Modern TRIZ

Michael A. Orloff

Modern TRIZ

A Practical Course with EASyTRIZ Technology

With 14 learning modules
and more than 340 examples

Springer

Michael A. Orloff
Modern TRIZ Academy International
Berlin
Germany

ISBN 978-3-642-25217-4 e-ISBN 978-3-642-25218-1
DOI 10.1007/978-3-642-25218-1
Springer Heidelberg Dordrecht London New York

Library of Congress Control Number: 2011944971

Printed on acid-free paper

Springer is part of Springer Science+Business Media (www.springer.com)

For Valentina

– my dear wife, friend
and comrade-in-arms

11.10.2010

Do you remember how it all started?

How new it was…

How we built boats and named them
"Faith", "Hope" and "Love"…

How we hacked off mooring lines
and watched
the earth slowly disappear
behind the horizon…

And
if the final destination remains the same
in good times and in bad times,
those
who have not lost their courage
and have not dropped their oars
will eventually find
their promised
land. *

* free translation from Russian; lyrics by Andrey Makarevich (b. 1953), co-founder of the Russian rock band *Time Machine* and creator of the *bard-rock* style; song title: *To Those Who Are At Sea*, 1982

EASyTRIZ: As Simple As Possible, But Not Simpler[1]

Book Presentation by Prof. Seung-Hyun Yoo

The headlined quote from Albert Einstein, eminent scientist and productive inventor, provides an apt description of the aim that Professor Michael Orloff pursued when writing his book.

We have known TRIZ for more than 10 years, and have successfully used it in engineering consulting and student training. We realized that TRIZ users needed to develop reliable training methods for real-life problem solving, streamline the *Algorithm of Inventive Problem Solving* (ARIZ) study, and devise a new way to organize theory models.

And we saw a new TRIZ at lectures and workshops delivered by Professor Orloff in the Republic of Korea before experts working for SAMSUNG and other companies, and before faculty and students of several universities.

The author designed a simple and efficient crash course in primary TRIZ models based on the method of *extraction*, or *retrieval of invention models from any real-life objects (artifacts)*.

He suggested a simple *standard model* presenting the problem solving process as a four-step ARIZ: *Meta-Algorithm of Invention* (MAI). MAI radically improves training methods, and can be efficiently used to solve most real-life problems.

The author developed the method of *reinventing, or "modeling the process of inventing", of any (!) artifacts on the basis of MAI.* Reinventing makes it possible to quickly master the inventing process and identify correct approaches to solving new problems.

Using reinventing examples presented in a standardized MAI format, the author created a TRIZ Knowledge Base, now known under a working title of *Modern TRIZ Idea Pool.* This is a real "creativity machine" that pool-members can use to exchange their inventive and innovative experiences, continue their self-perfection through ongoing study of new examples and pool digests, and advance TRIZ through new research.

The book offers a number of accessible and practicable general-purpose tools.

I wish you success with this book!

Seung-Hyun Yoo

PhD in Mechanics (Stanford University),
Professor* of Mechanical Engineering, AJOU University,
Director of AUCHRD (AJOU University Center for Human Resource Development),
Suwon, Republic of Korea

16.03.2010

---

[1] In consonance with "Everything should be made as simple as possible, but not simpler" by Albert Einstein (1879-1955) – great German-born theoretical physicist, 1921 Nobel Prize in Physics; http://en.wikiquote.org/wiki/Einstein

* In early 2000-es (for several years), Vice-President of Korea TRIZ Association

TRIZ Super-Task

Book Presentation by Dr. Anatoly Unitsky

With deep respect, I am presenting this book by Michael Orloff, expert in complex systems development management, researcher and global promoter of TRIZ – *Theory of Inventive Problem Solving* – which was discovered 55 years ago by Genrikh Altshuller.

For 15 years Michael – a considerate and empathetic person, a strong and independent individuality, and TRIZ expert – has been supporting my research and giving me the will to persevere, strength, and even inspiration to develop and build my inventions which, like TRIZ, must serve the entire mankind in the foreseeable future.

For more than 25 years, my ideas have always been fine-tuned with TRIZ and checked for compliance with TRIZ laws and models. And I see that our discoveries are aligned with TRIZ – and that in turn they reinforce TRIZ with "ideal" examples, corroborating the main TRIZ conceptual metaphors – "ideal solution" and "ideal machine".

Indeed, what can be more ideal than a string, if we need to build a straight and even speedway?! What can be more ideal than a self-ascending and self-returning string ring, if we need to create a transportation system linking the Earth's planetary and circumplanetary civilizations?

Michael has reopened to us that such ideas are alight with providence and even epiphany, and they engender feelings of beauty and wonder of creation!

He teaches us to discover a miracle of creativity in each artifact, in everything that surrounds us. Because all these things were first created by inventors and then implemented by designers and engineers.

Everything was invented, created by someone for the first time once!

And just the process of discovering beauty and wonder, I believe, is the super-task of TRIZ, of this book, of its author, Michael Orloff.

I think that it is also the super-task facing each of us, readers – to comprehend, through TRIZ and together with TRIZ, the miracle of creating things that are new, needful, efficient, unexpected, overwhelming and inspiring!

To perceive the wonder of inventing the future.

Anatoly Unitsky

PhD in transportation, inventor[2] of String Transport Unitsky (STU),

Moscow, Russia

02.04.2010

---

[2] Anatoly Unitsky (b. 1949) – author of the conceptual framework underlying string technologies, inventor, researcher and designer of planetary and space-planetary transportation systems; www.yunitskiy.com

Modern TRIZ

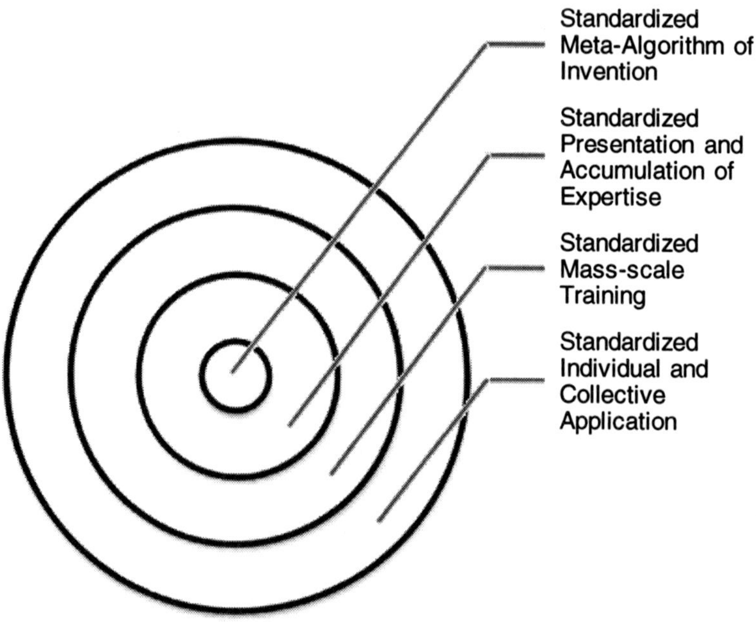

About the "EASyTRIZ" brand

The brand "EASyTRIZ" has two connotations:

1) it contains a reference to the adjective *easy*, and

2) it is based on a radiolocation abbreviation *EASy* which is deciphered as *Early Acquisition System* (radiolocation: target acquisition and locking-in so early as possible to get advance for efficient reaction) – and with respect to training objectives is interpreted as

EASyTRIZ – **E**arly **A**cquisition **Sy**stem for **TRIZ**

– to get right learning and skills in TRIZ as early as possible for efficient problem solving.

Who and How Taught Us to Think?

Author's Foreword

Photographic epigraph from the author:*
General Electric Building, Rockefeller Center,
New York, 29.10.2006

What was the starting point of your inventing experience?

Could it be the house that you once built with your favorite erector set? Or the sand castle that you made one sunny summer day on a small river? Or maybe the piece of pine bark that you turned into a small boat that you later launched down a short-lived rainwater stream? Or the toy airplanes that you manufactured first from newspaper pages or sheets torn out of your school notebook – and then from thin plywood, with wings of flimsy cigarette paper and engines of long tightly rolled elastic bands?

Or your drawings of bizarre flowers, trees and mysterious creatures that somehow always came out looking strangely human?

Or the feeling of awe inspired by the spring sunlight filtered through the arabesque filigree of maple leaves, the synchronic flight of doves, the myriads of stars dotting the fathomless dome of the night sky, the deafening roar of thunder and the blinding flash of lightning, the rainbow stretching above a humbled field, the weird world peering at you through a magnifying glass, the inscrutable attraction and repulsion of magnets, the magically slow emergence of pictures on sheets of photographic paper in the developing bath (now extinct "owing" to digital cameras and digital printing)?

Or could it be the village, complete with tiny houses and birch copses, that you built in the middle of the shiny wooden floor on the mountainous back of a good-natured "sea monster" made of a thick dark-green quilt? Remember how that "sea monster" carrying an entire living world on its back floated along under the sultry sun of exotic seas for what seemed to last an eternity… That is, until the evening came, and your parents returned from work, and your Dad asked you whether you had memorized the Pushkin' poem assigned by your teacher, whether there had been anything of interest in the paper that day, and whether you had thought up a move in the chess game you had started the day before…

---

* *Wisdom and Knowledge shall be the Stability of thy times*, Isaiah 33:6

I was about five years old.

Some of us were older, others younger – when at some point in time, somewhere in our inconceivably distant childhood, there happened something that led us, like the thread of Ariadne, through the labyrinths of life, something that prompted us, first young and inexperienced and then mature and sophisticated, to invent things that were exceptional, amazing, impossible – and at the same time very important, things that we would use doing our work, playing with our children, writing our lectures, consulting with our clients – engineers, managers, pedagogues.

I am not idealizing my childhood perceptions. On the contrary, more and more often I am haunted by realization of a strange, improbable and frightening fact: *nobody taught us to think.*

Indeed, did we learn to think while we memorized arithmetic rules or learnt theorem proving methods? And does anyone really remember those methods, or use them in his or her daily life? Well, definitely those who have become mathematicians, maybe some of those who have become engineers. A very important – or rather THE MOST IMPORTANT – component is conspicuously missing in our learning practices: we are not taught to "cogitate". None of our teachers told us about special *"cogitation"* methods designed to help us acquire skills required for *purposeful generation of ideas, resolution of intricate complex situations, conflicts and contradictions.*

The way I see it, we simply learned to live by *doing* things that we were required to do on an everyday basis at home, in the street, at school or at work. In a *certain learning environment* our brain apparently can master the ability for more or less *efficient, purposeful and inventive thinking.*

So we learned to think – widening the gap between ourselves and all other living things – animals and plants.

We had dreams and aspirations, we eventually amounted to something, achieved something, invented something – some of us wrote books and poems, while others composed beautiful music and imagined wonderful fairy-tale worlds.

And life went on – driven by the power of goodness and love, by the natural and sacred faith in the family, by the belief that we can become immortal through our children, and by the hope that we can change our way of life and make it more equitable and beautiful.

How and why did keeping that faith become the focal point of our lives and the lives of many other normal people who somehow still manage to keep the world going round? After all, all of us should have inevitably and inexorably become criminals totally devoid of any and all vestiges of morality – that is if we had succumbed to the infinitely iniquitous turpitude of movies and television shows, commercialized pseudo-music and pseudo-painting art, unrestrained squalor of the Internet, and persistent cruelty and amorality of many schoolyards…

I am surprised that, despite the overwhelming impact of evil and corruption spewed out by our TV screens and computer monitors, we are not turning into *mowglis* who are not destined to evolve into human beings despite Rudyard Kipling's benevolent fantasies. It is known that if a small child is not brought up by people, he or she *cannot become a human being.* Do we really want to spawn *mowglis*[3] who only look like people (and even that is not a given)? Do we want to let them walk among us, or control us?

It was only my pedagogical and psychological experience that helped me regain composure by identifying, in our collective socialization experience, certain prerequisites and indirect "techniques" which indicate *who and how taught us to think.* And there were many people who taught us to think. Some of them are mentioned in this book. But my "researcher gene" – the master of my subconscious – immediately and automatically triggered off a "reality check" alarm and demanded answers to more uncomfortable questions, such as these: Who and how taught us to resolve conflicts and contradictions? Who and how taught us to understand beauty? Who and how taught us to empathize and help other people? Who and how taught us to seek mutual understanding and engage in joint creative pursuits? Finally, what is the secret of survival of Personalities whose deeds keep humanity alive despite its stubborn endeavors to go to hell in a hand basket? Where are guarantees that our civilization will continue to develop and evolve?

It is difficult to answer all these questions. And there is yet another thought which vexes me – and justifies the emotional intensity of this address. It is the thought that arming people with inventive problem-solving methods fails to resolve the moral issues related to potential uses of this new "weapon". In other words, I am preoccupied with who and how will be exploring the new avenues opened by *efficient and powerful thinking.*

Contemporary history abounds in examples of situations where certain forces regard democracy exclusively as an irresponsibly generous, completely defenseless and basically gratuitous nutritional environment conducive to their amoral and criminal schemes. These forces, similar to democratic societies, organize into mimetic movements and associations which immediately declare that they intend to "fight for their rights". They rob normal people of benefits and resources and use those to disseminate evil and recruit new victims, e.g. through the drugs. As a result, we are being demoted to the bottom of the food pyramid which is topped by vice and iniquity. Our society is doomed to moral slavery because of our ignorance, indifference and cowardice – which for some inscrutable reason are called "tolerance". Well, be advised that evil is anything but "tolerant".

If some of you find this invective irritating, or believe that it distracts from their perception of their mission in life – which for many is as simple as the recipe for the proverbial Big Mac (incidentally, I have nothing against McDonald's technologies, with the proviso that their fast food cannot completely satisfy nutritional needs of a healthy human being) – this book is not for you; moreover, be warned that you may well be on the way to becoming both the victims and – *attention!* – agents of evil.

I am not trying to intimidate anybody – many are already scared of their own shadows having watched too many thrillers, but that does not stop them from gorging themselves on that drug because they know no other values or motiva-

[3] Abandoned "man cub" *Mowgli,* raised by wolves in the Indian jungle, is a main character of fable-stories *The Second Jungle Book* (1895), performed many feats of valour in jungle; created by Joseph Rudyard Kipling (1865 – 1936) – British poet and writer, Nobel laureate (1907)

tions, because they cannot imagine that life can be different, because they have become willing participants of a global – and very real – *Truman*[4] *Show.*

I have lived 60 years and mastered several crafts, and this experience has made me recognize the reality of active evil at work in the world, dispelled any illusions I may have had that good always triumphs over evil (evil can be prevented and defeated only by *active good*), and taught me to respect laws of system evolution and understand models designed to project future trends and study the driving forces of development. Naturally, this knowledge is imperfect and incomplete – as is any emergent knowledge – but it is enough to help me comprehend the *modus operandi* employed by evil, and understand (to my sincere regret!) the risks and dangers directly associated with the possible destruction and degeneration of our civilization.

Life would be impossible if we did not believe in goodness and fairness, in our friends and colleagues, in reasonability and inventiveness of the human race.

So – forward!

Keep doing your work day in day out – step by step, letter by letter, thought by thought, question by question.

Forward!

This is what I tell myself before making every step – however small – through the years of learning the world and making my own way.

This is what helps me push ahead even when my tormented eyes refuse to look at the monitor, when my joints and muscles scream for mercy after twelve hours of cruel and unusual punishment by intractable keyboard and recalcitrant mouse. Will these torture devices ever disappear from our lives?! I KNOW they will, and I know how it will happen, but they are still here… For the time being, special relaxation exercises remain my only solace. Write to me, and I will be happy to share my self-rescuing and pain-killing experiences.

Forward – all those who seek perfection and wish to make this world better.

Forward – starting with changing yourselves!

For a wise man once said[*]: *be transformed by the renewing of your mind.*

YOUR mind!

I wish you luck and well-being.

Michael Orloff

Berlin, Germany
May 9, 2010

Special thanks: I would like to express my respect and thanks to translator Val Prokhozhy for great pleasure of working together for almost a year, and for the nice fruit of that labor – the English translation of this book.

June 10, 2011

4 fiction-movie *The Truman Show* (1998); directed by Peter Weir, Paramount Pictures Corporation

* *Transformamini renovatione mentis* (Romans 12:2)

Table of Contents

Principles of TRIZ

While I was thinking through the milestones of my theory and its development, I was overcome with doubts... normal human doubts: if this is so important, why is it that over the course of many centuries so many clever people have failed to notice that, and have not done it without you? What, do you think you know better?[5]

Genrikh Altshuller

1 Introduction

1.1 On Creative Discipline and Standardization of Knowledge

1.1.1 Learn or Re-learn?

Many books have been written about TRIZ[*]. Accordingly, after a cursory glance at the cover the reader might assume that the idea of the book is not new, and that it is yet another repetition of something "about TRIZ". Especially with a heading like *EASyTRIZ*! Definitely a "popular book" for "dummies". Yeah.

In reality, it is very different from anything you may have encountered before. This is not just another "*imitation cum repetition*". And I hope that the reader will soon appreciate this fact.

This book has used a technology of modeling, training and problem solving, called *EASyTRIZ*, but it is not the same as "easy reading" or "easy doing". The title of the book is merely a reflection of the fact that this book cannot be made any easier. Any attempt to do that might result in oversimplification, loss of quality and, possibly, acquisition of wrong knowledge and skills.

And it is *much more difficult to re-learn* than it is to learn to do something right from the start! Accordingly, learning TRIZ today is still basically re-learning – even for schoolchildren, let alone grownups.

It is necessary to restructure *poorly organized* – in the words of Genrikh Altshuller, TRIZ founder – previous thinking experience.

---

[5] According to G. S. Altshuller, I. M. Vyortkin (1994) *How to Become a Genius. Life Strategy of a Creative Personality*. – Minsk: Belarus (in Russ.)

[*] as of the beginning of 2010

Such restructuring requires personal commitment and significant time, and even with those present, it sometimes remains incomplete. Because people have a tendency to return to their previous inefficient, hopelessly ineffective, but familiar experience of... *weak thinking.*

This book is for optimists, for people with a purpose. Studying this book, like any professional work, requires adequate labor, time and effort inputs to ensure that new skills are integrated into your personal experience.

1.1.2 About Modern TRIZ

First let me tell you what objectives I sought to attain writing this book, and what – *I hope and believe!* – may prove to be important and valuable for the reader.

So, let us first ask ourselves, what is it that we expect from TRIZ – or, rather, from **Modern TRIZ**?

Why is it that this technology designed to help find creative solutions to complex problems is becoming more and more attractive to practitioners and theoreticians, professors and managers, engineers and creative art psychologists? Why is it that TRIZ – with varying degrees of success – continues to expand its global footprint?

What needs to be done to help TRIZ enroll more supporters, both among developers and among users? What must happen for this theory to gain the recognition it deserves? And last but not least, here is a direct question that I keep asking myself (and expect to be asked by the reader): what is *most important* for TRIZ today?

Indeed, this is a very good starting point for any new book about TRIZ.

> My answer is as following: the most important for modern TRIZ is to ensure a mass-scale professional education in key, primary, fundamental models of theory.
>
> Therefore this book is designed to be used for education (including self-education) in these models as the point of origin, and the key element, of the Modern TRIZ universe.

Why do we need Modern TRIZ and what should it be like?

Extensive TRIZ teaching and application experience has shaped the convictions that the author needs in order to answer these questions. For many years, the soundness of these convictions has been confirmed by practice.

In addition to that, practice encourages further development of previous ideas. But even a brief answer will have to be broken down into several bullet points to reflect both the areas where TRIZ can be gainfully employed to perform practical tasks, and the prospects of its future development.

Well, here are the core objectives of our work:

a. Correct introductory education in key primary TRIZ models

The new TRIZ training technology must overcome the following known drawbacks:

- lack of an efficient introductory TRIZ training methodology,
- lack of generally accepted definitions for basic TRIZ concepts,
- lack of efficient standard examples to illustrate primary concepts and models.

Results of the "old-way" training include the following:

- fuzzy understanding of the simplest models,
- lack of skills required to use the simplest models,
- unrealistic expectations from theory and practical tools (particularly the TRIZ software package),
- uncertainty, fear or inability to independently resolve practical problems.

Pathetic inefficiency of the "old" TRIZ teaching methods erects insurmountable barriers in the way of practical application of new knowledge and skills by most students as soon as they finish "traditional" workshops and seminars. The absence of sustainable skills required for error-free independent operation hamstrings them, dilutes their initial enthusiasm, and deprives them of the possibility to efficiently apply basic TRIZ knowledge in real-life situations.

b. Standardized mass-scale education as an adequate reaction to new challenges created by global economic integration

Today not only large concerns have branches on different continents and in different countries. Small and medium-sized enterprises, universities, independent researchers and developers successfully form associations focused on research, design, production, marketing, education, consulting and other tasks.

Particularly complex problems are tackled by multinational Think Tank Teams.

Efficient operation of such Think Tank Teams is predicated on availability of a common problem-modeling language and a standard set of tools to generate ideas that can be understood by all team members.

Such language and such toolset can be created only on the basis of efficient structuring and standardization of TRIZ models. Conversely, failure to standardize TRIZ concepts and tools represents a major obstacle in the way of global dissemination of TRIZ, and hampers implementation of TRIZ programs in universities and other educational institutions. Meanwhile, students – those of them who are not infected with negative problem resolution stereotypes – might become the most appreciative and fruitful TRIZ users fully armed to implement innovative methods in their post-graduation work environments.

c. Accumulation of innovative experience by firms, associations and universities

Several years ago, the author undertook an analysis of innovations implemented at a major German enterprise over a period of almost 50 years. The findings were later verified by three reliable independent sources. The analysis convincingly demonstrated the following:

- efficient practical solutions are not analyzed to extract *objective models and methods which were employed to develop those solutions*, and which could be used to train young specialists!

- innovative experiences are recorded and codified only in the form of general engineering descriptions and technical specifications, and are *not presented as formalized creative models* similar to reusable TRIZ models which make it possible to avoid waste of time and effort on solving essentially identical problems!

- *failure to implement a workable paradigm for amassing and structuring innovative and inventive experiences* excludes the possibility of preservation and accumulation of invaluable intellectual capital; moreover, it amounts to wasteful and neglectful misuse of a legacy that may be irretrievably lost – for the company, for the young generation, for the society – when those who created that legacy retire or resign.

This state of affairs is global and typical. It is also lamentable and incomprehensible.

Without understanding this phenomenon, without creating a new technology for ongoing accumulation of knowledge and skills, how can we talk about the aims, objectives and functions of quality management, project management or personnel management? About innovative thinking culture and innovation management, about inheriting intellectual capital and ensuring corporate continuity? Continuity… in what?

This outrageous extravagance is valued at dozens of millions of dollars at any large enterprise. It is difficult to make a reliable estimate of the losses sustained by the industry as a whole, but we can safely assume that those losses may be as high as dozens – or even hundreds – of billions of dollars.

d. Exchange of experience among thousands of innovators and availability of ongoing "lifelong" self-training to all interested specialists based on deliverables produced by such exchange

Archaic TRIZ "transfer-of-learning" and teaching methods have the following drawbacks:

- poorly structured examples leading to unclear interpretations and inconsistent training applications which depend almost entirely on personal experiences accumulated by individual trainers,
- lack of objective criteria to evaluate model efficiency,
- grievously limited training in practical resolution of real-life problems.

All these drawbacks can be eradicated by creating centralized databanks containing information on standardized TRIZ models which have been successfully used to resolve real-life problems. Instead of being arbitrary collections of haphazardly selected examples, such databanks represent highly efficient "machines" assuring reproduction and productive transfer of TRIZ learning. They can be built on the basis of the following simple principle: the more participants a databank has, the higher its efficiency! The databank of TRIZ knowledge created by Modern TRIZ Academy (the Modern TRIZ Pool, or its company-specific implementations) will ensure efficient transfer of knowledge and skills among all members, enabling them to engage in ongoing self-improvement through daily training and review of periodically published digests.

e. Development of Modern TRIZ

Standardization of TRIZ models and accumulation of numerous examples of efficient solutions result in the attainment of a "critical mass" which generates new opportunities for Modern TRIZ improvement.

The Modern TRIZ Pool is rapidly becoming a treasure-trove of collective intelligence, while its empirical uses open up breath-taking applied research vistas.

Other important features of Modern TRIZ education include Internet availability of TRIZ knowledge and organization of mass-scale TRIZ basics remote training. It is this standardized education that is being developed by the Modern TRIZ Academy.

1.1.3 Standardization of Education – from Experience to Conviction

Can one "standardize"... organization of thinking? Or: what is a "standard model" in TRIZ which is designed to promote and organize "non-standard" creative thinking?

How do we answer these questions? And we do need to find those answers. Or we may wind up in a situation similar to what that occurred at a workshop that was held in Berlin for patent experts of a well-known German company specializing in cutting-edge space communication and telemetry technologies.

After I had demonstrated the standardized scheme and MAI-based reinventing examples, one of the attending patent experts said: "So, with your "standard" algorithm anyone can resolve any problem, or invent any device, say, a computer? But that means that such solution cannot be patented! There will be no know-how in it! If solutions can now be developed on the basis of objectively existing laws and rules, the role of the "subjective factor", of the MAN, is reduced to zero! Where is creation here?"

I had to explain that methodology is not a substitute for knowledge. That *"rational intellect"* models are not substitutes for talent. That success of problem-solving also depends on motivations, moods and preferences of a specific person, i.e. on what is called *"emotional intellect"*.

And, naturally, I had to elaborate on how TRIZ helps to develop talent, and that TRIZ radically improves application of knowledge and talent. That is why the bulk of my book is dedicated to "rational thinking" models – they "cover" most TRIZ methods!

The importance of *technical excellence* resulting from using the *right teaching methods* is best illustrated by examples from other areas.

Consider the famous boxing duet – brothers Vitaly and Vladimir Klichko. Here is the standard that they may use to organize their fights:

- correct ("beautiful") classical boxing stance,
- efficient tactically-justified movement around the ring,

- excellent defenses (evasive maneuvers come before blocks),
- excellent feeling of distance – for both defense and attack purposes,
- elaborate preparation for the decisive attack,
- effective winning punch.

This reminds me of Olympic champions Cassius Clay and Valery Popenchenko – the idols of my boxing youth. Over the last 10 years, such boxing has been demonstrated not only by the Klichko brothers, but also by other boxers in the world. It is next to impossible to mention them all. In 1996 Vladimir became the first post-Soviet super heavyweight Olympic champion. Ten years later, the entire victory podium in that most prestigious weight category was occupied by sportsmen hailing from the ex-Soviet school which became globally known as the *Russian Modern Boxing.*

New champions introduced *efficient sports thinking* instead of the brutal force show which involved two Behemoths slowly milling and shuffling in the ring – the epitome of obsolete professional boxing. At a World Boxing Organization conference in Fort Lauderdale, US, in November 2008, Vladimir was awarded the title of "Super-Champion 2008".

Where did this thinking come from? It came from the so-called "Soviet" boxing school – the *de facto* standard effective throughout the entire territory of the former Soviet Union. Now they would say: from "Russian" boxing.

And the same "branding" is growing now for TRIZ. Some people say that three Russian words will be the most known in the world from XX century: *sputnik, Kalashnikov* and *TRIZ.*

So, correct training in correct boxing! Preparation of elite sportsmen based on standardized baseline techniques!

New international professional boxing champions in many categories (and, most importantly, in the most prestigious "heavyweight" categories) were originally coached by using "Russian" school standards!

And now is a good time to recall the outstanding Russian composer and pianist Sergey Rachmaninoff who used to say: *If I don't play for one day, I notice that; if I don't play for two days, my family members notice that; if I don't play for three days, everybody notices that!*

Now, do not forget that he was a great pianist! This means that even masters need *ongoing standardized training.* If a master ceases training, he inevitably loses part of his mastery. Therefore, he has to train throughout the active phase of his professional career.

Indeed, at the decisive moment, when you need to be mobilized and concentrated, it is too late to ask yourself whether you have done everything that you could have done in the period leading up to that moment.

This philosophy underlies training in Oriental martial arts.

This philosophy of lifetime (!) training in the use of *"correct standard basic elements"* is the foundation of the world-renowned "Russian" ballet school and the "Russian" figure skating school, the Stanislavsky theater school and the Eisenstein cinema school, "Russian" musical culture and "Russian" ice hockey…

When half a century ago Russian ice hockey players started to beat Canadians and Americans, it was a sensation. When Russians won several world championships and Olympic games in a row, people started to learn from them. The results of that learning will be discussed at the end of this preface.

Clearly, professionalism is the fruit of numerous training sessions involving repetition of simple exercises which progressively become more and more complex. In any walk of life, acquisition of professional skills changes the person and his or her mode of thinking.

In turn, *changes in the mode of thinking and increased efficiency of thinking (in the relevant professional sphere) are both inherent attributes of mastership.*

Genrikh Altshuller, the founder of TRIZ, wrote[6]: "Each tool has retroactive effect on the person using it. TRIZ is a tool designed for subtle, audacious, highly organized thinking operations. Resolving one problem does not change the thinking style. But in the course of training the student resolves dozens, hundreds of problems. Then his thinking is gradually transformed". And so the world began to study TRIZ.

The TRIZ school encourages personal creativity.

And now let us recall what the father of TRIZ had to say about manageable creativity:

- for a very long time, inventive problems have been solved by enumeration of possibilities (What if we try this?...), and *that* process was impacted by a host of random factors that were very difficult (if not impossible) to factor in; that is, to all intents and purposes, the process was unmanageable. It was necessary to *advance to a different technology* yielding the same product (inventions), but within the framework of a completely different production system – a system which is manageable, well organized, and efficient[7];

- TRIZ essentially changes the technology used to produce new technical ideas. Instead of enumerating possibilities, TRIZ uses cogitative activities which rely on knowing the laws governing evolution of technical systems. The creative universe becomes infinitely manageable and, therefore, can be infinitely expanded. This creative revolution is probably as important as scientific, technical and cosmic revolutions[8];

- every component of the sci-tech paradigm is subject to change, with only one notable exception: the notion that the creative process cannot be managed. Moreover, it is widely believed that in the future – in, say, a hundred years or a thousand years – creation will remain the same. The strength of the old views on the nature of creation is colossal. That is why it is so difficult to see what apparently lies on the surface: technical systems evolve in accordance with certain laws which can be cognized and utilized to develop a new creative technology[9];

- the fundamental concept of a bad school is: prepare students for memorizing the material. In a good school, students are trained to understand the material and operate with it.

---

6 Altshuller, G. S. (1987) *Activation of the Human Factor in Education and Upbringing*. – Moscow: Knowledge Publishers (in Russ.)

7 Altshuller, G. S. (1979) *Creation as a Precise Science*. – Moscow: Soviet Radio Publ. (in Russ.)

8 Altshuller, G. S. (1986) *To Find an Idea*. – Novosibirsk: Science Publishers (in Russ.)

9 *Ibid.*

And only very few Teachers (there is still no mass-scale super-school of that type) can (or at least understand this need and try to) prepare students for *creating* new material.[10]

The founder of TRIZ elevates talent to a height not accessible in the old paradigm where creation is regarded as a basically random process which is ultimately reduced to enumeration of possibilities and brainstorming;

- the algorithm does not negate the need to think, it simply guides the thinking process, protects from mistakes, and forces the user to perform unusual – talented – thinking operations. [11]

So, is it possible to standardize the creative process? I am inclined to say "yes". Then what is this "standard"? What is "thinking discipline"?

Standard makes it possible to save effort when there is a need for correct "automatic" action, and to use the energy released to fuel creating thinking.

Standard reinforces thinking with tested methods which involve concentration, cautious maneuvering, and audacious construction action.

Standard helps to achieve the goal by the most efficient trajectory with minimum expenditure of energy.

The notion of "standard" is well known in engineering. The building of the engineering temple is underpinned by standards – ranging from the theory of "tolerances and calibers" introduced by F. Taylor in the USA in 1905 to quality assurance theories, such as *Six Sigma.* Incidentally, F. Taylor with his *PDCA* "algorithm" (Plan, Do, Check and Act) can be regarded a *Six Sigma* forerunner [12].

The merger of "standard" and "quality", as these notions apply to the theory of creation, the theory of invention, produces a WINNER. Let us say it again: **"standard"** and **"quality"**.

Indeed, what lessons can we draw from the three very different examples that we discussed above: excellent boxer + outstanding pianist + inventor, writer, teacher, founder of TRIZ?

First: the road to outstanding results starts with correct and standard (!) organization of initial education.

Second: elite arises from a large starting group tempered by correct training.

Third: professional growth continues on the basis of ongoing self-training, self-control and self-discipline "until" and "when" mastery is attained.

Nature produces many extraordinarily talented people who show a lot of promise, but it is common knowledge that incorrect initial training can prevent such people from reaching their full potential – or even arrest their development.

Starting this book, I relied on my personal TRIZ training experience (I coached more than 3 thousand TRIZ specialists in several countries), and on the learning and teaching experience I acquired at my *alma mater.* That experience rests on a firm foundation of deep systemic organization of subjects taught, and continuity of expertise and mastery.

---

[10] Altshuller, G. S., Vertkin, I. M. (1994) *How to Become a Genius. Life Strategy of a Creative Personality.* – Minsk, Belarus (in Russ.)

[11] Altshuller, G. S. (1979) *Creation as a Precise Science.* – Moscow: Soviet Radio Publ. (in Russ.)

[12] See the DMAIC "algorithm" in the description of the *Six Sigma* methodology: Define, Measure, Analyze, Improve and Control.

My *alma mater* – Minsk Radio-Technical Institute (now promoted to university status[13]) to which I dedicated more than 20 years of my life, both as a student and as a professor – boasted an extremely efficient academic organization which included one-hundred-percent attendance of all classes, strict supervision of work assignment completion, numerous mandatory field practice tours at "real-life" enterprises, massive involvement of students in research projects, and encouragement of students' inventive and creative efforts.

The same is true about Moscow State Technical University named after N. E. Bauman[14] where I undertook a course of advanced training, and with whose experts I spent several years designing specialized computing devices and systems.

A powerful training system reinforced by military discipline was in place at the Military Space Engineering Academy named after A. F. Mozhaysky[15] in Saint-Petersburg, where I worked for 10 years as an expert and co-developer of mathematical and technical methods of multi-level and multi-aspect modeling of large-scale dynamic systems and their components. That work integrated system analysis, high-class engineering, sophisticated applied mathematics, Lefebvre model psychology[16], AI event recognition systems from many Soviet schools, including the D. A. Pospelov[17] school, dynamic conflict resolution according to D. S. Kontorov[18], system evolution management, and... TRIZ which the author applied to the extent possible.

Similar powerful potentials were *accumulated and cultivated* in dozens of industry research institutes where I was lucky to work. Outstanding results were produced on the basis of high professional culture and strict executive discipline regulated, throughout the entire country, by uniform research, engineering and academic performance and evaluation standards.

So, TRIZ creates talents. Is it not evident, then, that TRIZ deserves the same measure of success as that achieved by "Russian" boxing in the professional world – that it is worthy of global recognition and dissemination?

The only difference is that boxing champions come and go, while TRIZ will always be there because it is constantly evolving!

Conditions precedent include the following:

- extraction and application of standard models,
- availability of TRIZ to the general public through promotion of textbooks, software packages and training programs on the basis of standard modeling!
- possibility to accumulate experience in solving real-life problems and modeling known solutions (reinventing) on the basis of standard models,
- possibility to exchange user experiences on the basis of global, regional and industry-specific databanks (pools) of standard examples – samples of efficient solutions reinvented in the *Meta-Algorithm of Invention* format,

---

13 Belarus State University of Information Sciences and Radio Electronics

14 Previously, Moscow Higher Technical School named after N. E. Bauman

15 In the past, the Academy has been known under several different names

16 V. A. Lefebvre, author of the reflexive management theory, currently (2007) researcher at the University of California, Irvine

17 D. A. Pospelov, outstanding Russian AI scholar

18 D. S. Kontorov, outstanding Russian system engineering scientist

- possibility for thousands of *Modern TRIZ Pool* members to engage in ongoing self-training in TRIZ modeling,
- communicating the basics of standard Modern TRIZ to teachers, professors and managers to ensure that they are able to integrate TRIZ models into their academic and training curricula.

Then TRIZ will truly become both Modern TRIZ and "mass-scale" TRIZ – or, for short, MTRIZ[19]. It is to attain that objective that we started our *Academy International for Modern TRIZ (AIMTRIZ)*[20] more than 10 years ago.

1.1.4 Modern TRIZ herbie!

The book you are holding in your hands gives you the standard professional minimum knowledge to qualify for MTRIZ Expert certification under the *EASy-TRIZ* program. But that alone is not enough – you also need training and character to master and use that knowledge and keep going despite possible failures.

In conclusion of this preface, I would like to tell you a story[21] that not only confirms the importance of standard training and vigorous practice, but also reminds us that in anything we do our success depends, to a large extent, on motivation, perseverance and will to pursue and attain our goals.

Herb Brooks[22] was appointed to coach the US national ice hockey team for the 1980 Winter Olympics in Lake Placid, USA.

He started off by saying to his boys: *You think you will win with only your talent... Gentlemen, you don't have enough talent to win on talent alone.*

Just think about it – he said THAT to *outstanding American players* hand-picked to become one team, one fraternity, one family!

To boost the starting speed and increase speed endurance of his players, he invented and implemented a method which was later dubbed *herbie* based on the diminutive version of his first name.

The method basically boiled down to this[23]: during each training session, the players had to start from their goal line (fig. 1.1), sprint for the blue line on their half of the playing field, stop dead in their tracks as soon as they reached it and skate back to the goal line; then stop, turn around and immediately run for the middle red line and back to their goal line; then stop, turn around and immediately run for the blue line on the other half of the playing field and back to their goal line; finally, stop, turn around and immediately run for the other team's goal line across the entire playing field – and then back to their goal line!

---

19 Inasmuch as the words "modern" and "mass-scale" start with the same letter, it will suffice to leave only that one letter in the name of the future "state" of MTRIZ.

20 www.modern-triz-academy.com

21 The author started this book concurrently with his book *ABC-TRIZ.* He deemed it possible to recount this story in both books as he believes it to be very significant.

22 Herbert Brooks (1937-2003) – outstanding US ice hockey coach

23 Herb Brooks quote: *Red line, back. Blue line, back. Far blue line, back. Far red line, back. And you have 45 seconds to do it. Get used to this drill. You'll be doing it "a lot". Why? Because the legs feed the wolf, gentlemen. I can't promise you we'll be the best team at Lake Placid next February. But we will be the best conditioned. That I can promise you.*

And they had to do it hundreds of time!

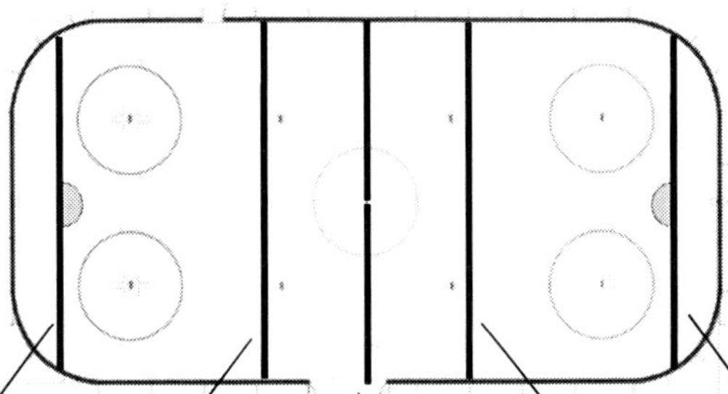

fig. 1.1. The ice-hockey rink is 30 meters wide and 60 meters long!

Just ask yourselves, would you be able to withstand such drill?!

If the answer is *no*, then maybe this book is not for you, and you need to find something else to do with your time – maybe something simpler.

If the answer is *yes*, then rest assured that the *EASyTRIZ* program is not all that difficult for those who are able and willing to work hard.

Prior to his appointment, Herb Brooks said to the members of the US Olympic Committee: *The Americans have to change their style of playing and adopt the hybrid style of the Canadians and the Russians!*

In the semi-final match with the invincible (!) USSR team, his boys scored a 4:3 victory. Many remember that match as if it were the final. Because it was more memorable, even though the US-Finland final (4:2) was equally dramatic. The US national ice hockey team became the 1980 Olympic champion.

Herb Brooks wrote in his memoirs[24]: *We were a fast, creative team that played extremely disciplined without the puck.*

> **Thus the victory was made *before the game*!**
> **And only then – *during the game*!**

The match (popularly known as the *"Miracle on Ice")* became an outstanding sporting event because it showed what could be achieved by individual mastery and team work based on *well-designed intensive practice regimen* multiplied by single-mindedness and purposefulness.

I remember that match…

Individuals and teams which understand and skillfully apply Modern TRIZ are capable of similarly great achievements.

The *EASyTRIZ*-training program can become a catalyst of success for anyone who is willing to succeed, and is prepared to go through *MTRIZ-herbie* to train to be successful.

I sincerely wish you luck!

---

[24] www.herbbrooksfoundation.com

1.2 What is an "Inventive Problem"?

In most cases, assignments which you find in this textbook have once been solved without TRIZ.

Thus, it can be possible for you too, to solve the problems or sometimes to know the solution, no matter what skills, experience or creative talent you have.

Now, the question is: Can we extract any creative experience from solving these or similar problems? Can these experiences then be generalized and furthermore represented as models or practical recommendations for solving similar tasks in the future?

TRIZ – *and only TRIZ!* – says *Yes.*

If we extract creative models from formerly solved problems and thereby learn how to solve similar ones, doesn't it lead to the conclusion we are facing the beginnings of a new theory of creativity, namely the theory of inventing?

To this question TRIZ again gives a yes for an answer. As TRIZ per definition is indeed the theory of inventing. TRIZ is constantly developing new models, definitions and ideas which support the further development of the theory. Still, there is a *primary knowledge* with which TRIZ once started out and without which TRIZ as a theory cannot exist. Everyone who wants to acquire skills in innovation and inventive thinking needs this knowledge.

The study of the primary (basic and advanced) models is absolutely necessary for the acquirement of the skills to apply these models.

The models are universal and independent from the field of their application, though some of the models are still too "technocratic".

They can be adapted to any kind of situation or problem.

Now let's shift our attention to the assignments. We would like to kick off with a very beautiful creative assignment, one of those we call "benchmark assignments" – assignments whose solutions can serve as models in terms of their effectiveness and exquisiteness.

We should remember that none of these problems would have been solved without the effort and creative talent of the ones who did solve them. Many of the names of such solution finders were lost throughout our civilization's history. Still though, we are making use of the fruits of their creativity.

Meanwhile, we also offer assignments which are still far from solved. And you can apply your talent and knowledge – as well as you TRIZ skills – to deal with the problems you will find here.

Problem P1. How Do You Design the Ruby Stars of the Moscow Kremlin?

The ruby stars of the Moscow Kremlin have become famous throughout the world as symbols for the Soviet Union. They are now symbols representing Russia and Moscow. In 1937 the ruby stars had been fixed on all of the main towers of the Moscow Kremlin, the tallest of those topping out at more than 60 meters – even without the star.

The stars are made of special ruby-colored glass containing selenium and other additives ("selenium ruby") and semitransparent milk-colored glass fastened with a special stainless steel mesh. The steel strips of the mesh follow the contour of the stars and make them extremely robust.

The main stars have a diameter of approximately 5 meters and a surface of 6 square meters. In other words, in a strong wind there is great probability that the star could fall off the tower.

With the star weighing about a ton, it was a serious task for the engineers who planned and constructed the stars to guarantee their reliability and safety in strong wind conditions.

Assignment: How can one guarantee reliable protection of the stars in case of heavy storms, and at the same time keep the construction simple and the expenditure of energy efficient?

Problem P2. How can we hold glass without touching it?

Once, when in Berlin, I was asked to consult a small company with a staff of some 30 people. The company was making sun roof glass units for Mercedes, BMW and other passenger cars. Small though it was, the company still faced a plethora of technical problems which reduced its efficiency and resulted in losses from deficient products and equipment downtime due to the need for repairs and adjustment.

The manufacturing process started with intake and storage of huge (~ 4 x 6 m) glass sheets. The sheets were then extracted from the pack one by one (a feat in itself!), placed on the table, and cut into blanks. Then the blanks were fine-clipped, contoured and – last but not least – somehow shaped into incurved glass units.

Incidentally, each sheet came to the bending machine out of a 700°C furnace speeding along a roller conveyer. And then the magic started! The sheet slid off the conveyer belt to go under a massive platform where – look, Ma, no hands! – it hung right below that platform (see figure).

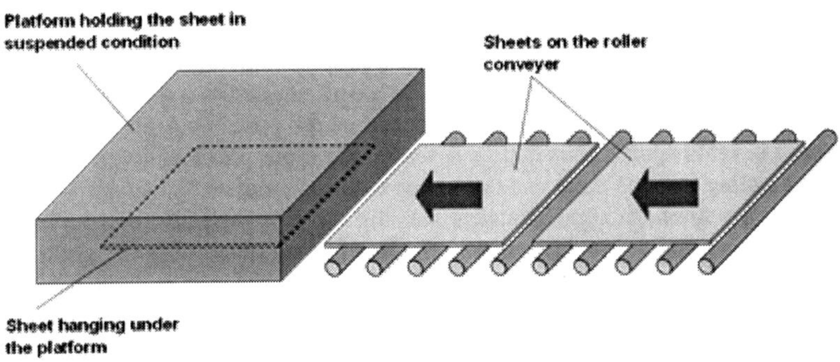

Then the hanging act ended with the sheet free-falling into a well from a height of about 30-40 cm to hit a wooden frame (see figure below). In the process, the hot-softened sheet bent in the middle and assumed the shape fitting the roof profile of a certain car model! To pull that off, the sheet had to fall off a certain height.

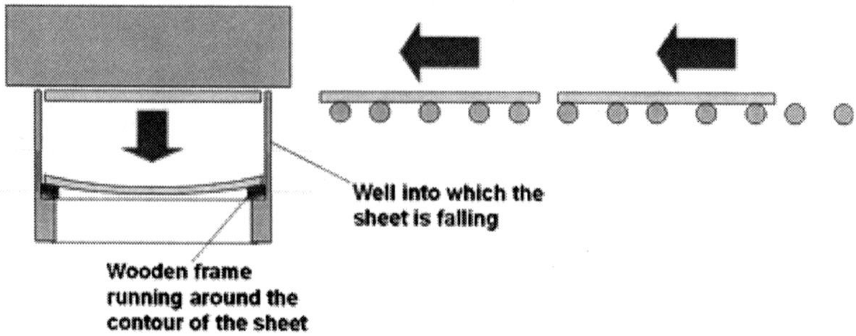

Assignment: How can we hold the sheet under the platform above the well, if neither metal nor wood may have contact with the sheet from above?!

Problem P3. How can we stop drops from spilling over when pouring liquid from a bottle?

When pouring colored liquids, such as juices, sauces or wine, you often get a nasty surprise when drops of that liquid mar the pristine beauty of your snow-white tablecloth or ruin your new expensive clothes.

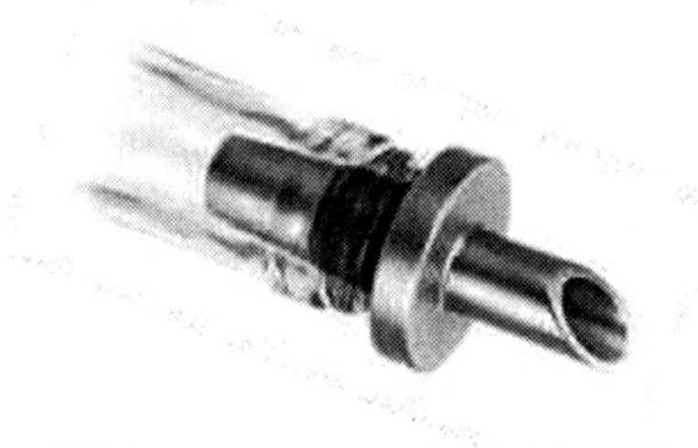
The Vinotto Bottle Pourer

The drops fall off the neck of the bottle or any other vessel from which the liquid is being poured. The edge of the neck is usually thick and rounded, making it an ideal breeding ground for large drops which proceed to effortlessly roll down *en bloc* or in parts.

To avoid this, you can stop the bottle with a cork which has a thin little pipe running inside down the middle. The outlet end of the pipe is cut off at an angle to make a beveled spout. This is how it works: the spout prevents drops from forming by cutting through them and not letting them accumulate on its edge.

There's a hitch, though. The cork housing the pipe (and also part-timing as a bottle-stopper and a liquid dispenser) fits in a limited range of bottle necks. Both broad-necked and narrow-necked bottles require special corks. As a result, you have to keep different corks for different bottled which – you must agree – is a nuisance.

Assignment with a prompt: What can you do to make the "drop terminator" fit a larger range of bottle necks, while retaining the ease of manufacture, storage – "Do you have pocket-size stoppers?" – and use?

Problem P4. How do you make automotive civilization more humane?

The advent of the 21st century is marked by rapid exacerbation of global environmental and security issues. In particular, automobiles – due to their massive use – may have become the most lethal of human inventions.

Car accidents kill

one and a half million – 1,500,000 – people

EACH YEAR.

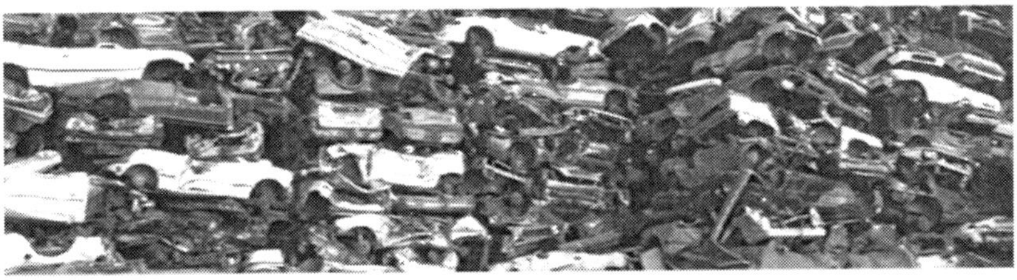

Besides, about 50 million people suffer injuries turning them into cripples and invalids – which is unacceptable from the position of humanism and sustainable development of our civilization.

The total length of communications spanning the planet now stands at about 35 million (mln) km, with motor roads, rail roads and pipelines accounting for 32 mln km, 1.2 mln km and 1 mln km, respectively.

These communications are built on about 60 mln hectares of land, which is equal to total area of such countries as Germany AND Great Britain.

This land does not breathe, nor does it produce oxygen, as it is devoid of plants and stripped of top soil and humus, the product of many million years of natural creation. Areas adjacent to major transportation routes are characterized by abnormal patterns of migration of large and small domestic animals and wild animals, with more than one billion of them ending up as road kill every year.

Within the territory

ten times the size of the area mentioned above

soil, as well as everything that grows and feeds on it, is contaminated with carcinogens and more than 100 other deleterious substances representing fuel burning products, tire and road-top wear-and-tear residue, de-icing salt mixtures, etc.

Assignment: How can we prevent mass murder of people by cars, "peaceful" technogenic weapons – each of which is, incidentally, steered by a human being?

Problem P5. How can we build a circumterrestrial space civilization?

Let us peek into the immediate (as compared to the existence and development of the terrestrial civilization) future of our planet, and imagine the dawn of a new space civilization. It may be marked by creating a man-made industrial/residential ring at an altitude of 300-400 km above the equator.

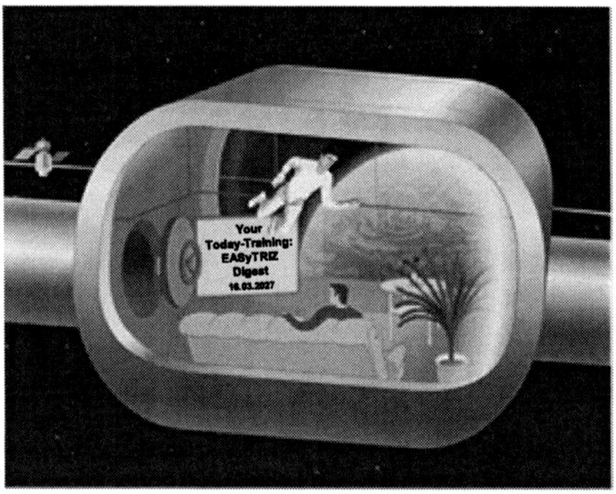

Let us assume that the weight of materials to be uplifted to the ring will "soon" reach 10 mln tons per year (this is not too much: just about 10% of the annual cargo turnover posted by the 10,000 km Russian Trans-Siberian Railroad).

Let us further assume that each rocket can carry 100 tons per launch (such heavy-duty rockets do not exist, but they can be built). Then each year we need to make 100,000 launches, or almost 300 super-rocket launches… per day! This avenue is apparently a dead-end, and not only because of the mind-boggling power consumption and related costs, but also because it will completely destroy the planet's protective ozone layer killing all life by hard UV solar radiation.

Now, this scenario does not even start to mention reverse transportation issues – and that is not a walk in the park, either!

Bottom line: the transportation system linking the Earth with the industrial/residential ring hovering above the planet *must* support annual cargo lifts of up to 10 million tons, and *must not* (*cannot* with existing rocket systems) lift that much if we are to prevent annihilation of the planet's protective ozone layer.

Assignment: How do you build a transportation system to support massive cargo turnover between the Earth and its circumterrestrial sister civilization?

1.3 EASyTRIZ: Constructive Learning Technology of Modern TRIZ

The proposed Modern TRIZ learning technology[25] (brand name: EASyTRIZ) has been developed over an approximately 10-year period (starting in 1995) of work on TRIZ-related questions of varying degrees of complexity. Examples: What is an *"invention technique" (transformation models)*? What is a *contradiction*, and is it possible to more accurately and universally define a *contradiction type*? How does one devise *the simplest and at the same time most constructive "invention algorithm"*? How does one *increase the speed and accuracy of the teaching process and form reliable practical skills* that can be used to solve new problems?

> The *constructive teaching technology* has been developed by the author for the first time in TRIZ history.
>
> It ensures correct, consistent and adequate (for the beginners) teaching of primary TRIZ models, that is possible by the following features of the new technology:
>
> - *creation of new teaching methods (extracting and reinventing – to examine efficient solutions and model the invention process);*
>
> - *development of a simple standard format to represent reinventing results – Meta-Algorithm of Invention T-R-I-Z (MAI);*
>
> - *use, to the extent possible, of the simplest available objects (artifacts) to explain TRIZ models and complete practical exercises.*

These features help to reduce the level of complexity, assure universality of acquired knowledge, boost credibility of the proposed method, and accelerate formation of independent TRIZ application skills.

Over the last 6 years, the number of people who have benefitted from this technology has reached several thousand, with training provided in several countries to large and small groups comprised of students coming from all walks of life, of different ages and with different interests. Also, there is an association of partners in global dissemination of Modern TRIZ distant training methods which has been successfully recruiting new members for more than 3 years.

And now to my answer to a frequently asked question: What specifically are the main distinctive features of your "constructive technology" as it applies to teaching TRIZ basics, and why do you call your approach (if it may be called that) a "technology"?

The answer is: constructive Modern TRIZ basics teaching technology is based on 5 radically innovative paradigms:

I. **The "Artifact" Paradigm:** it is explicitly postulated that the *scope of learning is unlimited and includes any artifacts*, not just technical patents, as well as *any sources of information* offering efficient ideas and solutions.

---

[25] This section is written primarily for higher school teachers and university professors who are interested in TRIZ, and for Modern TRIZ followers who are using MTRIZ independently.

The purpose of the paradigm is to examine transition of any artifact from the state "was" to the state "has become" accompanied by resolution (elimination) of systemic contradictions which existed while the artifact was in the state "was"; the methodology used essentially involves reproduction, reconstruction, repetition of the process by which an efficient idea was originally created by comparing "constructions" of the prototype and resultant artifacts performing similar functions.

This paradigm generalizes and reinforces, by transforming it into a methodological axiom, the available experience stemming from examination of innovative changes not only in technology, but also in non-technical spheres, such as marketing and management, design and art (any kind of art), military and business operations, education and politics. There are no observable factors which might restrict application of the paradigm.

II. The "Extracting" Paradigm: structuring *the research method to identify models* of transformations (creative, inventive methods used to generate ideas) and contradictions (reasons for creating changes, innovations, inventions), and using it as an *intensive teaching method.*

III. The "Reinventing" Paradigm: organization of efficient training based on *modeling the complete cycle of creation (invention) of a constructive idea*, as a given artifact is transformed from the state "was" (prototype artifact) into the state "has become" (resultant artifact), *with a view to accelerate acquisition of correct practical skills for subsequent generation of new ideas.*

Extracting and reinventing can both be described as trail-blazing methods comprising the constructive teaching technology as it is used to promote systemic creativeness in general, and Modern TRIZ, the author's "pet project", in particular.

IV. The "Meta-Algorithm of Invention T-R-I-Z (MAI T-R-I-Z)" Paradigm: *standardization of training, and of subsequent idea generation process,* based on the four-stage Algorithm of Inventive Problem Solving (ARIZ), the four stages being *Trend – Reducing – Inventing – Zooming*, or T-R-I-Z.

Development of a systemically organized standardized methodology to model the transition (transformation) of a *known* prototype artifact in the state "was" into a *known* resultant artifact (heir artifact) in the state "has become" is a radical step towards assuring reproducibility and efficiency of training.

Such modeling eventually evolves into *efficient generation of ideas* which is used to resolve existing systemic contradictions in a known artifact in need of improvement. The artifact then becomes the prototype in the state "is", and we need to invent the principle and "construction" of the resultant artifact in the state "needed" which possesses the required new properties, and is free of the contradictions inherent in the prototype artifact.

V. "The Intelligence Amplifier"[26] Paradigm: *technology of mass-scale training* as realization of two processes –

---

[26] In favor of the famous Polish writer, philosopher and futurologist Stanisław Lem (1921-2006), the author of a futurological treatise *Summa Technologiae* (I have used the title of one of the chapters of that book published in Russian in 1968 – M.O.) and several novels, including *Solaris*, *Powrót z gwiazd (Return from the Stars)*, etc.

1) distant learning supported with the books and software provided at different training levels:
2) accumulation and transfer of extracting and reinventing examples presented in MAI T-R-I-Z format to amass innovative and inventive experience which can be used to train new generations of innovators and to develop new solutions.

The constructive technology has unlimited prospects of functional improvement. It also has a reliable technical platform which is being prepared for publication, and is designed to ensure future efficient implementation (individual instrumental software, TRIZ-knowledge databanks and network systems).

As new technologies evolve in the future, they will encourage further advancement of TRIZ. This is assured – and required – by the theoretical platform of Modern TRIZ (now under development).

Let us consider, as a small illustrative example, two artifacts hailing from an area which immediately takes me back to my youth, when I was of the same age as senior schoolchildren, the intended recipients of the this book, and when I first partook of the thrill of generating my own inventive ideas.

Demo-example. This is KOH-I-NOOR!

There is no doubt that this mysterious name is well-known to many of those who like drawing with a good pencil! Yes, this example is about Czech-manufactured[27] KOH-I-NOOR pencils which the 17-year-old author used at the dawn of his career when he was a design engineer at one of the then-new enterprises of future electronic industry.

Ordinary pencils were first manufactured in 1802, mechanical pencils – in 1890, almost 100 years later. The prototype artifact has a functional deficiency: it needs to be repeatedly sharpened to enable drawing thin lines. Clearly, this results in time loss and massive waste of lead.

Incidentally, the body of the pencil is made of fine wood – Siberian pine.

Without going into details, we can say that the heir artifact makes it possible to draw thin lines continuously, as the lead has the required diameter.

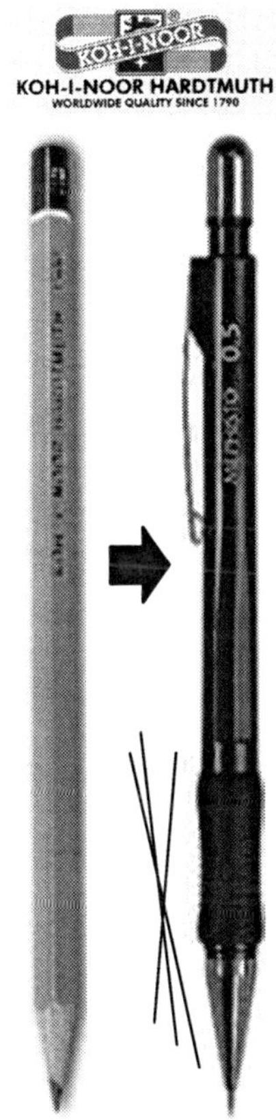

fig. 1.2. Ordinary and mechanical KOH-I-NOOR pencils

[27] The factory manufacturing pencils under the global brand of KOH-I-NOOR was founded in Vienna in 1790 by Josef Hardtmuth, the inventor of graphite-clay leads suitable for industrial production. In 1848, the facility was transferred to Czechia. The name is translated from Persian as "Mountain of Light", and is derived from the name of the largest (at that time) Indian diamond with a dramatic history spanning several centuries.

Besides, the stock of leads inside the mechanical pencil lasts much longer than the single lead inside the ordinary pencil. The body of the mechanical pencil can serve for many years, and no wood of any kind is used.

This is what we want to know: what problems are solved with the emergence of the new principle and new design? What creative models are objectively present in these well-known and widespread objects?

Let us focus on just one problem inherent in the ordinary pencil – the need to sharpen the lead.

Apparently conceptual design and functional properties of the ordinary pencil contain the following contradictions:

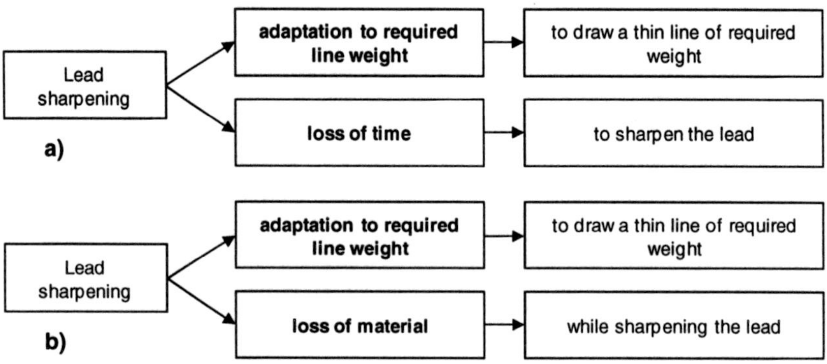

fig. 1.3. Examples of standard contradictions

Let us note that similar contradictory properties can be identified in any object; accordingly, this type of contradiction with two different contradictory properties can be called *standard*, i.e. present in all situations and artifacts as a "template", a "standard".

We can aggravate the conflict by formulating another contradiction:

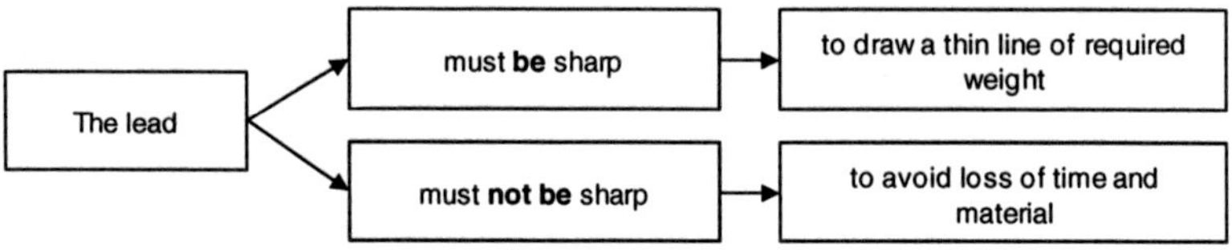

fig. 1.4. Example of a radical contradiction

Indeed, in any situation there are two possibilities: "to be" or "not to be" (sharp), "to have" or "not to have" (the property of "sharpness"). Each possibility leads to diametrically opposite and (apparently) incompatible results. This is a *radical* contradiction, the reason, the root of the problem.

The new mechanical pencil uses the following technical idea: we make thin leads of the requisite size, place them into a chamber inside the pencil, extend lead from the pencil long enough to enable drawing, replace spent lead with a new one from the inner chamber. When all leads have been spent, we load a new set into the chamber.

In this case regardless of the authors' will, i.e. *objectively* (!), there exist several dominating (critically important) transformation models whose names (according to TRIZ) reflect their constructive nature:

- "preliminary action" – the leads are "sharpened" in advance to a certain degree, and placed into the chamber in advance for future replacement,
- "segmentation" – separation of the pencil into parts (the body is separated from the lead, the lead is separated into several smaller leads),
- "replacement of mechanical matter" – the prototype construction with a solid one-piece core is replaced with a system that has special structure, mechanism, and moving parts,
- "dynamization" – the lead is made movable, there is a new mechanism which moves the lead and fixes it in working position, etc.,
- "inverse action" – do not sharpen the lead!
- "discard and renewal of parts" – replacement of leads and replenishment of their stock,
- "matryoshka (nested doll)" – use of the inner chamber to house spare leads.

Leaving out the general deficiency associated with the loss of remaining lead, the outcome is this: lead loss – 0 (zero), time loss – virtually 0! Ideal result!

The actions we have just reviewed reflect the nature of certain operations, such as selection of artifacts and extraction of transformation models and contradictions.

Now that we have extracted transformation models and eliminated contradictions, the entire process of inventing the mechanical pencil can be squeezed into four major stages:

Stage 1. Trend. Identify deficiencies in the prototype artifact, and determine the direction (trend) of future changes. For example, what do we do to avoid sharpening the lead? This, as a minimum, points to the objective of improvement.

Stage 2. Reducing. Identify causes of the problem in the form of contradictions (see above). It becomes clear that attainment of the stated objective is contingent upon elimination of contradictions. Indeed, if we do not have to sharpen the lead, no time or material will be lost! Ideal Final Result!

Stage 3. Inventing. Key idea: make lead of the required diameter in advance! Also, structure the pencil so that the user can extend the lead as necessary, and replace it with a new lead when it is spent.

Stage 4. Zooming. Now, in the same way as we change focus and zoom in a camera, we need to review the result from various positions and levels of scrutiny. For example, at the level of the working element: the lead does not have to be sharpened! At the level of the pencil: the original contradictions have been eliminated. At the level of production: it is much simpler to manufacture the mechanical body and the leads separately. At the level of the user: time is saved, line weight is guaranteed, it is cheaper to use the mechanical pencil than keep buying ordinary pencils. At the level of the environment: precious wood is not wasted.

The first letters of the names of these stages make the abbreviation **TRIZ**, or **T-R-I-Z**. And the four stages comprise **MAI T-R-I-Z**.

The aim of TRIZ is, and has always been, to learn to apply such models deliberately (as and when the authors want it!), in a purposeful and efficient manner.

The aim of Modern TRIZ is to assure mass-scale dissemination of TRIZ basics and efficient standardized application of TRIZ models, improve existing TRIZ models on an ongoing basis, and develop a general systemic and mathematical platform for the presentation of TRIZ theory.

fig. 1.5. There are 1,000 attendees in the audience!

Upon completion of an induction course provided by this book, this technology can be successfully reproduced and implemented by all Modern TRIZ practitioners: trainers, teachers, applied experts and, naturally, senior schoolchildren and college students.

Efficiency of the technology can be illustrated by the following example: after a two-day workshop in Harbin (China), none (!) of the individual reinventing papers written by the 1,000 participants had any mistakes. It was a TWO-DAY workshop!

Distant training is even more efficient, as trainees complete considerably more practical assignments. Incidentally, most certification works focus on areas which are of professional interest to the trainees.

In conclusion of this section, the author invites all interested parties to cooperate in TRIZ development, dissemination, mass-scale learning and application.

And last but not least: We get used to everything around us.

And we stop wondering at the talent of the maker and the beauty of the idea inherent in almost any artifact.

Therefore, this course is an invitation to boost your power to discern:

- *the simple in the complex,*
- *the beautiful in the practical,*
- *the amazing in the habitual.*

The emblem of our Modern TRIZ Academy bears a biblical maxim[28]:

Be transformed by the renewing of your mind.

This textbook is invitation to change, to become more inventive and powerful, so that you learn to overcome obstacles, transform the "impossible" into the possible, and simply rejoice in creating beautiful ideas and useful solutions.

---

[28] Romans 12:2; the original Latin text is: *Transformamini renovatione mentis. Source: Nova Vulgata, Apostoli ad Romanos Epistula Sancti Pauli, 12 (2).* The fact that we quote the Bible does not mean that we are biased towards any particular religion, nor are we trying to proselytize the reader. It simply means that we are drawing wisdom from a specific historical literary source.

2 Original Concepts

The list of standard methods – this is some kind of inventor's desk reference book, but a special kind of reference book: the inventor must treat it as a base that is necessary to replenish yourself from the new technical and patent publications.[29]

Genrikh Altshuller

2.1 Contradiction

PRIMUS INTER PARES
First among peers (Lat.)

2.1.1 Definition of Inventive Problem

The notion of *"contradiction"* is the most fundamental and central of TRIZ concepts. Another important concept is that of *"transformation model"*.

A contradiction exists where certain system properties prove to be incompatible, or inconsistent, or contrary to the purpose of the system.

Definition of "**Inventive Problem**"	**"Inventive Problem"** (Inventive Assignment) – a problem containing *a contradiction* in the form of incompatibility of requirements and-or properties which are appeared through an irregular development of parts of the system or of the given system and its surroundings and which possesses no adequate methods and means to solve the problem situation.
Addition 1 to definition	Resolution of problem-based contradictions requires application of special creative models and methods to constructively reorganize available professional knowledge and/or acquire new knowledge with a view to synthesize an efficient idea.
Addition 2 to definition	Resolution of the contradiction is an indispensable condition of removal of the relevant inventive problem.

[29] According to Altshuller, G. S. (1973) *Algorithm of Invention*. – From Russian: Г.С.Альтшуллер (1973) *Алгоритм изобретения. – Москва: Моск. рабочий, с.141-177*

Example 2.1. Original Issue in Problem P1. Moscow Kremlin Stars.

Architects wanted Moscow Kremlin Starts to be big. However, with an arm span of 5 meters, the star has a large wind area (surface exposed to the wind). This means that strong frontal wind can break the star or its support, and throw the star off the tower. Towers have a height of 60-70 meters.

The star is very heavy, weighing in at about one and a half tons. An attempt to strengthen the construction with additional elements to reinforce its structural integrity by definition increases the star's weight, which is undesirable.

The original problem can be shortly formulated as follows: structurally, the star must withstand exposure to strong winds. This is a *general contradiction* which can usually be "framed" by asking the question: *"What can we do to...?"* It is naturally assumed that we do not know.

Indeed, at the time when they came up with the architectural idea, designers of Kremlin Stars did not have a ready answer to that question.

Example 2.2. Clarification of the Issue in Problem P1. Moscow Kremlin Stars.

The key to solving this problem lies a more detailed definition of contradictions inherent in the original problem situation.

In this case, the star must be large and, consequently, must have a *large surface area*, so that it can be seen from a great distance; on the other hand, an increase of the surface area of the star increases its windage and reduces its resistance to strong winds, and, accordingly, detracts from *reliability* of construction. This is a typical *standard contradiction* where one factor (which is usually placed, or named, first) requires improvement, while the other factor either deteriorates concurrently with improvement of the first factor, or hinders such improvement. Both developments are unacceptable, and must be prevented.

The original contradiction may also be escalated using the following formulation: the star *must* and *must not* be large – for different reasons! This is a *radical contradiction* where incompatible or, to be more precise, directly opposed requirements are imposed on one and the same property.

2.1.2 General Contradiction

General Contradiction	**General Contradiction** (in classical TRIZ this contradiction is called *administrative*) is a systemic requirement which merely reflects the general need to attain a certain property (or state) or remove an obstacle preventing the system from operative as desired.

Problem 2.3. "Clever" Gate (beginning). Not too far from the downtown area, in the south-west of Berlin, there is an agricultural experimental field run by the Free University (Freie Universität Berlin) – a plot of land with an area of about 20 hectares planted with potatoes, wheat, strawberries (!) and many other things.

The field consists of smaller sections, fenced in by inconspicuous but very durable wire, with small barns housing sheep, horses, pigs and cows. A real "natural" pastoral nook complete with dirt roads and tracks, roadsides covered with wild bushes and grasses, and even a small bog with nocturnal frog choir presentations.

A place which enjoys unflagging love and devotion on the part of many local residents. Entrance fee: zero.

Incidentally, in Germany such "natural" places are very few and far between – all roads and paths leading to houses are covered with asphalt or tiled. And every house strives to present, in a somewhat theatrical fashion, fabulous fusion of human residence with Nature, and has a well-tended flowerbed. Flowers, many different flowers, are everywhere, as are manicured lawns, shrubs and trees!

Well, with all this talk about cleanness and flowers, I cannot help myself and not tell you a totally "irrelevant" story. Incidentally, this story also contains a "general contradiction" that we can discuss or, better yet, use as the title of this example.

Example 2.4. General Contradiction: How Do We Keep It Clean?

fig. 2.1. No comment…

I was amazed (fig. 2.1) to see, amid a huge mountain meadow, a post carrying a notice to the effect that visitors are kindly asked to clean after their dogs so that other people can walk around and lie down without running the risk of stepping into… well, you know… Right under the notice there was a small box filled with disposable plastic bags that one could use to pick up "pet products", and next to the post there was a big box where one could dump those products. The big box was firmly embedded in the ground, and its lid was weighed down with a heavy brick to prevent it from being accidentally opened by the wind.

Naturally, the solution of the general contradiction described in this example does not boil down to "provide the box and the bags and display a polite notice." The solution is to accustom people to cleanness from childhood. Unfortunately, it does not work for all people…

Problem 2.3. "Clever" Gate (continued). But let us get back to the Free University experimental lot. It can happen that by some miracle a particularly curious animal may escape. Meanwhile the gates that people use to enter and exit the plot may remain open, and the animal may find itself in the middle of a street with cars zooming by, next to an underground line (U-Bahn) which – in this particular place – runs above ground.

What can we do? - Ah, there it is, the general contradiction!

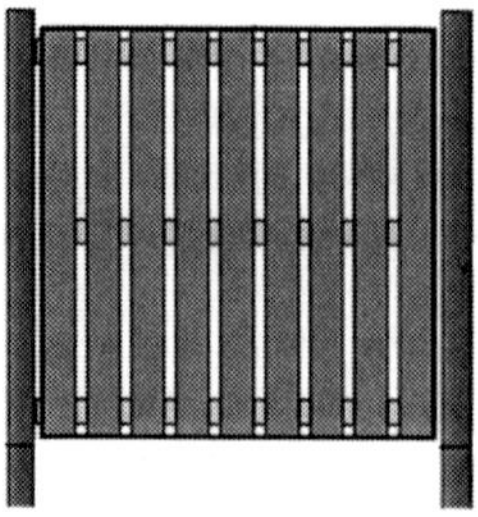

fig. 2.2.

We can fit the gate door with a spring – OK, not bad. We can use a weighted cable – that is a good one, too! We can install a special mechanism – but then it will lose its "rustic" charm. Alright, can we make a "clever" gate that would close ITSELF, and that no animal, however smart, would be able to open?

Let us take a closer look at the situation (fig. 2.2): normally, a gate hangs on hinges, on a small post, and - when closing - is stopped by another post. The gate hangs vertically, and without external influence can stop in any position between "opened" and "closed". Apparently, the *general contradiction* does not indicate the possible solution – which, indeed, remains unknown.

So the question "How do we make a simple self-closing gate?" remains.

Problem 2.5. "Clever" Platform. Two large (2 m x 2 m) sheets of glass are joined in a vacuum chamber. The sheets are very thin – less than 1 mm. They are placed horizontally. A robot with almost 100 "suction-cup fingers" puts the first sheet on the lower metal platform, then grabs the second sheet, turns it over and raises it from below to the underside of the upper platform. The upper platform must "capture" and hold that sheet. The robot then extracts its "tentacles" from the chamber, and the chamber is closed. In the past, such platforms used to hold onto the sheets "by suction": they had openings and channels through which the air was pumped out until the pressure became lower than residual pressure inside the chamber.

At this point we need to explain the situation in more detail (fig. 2.3). The sheet must adhere to the platform uniformly throughout its entire surface area. Then the upper platform slides down (see the arrow) until the distance between the two sheets becomes about 100 microns. The horizontal position of the lower platform is adjusted (see two thin arrows) so as to align certain markers placed on the sheets. Then the upper sheet is pressed to the lower sheet.

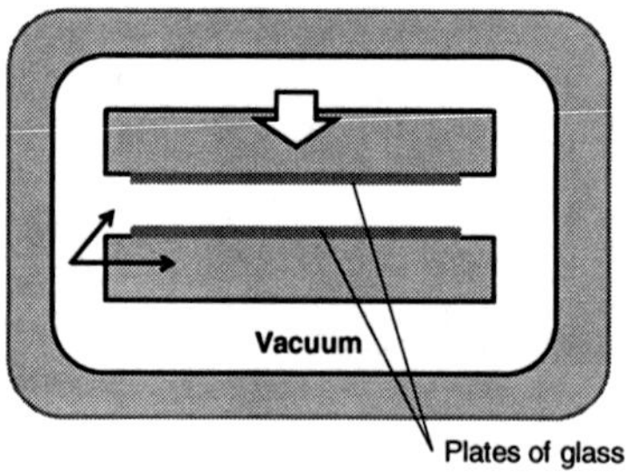

fig. 2.3. Problem: how do we keep a sheet of glass under the upper platform in a vacuum chamber?

The sheets are glued together, because epoxy adhesive is applied to the upper sheet in advance along an outline tracing locations of future monitors (the lower sheet houses liquid crystal fields).

Finally, the chamber is depressurized, and atmospheric pressure "fine-presses" the sheets into a monolithic construction. This is one of the technologies used to manufacture liquid crystal displays for computer monitors and television sets. The resultant monolithic sheet is cut into individual monitors.

General contradiction: how do we ensure that the upper sheet holds onto the platform after deep vacuum is created in the chamber (i.e. after the air is pumped out, and holding "by suction" becomes impossible)?

Problem 2.6. Tricks and "Illusions". Here is a simple task[30]: three glasses are placed on the table at distances exceeding the length of a table knife (fig. 2.4).

We need (general contradiction!) to use the glasses and the knives to construct a "bridge" so that an imaginary "ant" would be able to travel from one glass to another, and so that the bridge would be sturdy enough to hold, for example, a glass of water.

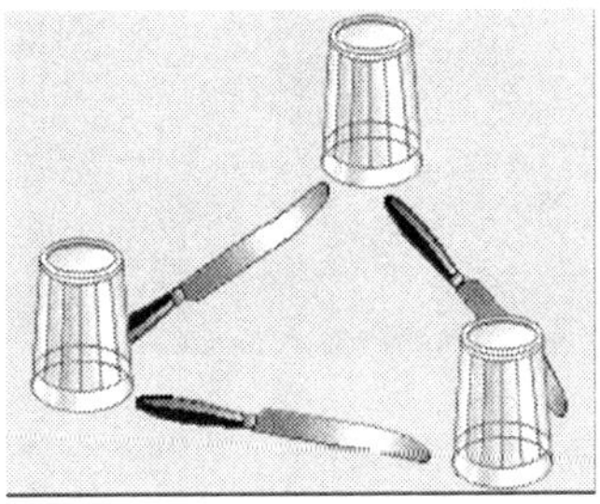

fig. 2.4. Problem: how do we build a "knife bridge" that will join the glasses?

This is a game problem meaning that there is at least one solution. In general, as long as we are dealing with tricks and "illusions", it is important to recall one of the fundamental TRIZ systemic analysis postulates: if a phenomenon exists, all resources required for its realization are available.

This is true for puzzles, too, except that puzzles may have only one solution based on a very subtle secret which is very difficult to discover. You need to *know*. And sometimes it may be impossible to figure out (reinvent) the solution! (Do you agree, by the way?)

2.1.3 Standard Contradiction

Standard Contradiction (1)	***Standard Contradiction*** (in classical TRIZ: *technical contradiction*) – binary (of 2-factors) model which reflects incompatible requirements between *two different functional features* of an object (or several conflicting objects).
Standard Contradiction (2)	***Standard Contradiction*** – 2-factors-model in which one of the factors corresponds with and supports the most important feature of the system (positive *Trend-factor* or *Plus-factor*) while the other factor does not correspond this feature or counteracts it (negative *Problem-factor* or *Minus-factor*).

30 This problem was published in 1913 by Yakov Isidorovich Perelman (1882-1942) – outstanding Russian pedagogue; Edward de Bono (b. 1933) – renowned psychologist and educator, developed a series of instructive versions of this game in the early 1960-es

fig. 2.5. "Innovative" leaf transportation devices!

Problem 2.7. "Short Arms".

In autumn, when cleaning fallen tree leaves, we face a problem: to carry the leaves from the heap to the compost pit, we have to resort to "complex" devices – carts, shovels or pitchforks – which are often not readily available.

Of course, we can try and *grab* as much as possible, but that usually does not work – our arms are "too short!

Maybe that is why the "cleaners" in fig. 2.5 had to come up with some really weird leaf transportation devices!

Let us present the standard contradiction (SC) in different notation systems.

Option 1 (Text): Cleaning fallen leaves involves the need to move large quantities of "material", but devices normally used to do that are complex and inconvenient.

Option 2 (Formula): Leaf Cleaning ► Copious Material **VS** Complex Devices

In this option, we have used a special sign (►) meaning that the situation (process, state, action, object, etc.) is "modeled" as a contradiction. Its choice is completely arbitrary, and it can be replaced with any other sign.

The sign **"VS"** is an abbreviation of the Latin (and English) word ***versus*** meaning *against, with respect to, in comparison with.*

Option 3 (Graphic): All three options represent an *informal* standard contradiction (we use the epithet "informal" because this contradiction does not use any TRIZ models).

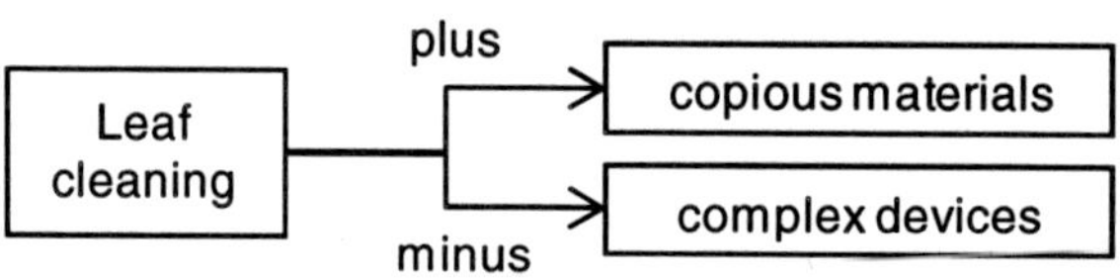

We also say: informal contradiction is not *reduced* to TRIZ models.

Other options are available, too. This is not the point. The point is we have singled out, from the description of the problem situation, the "main" conflicting properties (demands, states, actions, etc.).

Problem 2.8. That's a Lotta Line!

fig. 2.6. Mooring at Key West, Florida

I made these shots in October 2006, when our cruise liner *Majesty of the Seas* was nearing the isle of Key West, the southernmost point of the USA and the State of Florida, a place that Ernest Hemingway[31] visited over and over again, a place where he lived for many years.

fig. 2.7. The vessel is successfully moored

I was filled with the thrill of anticipation, as I was about to step over the threshold of the house of Hemingway who was one of my two most favorite writers when I was a young man (the other one being Jack London[32]). In the year when I turned 18, I won the boxing championship of Minsk, Belarus, among light middleweights, scoring the remaining points that I needed to be awarded my first sports rank (to be followed by Candidate Master of Sports and Master of Sports – but I never got those, as I dedicated all my time to work and study (at that time I was doing an evening correspondence course).

---

31 Ernest Miller Hemingway (1899-1961) – well-known US writer, 1954 Noble Prize winner, author of "Farewell to Arms", "For Whom the Bell Tolls", "The Old Man and the Sea" and other works.

32 Jack London (1876-1916) – well-known US writer, author of "The Sea-Wolf", "Martin Eden" and numerous short stories about courage, fidelity, will to live and win; my favourites are "White Fang", "A Piece of Steak", "The Mexican", "Grit of Woman" and "Love of Life"

Both writers went in for boxing, and wrote many wonderful stories related to box. And now I was going to have a meeting which would surpass even my wildest dreams…

But first things first! First we need to moor, and mooring involves delivering several lines (one of those is marked with an arrow in the photo) from a height of about 20 meters to the berth below. This is what three sailors are doing (see the circled area in fig. 2.6, a and b). But therein lies a contradiction:

Line ► long **VS** heavy, sailors are not strong enough to swing it,
especially from above, so that it reaches the berth

All the energy being applied is consumed by the line! Have you maybe remembered the thin-line problem from the kindergarten described in the author's other book, *ABC-TRIZ*?!

Problem 2.9. On the Way to a New Automotive Era: Electric Car. Many electric car models have been suggested over the 100+ years since the first such cars were designed – and even broke speed records outstripping the first cars equipped with internal combustion engines! Did you know that? Amazing, is it not?

However, even today the main problem persists: low efficiency and high weight of the onboard accumulator batteries and excessive power losses.

Standard contradictions – possible variants:

1) Electric car ► time in motion between battery recharges
VS power expenditure;

2) Electric car ► time in motion without battery recharge
VS excessive weight due to heavy battery;

3) Electric car ► time in motion without battery recharge
VS power losses at every stop, for example,
at a crossroads, if the engine is left running.

Problem 2.10. Back to the Future! It is a well-known fact that ocean-going vessels use up huge quantities of fuel and have to "carry about" fuel stocks instead of more payload. Also, the higher the speed, the higher the fuel consumption rate. Naturally, intensive fuel consumption results in intensive environmental contamination (ocean water).

Standard contradiction (it is truly "standard" for all means of transportation!):

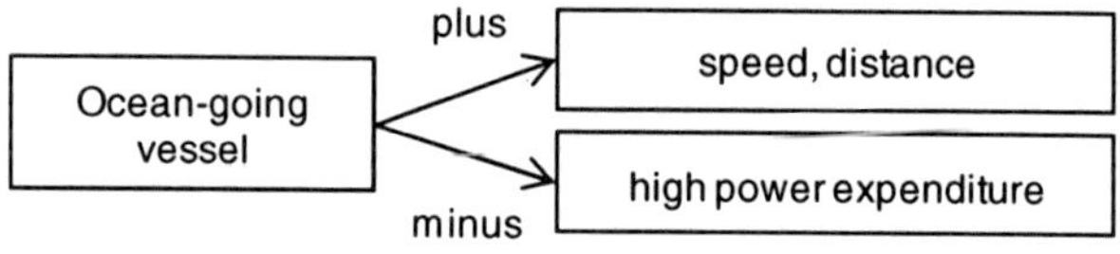

Problem 2.11. Lord of the Wheels. In section *3.6 The Invisible in the Obvious*, discussion of Problem 30 referred to the fact that the Honda U3-X scooter can move in any direction!

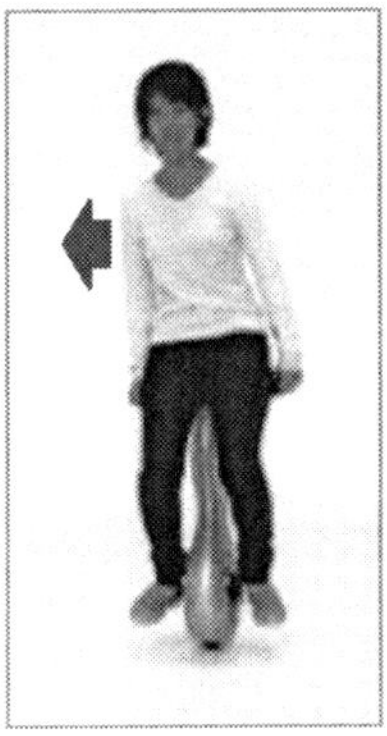

fig. 2.8.

Literally in any direction: to change direction, the rider can stop and start moving, for example, at right angles to the previous direction... without ever turning the wheel! How can this be?

In known prototype scooters, such change of direction clearly detracts from comfort of operation, as the rider has to twist his or her body like a circus contortionist to make the steering, or main, wheel move along a rounded transitional trajectory.

Conversely, with Honda U3-X the rider only has to lean his or her body in the required direction (fig. 2.8)!

There arises the following informal standard contradiction:

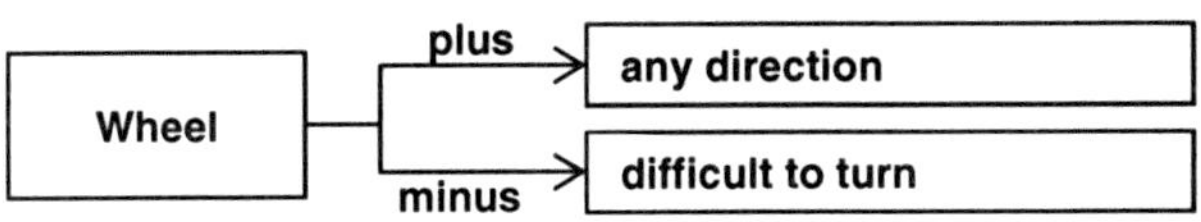

It is easy to see that in all such situations there are two different properties at war (fig. 2.9):

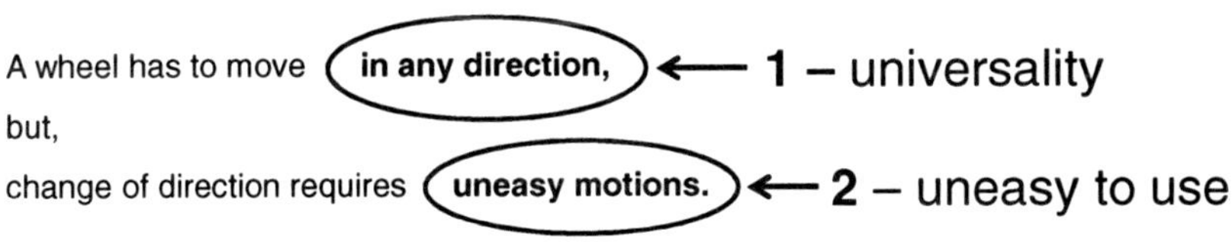

fig. 2.9. Standard contradiction for standard prototype scooters at the time when they turn

A similar conclusion can be made for any standard contradiction, for example, for the "mooring line" model (fig. 2.10):

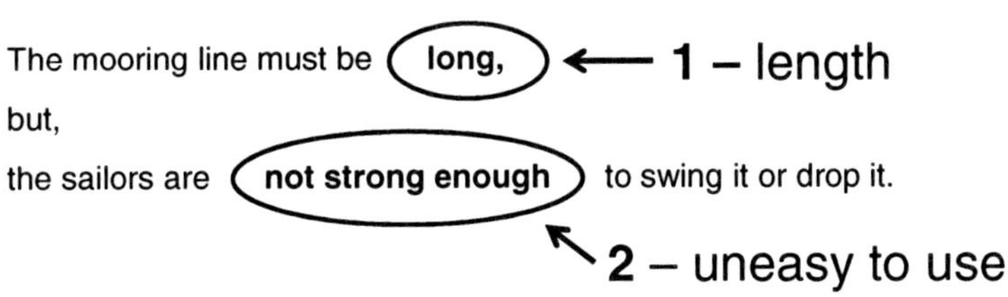

fig. 2.10. Standard Contradiction for the mooring line

Problem 2.12. "Clever" Mailbox.

Private residences are fitted with mailboxes (fig. 1.11). This is Long Island, New York. And this is very traditional for America. What can we do to enable the owner to learn that there is mail in the box without leaving home?

You can define the standard contradiction by yourselves. Very soon you will develop a taste for building similar conflict situation models.

fig. 2.11. These are not birdhouses – these are mailboxes!

Still, take a closer second look at this short section, compare your versions with those suggested by the author – and maybe yours will prove to be more accurate, and will better approximate the nature of real-life conflicts inherent in the situations described, the contradictions existing in these prototype artifacts.

Write to me about your models.

2.1.4 Radical Contradiction

Radical Contradiction (1)	***Radical Contradiction*** (in classical TRIZ: *physical contradiction*) – binary contradiction model in which contrary, excluding one another, requirements are being requested from one and the same feature of one and the same construct (component, resource, function, effect, condition etc.).
Radical Contradiction (2)	**Radical Contradiction** – binary two-factor model where the first factor reflects one demand ("plus-factor"), and the second factor reflects the other demand ("minus-factor") so that both factors represent the same property of the same construct (*component, resource, function, action, state, etc.*), but are incompatible.

Problem 2.13. Parking Lot. Large parking lots, such as those found next to exhibition centers and shopping malls, often look quite depressing – huge gray concrete or asphalt fields. People would definitely find it more pleasant and useful to have, instead, grass fields – better yet, grass fields dotted with trees! Alas, heavy cars would soon grind the grass into the ground and eventually turn those fields into dirty wastelands covered with weeds.

General contradiction: How can we improve the external appearance of parking lots?

Standard Contradiction: parking lots must have *durable coating* over the soil, but because of that they *look bad*, they *overheat*, and they *get overrun with rainwater, so that sometimes you cannot get to your car without wetting your feet.*

Radical Contradiction: the parking lot must *be concrete* to bear the weight of the cars, and it *must not be concrete* (or: it must be grass) to look beautiful, not overheat, and have better rainwater removal properties.

Graphically, this conflict can be presented as follows:

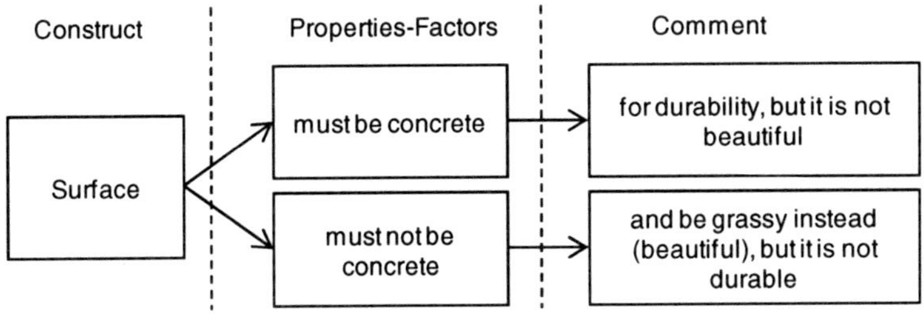

Radical contradiction formula (variants):

"concrete" **VS** "non-concrete",
"grass" **VS** "non-grass",
"durable" **VS** "non-durable",
"beautiful" **VS** "non-beautiful".

Note. If you found a solution outside of the TRIZ universe, or knew an efficient solution for the start, present it using TRIZ models.

Problem 2.14. Bridge Across a Navigable River. Radical contradiction: the bridge *must be* to join the banks so that pedestrians and vehicles can cross the river, and the bridge *must not be* so that vessels with high masts and decks can navigate the river. (See note to the previous problem.)

Problem 2.15. Ads on the Bus. Would you trade places with the passengers of this bus pretending to be an airplane (fig. 2.12)? I would not – the windows are "painted over". They probably can see nothing from the inside?!

Now, that is a truly radical contradiction:

the ad *must be* on the bus, particularly when it is so impressive – an airplane rolling along the street with its landing gear legs propped against the wheels of the bus will DEFINITELY draw stares from the amazed onlookers! –

fig. 2.12. Airplane-Bus "hybrid"!

and it *must not be* on the bus, as the situation where you cannot see the sights on a sight-seeing tour is unacceptable!

Who wants to ride around the city "inside a suitcase"? So, you decide: Do you want to see things through a small porthole (at best) – or through large bus windows?

Problem 2.16. Comfortable Bus Step. How can you keep passengers comfortable when they are boarding and alighting the bus, especially when they carry heavy suitcases or travel with a baby carriage? You cannot lift boarding platforms to the bus-floor level at each bus station, now, can you?!

Radical Contradiction:

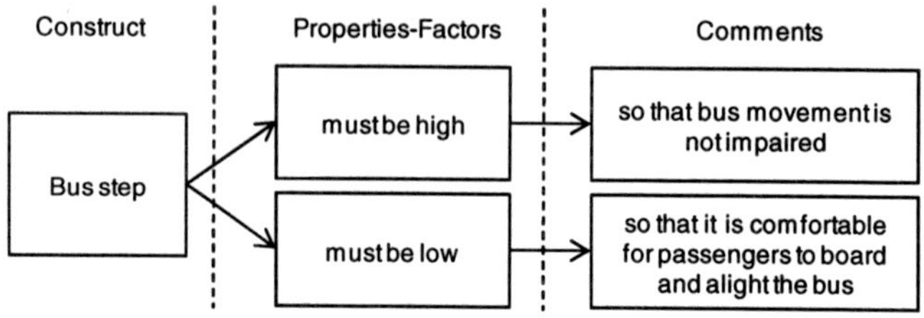

Formula: The step must be "high" **VS** "low", or the step ▶ "high" **VS** "low".

Problem 2.17. Like a Fairy Tale! You probably remember a clever (you might even say "inventive") girl which managed to comply with the king's demand to come before him "neither dressed nor naked, neither on horseback nor in a cart, neither by path nor by road"...

Incidentally, the fairy-tale lassie successfully dealt with a SET of radical contradictions!

Try to create formulas and graphic models for those contradictions.

Problem 2.18. Memory Eternal. What can be a symbol of remembrance – of "memory eternal"? It is important because people continue to live in us and through us while we remember them. They continue to live in the new generations while representatives of such new generations remember their families, their history, their motherland. This is the only way to preserve and pass on memories which hold together national identities, families and religions. To refuse to remember is to change or lose one's identity.

With time, monuments decay and fall apart. Memories, on the other hand, can be preserved forever while their carriers remain alive and pass on to their heirs things that need to be remembered. So what is the most "adequate" symbol of memory eternal?

Informal radical contradiction: the symbol *cannot (must not* – according to the laws of Nature) *be* eternal, as with time any "physical object" is destroyed, and it *must be* eternal to remind us about things that should not be forgotten.

Here is a more formal presentation: symbol ▶ "eternal" **VS** "non-eternal".

Problem 2.19. Sun Beds. And now back to the sea! These people are sun-bathing on board the Star Princess (photo by the author: Miami, 2007) – hundreds of sun beds (fig. 2.13) are set on deck each morning, and taken away each evening. But how and where do they go? The photo clearly shows that they occupy a lot of space!

Therein lies a radical contradiction:

sun beds ▶ "much space" – for use
VS "little space" – for storage.

How do we resolve this contradiction?

fig. 2.13. Here everyone has a place under the sun!

Problem 2.20. Window Washer. Here is a seemingly simple task: how do you wash a window if it does not open so that you are able to reach each and every spot on the pane from the outside? Well, you could always make, say, a special rod with a curve. But that may cause another problem: the rod may become too complex.

Accordingly, we have the following *general contradiction:* what can we do to wash (whether on a one-time basis or repeatedly) window panes from the outside?

We can also construct a *standard contradiction*:

the rod makes it possible to wash the window both from the inside and for the outside, ***but*** *the construction of the rod may become too complex.*

Last but not least, we can proceed to formulate a *radical contradiction*:

the rod must be outside the window (to wash the window pane from the outside) ***and*** *it must not be outside (as this is physically impossible).*

In the last two variants, we have named specific obstacles preventing the use of the main property of the rod to enable completion of the main useful function of the entire operation – namely, to wash the window panes (both from the inside and from the outside).

We can you visualize these models (fig. 2.14).

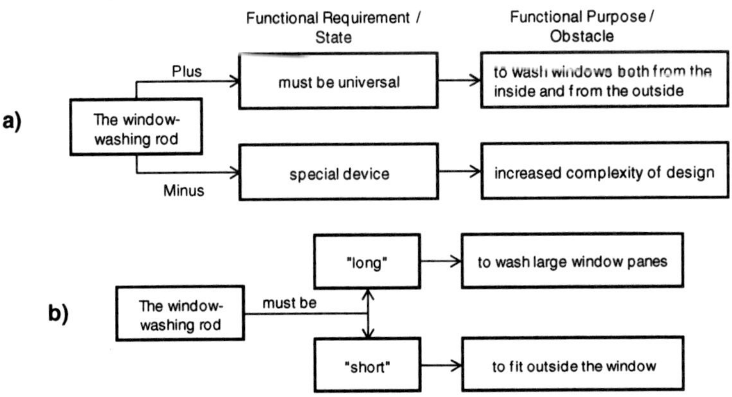

fig. 2.14. Graphic representation of informal Standard (a) and Radical (b) Contradictions for the "Window Washer" example

Also, to avoid psychological stereotyping, we recommend that the word "rod" be replaced with some conventional name for the future construction. For example, we can call it a "washer". Thus we finally formulate the trend that can be used to search for an idea for the solution: we need a "washer" that can wash windows from the inside and – importantly! – from the outside, and is not too complex!

Problem 2.21. Makeyev Missiles[33]. In the early 1960-es, with the commencement of construction of nuclear submarines, it turned out that to enable installation of more powerful missiles it would be necessary to build huge submarines.

Naturally, the missiles could still be transported in horizontal position (fig. 2.15), but before the launch they would have to be rotated and placed vertically on their launching pads.

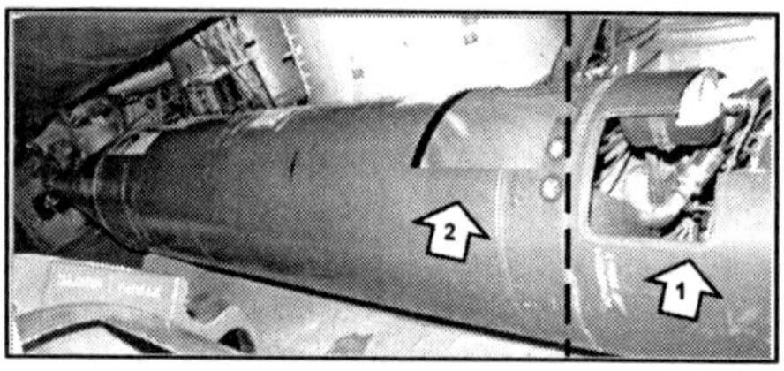

fig. 2.15. 11FM Missile (1950-es)

This, however, would necessitate time-consuming pre-launch preparations and exclude the possibility of submerged salvo launches.

The length of 11FM, the first one-stage sea-launched missile, was slightly more than 10 meters, its range – less than 200 km. The dotted line in the drawing separates the missile's engine (1) from its fuel tank (2).

To extend the range to 2,000-3,000 km and increase the payload, the missile would have to have two stages. Unfortunately, an attempt to add a second stage without changing the missile's construction would increase its length to more than 16 meters. As a result, the missile would simply not fit inside the submarine.

Radical contradiction with respect to the missile:

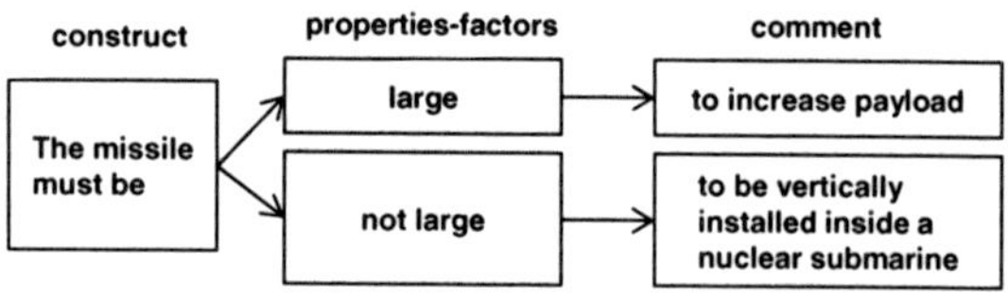

Radical contradiction with respect to the submarine:

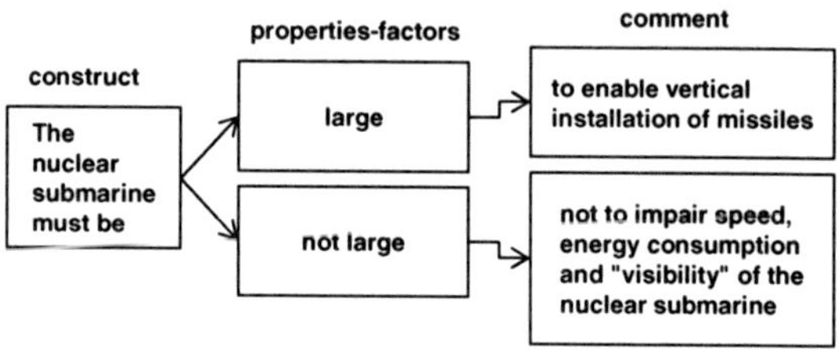

Which solution would you choose if you were the "Commander-in-Chief": a submarine of unprecedented height or a missile with inferior parameters?

---

33 Viktor Petrovich Makeyev (1924-1985) – outstanding Soviet rocket designer, creator of the Soviet submarine-launched ballistic missiles design school

Problem 2.22. Time is Powerless! The temple in Philae, Egypt, was built approximately in the 3rd century BC, and used to glorify Isis, a Pharaonic goddess. In the 6th century AD, Christians converted it into a church.

Question: what would you do to try and preserve the beautiful frescoes inside the temple for posterity in the hope that one day justice would prevail?

Radical contradiction:

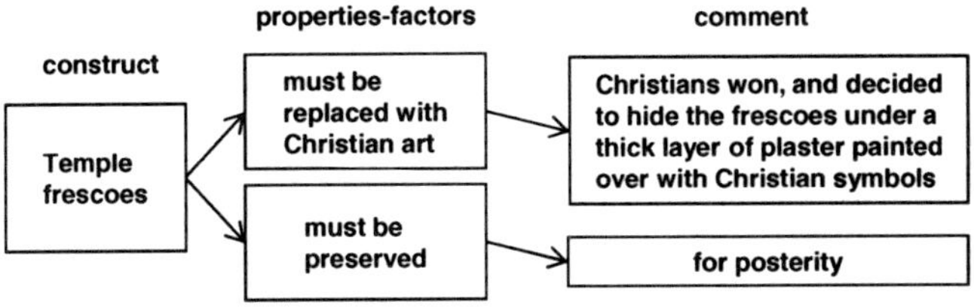

Problem 2.23. Art and Craft of Ice. One evening during a workshop that I gave in Harbin, China, in February 2008, my translator and good friend Sung Siandung (whom I called by Russian name Aleksey – according to the long-standing tradition and, naturally, with his consent) invited me to the famous ice festival.

A 1 km^2 plot was filled with magnificent works of art created of ice by sculptors from all over the world. Replicas of the London Tower Bridge, the famous Notre Dame de Paris, the Athens Acropolis, a mosque, an Orthodox Christian church and many other buildings were towering above the square covered with myriads of color lights (fig. 2.16 – color photos are presented at www.eastytriz.com).

Ice sculptures were made of large ice blocks ("bricks"). Each such "brick" had dimensions of 25 x 25 x 50 cm and weighed almost 30 kg. The builders must have used several hundred thousand "bricks"! But neither any single one sculptor, nor all of them together for that matter, would be able to manufacture enough "bricks" to build, say, the Acropolis!

How was that grandiose construction site supplied with building materials? What did the organizers and participants invent?

fig. 2.16. Harbin Ice Sculpture Festival in "miniature"

Radical contradiction:

"bricks" ► "many" – approximately 1.5 million **VS** "few" – the sculptors cannot make about 20,000 "bricks" needed for a relatively small sculpture, and then build it over a reasonable period of time.

2.2 Transformations

2.2.1 Transformations Pyramid

TRIZ continues to evolve. Today it still exists as a theoretical *background*, an empirical substantiation, a proto-theory. We still have to walk the long and winding road leading to the creation, within the structure of Modern TRIZ, of in-house "algebra" and "systemology" or, maybe, proprietary "systemogenetics" and "nanotechnologies". Ahead of us lie breathtaking discoveries whose fruits will need to be cemented through value- and purpose-oriented synthesis.

Besides, Modern TRIZ currently relies on proto-models awaiting future structuring, algorithmization, expansion and development.

In a first approximation, TRIZ models[34] can be presented within a single structure as a "pyramid" (fig. 2.17). The general principle underlying this pyramid is that the "lower" models can be regarded as more detailed versions of the "upper" models. The "upper" models, in turn, can be regarded as meta-models, or "capsules" containing the "lower" models.

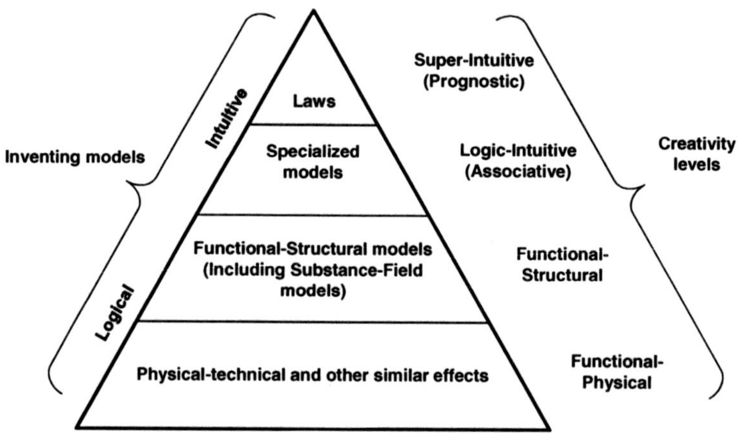

fig. 2.17. Pyramid of the TRIZ-models of transformations

Next to the pyramid are the "names" of these levels from the viewpoint of the role played by logic (left hemisphere), intuition (right hemisphere) and "super-intuition" (according to P. V. Simonov[35]), and their interactions. Each level assumes the dominating position depending on which type of model is in use. All

[34] M. Orloff (2006) *Inventive Thinking through TRIZ*, sections "Classical A-Studio Invention Navigators", "Strategy of Invention" and "Art of Invention". – Springer Verlag

[35] Academician Pavel Vassilyevich Simonov (1926-2002) – outstanding Russian neurophysiologist, biophysicist and psychologist, creator of the "information-need" approach to behavioural and creative studies.

levels of the pyramid are transfixed by three systemic entities which hold the pyramid together – and which we intend to explain at higher levels of Modern TRIZ training.

These notions are best introduced after there has emerged a relatively numerous and qualitatively diverse community of theorists (scholars) and practitioners (users). It should also be noted here that specialized models – which we are going to tackle in the following sections – are the simplest in terms of understanding and practical application.

2.2.2 Specialized Transformations

Here we intend to deal with the purely empirical aspects of the notions "transformation" and "transformation model". The algorithm which can be used to extract models from individual artifacts is presented in the section "Extracting".

Problem 2.24. See the Invisible in the Obvious (beginning). What do these things have in common: health insurance, a safety belt in a motor car, and a fuse installed in your apartment and protecting all electric devices and wiring from short circuits and overloads?

Let us look at the Main Useful Function (MUF) of each system and see what critically important and creative things have been introduced into it in comparison with systems that do not have such function.

Example 2.25. Insurance System. Let us assume that a certain individual (let as call him Mr. N) is not paying the relatively small annual or monthly insurance premium, and is not a member of any insurance program. Then it so happens that Mr. N finds himself in need of prolonged special treatment. Accordingly, to restore his health, Mr. N has to find a sizeable amount of cash to cover these unexpected expenses. Naturally, Mr. N will have to pay all treatment costs by himself. And what if he does not have the required amount?

So, in this case the MUF is health insurance. And Mr. N has failed to use this function

Let us imagine an opposite situation where Mr. N is a member of an insurance program. The program would cover all his medical costs! Let us also assume that the annual premium that Mr. N would have to pay is € 3,000, that the required course of treatment costs € 60,000, and that the tr eatment cannot be delayed or replaced, say, with a vitamin course. Therefore, all insurance program members contribute relatively small amounts to create a massive cash fund which can be tapped if any of them needs significant assistance.

Yes, insurance is a truly valuable invention!

Example 2.26. Safety Belt. It is a well-known fact that safety belts save numerous lives in car accidents – which occur when the driver becomes exhausted, or fails to notice a patch of ice or promptly react to an animal jumping into his headlights out of nowhere. Safety belts also help us win the war waged, both against themselves and against all others, by irresponsible individuals which violate traffic rules.

Rules where virtually each paragraph is written in blood. And what? See information in Problem P4.

So safety belts are a really useful invention! The MUF of the safety belt is to help the person survive a dangerous unpredictable situation which unfortunately may occur through somebody else's or his (alas!) fault.

Example 2.27. Fuse Guarding Against Short Circuits or Overloads. A short circuit happens when live wires come into direct contact. This generates an electrical charge resulting in overheating and inflammation of insulation and other nearby objects. A network overload may lead to inflammation from overheating of, or damage to, electrical devices.

Fuses installed throughout the house immediately cut off power supply and discontinue the charge when there is a short circuit, or the permissible aggregate load is exceeded. After the problem has been remedied, the fuse is restored to its working condition.

This invention has saved innumerable devices, houses, and lives.

So, a question for Trivial Pursuit aficionados: name the creative model which is common for all these three situations!

Problem 2.24. See the Invisible in the Obvious (solution). All three problem situations in Examples 2.25 – 2.27 relate to events which may occur in the future. The MUF *in advance* takes certain measures designed to *mitigate the negative impact of such events.*

It brings to mind a Russian proverb: if I knew where I would fall, I would put some straw there to cushion the blow! Wait, why not use a *cushion*? Interestingly, the *creative model* used in all three situations is called "previously installed cushion"! In TRIZ this model is represented as follows:

Previously installed cushions	Increase the relatively low security of an object with safety measures in advance.

The "specialized" principle underpinning the model can be used to find efficient solutions in situations where you need to protect yourself in advance from developments which are possible, yet undesirable. All other "specialized" models are defined similarly.

Clearly, the *technical implementation* of the principle in all three systems is completely different:

- cash fund in an insurance program;
- safety belt activated when the body moves with certain acceleration (try to rapidly lean forward in your car while your safety belt is fastened – even when the car is not moving – and the belt will arrest you progress!);
- fuse activated when the intensity of the current which runs through it exceeds a certain predefined level.

Now let us identify the differences between creative solutions used in these situations and artifacts.

Example 2.25 (supplement). Insurance System. The TRIZ creative principle used here is called "segmentation".

Segmentation	a) disassemble an object into individual parts; ... c) raise the degree of disassembly (reduction into parts) of an object.

Clearly, the overall insurance benefit paid upon the occurrence of an insured event is divided into small contributions paid by individual members of the insurance program.

Example 2.26 (supplement). Safety Belt. Several more creative models are used here, and one of them is called "mediator".

Mediator	a) use another object to transfer or transmit an action; b) temporarily connect an object with another (easily separable) object.

This basically is a direct description of the "creative technique" implemented in the invention called "safety belt".

Example 2.27 (supplement). Fuse Guarding Against Short Circuits or Overloads. For the sake of illustration, in this case we can demonstrate several (but not all!) transformation models making up the "composite creative picture" of this apparently simple solution.

Copying	a) use a simplified and inexpensive copy instead of an inaccessible, complicated, expensive, inappropriate, or fragile object

Indeed, the fuse is like a switch which terminates electric current under certain conditions. The fuse, therefore, copies and directly implements the switch function.

Inexpensive short-life object as a replacement for expensive long-life object	replace an expensive object with a group of inexpensive objects without certain properties, for example, long life.

Previously (and in many systems currently) structurally simple and inexpensive fuses were made of easily fusible materials, and were destructible and expendable. As a result, the live wire within the fuse melted and broke when the current exceeded a certain threshold value sufficient to destroy it.

Transform damage into use	a) use damaging factors, especially damaging influences from the environment to achieve a useful effect

In this case, strong current which could potentially damage electrical devices installed in the house effectively "commits suicide" by destroying its own circulatory system (the fuse).

> The minimal primary objective of TRIZ training and self-education is to learn to *see the unobvious* – see transformation models – in any artifact, in any situation where positive (or, sometimes, negative) changes have already occurred, or may occur in the future.

Let us look at more examples.

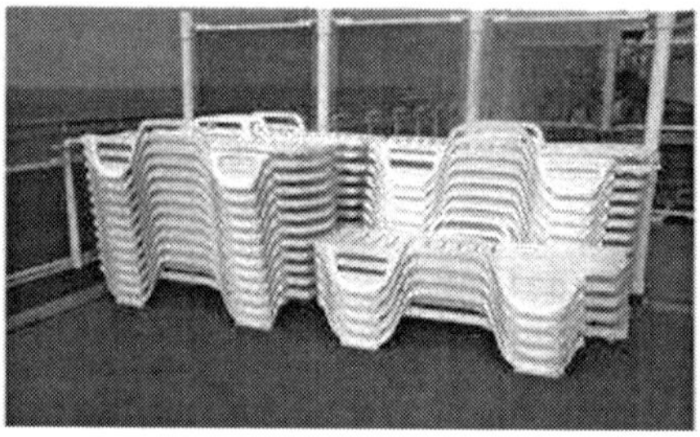

fig. 2.18. A sun beds pyramid!

Problem 2.18. Sun Beds (solution).
The solution of the radical contradiction in this assignment is provided by the special construction of the sun bends which makes it possible to stack them up one on top of the other (fig. 2.18). As a result, the sun beds occupy as little space as possible!

Example 2.28. From Wooden Pointer to Metal Pointer. A wooden pointer is usually about 1.0-1.5 m long (fig. 2.19).

fig. 2.19. Wooden pointer

Let us formulate the following radical contradiction:

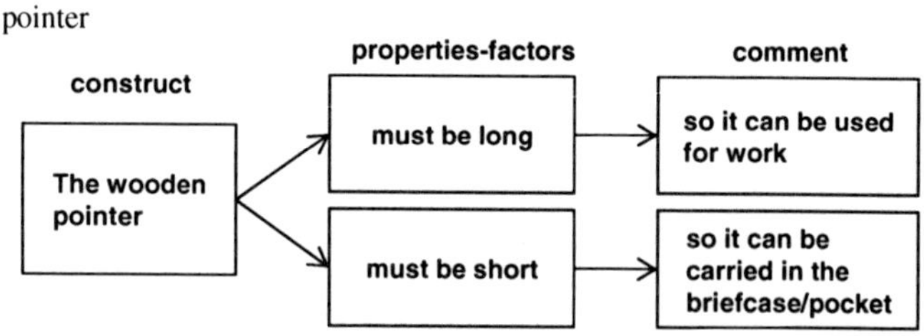

Solution of the problem: metal telescopic pointer (fig. 2.20). Such pointer can be extended to 1.0-1.5 m, but when collapsed, it fits inside a briefcase or even in a jacket pocket.

fig. 2.20. Metal telescopic (extensible) pointer

There are two leading, dominating transformations here. For now, we will name only one of them: make an object of several parts, each part inserted inside another! Have you noticed that this solution is in principle very similar to the solution of the sun bed problem?

Example 2.29. Shooting Tower. This military solution[36] presents interest to us only in connection with examination of the creative model objectively realized in it (fig. 2.21). We are not posing or discussing questions related to its functionality or efficiency.

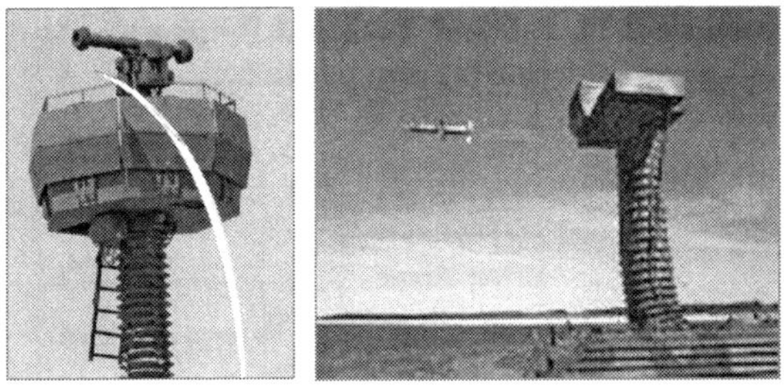

fig. 2.21. Elevating tower used to launch… no, not arrows – missiles!

In the opinion of the manufacturers, one of the intended uses of the tower is to destroy enemy tanks with self-guided missiles.

It is assumed that the tower can be covertly delivered to, and deployed in, the operative zone (in a forest or on broken terrain), unexpectedly assume the firing position, and attack the enemy.

The mast of the tower is made in the form of a telescopic construction. Thus, we already have three artifacts sharing the same key idea for resolving the main contradiction "be large VS be small" by using a similar approach – a transformation model which in TRIZ bears the name of "Matryoshka", or "nested doll".

fig. 2.22. Matryoshka doll.

Matryoshka (nested doll)	a) an object is inside another object that is also inside another, etc.; b) an object runs through a hollow space in another object.

Let us look at more examples.

Problem 2.13. Bridge Across a Navigable River (Solution).

One of the known solutions involves the use of a draw bridge. In other words, the construction is no longer static, as is now incorporates a dynamic component.

fig. 2.23. Palace Bridge in Saint-Petersburg

[36] MSLP, or Missile Stand Lifting Platform, is an antitank system designed by the Danish company Falck Schmidt Defence Systems.

During one (longer) time interval, the bridge is "closed", and is used in accordance with its main useful function, linking the banks of the river so that it can be crossed by pedestrians and vehicles, and during another (shorter) time interval it is "opened" for the passage of high vessels.

Example 2.30. B-1 and Tu-22M3 Bombers[37]. At one time designers of jet aircraft faced the following contradictions:

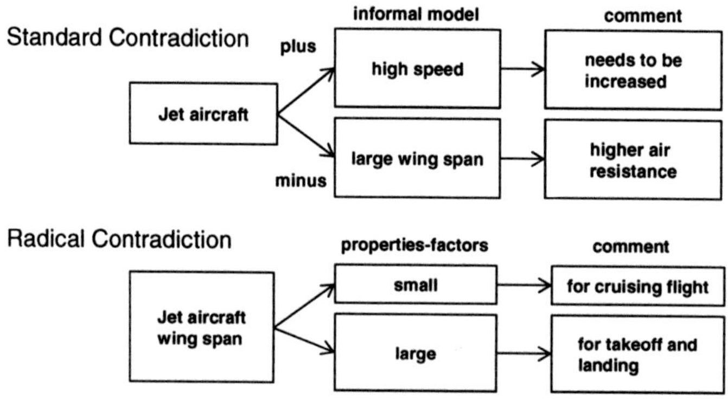

Solution-invention: use of variable-sweep wings!

In fig. 2.24,a B-1 is preparing to land, and the wings are unswept (spread out).

In fig. 2.24,b B-1 is performing a high-speed flight, and the wings are swept (pressed to the fuselage).

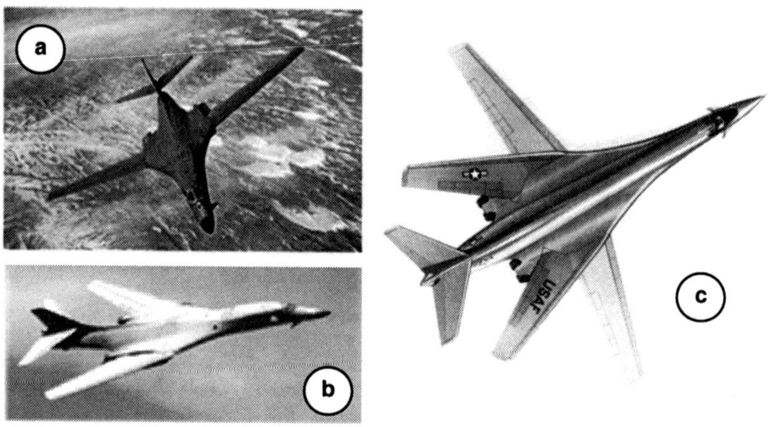

fig. 2.24. B-1 – currently (2010) the only US combat aircraft with variable-sweep wings.

[37] The Rockwell International B1 Lancer bomber has been used by the USAF since 1986; the Tupolev Design Bureau began to design the Tu-22M3 (Backfire-C) bomber in 1974; the Soviet Air Force started to use it in 1984 (officially, in 1989)

The schematic representation of wing sweep variation in the Tu-22M3 aircraft is shown in fig. 2.25.

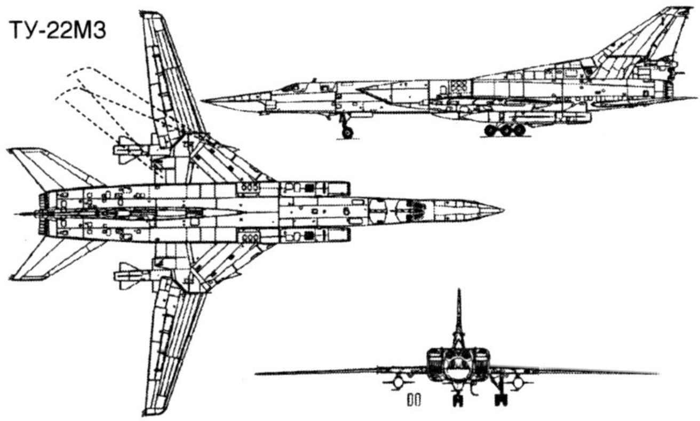

fig. 2.25. Variable-sweep wings of the Tu-22M3 aircraft

Example 2.31. Su-24 Attack Aircraft[38]. In the 1960-es, the Special Design Bureau headed by P. O. Sukhoi[39] was instructed to create a multi-purpose jet aircraft meeting certain opposing requirements which can be represented in the form of the following standard contradiction:

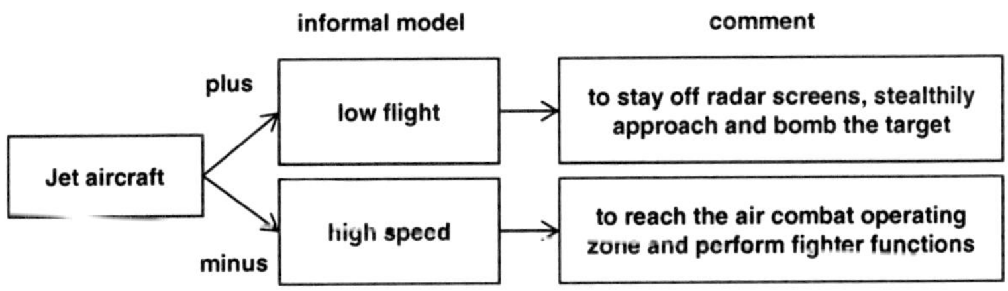

The key TRIZ principle underlying the solutions in Examples 2.13, 2.30 and 2.31 is called "dynamization".

Dynamization	a) the characteristics of an object or an environment are changed to optimize every work procedure; b) disassemble an object into parts that are moveable among each other. c) make an object moveable that is otherwise fixed.

38 The Soviet/Russian Sukhoi Su-24 (Fencer) aircraft has been used by the Soviet/Russian Air Force since 1975; its production was discontinued in 1993

39 Pavel Osipovich Sukhoi (1895-1975) – outstanding Soviet aircraft constructor and designer, one of the founders of the Soviet jet-propelled and supersonic aviation

fig. 2.26. Su-24 bomber with variable-sweep wings

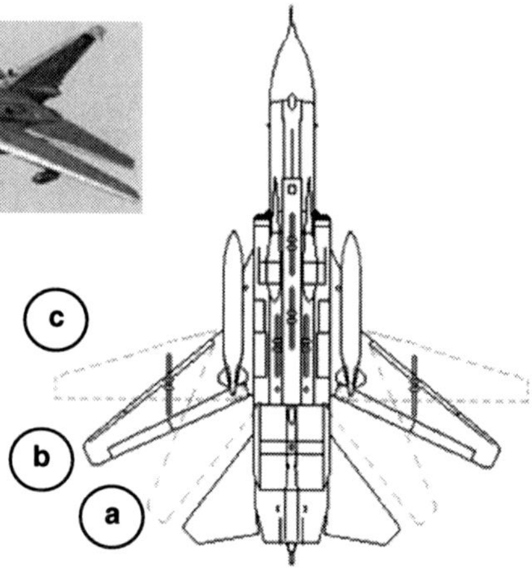

fig. 2.27. Su-24 three wing positions

And in conclusion, let us "transform" the telescopic pointer to illustrate other transformation models.

Example 2.32. From Telescopic Pointer to Laser Pointer. Let us assume that we need a pointer, say, 3-5 m long. Clearly, a wooden pointer of that length – even if it could be manufactured – would be very heavy.

Standard contradiction for this example:

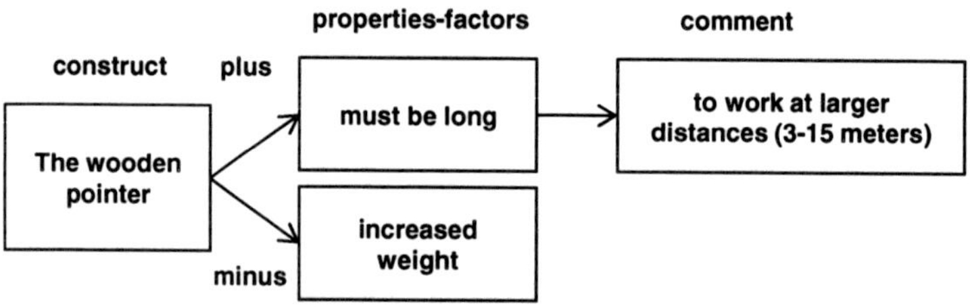

For this problem, TRIZ suggests a solution based on the "counterweight" model – even though for distances of 2-4 m we still might use a collapsible pointer (I once saw such pointers in a military school – they were sectional and consisted of many links, like fishing rods) or a telescopic pointer based on the "dynamization" model.

With the dimensions specified, the telescopic pointer encounters the following standard contradiction:

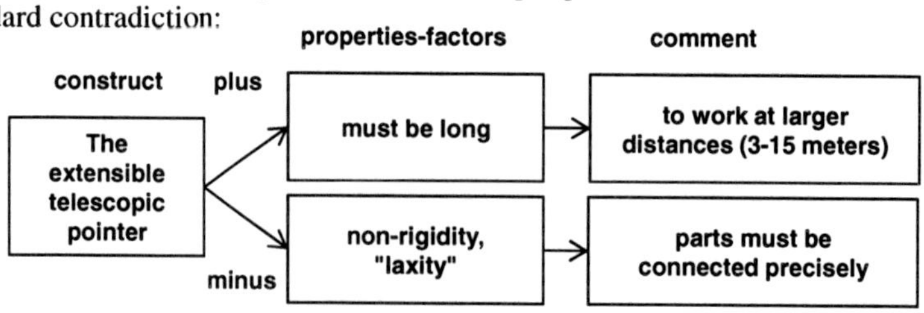

Even a cursory inventory of the TRIZ arsenal shows that this contradiction is best resolved by the "replacement of the mechanical matter" model – *a) replace mechanical structure with optical structure.* And *voila* – many speakers use laser pointers (fig. 2.28).

Specialized transformation models were the first TRIZ tools to be designed – and will be the first to be learned.

fig. 2.28. Laser pointer

2.3 Operative Zone

The Operative Zone (OZ) is one of the fundamental TRIZ notions.

Figuratively speaking, OZ is the *hypocenter of the problem*, unlike the *epicenter of the problem* which refers to external manifestations of the problem represented, in particular, in the form of contradictions.

This metaphorical definition can be explained with the following simple scheme (fig. 2.29).

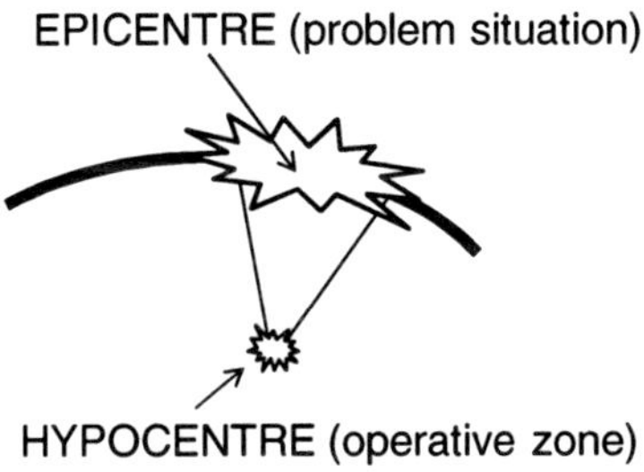

fig. 2.29. "Epicenter" and "hypocenter" of the problem

The earthquake starts in the *hypocenter* because of the shifts or other changes that occur to the Earth's crust at some depth.

The *epicenter* is defined as the area where those subsurface changes manifest themselves on the surface of the Earth. In other words, "invisible" processes evolving in the hypocenter are the cause of the "visible" observable problems (destruction and other catastrophic consequences) that we see in the epicenter.

When analyzing a problem situation within the TRIZ framework, we formulate models of the problem (or multiple problems) in the form of contradictions. Problems (regardless of the form of their presentation), conflicts, contradictions – all these are things that lie on the surface, in the "epicenter".

But what causes those contradictions? What subsurface processes create problems, conflicts, contradictions? What elements act as "initiators" or participants of those subsurface processes?

The common answer to all those questions can be located within the operative zone – the "hypocenter" of the problem situation.

The formal definition of the OZ gives only a marginal idea of this notion. To master it, you need to accumulate ample practical experience. Still, we will use examples provided in the following sections to review situations where identification of the OZ is a relatively straightforward and intuitive process.

In the simpler examples, the hypocenter and the epicenter may be deemed to be virtually the same, as the relevant contradictions directly reflect the conflict between the properties inherent in OZ elements.

<table>
<tr><td>Definition
"Operative Zone"</td><td>**Operative Zone (OZ)** is the totality of system components and, in some cases, system environment elements which are directly related to, and lead to the emergence of, the contradiction.</td></tr>
<tr><td colspan="2">OZ is perceived as a certain functional/structural model reflective of the composition of, and the connections and interactions between, the elements within a certain modeling system (language).</td></tr>
<tr><td>**"Occam's Razor"**
- addition to OZ definition</td><td>OZ consists of elements which are necessary and sufficient to fully describe the processes (as well as the causes and consequences) leading to the emergence of contradictions.</td></tr>
<tr><td colspan="2">The necessity condition sets forth the requirement in accordance with which the OZ consists of such elements that, if even one of them is excluded, it will make it impossible to fully describe the processes (as well as the causes and consequences) leading to the emergence of contradictions.
Such OZ description is *a priori* non-excessive and, therefore, minimally complicated.</td></tr>
<tr><td>**"Hypocenter of the problem"**
– addition 2 to OZ definition</td><td>For relatively more complex systems, OZ delineation (location and structuring (modeling)) often involves development of several intermediate models, and for each such model it proves possible (or necessary) to find an even "deeper" OZ.</td></tr>
<tr><td colspan="2">If there are several successively linked OZs, any of them can be regarded as the epicenter for a "deeper" OZ which, in turn, is the hypocenter for the previous, more "superficial" OZ.
Any OZ can have, as its epicenter, reformulated contradiction models related only for those processes and consequences that occur in such OZ. Those contradictions reflect the requirements posed by the superior OZ to the "deeper" OZ.
The "deepest" OZ can be defined as the primary OZ-hypocenter for the entire system, entire problem situation, specific problem, conflict or systemic contradiction under analysis.</td></tr>
<tr><td>**" Dynamiztion of OZ"**
– addition 3 to OZ definition</td><td>Creation of a chain of operative zones (or creation of a functional/ structural model of any OZ) can be regarded as a dynamic process evolving not only "depth-wise" (determination of the "hypocenter of the problem"), but also "width-wise", with the final result depending on the composition and source of resources used by the model.</td></tr>
<tr><td colspan="2">OZ models can widely vary in terms of composition, structure and processes depending on the selected aspects of analysis, with type of resources by which core analysis is being performed and the nature of anticipated change (Ideal Functional Model) being the most critical aspects.</td></tr>
</table>

Example 2.33. Operative Zone in the Construction of Moscow Kremlin Stars

Let us recall the standard and radical contradiction as they are defined based on the description of the problem situation:

Standard Contradiction: the star must have *large surface area* so that it can carry many lamps and be visible during the night; however, this increases windage and reduces *reliability of construction* when it is exposed to strong winds.

In this case, there are two conflicting properties: *area* (of static object) and *reliability*. Such conflict is normally present in virtually all problem situations: something needs to be reinforced or developed – but such reinforcement of development detracts from another aspect (property). There emerges a general trend (direction towards the goal, and the goal *per se*) where the desired solution is synthesized subject to the following requirements: the first property (*plus factor*) must be improved, while the second property (*minus factor*), as a minimum, must not be impaired.

The **radical contradiction** is also quite simple to build – it is a set of conflicting requirements related to one and the same aspect (factor, parameter, element): the *area* must be *large* (so that many lamps can be installed throughout the surface of the star), and it must be *small* (ideally, equal to "zero", so that the wind cannot bring pressure to bear on such "zero" surface).

In this case, the operative element is the entire frontal surface of the star. The opposing requirements (attainment of different parameter values and, consequently, of different properties) are posed to the "area" of that surface: it must be *large* and *small.* The critical parameter here is the measurable surface area.

These contradictions are "informally" presented in the table below (fig. 2.30).

<table>
<tr><td>Problem P1. Moscow Kremlin Stars</td><td colspan="2">Standard Contradiction</td></tr>
<tr><td rowspan="4">A large star on a tall tower can be damaged by strong wind.</td><td>(+)-factor</td><td>(-)-factor</td></tr>
<tr><td>Large surface area, so that the star can be seen from a large distance</td><td>By large area, the star could be destroyed with a strong wind</td></tr>
<tr><td colspan="2">Radical Contradiction: The star must have –</td></tr>
<tr><td>a large area to house many lamps</td><td>a small area to weather a strong wind</td></tr>
</table>

fig. 2.30. Moscow Kremlin Stars' OZ contradictions

Operative Zone: the OZ is represented by the entire construction, as the hypocenter and the epicenter (in this simplified description) are the same. The OZ consists of the following elements: the star and the support in the form of a pointed summit (spire) of the tower. To put it simply, the problem in the OZ arises because strong frontal winds create significant pressure on the star, and the star, or the support, or the junction between the star and the support can break.

It is necessary to change the composition of construction elements, or the structure of one or several elements, so as to eliminate the existing contradictions.

2.4 Invention

The notion of "inventive problem" in TRIZ was defined as a problem whose efficient (acceptable) solution still awaits to be created because obvious solutions would be unsatisfying or they do not exist and would be impossible to obtain by using existing professional methods.

It is the emergence of contradictions in the course of evolution of all systems that is the source of problems and problem situations. Contradictions are eliminated by inventing efficient ideas which change the operative principle and/or the construction of the system. Such construction change promotes the system to a new qualitative level, and creates a set of properties which are free of the original problem and related contradictions.

Definition of **"Invention"**	**Invention (inventive idea, inventive solution)** – idea that abolishes the contradiction and thereby the problem containing that contradiction.
Effectiveness – "system" addition 1 to the definition "Invention"	The inventive idea excels itself by its high effectiveness due to the following preferences: 1. clears the problem which hinders the system's development; 2. creates the object with new properties that are necessary for higher systems; 3. opens the opportunity for a further development of the system and-or of the surrounding systems; 4. usually solves the problem with the least amount of means (resources) possible; 5. requires the least amount of small changes within the system itself or within bordering systems.

The characteristic of the idea's effectiveness is in its core, according to addition 1, a complex of important demands on the quality of the inventive idea.

Let us consider the notion of "invention" using one of our first benchmark inventions – the Moscow Kremlin Stars.

Example 2.34. Moscow Kremlin Stars. The Secret of Invention!

In the absence of a solution inventing system, such as TRIZ, all we really can rely on is professional knowledge and talent of the designers. And for the Kremlin Stars this talent created a truly wonderful solution! Solution which today we will use to learn to invent.

The key idea of the solution is to turn the star into a… wind vane (fig. 2.31)! To that end, the axis of the support is displaced relative to the star's axis of symmetry so that one part of the star has a larger surface area than the other. Then strong wind will have to do useful work, turning the star so that it positions itself with its "side" – its shorter arms – towards the wind. The stronger the wind, the more sustainably will the star be fixed in this "correct" – safest – position! This means that *harmful impact has been made useful.* And this is exactly how inventions are created in TRIZ! Another principle which was actually used when inventing vane-stars is the principle of *asymmetry.*

And now let us review this solution as if it had been made according to TRIZ. In Modern TRIZ such modeling is called *reinventing*. Reinventing is the fundamental method to practice inventing thinking which can then be used to solve any problems of similar or even higher complexity.

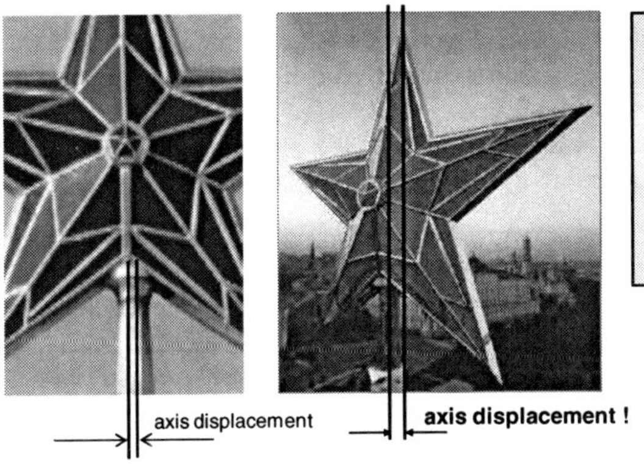

a) frontal view b) back view at an angle

fig. 2.31. Ideal solution – the construction PROTECTS ITSELF!

fig. 2.32. You do remember what a wind vane looks like – and how it works?

Example 2.35. Reinventing of Moscow Kremlin Stars According to TRIZ

To resolve such contradictions, TRIZ has a wide gamut of 40 models, such as "*inverse action*", "*copying*", "*dynamization*", etc. To select the models suitable for the given situation and given contradiction, a special table [40] – A-matrix[41] – is used.

To solve the problem under consideration, we use, in A-matrix, two factors from the standard contradiction: the first factor corresponds to the positive objective of problem solution – the star must have a large surface area, which corresponds to A-matrix plus-entry Number *18. Surface of the Fixed Object;* the second factor corresponds to minus-entry *04. Reliability*, as it is reliability of the star that deteriorates as its size and surface area increase.

At the intersection of Line 18 and Colum 04 of the A-matrix we find a cell containing the numbers of transformation models – 01, 09, 17 and 24 – which statistically have been frequently used in practice to solve contradictions represented by Factors 18 and 04. The notation form used is: 18 **VS** 04.

To generate an idea, we need to look up descriptions of these models in a special As-Catalog, and attempt to interpret them as they apply to solving the problem at hand. In this case, the highest "creativity" potential is displayed by Navigator *24 Asymmetry: (a) change the form of the object from symmetrical to nonsymmetrical.* Objectively, there is also Navigator 07 *Dynamization*: (c) if the object is im-

---

40 See also: M. A. Orloff (2006) *Inventive Thinking through TRIZ.*

41 The terms "A-Matrix" and "A-Catalog" were suggested in honor of G. S. Altshuller by the author of this book in his first textbook – a TRIZ summary based on the works of Genrikh Saulovich Altshuller: M. A. Orloff, A. M. Shirokoff (2001) *Contradiction. Invention. Development. Selected Pages from Classical TRIZ.* – Minsk: Modern Knowledge Institute– 210 p. (in Russian)

mobile, make it mobile, motive. Application of these transformation models yields an idea which leads to the following wonderful solution: make the star in the form of a "wind vane" (fig. 2.32)!

A quick test of the solution shows that all contradictions are removed! The design does not become more complex, and there emerges a "bonus" useful effect – we save energy, as the wind does all useful work by itself, turning the star into the safest position. Accordingly, the solution objectively implements yet another model, *21 Transform Damage into Use.*

Now we can supplement the brief description of the problem, as presented in fig. 1.30, with a more detailed one incorporating the solution (fig. 2.33).

Problem P1. Moscow Kremlin Stars	**Standard Contradiction**	
A large star on a tall tower can be damaged by strong wind.	(+)-factor	(-)-factor
	Large surface area, so that the star can be seen from a large distance	*By large area, the star could be destroyed with a strong wind*
	Radical Contradiction: The star must have –	
	a large area to house many lamps	*a small area to weather a strong wind*
Time: 1936	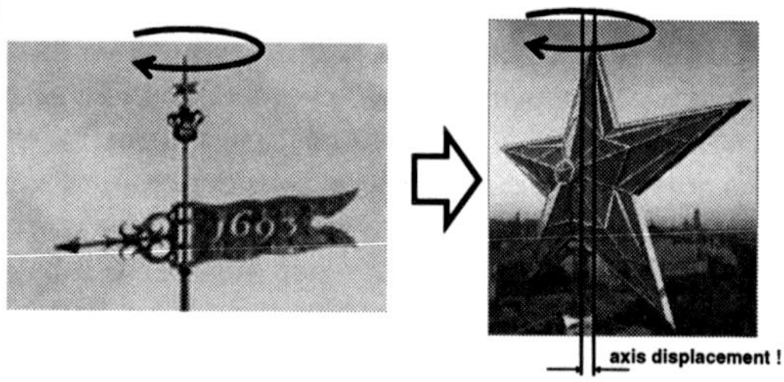	
Solution: the stars are made in the form of wind vanes. Axis of rotation (axis of support) is displaced relative to the axis of symmetry of the star. **TRIZ Models:** "Asymmetry", "Dynamization", "Transform Damage into Use", "Copying"		

fig. 2.33. Brief presentation of solution for Problem P1. Moscow Kremlin Stars

Reinventing of this invention is presented in a standard blank (to work in the class without software EASyTRIZ™) shown in fig. 2.34.

Additional examples of similar application of dynamization and asymmetry models are provided in fig. 2.35-2.37 (all photos are made by the author).

So, you have learnt initial TRIZ concepts: contradiction, transformation model, and invention.

Now we will proceed to an in-depth review of these and other TRIZ notions. In doing so we will – as you may have already noticed – rely, to the extent possible, on real-life artifacts. This entire course will be accompanied by practical examples similar to that of Moscow Kremlin Stars (fig. 2.34).

We wish you success – and yes, perseverance! – in mastering "easy" TRIZ!

TREND

The main Moscow Kremlin Stars are large – more than 5 meters tall. Each star weighs 1 ton or more. All stars have large surface area and, consequently, large windage. As a result, in high winds there is a risk that they may be thrown off their respective towers. PROBLEM: how do we redesign the stars to make them reliable and resistant to strong wind?

REDUCING

FIM: X-resource, together with available or modified resources, and without making the object more complex or introducing any negative properties, guarantees attainment of the following **IFR**:

[*the star is reliably steady in strong wind*].

Standard (Technical) Contradiction

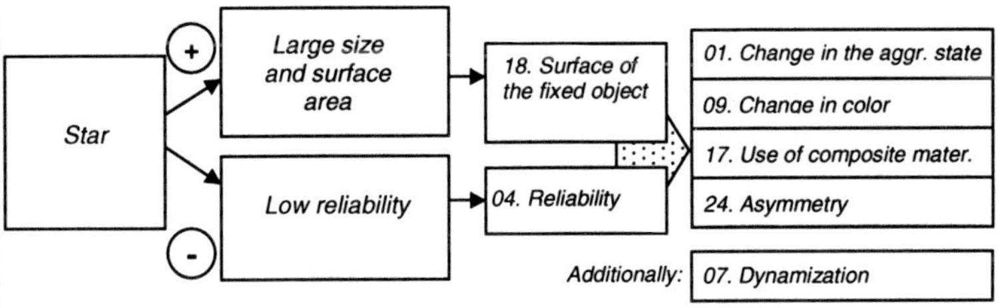

Radical (Physical) Contradiction

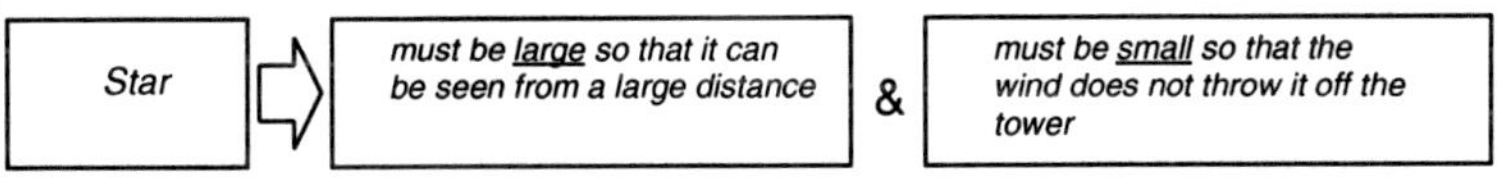

INVENTING

Key model: 24 Asymmetry – a) move from a symmetrical shape of an object to an asymmetrical one. What does a "nonsymmetrical star" resemble? A wind vane! So turn the star into a wind vane! This is a clear case of application of Model 07 Dynamization.

ZOOMING

Have the contradictions been removed? – Yes. – ~~No~~.

Super-effects: *the wind ITSELF aligns the star with its own direction. The solution objectively implements Model* 21 Transform Damage into Use, *as the stronger the wind, the better it keeps the star in a position where its smallest section "cuts through" the air flow, while its largest section "rides with" the air flow. The design is simplified. No energy is consumed.*

Negative effects: *none*

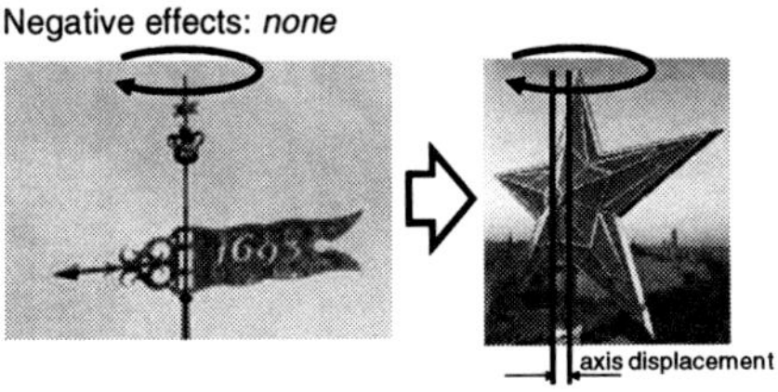

BRIEF DESCRIPTION

To simplify design and reduce energy expenditure, Moscow Kremlin stars are designed so that their axes of rotation are displaced relative to their axes of symmetry in accordance with Navigator 24 Asymmetry*; model* 07 Dynamization *is also realized.*

fig. 2.34. Reinvention of the Moscow Kremlin Stars

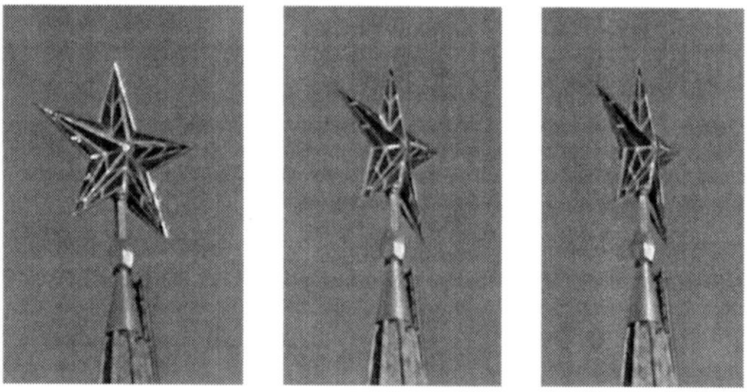

fig. 2.35. Rotation of the star on the Troitskaya Tower of Moscow Kremlin

fig. 2.36. Rotation of the coat of arms on the tower of the Museum of History in Moscow as viewed from the Alexander Garden

a) view in the morning from the Palace Square; the fountain is to the left

b) view in the evening from the fountain opposite the main entrance; the Palace Square is to the right

fig. 2.37. Rotation of the ship wind vane on the spire of the Admiralty Building in Saint-Petersburg
(two positions of the wind vane with different wind directions)

Technical solutions used in the making of the Kremlin Stars inspire admiration, and can serve as examples of inventive brilliance and audacity. These solutions are simple and highly efficient. Some of the photographs provided are made by the author. And one more thing: the same principles – "dynamization", "asymmetry" and others – that can be extracted from these solutions are used in other constructions, for example, in the rotating "eagles" towering above the Museum of History in the Red Square in Moscow, and in the ship-shaped wind vane on the Admiralty Spire in Saint-Petersburg. All photographs are dated July 2009.

3 Red Thread of Victory

It happened so that, before leaving Crete, Daedalus[42] gave Ariadne a magic ball of thread and told her how she could enter and exit the Labyrinth. She had to open the door and tie the loose end of the thread to the door post. The ball then would start rolling ahead of her, becoming smaller and smaller with every step, and would finally take her, through many tricky turns and passages, to the heart of the Labyrinth, the dwelling place of Minotaur. Ariadne gave the ball to Theseus and instructed him to follow it until he would come upon the sleeping monster... He would find the way back by rolling up the thread.[43]

3.1 Algorithms of Inventive Creativity

The third fundamental concept of TRIZ is the *Algorithm of Inventive Problem Solving* (ARIZ). This is the thread of Ariadne, the *red thread of victory* blazing the way through the Labyrinth, the way to solving the problem.

The aim of TRIZ is to create efficient models for inventing efficient ideas. Solving problems with embedded contradictions requires special supra-professional creative knowledge and skills which to date have been presented in an integrated form, i.e. as a certain unity of concepts, theoretical models and practice-oriented technologies, only in TRIZ.

No system (product, technology, organization, conflict situation) can develop without resolving contradictions. That is why problems defined in terms of developmental contradictions always represent a challenge for creatively-minded engineers and managers, and require full mobilization of their knowledge and creative abilities, special psychological concentration.

Application of TRIZ can be illustrated as follows (fig. 3.1).

The original problem situation arises because the prototype artifact requires an improvement, but an attempt to make such improvement creates a certain contradiction. For example, we may need to build a large star, and install it on a high tower; however, the star has large surface area, and may be thrown off the tower by strong wind. Contradiction is a shorthand description of the problem situation model. One could also say that description of the problem is formally reduced to a contradiction.

Once that is done, it is necessary to determine the purpose – the *Ideal Final Result (IFR)*, and formulate one or several *Functional Ideal Models* (FIMs) setting the direction towards the IFR (the "meta-trend"). For example, here is a possible

---

[42] Heroes of ancient Greek mythology: Daedalus, artist, engineer and inventor who invented, among other things, wings that he used to escape, together with his son Icarus, from captivity (Icarus died because he did not heed his father's warning, and rose too close to the Sun); Minotaur, part man and part bull who lived on the Isle of Crete and devoured youths sacrificed to him; Theseus, Athenian hero who slew Minotaur; Ariadne, daughter of a Cretan king who helped Theseus to kill Minotaur and escape the Labyrinth with the help of a ball of red thread.

[43] Based on: Robert Graves (1992) *Myths of Ancient Greece*. – Moscow, Progress Publishers

wording of the IFR: the star must be large and reliable. To attain that result, it is necessary to realize the following FIM: the star must BY ITSELF interact with the wind so that it will always position itself "sidewise" to the direction of the wind.

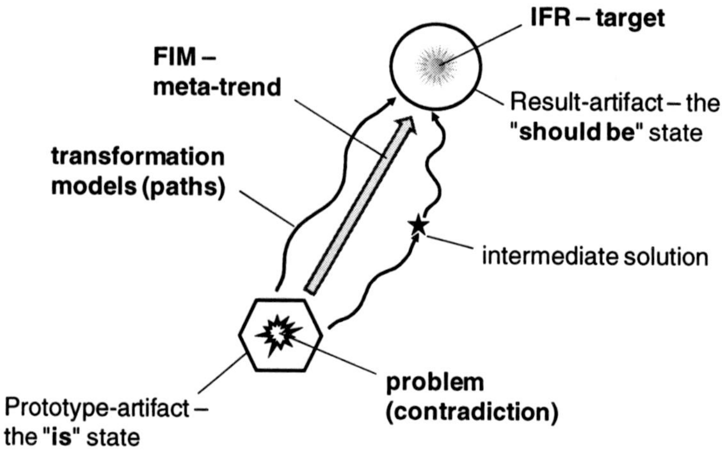

fig. 3.1. Resolving the problem according to TRIZ

Then we need to select transformation models – specific pathways that will take us to our destination. For the various formulations of the contradiction described above, TRIZ offers the following transformation models: *dynamization, asymmetry, copying, transform damage into use*, and some others.

Last but not least, we walk one or several of these paths – invent specific solutions and modify resources of the prototype artifact so as to finally arrive to a new target prototype artifact consistent with the sought IFR. At this stage we need to make use of our imagination, associative thinking, experience, knowledge, character. In this case, if we follow the "prompts" provided by TRIZ in the form of thinking navigators, emergence of an idea to copy a weather vane seems quite logical and predictable!

And, naturally, in the end of the solution generation process, we need to verify the idea, check it for implementability and efficiency.

Therefore, invention according to TRIZ is the path from the available state of the artifact ("is") to a future state ("should be", or "needs to be") with the help of transformation models acting as thinking navigators.

This process can be described in more detail with the following chart (fig. 3.2):

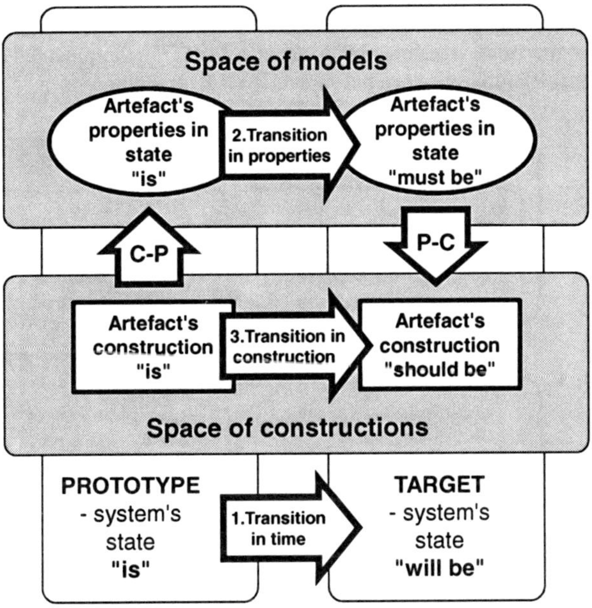

fig. 3.2. Process of invention as sum total of system transitions

Any invention involves three transitions:

1) *transition in time*: the prototype in the state "is" is modified in the required direction (in line with the specified trend) to arrive at the state "needs to be"; the trend is defined by the growing needs and restricted by available resources, including the informational resource – the *sum total of knowledge* required to create the new artifact;

2) *transition in properties*: while the prototype remains in the state "is", it always has contradictions that need to be resolved; when its state changes to "needs to be" or "will be", those contradiction are eliminated;

3) *transition in construction*: in the state "is", the existing prototype construction fully determines all its properties (direction "Construction → Properties", or C-P); for the state "will be", it is necessary to create a new, currently unknown target construction which will have the required properties, i.e. the direction is changed to its opposite – P-C, or "Properties → Construction".

> In terms of *technical implementation*, invention is a transition in construction. In terms of *systemic development*, invention is a combination of three transitions: transition in properties, transition in construction, and transition in time.

Practical realization of such transformations of the original system is supported by ARIZ. Modern TRIZ offers for this purpose the simplest version of ARIZ (fig. 3.3) either in the form of the so called Meta-Algorithm of Invention[44] (MAI T-R-I-Z), or in the more instrumental form of START[45] (Simplest TRIZ Algorithm of Resourceful Thinking) – see below section *6 Reinventing*.

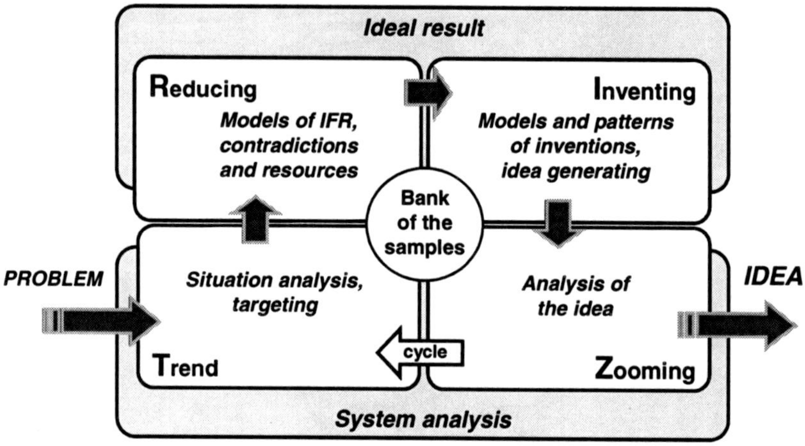

fig. 3.3. Meta-Algorithm of Invention (MAI) T-R-I-Z

MAI T-R-I-Z is a methodological scheme containing practical TRIZ instructions that must be followed to ensure transition of the system being improved from the state "is" to the state "needs to be". MAI T-R-I-Z exists in an algorithmic form, and consequently is the most compact of all modern ARIZ – which is basically "collapsed" to the format of the first ARIZ (fig. 3.4) in the history of TRIZ.

ARIZ – Algorithm of Inventive Problem Solving – is a method[46] which contains practical instructions for the transformation from a state of "is" to the state of "should be" (fig. 3.4) whose form and character are algorithmic. The first ARIZ was published in 1956. Back then Genrich Altschuller and Rafail Shapiro published a 3-phase scheme, the first "algorithm of invention" ARIZ–1956, which became not only the foundation of contradiction and transformation models but also the core of growth for the future TRIZ.

The most prominent difference of this *meta-navigator of thinking* to other similar schemes and theories consisted of its *instrumentality* and *constructiveness*.

[44] Orloff, M. (2006) "Inventive Thinking through TRIZ", part 7.2. Meta-Algorithm of Invention
[45] Orloff, M. (2011) "ABC-TRIZ"
[46] *The Psychology of Inventive Creativity*. – Journal "Questions of Psychology", No. 6, Moscow, 1956 (in Russian); co-authors: Genrikh Saulowitsch Altshuller (1926-1998) and Rafail Borisovich Shapiro (1926-1993)

Instrumentality means that the three main phases of ARIZ-1956, the analytical, the operative and the synthetic, were already then equipped with the first models and recommendations for practical changes (transformation) of any object which at the same time allowed the *constructiveness* (goal-oriented and resultant) for ARIZ-1956.

TRIZ developed from ARIZ-1956 towards a strong key idea – towards installation and comparison of certain kinds of contradictions (of certain kinds of conflicts) corresponding a certain group of adequate and effective models of transformation. This idea wasn't formulated in the ARIZ-1956 but soon appeared to be a natural and logical discovery that gives the models of TRIZ a constitutional similarity to models from physics, mathematics and all sciences in general.

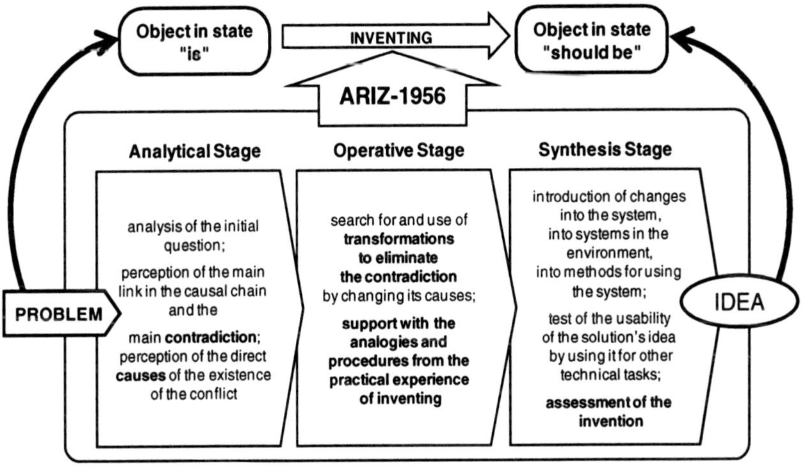

fig. 3.4. The first Meta-Algorithm of Invention ARIZ-1956

Over the years ARIZ changed and it became more complicated, then by 1985 it developed into one of the most complicated versions and contained a double digit number of steps and transitional recommendations.

> To study ARIZ-1985 is practically impossible for beginners as it is necessary to first of all master the instrumental (fundamental) models of TRIZ. Genrikh Altshuller also stated that ARIZ is designed for a mastership level.

For a mass application we need more simple versions of ARIZ – specialized for example according to the level of education of students and users or even according to the specialized language of the specific occupation and its application.

That is why the experiences of the TRIZ education have so far proven that learning the fundamental TRIZ models on the basis of simple schemes, such as the Meta-Algorithm of Invention T-R-I-Z and others which have been elaborated in Modern TRIZ to be a necessity.

You have already seen the results of MAI T-R-I-Z application in the previous section (see fig. 2.34) to reinvent the Kremlin stars.

However, MAI T-R-I-Z can be used to create ANY new invention.

This can be explained by the fact that every process of generating an idea to solve a problem contains the following four steps:

> **1)** analysis of the initial problem situation and setting a goal (**Targeting and Trend**);
>
> **2)** construction of a preferably simple model of the problem (**Reduction**);
>
> **3)** creation of an efficient idea (**Invention**);
>
> **4)** examination of the idea from different scales and surroundings in order to evaluate the effectiveness (**Zooming**).

Let us present this scheme in a more detailed way (fig .3.5).

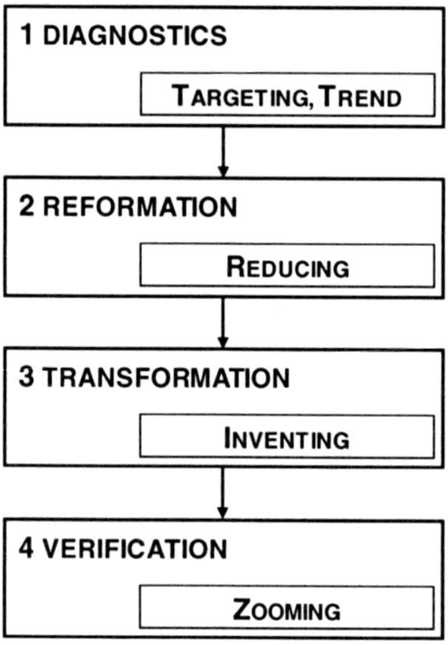

fig. 3.5. Main stages of the MAI

In the diagram the names of the stages are linked to the terms which have a broader meaning than the terms used in the preceding example. The reader can autonomously find differences and similarities in every pair of names.

For that reason the diagram in fact unites two variants of Meta-Algorithm: MAI T-R-I-Z and MAI D-R-T-V (for "Diagnostics – Reformation – Transformation – Verification").

We shall again remark that the part "Meta" designates the very abstract and generalizing level of MAI; it defines only the widest steps of problem solving and doesn't contain specific models or instructions for every step.

The steps themselves though, are obligatory and characteristic when solving any problem or creating any invention.

When fulfilling each of MAI's steps with specific models and recommendations, MAI is functionally unravelling and transforming into a specific practical algorithm of invention. The type (designation) of practical algorithm is being defined by what type of model has been employed. Still in every practical algorithm the four important steps of MAI can be singled out precisely:

Definition **"Meta-Algorithm of Invention"**	**Meta-Algorithm of Invention (MAI)** – *standard fundamental model used to organize the process of making an invention (and, more broadly, any creative idea)* on the basis of a four-stage structure which includes the following: - ***Diagnostic (Trend, Targeting),*** - ***Reformation (Reducing),*** - ***Transformation (Inventing),*** - ***Verification (Zooming),*** as necessary and sufficient stages of the process of solving any problem.
Addition 1	When equipping the stages of MAI with models and methods on the basis of TRIZ instrumental (practical) algorithms of TRIZ emerge.
Addition 2	When some stages are unified, MAI can be presented in a "compressed" structure which contains less than four components.
Addition 3	When some stages are separated MAI can also be presented in a "decompressed" structure which contains more than four components.
Statement 1	For every artifact and for every invention (from an infinite range of artifacts and inventions) a reinventing on the basis of MAI can be performed.
Statement 2	Every invention can be created on the basis of MAI.

The fundamental value of MAI is that with its help the important stages of TRIZ based problem solving are easily made clear and understandable on first sight and already in the first lesson.

MAI being equipped with TRIZ's instrumental arsenal turns into an effective navigator of inventive thinking, even for beginners. MAI can be effectively used to solve both training problems and real-life practical problems. We will note, at the end of this section, that conceptually MAI is the closest to the very first algorithm for solving inventing problems, ARIZ-1956.

Evolution of the ARIZs are shown in fig. 3.6.

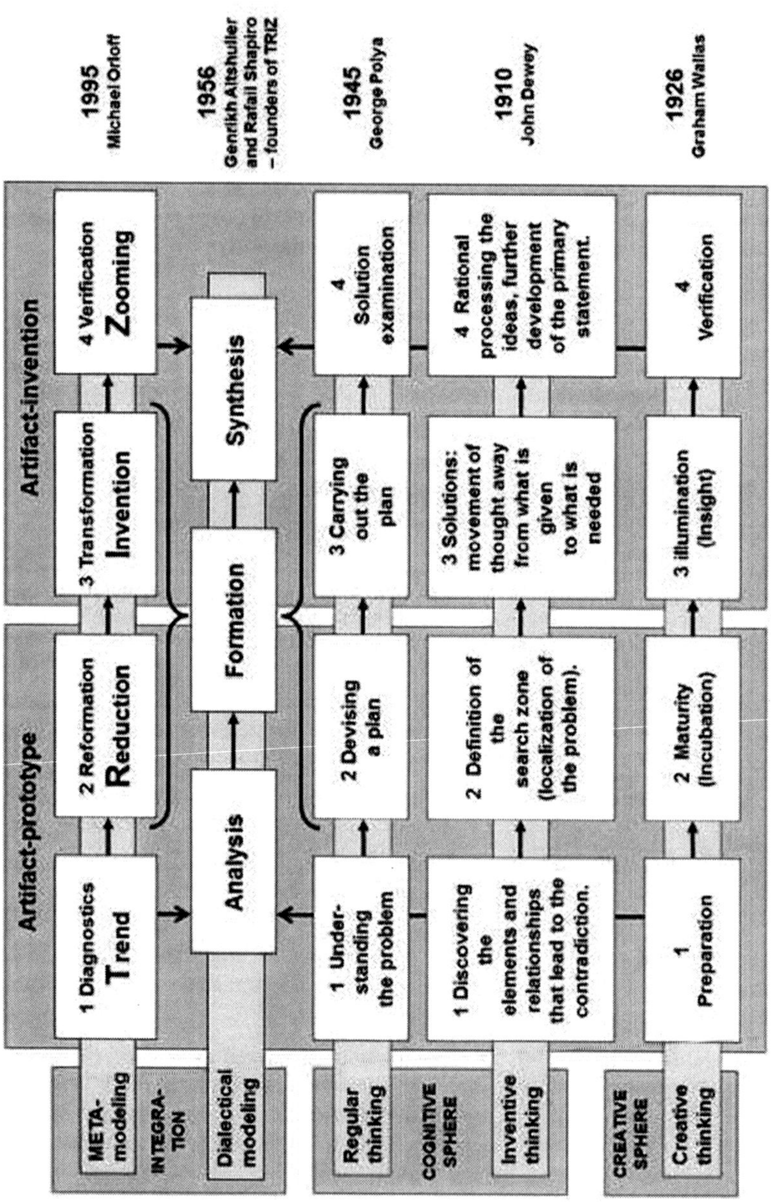

fig. 3.6. Evolution of the ARIZs

ARIZ is the main "navigator" used to develop TRIZ-based solutions. This is demonstrated by the chart (fig. 3.7) illustrating the structure of classical TRIZ.

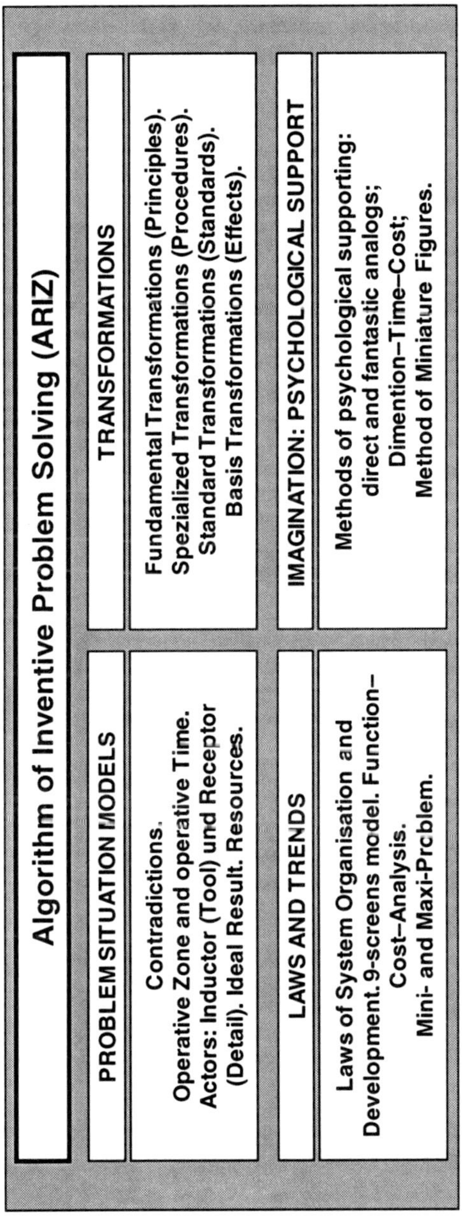

fig. 3.7. Structure of classical TRIZ

And in conclusion of this section, let us look at the detailed definition of TRIZ (fig. 3.8), and go through the key notions promulgated by TRIZ founders.

Definition of **TRIZ**	***Theory of Inventive Problem Solving (TRIZ) – empirical, constructive, qualitative, universal methodology*** *for generating ideas and solving problems, primarily when projecting engineering systems*, on the basis of contradiction models and methods to solve them that were extracted from known inventions.
Empirical methodology – addition 1 to definition	TRIZ is an *empirical methodology* which postulates possibility of resolving any contradiction on the basis of models and methods extracted from numerous examples of efficient solutions.
Constructive methodology – addition 2 to definition	TRIZ is a ***constructive methodology*** for it includes practically reproducible models and methods which allow the development of new inventions and the teaching of the process, of the models and the methods of creating inventions.
***Qualitative* methodology** – addition 3 to definition	TRIZ is a *qualitative methodology*, as its models and methods have the nature of recommendations which are based on *similarity of qualitative (as opposed to quantitative) properties of models* created for the artifacts being transformed or developed, and are *not* based on rigorous mathematical axioms, or strict mathematical rules, used to deduce consequences (solutions) from certain original data.
Universal methodology – addition 4 to definition	TRIZ tends to become the *universal "supra-professional" methodology of inventive generation of ideas and resolution of problems* in any applied area where efficiency of TRIZ can be limited only by the lack of basic scientific knowledge specific to such area.

fig. 3.8. Detailed definition of TRIZ

In their first article entitled *The Psychology of Inventive Creativity*, G. Altshuller and R. Shapiro formulated three fundamental premises which are found in each creative solution of each new problem:

> 1. Formulation of the problem and determination of the contradiction which prevents its resolution by applying the usual methods already familiar to engineers.
>
> 2. Elimination of the cause of the contradiction with a view to attain a new – higher – technical effect.
>
> 3. Transforming the other system elements to bring them into conformity with the modified element.

Subsequently G. Altshuller developed the following principles governing resolution of inventive problems:

> 1. Identification and elimination of the systemic contradiction is the key to resolving the problem!
>
> 2. "There is an endless multitude of inventive problems, while the number of systemic contradiction types is relatively small. There are typical systemic contradictions – and typical techniques used for their elimination[47]." Problem resolution methods (techniques) can be identified by analyzing great inventions.
>
> 3. The strategy and tactics of directed problem resolution must rely on laws governing evolution of technical systems.

TRIZ rests on the systemic foundation of the following basic provision:

System Platform	*Technical systems evolve in accordance with certain laws which can be cognized and applied to designing a new creative technology... thereby turning development of technical systems into a precise science*[48].

And here is yet another provision which is critically important for understanding TRIZ fundamentals:

Psychological Platform	*Invention methodology is based not only on the laws governing the overall progress of the engineering science and generalization of experiences accumulated by inventors. It also takes into consideration human psychology. It assumes that... each human being has certain talents. The methodology helps develop those talents and put them to good use*[49]. *Directed search does not exclude intuition. On the contrary, normalization of thinking processes creates a special "attitude" which is conducive to manifestation of intuition*[50].

[47] Altshuller, G.S. (1961) *How to Learn to Invent.* – Tambov Book Publishers (in Russian)
[48] Altshuller, G.S. (1986) *To Find an Idea. Introduction to Theory of Inventive Problem Solving.* – Novosibirsk, Nauka Publishers (in Russian; there are also later editions)
[49] Compiled by the author (M.O.) based on: Altshuller G.S. (1961) *How to Learn to Invent.* – Tambov Book Publishers (in Russian)
[50] Altshuller, G. S. (1973) *Algorithm of Invention.* – Moscow Worker Publishers (in Russian)

3.2 Noospheric Modeling Concept in TRIZ*

Psychology of inventive creation as a scientific discipline is unique in that there is a need to concurrently take into consideration both objective laws of technical development and subjective, psychological factors.[51]

Genrikh Altshuller and Raphail Shapiro, 1956

In this section, we will take a brief look at the models explaining certain key creative thinking mechanisms.

Based on conclusions drawn by a number of psychologists and neurophysiologists, primarily Prof. P.V.Simonov[52], we suggest the following model of "allocation of responsibility" for ideation processes and products among certain parts of the brain (fig. 3.9):

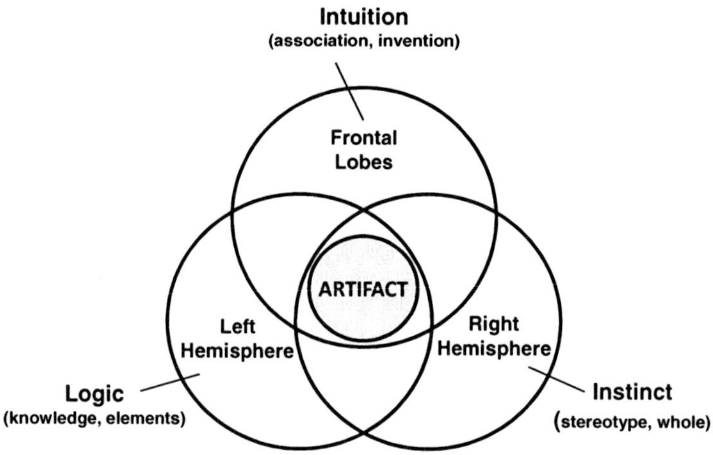

fig. 3.9. "Topological" model of the creative brain

Human creativity can be regarded as consisting of two components relying on specific knowledge – instinct and intuition.

* This paragraph is not part of certification, but plays a very important practical role.

51 *The Psychology of Inventive Creativity*. – Journal "Questions of Psychology", No. 6, Moscow, 1956 (in Russian); co-authors: Genrikh Saulowitsch Altshuller (1926-1998) and Rafail Borisovich Shapiro (1926-1993)

52 P. V. Simonov (1993) *Creative Brain. Neurobiological Foundations of Creation*. – Moscow, Science Publishers (in Russian)

Instinct - skills, automatisms, whether inherited or acquired through training. The set of creative models which are inherited and acquired through usual experiencing and learning is relatively narrow and very general. It is used in brain storming, and includes, *inter alia*, the following procedures: increase-decrease, unite-disunite, make by analogy (copying, imitation), use a mediator, combine, recombine, insert, extract, replace part or whole, do the reverse, etc.

Intuition - the talent of inventing new things. Naturally, it is an inherited property. But can you train it? Can you enhance talent by imparting the knowledge of certain models and methods that can be used to generate efficient ideas? TRIZ says "Yes, you can." TRIZ radically expands the creative toolset of the person, and points to efficient solution search strategies which are either absent from, or underutilized in, other creative theories.

Artifact - this term is used here to refer to any "artificial object", i.e. an object created, modified or studied by man. It can be a material object, such as water, an automobile, a computer, or an "immaterial" object, such as a thought, whether expressed in written or oral form or still existing only in the individual's head.

Next we are going to need the term "noosphere."

Noosphere - this term[53] is generally understood here as the "sphere of intelligence". The first part of the word ("noo") comes from the Greek "νους" – *mind.*

The term can be applied to notions of different scale. For example, according to V.I.Vernadsky, it reinforces recognition of a new period in the evolution of the Earth – a period which is determined by participation of an intelligent human civilization capable of producing, starting with the 20th century, effects comparable with those created by natural forces. Two preceding periods are defined as "geosphere" and "biosphere".

Noosphere can also be understood as the global body of knowledge accumulated by the entire human civilization. This is only one step away from applying the term to any artifact as a vessel containing accumulated materialized knowledge stemming from an even more global or even cosmic noosphere.

Let us now introduce the notion of "artifact noosphere" to denote both the totality of knowledge implemented in any artifact, and specialized spheres comprising the "artifact noosphere" and belonging to certain specific areas of human knowledge.

---

[53] The term was coined and introduced into general scientific circulation by Vladimir Ivanovich Vernadsky (1863-1945) – an outstanding Russian natural scientist and thinker of encyclopaedic stature

Definition of the term **"Artifact Noosphere"**	**Artifact noosphere** is the totality of knowledge accumulated and materialized in an artifact.
Addition 1 to the definition	**Artifact noosphere** is the totality of specialized "spheres" comprising the "artifact noosphere" and belonging to certain specific areas of human knowledge.
Addition 2 to the definition	**Artifact noosphere** is organically linked to numerous specialized "spheres" of the general noosphere of the entire human civilization (the largest artifact existing on the face of the Earth).

The noosphere of the human civilization consists of numerous interdependent "specialized" noospheric structures. The noosphere of each artifact is linked to many such structures.

Such links may include functionalities, process-time structures (relations), space-matter-energy structures (relations), consumer preferences, finally, knowledge existing "within" each artifact – collective knowledge accumulated throughout the history of the civilization, but carried by a relatively limited number of individuals which possess such knowledge in each specialized noosphere.

Based on the above definitions, the process of generating an idea (solution) can be represented as follows.

To create the concept of a new artifact, or prepare the decision to modernize an existing artifact, we need to identify key noospheric aspects and relations which are critical to inventing the idea of the solution.

As a minimum, ideas are generated with the participation of the following three "specialized" noospheres:

1) cognitive (gnostic, applied) sphere;

2) creative sphere; and

3) psychological sphere.

To understand the nature and extent of each sphere, we will resort to "reverse" analysis. We will show that to generate an idea the individual must first overcome three fundamental mental "barriers" (fig. 3.10):

- *cognitive* barrier: lack of requisite applied (specialized, professional) knowledge and/or established skills of application of such knowledge in novel situations,

- *creative* barrier: lack of methodological knowledge or established skills required for problem resolution,

- *psychological* barrier: lack of adequate motivation and mobilization to resolve a specific problem.

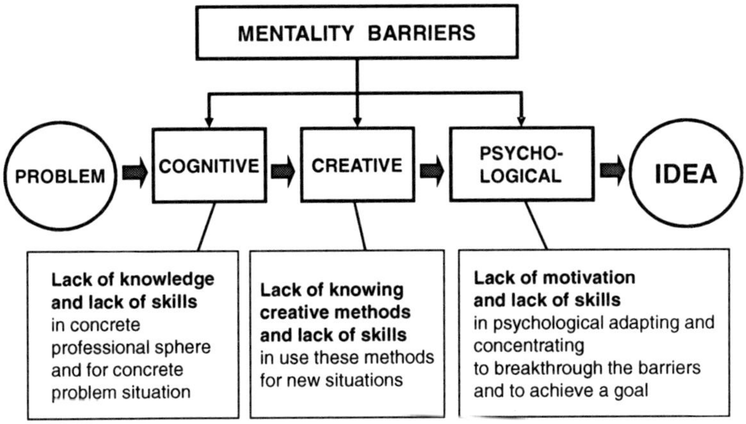

fig. 3.10. Mental barriers in problem situations

Description of the Cognitive "Barrier":

1. KNOWLEDGE IS IRREPLACEABLE (relevance condition): lack of specific knowledge regarding methods of modification of the object to achieve a certain desired outcome cannot be compensated by other means and abilities in most situations;

2. KNOWLEDGE IS INCOMPLETE (non-sufficiency condition): availability of knowledge, however complete, regarding methods of transformation of objects existing in a given applied sphere does not guarantee a "simple" logical identification (generation) of an efficient idea resolving a problem situation, as a minimum, because of the existence, in any such situation, of the two other "barriers".

Description of the Creative "Barrier":

1. CREATIVITY IS IRREPLACEABLE (relevance condition): lack of specific knowledge regarding methods of generation of creative ideas to achieve a certain desired outcome [almost] cannot be compensated by other means and abilities in *complex, intricate situations* which contain *acute "irreconcilable" contradictions*;

2. CREATIVITY IS MOTIVATED (non-sufficiency condition): availability of knowledge and skills required to generate creative ideas does not guarantee "automatic" synthesis of an efficient idea to resolve a specific problem situation, as a minimum, because of insufficient motivation and volition, i.e. because of the existence of the psychological "barrier".

Note: the bracketed word "almost" (as used above and below) means that in certain relatively simple situations, and subject to certain special conditions, there have been precedents of "insights" and "visions" producing ideas for solutions. To all intents and purposes, practical replication of such conditions is impossible.

Description of the Psychological "Barrier":

1. MOTIVATIONAL DOMINANT (relevance condition): lack of willingness to attain the objective cannot be compensated by other means and abilities;

2. EQUIPPED DOMINANT (sufficiency condition):
- availability of creative modeling and idea generation knowledge and skills increases personal confidence and motivation to invent efficient solutions;
- availability of psychological self-regulation knowledge and skills increases personal sustainability and successfulness in generating efficient ideas in situations with "irresolvable" problems.

To overcome the above "barriers", one has to first acquire knowledge and skills in the "three spheres of creative excellence" which contribute to successful resolution of complex problems (fig. 3.11).

Resolution of a problem can be presented, in a simplified way, as follows:

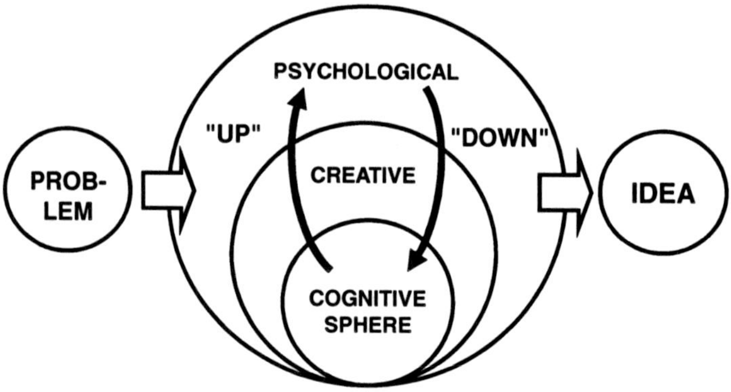

fig. 3.11. Three spheres of creative excellence

"Up" movement:
- collection of primary information about the problem situation (cognitive sphere);
- attempted transformation of the problem situation with customary means (creative sphere);
- emotional evaluation of the situation and first attempts to resolve it (psychological sphere).

"Down" movement:
- additional mobilization of psychological resources: increased focus, active use of memory and concentration, intensive associative thinking, stronger volition (psychological sphere);
- active use of efficient creative instruments (creative sphere);
- active use of, and search for, broader applied knowledge, synthesis of new knowledge (cognitive sphere).

Creative excellence in these three spheres metaphorically corresponds to the well-known triad (fig. 3.12):

Know + Can (Be Able) + Want

Incidentally, this triad also provides a laconic and rather apt presentation of the *ideal end result of any learning process.* To achieve such end result, it is necessary to:

1) KNOWLEDGE: continuously improve general and specialized knowledge and skills in the selected professional area,
2) SKILL: continuously improve creative problem resolution and efficient ideas generation knowledge and skills in the areas of creative problem resolution and generation of efficient ideas,
3) MOTIVATION: continuously improve knowledge and skills in the area of self-control and self-management with a view to be able to mobilize individual psychological resources at the requisite time in the requisite place ("here and now").

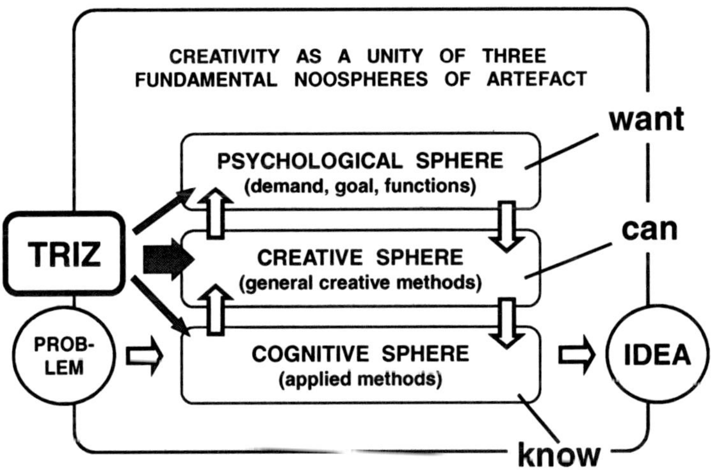

fig. 3.12. Noospheric model of integration of knowledge and skills in the process of inventing an idea

To successfully resolve a problem situation, it might be useful to take a closer look at the reasons which make the situation *difficult-for-the-resolver* and, as a minimum, learn to determine the *dominating sphere* for the purposes of synthesizing an idea.

Thus, if there is a lack of requisite applied knowledge, the problem may prove to be too complex, and it may be necessary to conduct *additional research* to study the processes and/or materials in the system in need of improvement, while selection of methods of such research may turn out to be difficult in the context if available applied knowledge.

Such problems may include, for example, the following: increase the power efficiency ratio of solar batteries; eliminate traffic jams in large cities and highways; h prolong human life to, say, 200 years; effect long-distance thought transmission, etc. – you can extend the list yourselves.

If there is a lack of creative knowledge and skills, things can be greatly improved by studying creative techniques, first and foremost TRIZ, concurrently with methods helping to evolve associative thinking, memory, attention, concentration, artistic taste, etc. Finally, if there is a dominating psychological barrier (for example, negative thinking stereotypes and inadequate attitudes towards the problem situation in question), it is possible to galvanize personal psychological resources through preliminary training and mastering of at least one self-mobilization technique based, e.g., on *neurolinguistic programming*.

Clearly, creation of an idea is not the same as its practical implementation. All we can safely assume is that the idea has been verified by modeling, estimation, etc. It should be borne in mind, though, that sometimes it is hard to make the leap from the ideal to the material even with the most brilliant ideas. The reasons for that are many.

The most general reason is resistance put up by large systems (for example, the society at large acting through its specific structures – firms, bureaucratic establishments, markets, etc.) against changes – despite widespread calls for innovation and fashionable avowals of commitment to innovative corporate culture.

Such resistance is a manifestation of one of the fundamental laws governing the operation of large systems – the *law of self-preservation* (through conservation of status quo). The famous maxim to the effect that "the best is the enemy of the good" is a somewhat simplified but, unfortunately, accurate reflection of production and market pragmatics: changes increase the risk of unknown consequences, generate additional technical refurbishment costs, often undermine existing market-sharing arrangements, etc. An individual inventor may run into trouble as soon as he attempts to patent a promising idea.

On the whole, implementation of a creative idea requires activation of at least two more "spheres":

- *resources*: securing funding, materials, technologies, etc. required to implement the idea;
- *activities*: organization of support and implementation of the idea in a real-life artifact, promotion of the newly-created object among potential users, recoupment of invested capital, receipt of profit and/or other awards.

Accordingly, for any practicing inventor his creative excellence must be completed with the following dyad:

Have (Resources, Possibilities) + Make.

Incidentally, this dyad also provides a laconic and rather apt presentation of the *ideal end result of any "production" process*.

Also, to implement the idea, i.e. to find the required resources and possibilities and organize "production", one also needs to have talents residing in the first three spheres – but as they apply to the practical implementation stage:

1) it is necessary to know (where and what), be able (how) and want (what for) *to find resources and create possibilities;*

2) it is necessary to know, be able and *want to "make", i.e. seek to practically implement* the idea.

Symbolically, the full cycle of creating and implementing an idea can be represented as a chain of characteristic verbal expressions:

Know + Can + Want + Have + Make.

In conclusion of this theoretical essay, let us take a look at an example[54] illustrating decomposition of the original problem situation into the first three "spheres".

Problem 3.1. Lesson Learnt from Mistakes Made by Engineer Ernest Pavlovich Shchukin.

If you wish, you can independently relate the example in full-text form or, better, in the reinventing format.

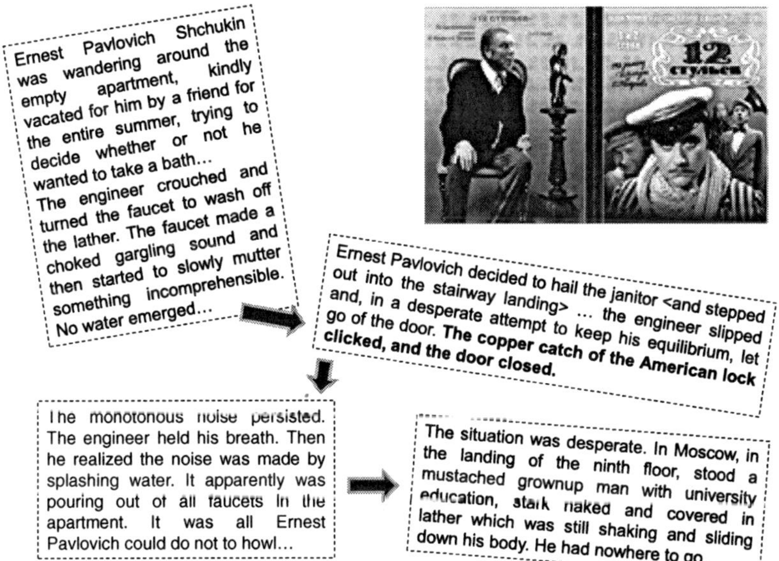

fig. 3.13. Problem situation

Of course, the readers have already guessed how this problem can be prevented from reoccurring in the future. We will give you but one example: you have to buy, or prepare in advance, a piece of cord, and then make loops at both its ends so that it can be slipped onto the door handles (see fig. 3.14) when you need to leave the apartment for a short time.

[54] Il'f, I. and Petrov, E. (1979) *Twelve Chairs. Golden Calf.* – Moscow, Hudlit (in Russian; www.hudlit.com). See *Twelve Chairs*, Chapter XXIII "Absalom Vladimirovich Iznurenkov", and denouement in Chapter XXV "Dialog with a Naked Engineer"

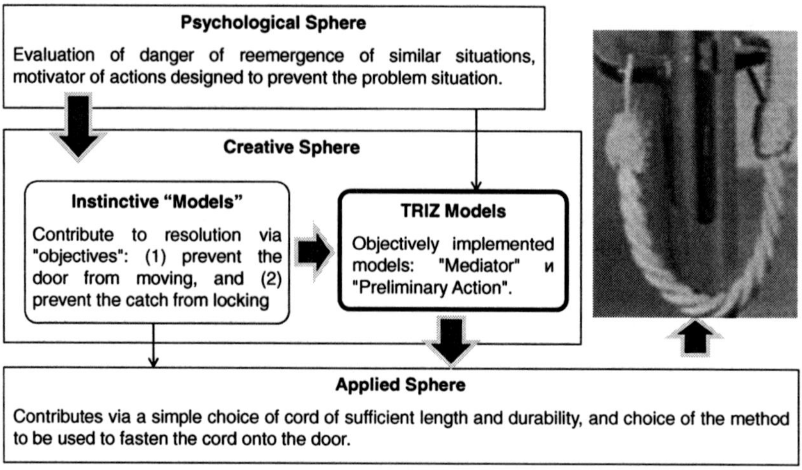

fig. 3.14. Reinventing of the cord door stopper

The main TRIZ-resolution path is shown in fig. 3.14 with dark arrows. We would like to draw your attention to the fact that, even when suitable TRIZ models are available, you need imagination to invent a specific construction that will resolve the problem.

Incidentally, specific solutions may be very different depending on the resources you use: an insert placed inside the groove in the doorframe to prevent the catch from slipping in; a wedge placed under the door to prevent it from moving, etc. You can come up with even more interesting solutions.

But see also that, more likely than not, the TRIZ models listed above will be objectively – and all together – manifesting themselves in, or contributing to, most of your solutions.

And one last thing: to resolve this (and, naturally, any other) problem, TRIZ has other model types whose study lies beyond the scope of this book.

When resolving real-life problems, you need to identify their causes and deal with them using appropriate tools. TRIZ is designed, first and foremost, to overcome the creative barrier; however, it cannot replace specific knowledge or compensate the shortage of patience and will.

That is why you always need to be prudent and persistent in verifying whether you have sufficient information about the problem and (at least potentially) sufficient resources to resolve the problem (see fig. 3.10).

In conclusion, let us consider one more addition to the definition of "invention" provided above in section *2.4 Invention.*

Disappearance of the "Miracle" Phenomenon – "psychological" addition (2) to the definition "Invention"	Inventive Idea is: 1. not obvious until it has been discovered; 2. perceived as an exceptional and shining phenomenon, a "Miracle", when being discovered; 3. perceived as an ordinary and simple phenomenon after it has been explained.

To understand the particularities of the emotional reception to an idea, according to addition 2, is a warning and a protection against at least two dangers:

1) too high (very rarely – too little) evaluation of the idea discovered by its author; as a result, the author often becomes obsessively persistent in protecting his idea (which sometimes proves to be weak and ineffective), and impervious to objective criticism and even friendly recommendations. One shouldn't forget that objective evaluations take place only after the practical realization of an idea, after the real experiment;

2) too little (very rarely – too high) evaluation of the idea through people (society); this phenomenon is encountered even more frequently than the first one and, conversely, requires the author to be tenacious and consistent in protecting and promoting his idea. Of course a better evaluation depends on a successful use of the idea in practice.

Finally, this is a good place to recall a famous maxim[55] which may help us balance the two approaches:

> *"Lord grant me the serenity to accept the things I cannot change, the courage to change the things I can, and the wisdom to know the difference".*

And my very last and absolutely sincere recommendation: find and read a book by Genrikh Altshuller and Igor Vertkin called *How to Become a Genius*[56]. It is hard to read, it may be not too optimistic, but it is definitely truthful – and this alone makes it a valuable asset for anybody who wishes to find a suitable answer to the dilemma described in the previous quote.

I wish you all the best!

[55] Quotations from Saint Francis of Assisi (1181-1226) – an Italian and the Roman Catholic monk who founded the Franciscan Order of friars

[56] G. S. Altshuller, I. M. Vertkin (1994) *How to Become a Genius. Life Strategy of a Creative Personality.* – Minsk, Belarus (in Russian)

3.3 The Big Through the Small

It all started in January 1965.

I, a 17-year-old student of the Minsk Polytechnic majoring in *Control and Measurement Instruments* and doing a field placement, was transferred from the Minsk Motor Plant (where I had spent the last six months working as a fitter, turner and miller in the most "universal" of the plant's shops – the mechanical repair shop) to a new enterprise, a special design bureau which was rather unimaginatively called "P.O. Box 19".

It turned out that the "box" (as it was familiarly referred to by its employees and the all-knowing townspeople) was sitting behind a barbed wire fence in an old storage facility of the former tire factory, and that it was guarded by armed sentries around the clock.

The winter was snowy, and while I was being escorted to my new workplace, staggering though a vicious snowstorm, slowly negotiating the external iron staircase to the second floor, it was hard to believe that right behind that heavy door there was a large well-heated and brightly lit hall lined up with several rows of Kuhlmans[57] manned by young people who were engrossed in drawing or engaged in lively conversations.

Later I learned that the ground floor housed several shops "inhabited" by masters who were capable of feats even more surprising that that performed by the famous Leskov's Leftie[58] (who, as you may well know, shod a flee). Because their microscopes were much more advanced than Leftie's.

P.O. Box 19 was the pioneer designer of the *first Soviet* semiautomatic devices for the *first Soviet* process lines manufacturing the *first Soviet* electronic chips.

Virtually all technical processes were conducted under strong microscopes which made it possible to discern minute parts embedded into silicon crystals and connected with thin golden wires with diameters of 40 or even 20 microns – thinner than human hair.

By then I already had some idea of inventing methods, having read my first book on that topic. The book was written by Genrikh Saulovich Altshuller and called *How to Learn to Invent.*

I had bought it back in 1963 at the recommendation of Stanislaw Antonovich Baltsevich, the supervisor of my student group at the Polytechnic. He taught *"Metal Smelting and Processing"* and *"Machine Tools and Instruments".*

After classes and in the evenings, he also played chess with some of us, asking about how we lived and trying to talk some sense into us.

Once I said that I wanted to create "something unforgettable and never-before-seen", and he told me about this small book. He ALREADY KNEW about it!

I carefully preserve this book, now tattered and repaired time and again (fig. 3.15), with two added pages describing my first attempted invention "according to TRIZ" (even though back then this abbreviation did not even exist).

---

57 A huge drawing board with adjustable tilt and height

58 Nikolay Semyonovich Leskov (1831-1895) – famous Russian writer. The story of Leftie was written in 1881

The cover features a picture of a Kulhman drawing board. At that time, drawings were normally made on large sheets of shining white thick paper, the so-called Whatmans[59], with top-of-the-line KOH-I-NOOR pencils.

Anyway, I received my first assignment in the "box": I had to improve the assembly[60] responsible for snipping off golden wire after it had been welded to the microchip lead (fig. 3.16).

fig. 3.15. First book by G. S. Altshuller, 1961 (part of M. Orloff's collection since 1963)

Example 3.1. How to Cut Off a "Golden Hair"? Or: Previous Experience – Solution "By Copying" and "By Analogy"

The body of the electronic microchip, approximately 15 mm long, (1) is fitted with a silicon crystal (2). The lead (3) must be connected to the contact pad (4) of the crystal with golden wire (5). The body is heated, and when the wire is forcefully pressed to the lead they become firmly connected (welded). The same happens at the juncture between the wire and the contact pad. Because of the combination of high temperature (hundreds of degrees) and considerable pressure, this type of welding is called "thermal compression welding".

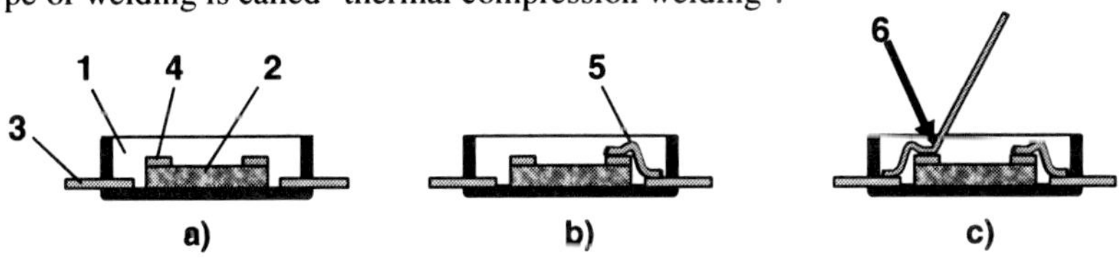

fig. 3.16. Connecting the microchip lead to the crystal-mounted contact pad with a golden wire

After the wire is welded to the contact pad, it must be cut off at the point indicated by the arrow (6) (fig. 3.16.c).

To cut the wire, I designed semiautomatic shears about 2 cm long (fig. 3.17) which were later improved and actually manufactured by experienced specialists. Those shears did well with wire 40 microns in diameter, but their performance took a drastic turn for the worse when wire diameter decreased to 20 microns.

---

59 This world-renowned type of paper was invented c. 1750 by the English leather dresser James Whatman

60 Note for experts: here and in the following sections, examples have been adapted for TRIZ teaching purposes

Drawbacks: 1) in many cases, instead of being cut off, the extremely thin wire twisted between the blades of the shears (fig. 3.17.c), as because of the high temperature the distance between the shears changed from the required **d** something else; 2) the cutting operation took a lot of time, and the shears had to be very carefully positioned in the correct position; 3) another general drawback was that in the end there remained a rather long "loose end", which could short-circuit when the body was being closed and pressure-sealed. The length of the shears as shown in fig. 3.3 is two times more than in reality, while the wire diameter has been increased many times.

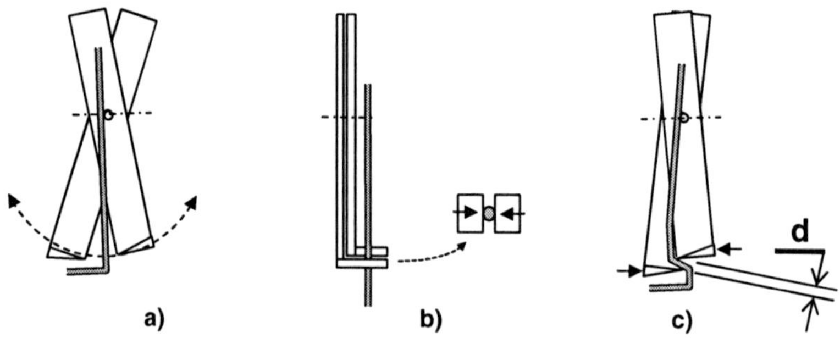

fig. 3.17. Cutting off the "golden hair" ... with "mini-shears"

There arose numerous contradictions (fig. 3.18):

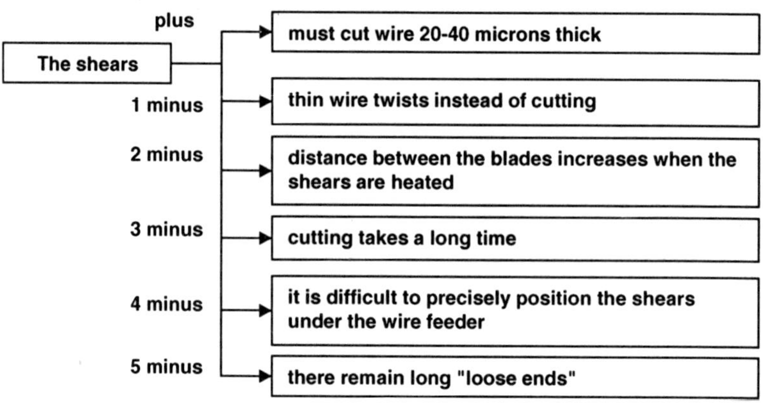

fig. 3.18. Contradictions emerging during the cutting of a thin golden hair

How can these contradictions be resolved? Should we continue to "improve" the shears?

Example 3.2. Craftsman's Instinct (Example for Proponents of Synectics!)

Once one of the "Lefties" came up to the designers and said: "Why are you so stuck on cutting? Why not just tear it off! The wire is so crushed by the instruments during the welding procedure that it can be easily detached!"

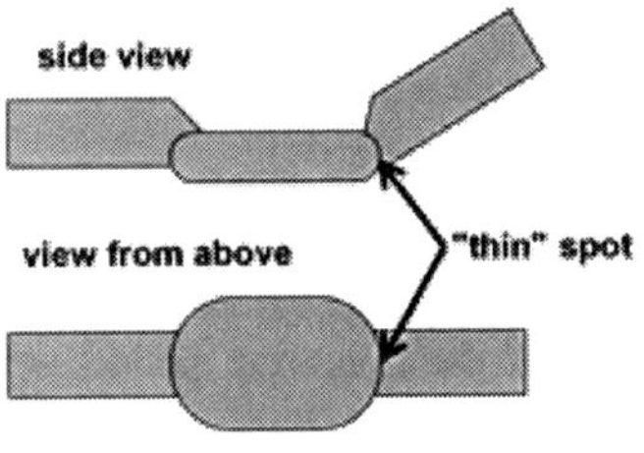

fig. 3.19. Contradictions emerging during the cutting of a thin golden hair

Indeed, the wire at the welding point looks like a "pancake", with a new section leading out towards the feeder from that "pancake" or, rather, from the adjacent attenuated isthmus (fig. 3.19).

And a well-known Russian proverb says: *It usually tears off where it is the thinnest*!

So we pressed the wire down at the feeder outlet and pulled – and it worked!

But this new solution produced a new contradiction!

The wire feeder (fig. 3.20) consists of a transparent plastic body (1) housing a wire spool (2), just like in a sewing machine. The end of the wire (5) goes through an opening in the body and then through the needle (4) which doubles as the clamp pressing the wire to the welding point.

The wire spool hangs on and rotates around an "air pressure bearing"! The shaft of the spool (3) has numerous radial openings (an enlarged drawing of this elements is provided in the left) through which air is pumped into the feeder (see arrows). The air exerts uniform pressure on the inside surface of the spool, holding it in suspension. When the spool is rotating, there is almost no friction! This is the basic operating principle of the air pressure bearing. The "thickness" of the bearing is about 100 microns.

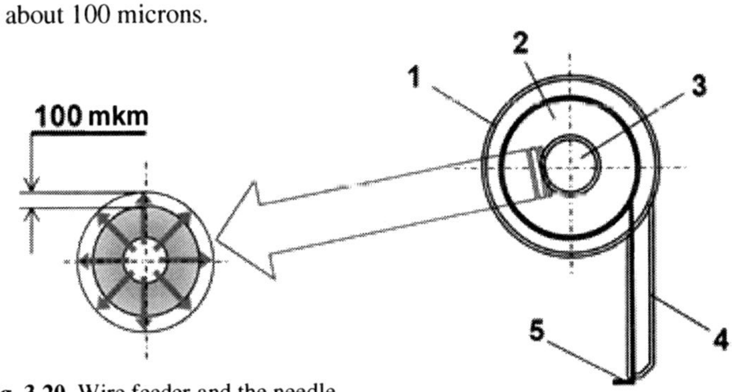

fig. 3.20. Wire feeder and the needle pressing the wire to the welding point

But this advantage hid a new problem! There arose a standard contradiction (fig. 3.21): when the tearing force was applied, the wire simply reeled out of the spool without breaking off at the welding point! There was nothing to arrest the spool!

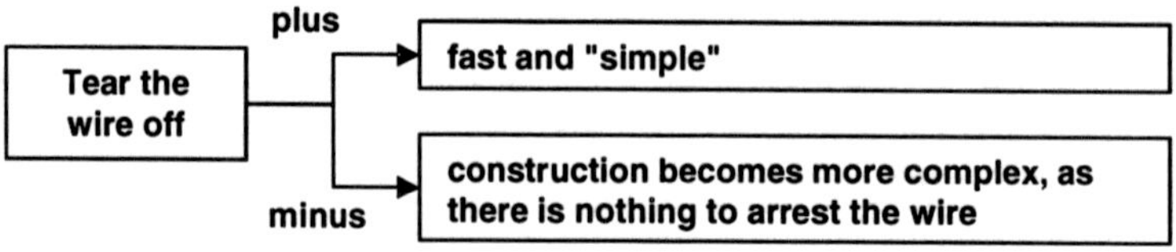

fig. 3.21. Contradiction arising in the wire feeder when the wire is being torn off

TRY IT!

> Propose an idea for an "arresting mechanism" which would meet the "ideal solution" requirements according to TRIZ: "nothing" changes in the operating zone, and the movement of the spool is reliably arrested for the duration of the operating period – the short interval of time when the wire is being torn off!

Soon our "sector" (this term applied to a small group of experts working on the same general topic) resolved this problem and learned to reliably arrest the movement of the spool!

But the problem – like a bad penny – turned up in a different place (fig. 3.22)! Sometimes the weld became damaged, and the contact pad was detached from the crystal together with the wire! The arresting force turned out to be too strong!

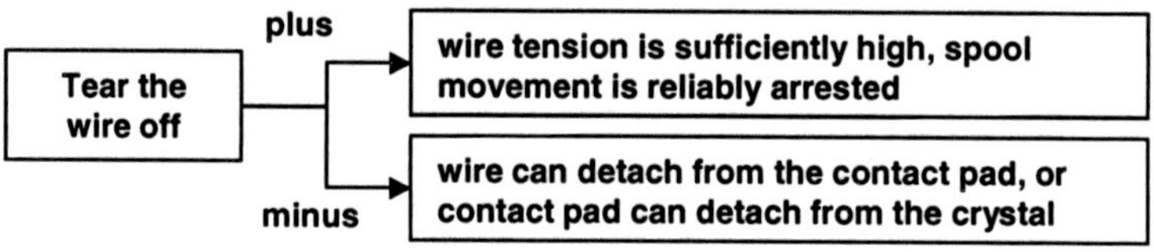

fig. 3.22. Key contradiction of the wire tearing method

Example 3.3. "Amelioration" of a Non-Ideal Solution – New Problems

To improve the situation, we tried to make needles with "beaks" to almost completely "crush" the wire after the welding was complete.

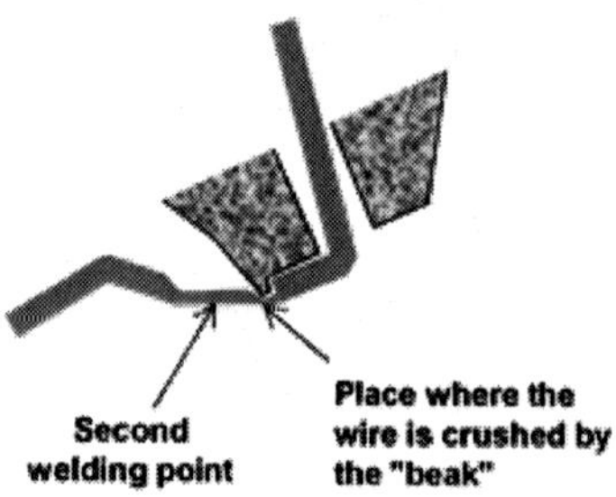

fig. 3.23. The wire is crushed by the "beak" before being torn off

Here is a simplified description of the new solution:

After the second weld was finished, the needle slid a little to the side and away from the contact pad, turned so that the "beak" jutted down, and then descended crushing the wire right beyond the welding point.

This, however, gave rise to new defects: the lengthening of the "beak" prevented the wire from tightly adhering to the pad during the second (and the first!) welding operation.

The underlying radical contradiction is shown in fig. 3.24:

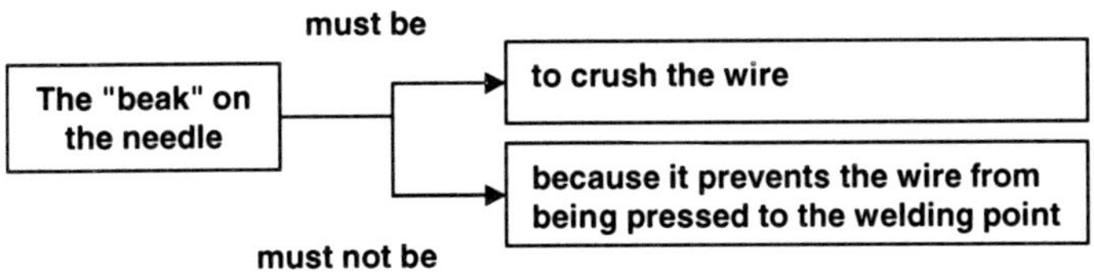

fig. 3.24. "Beak" contradiction

Example 3.4. Bad Solution Turned Good

A new welding technology was invented – ultrasound thermal compression welding. The needle was made to vibrate with a small amplitude at a frequency of more than 20 kHz. It became possible to tear the wire off without first crushing it with the "beak". The needle now had a simpler construction, but as contact pads became smaller and smaller, we again had the same old problem – the pads tore off from the crystal together with the welded wire (see contradiction formula in fig. 3.25).

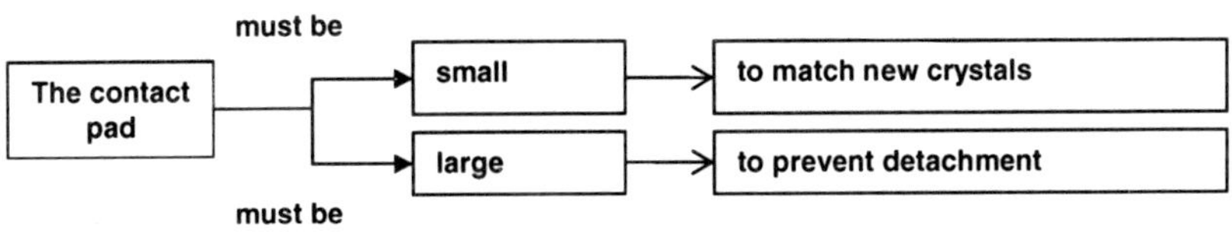

fig. 3.25. "Contact pad" contradiction

Example 3.5. Strong Solution – Change of Operating Principle

It was suggested that the wire could be burned off with an electric impulse. The needle was divided into two parts (fig. 3.26) isolated from one another inside along their entire length except for the top (insulation is shown by darker strips).

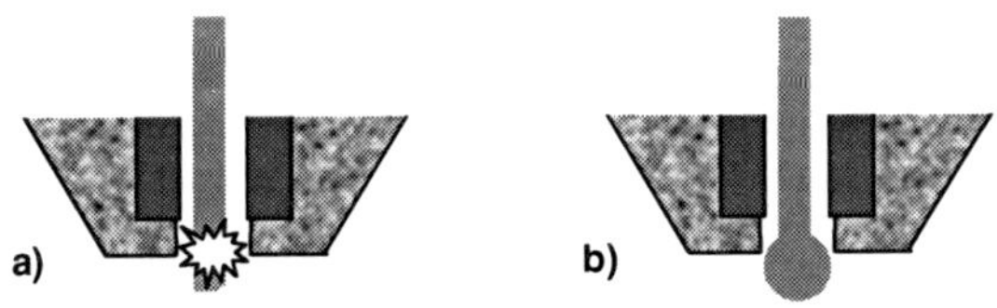

fig. 3.26. Wire severed by electrically-generated fusion

When electric current was applied to both halves of the needle, there emerged an electric discharge which hit the golden wire running through the needle (fig. 3.26,a). The wire was easily fused and torn! As an extra bonus, at the end of the wire formed a small "ball" (fig. 3.26,b) which prevented the wire from being pulled inside the needle, and increased the contact area during the next welding operation.

In other words, we received two unexpected positive "super effects" at once! Incidentally, such unexpected additional advantages often serve as telltale signs of a very good solution!

Unfortunately, this was not the end of our woes (fig. 3.27). The needles were becoming thinner and thinner to match miniature dimensions of crystal-mounted contact pads. However, the thinner the needle, the sooner it was worn out (burned) under the influence of electric discharges.

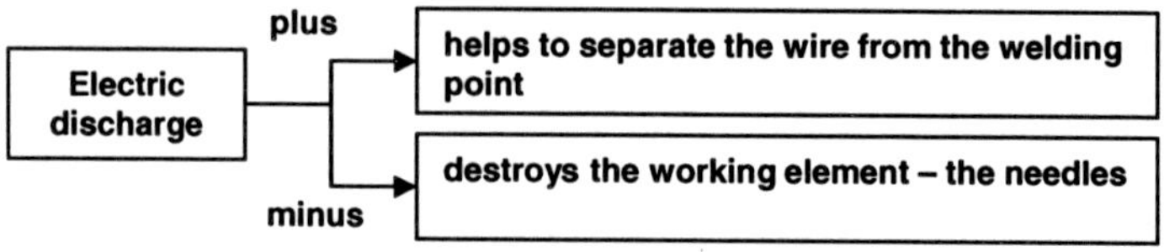

fig. 3.27. "Electric discharge burning" contradiction

Example 3.6. Revolutionary Change of the Operating Principle!

These radical problems were resolved in a radical fashion: it was suggested that contact pads be connected by using the "spider" method! In a very effective operation, all contact pads were joined to all leads "in one fell swoop" without intermediation by golden wire (fig. 3.28,a) – a truly beautiful solution!

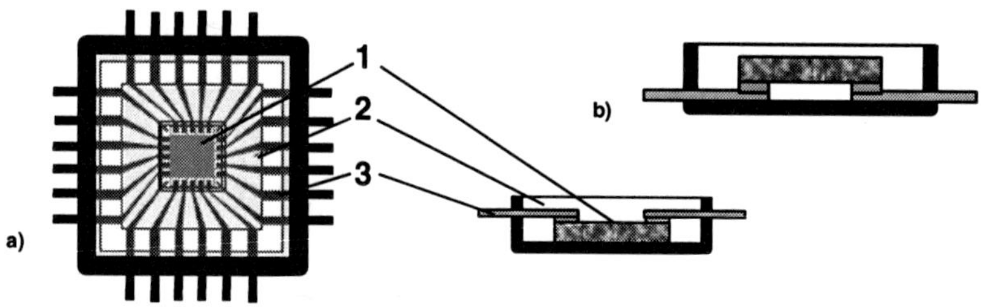

fig. 3.28. Connecting crystal-mounted contact pads with leads in the body of the microchip

The only inconvenience was that the leads were located in the upper part of the body, on a level with the top surface of the crystal. And designers found another simple and efficient solution – the "overturned crystal" method (fig. 3.28,b).

In TRIZ, this creative method is called "inverse action". A beautiful solution: a well-tuned automaton does not really care how it installs the crystal – with contact pads facing up or down!

Some impatient readers may object that these inventions are not "strokes of genius"! They have not affected development of the entire human civilization! They may even say: "We are pining for something global, better yet – cosmic!" OK, I will answer that objection in the next two sections.

Meanwhile, here is another story for you: In 1965, at the famous International Industrial Fair in the German town of Leipzig, the Republic of Belarus harvested an unprecedented five golden medals. Three of those were earned by the denizens of P.O. Box 19 – which was naturally exhibiting its wares under a different name!

It is with great warmth and gratitude that I recall the first general manager of P.O. Box 19, Ilya Mikhailovich Glazkov[61], the eminent leader of our collective.

The first thing he did when appointed to that post in 1962 was to visit several international exhibitions. He brought back a heap of advertising booklets, gathered his young wards and said: we must do this as well or even better! And 3 years later, in Leipzig, the first *Contact* machines received outstanding awards.

I will never forget my closest colleagues and teachers who did it: Valery Semyonovich Galkov, Guennady Pavlovich Kuzmichyov, Vitaly Trofimovich Kuleshov, Aleksandr Danilovich Kiselyov. And they all were 25-28 years old!

I also remember Evgeny Evgenyevich Onegin, the 33-year-old head of our department, who went on to become the general director of PLANAR corporation, heir to the miniscule P.O. Box 19.

ATTENTION!

And now please pay attention: the basic operating principles – and even constructions – of "assembling" machines have not changed over the last 40 years (I am writing this in December 2009)! Naturally, new welding methods have been invented: microplasma welding, laser welding, electron-beam welding, diffusion welding, "cold" welding eliminating the need to heat up the entire microchip…

Each year such machines put together hundreds of millions of microchips throughout the world!

So you tell me – are these inventions "small" or "big"? Think about how the *"big"* started with the *"small"*, how it went through hundreds of inventions and thousands of minor innovations, survived for almost half a century, and keeps on working!

3.4 The Unbelievable in the Routine

There is something more powerful than all the armies in the world: it is an idea whose time has come.[62]

Victor Hugo

What can be more familiar and even boring than a rail, or a steel wheel of a railway car rolling along that rail? Well, maybe a motorcar wheel rolling along a concrete highway or a battered dusty country road.

And now let us consider the five (according to the author) wheel revolutions in the history of the human civilization:

1st Revolution (6,000 years ago, Sumerian state in Mesopotamia): invention of the wheel as such – first made of wood;

2nd Revolution (5,000 years ago, Egypt; 4,000 years ago, Europe): invention of the pair "wheel – manmade road" (road made of wood, stone, etc.);

---

[61] Ilya Mikhailovich Glazkov (1922-1996) – prominent public figure of the Republic of Belarus, Deputy Chairman of the Council of Ministers (1974-1982)

[62] Victor-Marie Hugo (1802-1885) – famous French writer, poet and playwright

3rd Revolution (500 years ago, ore mines in England, Ireland, Russia): invention of the pair "cast-iron wheel – cast-iron rail";

4th Revolution (160 years ago, Robert Thomson[63]): invention of the pair "pneumatic wheel – road" (dirt road, stone road, concrete road, etc.);

5th Revolution (30 years ago – late 1970-es, Anatoly Unitsky[64]): invention of the pair "steel wheel – steel string rail".

To be more exact, Anatoly Unitsky

INVENTED A COMPLETELY NEW... TYPE OF ROAD!!!

Manmade road of a new type using cast-iron rails did not have a prototype in Nature! Anatoly Unitsky's string rail has prototypes both in technology *and* in Nature – for example, cables or spider web threads – but it is not identical to them, as its structure and properties are wonderful and truly revolutionary!

Some people may say: "Well, it's not like you've invented a bicycle or a wheel! What's so special, so unbelievable about it?"

Example 3.7. Automotive Civilization (Solution for Example P4)

Let us consider the key contradictions inherent in the pair "wheel – road" (fig. 3.29).

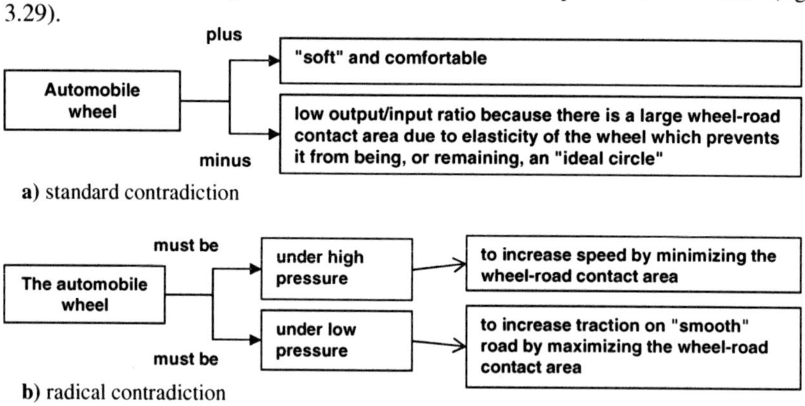

fig. 3.29. Major contradictions inherent in the pneumatic automobile wheel

Because of the road profile and the ensuing low efficiency of the pair "wheel – road" it is impossible to universally support movement even at speeds ranging from 150 to 200 km/h. If you are taking an autobahn from Berlin to Munich, you will move at an average speed of 120 km/h even if traffic conditions are favorable (i.e. there are no traffic jams, construction works, detours, etc.)! That is progress for you! And it is a tall task to get from Berlin to Munich in 7 hours – it all depends on traffic jams, road repairs and construction works!

---

[63] Robert William Thomson (1822-1873) – Scottish inventor, author of the idea for the pneumatic tire which he patented in 1846-1847

[64] Anatoly Eduardovich Unitsky (b. 1949) – inventor of surface and circumplanetary string transportation systems; www.yunitskiy.com

In addition to that, the motor vehicle is the worst environmental pollutant and the worst "legitimate" murderer killing **1,500,000** people and maiming about **5,000,000** people throughout the world every year!

Conclusions: 1. Motor vehicles must not be used for mass-scale transportation of people and cargoes for distances exceeding several hundred kilometers!

2. In the future, the rules governing city traffic, as well as conditions and scale of parking in cities, must be radically changed!

And now let us consider the radical problem affecting the railroad pair "steel wheel – steel rail" (fig. 3.30).

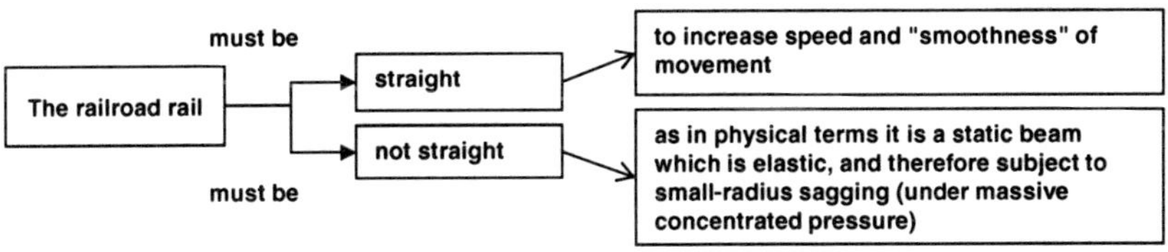

fig. 3.30. Radical contradiction inherent in the "steel rail"

The existence of a sag, however small, means that the railroad wheel is rolling along a road which is not "ideally" straight, that it is constantly trying to get out of a depression – the sagging rail! And that goes for several dozens or hundreds of wheels in each train! Clearly, the need to overcome ongoing resistance results in huge overconsumption of energy! Monstrous waste of energy! In other words, the pair "steel wheel – steel rail" also has a very low output/input ratio!

Accordingly, the maximum speed that can be achieved by trains without creating potentially lethal complications is limited by 200-300 km/h. And the situation is further exacerbated by immense railway construction costs!

And now I will tell you about one surprisingly simple invention which is unbelievably efficient! This is the new planetary transport of the future!

The essence of this invention authored by Anatoly Unitsky is the completely new rail construction (fig. 3.31).

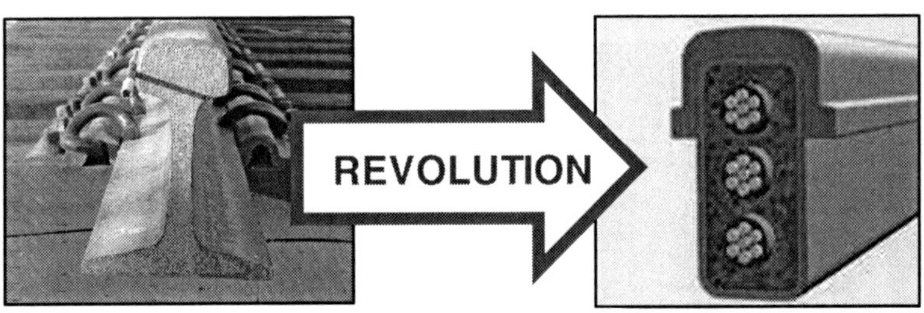

fig. 3.31. String rail construction

Pursuant to the TRIZ model that we know as "Matryoshka" (named after the famous Russian nested doll), the new rail is made of dozens and hundreds of wires. Each such wire is tight as a string, and all of them together make one powerful string which is completely straight!

In fig. 3.32 you can see the relative "sag" radii of motorway surface (Ram = 1, used as the base unit), high-speed railroad rail (Rrr = 3), and Unitsky string rail (Rsr = 10).

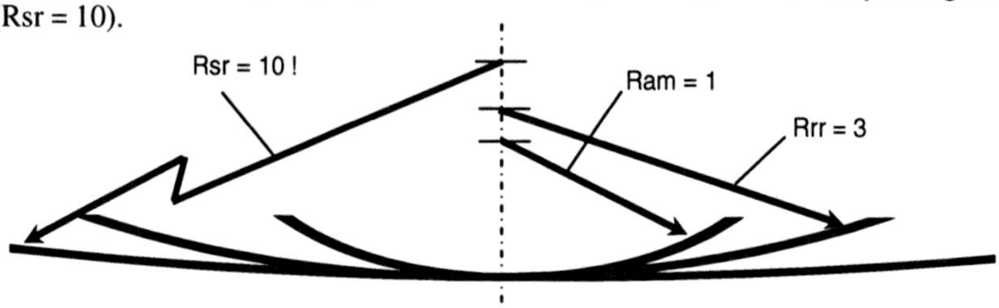

fig. 3.32. Rigidity and straightness of Unitsky road is 10 times better than those of motorway road surface, and 3 times better than those of a high-speed railway rail

The second dominant TRIZ model which is *de facto* realized here is "Transition into another dimension" – in all senses of the word! Technically, this means that the entire transportation structure is raised above ground (fig. 3.33) to a certain height in order to minimize up and down sloping, and to assure that the road is completely straight for dozens and hundreds of kilometers and that flight in a unibus while contemplating the beauty of Nature gives pleasure to each and every passenger!

a) STU station

b) high-speed transportation vehicle looks organic in any

fig. 3.33. Examples from engineering projects designed by Anatoly Unitsky

Indeed, land that can be used by surface transport, particularly in cities, is becoming more and more scarce.

And this means that further development – historically evolutional! – is possible only if we move up, make use of the "height" coordinate by raising the transportation infrastructure above the land surface. And while we are at it, we might just as well "correct" its "defects", the so-called "topographic inequalities" causing the roughness of terrain.

Still, the secret of this invention lies in the physical-technical properties of the string! These properties are completely different from the properties of a beam resting on its ends. All other things being equal, the beam sags much more!

As a result, the maximum speed that can be achieved by Unitsky transportation modules (unibuses) without complications is 200–500 km/h.

ATTENTION!

The string is an "intelligent" system which is capable of self-restoration and is exclusively – to the maximum extent possible! – straight and even along the way.

That is why Unitsky's invention is the last revolution in the development of the pair "wheel – rail": it cannot be surpassed or improved, except in terms of efficiency of materials used for its construction! But NOT in terms of its operating principle!

Also, like any other powerful invention, String Transport Unitsky (STU) creates extraordinary super-effects:

DIRECT ADVANTAGES!

- energy (fuel) consumption is 20 times less compared to motor vehicles!
- energy consumption is 4 times less compared to railway transport!
- STU construction costs are 20-30 times less compared to those of a motor highway of comparable length!
- STU construction costs are 10-15 times less compared to those of a railroad of comparable length!
- Passenger and cargo transportation costs are 10+ times less!

And last but not least, STU has certain systemic properties that are extremely promising.

There is a well-known saying: *states grew around roads.*

STU can link locations within any country by high-speed, comfortable, safe and reliable means of transportation. STU can link countries and continents, for example, Europe and Japan, China and South-East Asia, Africa, the Americas, even Europe and America, etc. Accordingly, STU can become a planetary transportation system (see example in fig. 3.34).

fig. 3.34. Eurasian STU-based transportation internet

There is no end to what I can tell you about STU. But its construction has not started yet. STU is still waiting for investors and statesmen who can think globally and historically, and who are capable of exerting strong political will.

And to crown it all – a little "back-to-the-future" turn: let us talk about *new old* values associated with living on this planet.

Throughout the last century, countries rapidly (like the notorious pebble leather[65]) coalesced around cities, gigantic metropolises inhabited by millions of people! For many, "nature-less" existence turned into chasing material well-being, while morality and dignity began to shrink like the magic talisman made of pebble leather and presented by an old storekeeper to the protagonist of the famous novel by Honoré de Balzac.

Vast expanses of many countries – especially Russia – remain unpopulated! Urbanization – what a terrible word! Man has retreated from Nature. This is one of the reasons why man has lost the emotional skills to perceive and appreciate Nature.

STU makes the impossible possible: it enables a mass-scale reverse migration from the cities to the Nature! Migration to wonderful living and working conditions created by a new strategy and a new conceptual approach to organizing new settlements in the most picturesque locations in any country, in any spot on the planet! Because STU makes distances irrelevant.

People will no longer be wasting their lives in metropolises, if given the chance to live far (by modern standards) from large cities, but still be able to get there in a matter of minutes, not hours, without getting stuck in traffic jams, without suffering nervous breakdowns, without destroying Nature.

65 "Pebble Leather" (Fr. *La Peau de Chagrin*), a novel (1830-1831) by Honoré de Balzac (1799-1850) – famous French writer

The new systemic conceptual framework[66] envisaging establishment of "line" settlements along streamlined STU networks is illustrated by fig. 3.35. Ability to use clean transportation infrastructure and noiselessly moving unibus trains will make it possible to build compact cities, existing in harmony with their natural environment, along STU lines and within walking distance from them.

These will be the growth points of a rationally organized society. It will also be very profitable, because eventually the possibility to live in normal natural and social conditions will be valued higher than possession of any thing, however precious and prestigious. Gradually, there will emerge a new civilization living in unity, rather than in confrontation, with Nature.

And now about the hard lot of pioneers and discoverers.

fig. 3.35. Starting point of systemic growth of a new civilization

"A new worthy purpose[67] is usually so much ahead of its time that people frequently perceive it as heresy. Life in creation does not start with laurel wreaths. You have to learn to fight, to make war. Creative personalities do not turn away from the worthy creation roads they have chosen. They cannot do it because of their sense of duty. Zander[68] called it *duty before mankind.*"

For the last 30 years, Anatoly Unitsky has been *doing his duty before mankind*, and not only does he dream of seeing his transportation system in action, but he also does everything he can to make it happen. And still ... people are not willing or able to understand – *heretic*!

Perception of *pioneers* by politicians, officials and even investors has not changed over the entire history of civilization.

"It is known that it is easier to look for treasure under a lamppost where it is light, rather than in the dark and impassable forest. A little more than 100 years ago, when the first cars had already begun to move along the streets, the Greater London Municipal Council gave credence to conclusions drawn by a special investigation committee in a report on the prospects of development of the city's transportation system: "In 100 years (*i.e. today – A.U. and M.O.*) London will have 2 million residents and 4 million horses; there will be more stables than residential houses; all adjacent lands will be used to grow grass and oats, but there

---

66 See *STU in Questions and Answers* at www.yunitskiy.com

67 Based on: G. S. Altshuller, I. M. Vertkin (1994) *How to Become a Genius*. – Minsk, Belarus

68 Friedrich Zander (Fridrikh Arturovich Tsander) (1887-1933) – Russian scientist and inventor, one of the rocketry pioneers who published his first work on future space flights in 1908 (when he was only 21 years old!). His personal motto was: "Onward, to Mars!"

will still be not enough land and fodder for horses; and in some places horse manure will be lying in heaps half a meter thick".

Current projections for the next 50-100 years are very much alike if we replace certain terms in the above analysis as follows: *horse* with *motor car*, *stable* with *garage*, *oat* with *gasoline*, *land* with *asphalt*, and *manure* with *exhaust gases*. Both then and now specialists and experts make the same mistake – 110 years ago nobody was ready to take seriously the first simple and non-presentable automobile which changed the world in the 20th century; now nobody is ready to take notice of the first simple and non-presentable STU concept built in the settlement of Ozyory, Moscow Region, in 2001, a device which will change the world even more drastically in the 21st century, making it a safer, more ecological and more comfortable place to live.[69] "

The time to build *String Transport Unitsky* has come!

3.5 The Impossible Is Possible

This invention – the one made by Anatoly Unitsky! – is very big in all senses! It is based on an extraordinarily simple idea, it is unbelievably difficult to implement, and it is exclusively efficient and fantastically beautiful!

And it is also something that our human civilization needs very much. And this means that mankind is capable of realizing it! This has always been the case when mankind was faced with a new purpose.

Example 3.8. Space Civilization (Solution for Example P5)

- So you insist that man can raise himself by the hair?

- Absolutely! *Every thinking human being simply must do it from time to time.*

Baron von Münchhausen

This statement by the famous Baron von Münchhausen[70], as brought to us by the actor Oleg Yankovsky, has become an unforgettable aphorism for many who saw this movie[71] and believed in the truthfulness of the story that it related (fig. 3.36).

fig. 3.36. *That Very Same Münchhausen* (copy of DVD cover)

---

[69] See *STU in Questions and Answers* at www.yunitskiy.com

[70] Karl Friedrich Hieronymus Freiherr von Münchhausen (1720-1797) – German Baron, descendant of a famous family of noblemen from Lower Saxony, cornet in the Russian army, historical and literary character. His name later came to be used to refer to people prone to telling tall stories.

[71] *That Very Same Münchhausen* (1979), a brilliant movie directed by Mark Anatolyevich Zakharov (b. 1933) – Russian theater and cinema director; the script was written by Grigory Israilevich Gorin (1940-2000) – Russian playwright and satirical writer; the title role was played by Oleg Ivanovich Yankovsky (1944-2009) – Russian theatre and cinema actor

One of the particularly "impossible" tales told by the Baron was the one where he pulled himself AND his horse out of the bog... by the hair made into a pigtail! Well, who knows, maybe it is really possible! After all, this event has become the subject of at least one painting[72] (fig. 3.37)!

And if someone tried to argue with him from a purely scientific position, the Baron would probably retort: *"I see what your problem is: you are too serious. But a clever face, gentlemen, is not necessarily a sign of intelligence. All stupid things in the world are made by people wearing that expression on their faces."*

fig. 3.37. *The "reliable evidence" of possibility what has been done by Münchhausen*

"And now – attention – question for the Cognoscenti[73]*:"*

Is there a practically implementable idea for an antigravity engine or propulsion unit?

First, some definitions (samples from various encyclopedias):

1) engine is a device which transforms some type of energy into mechanical energy; for example, an electric motor transforms electric energy into shaft rotation;

2) propulsion unit is a device which transforms the energy generated by an engine or some other external source into useful work by moving a transportation vehicle. For example:

- *wheel* – motor cars, locomotives, bicycles, etc.
- *caterpillar* – track tractors, tanks, certain types of all-terrain vehicles;
- *auger* – auger-wheel machines, certain types of all-terrain vehicles (auger-wheel cross-country cars);
- *limbs* – walking mechanisms: androids, exoskeletons, walking excavators;
- *flagellum* – main propulsion unit of microorganisms and spermatozoa of highly organized animals;
- *screw* (propeller) – aircraft, dirigibles, air-cushion craft;
- *water propeller*, ship rotating blade propeller – vessels and ships;
- *jet propeller* (water jet) – small vessels;
- *sail* – sail vessels;
- *electric rocket propulsion unit* (ion stream) – spacecraft, ion sources;
- *jet propulsion unit* (gas stream) – aircraft, rockets.

---

72 Theodor Hosemann (1807-1875) – well-known German genre painter, illustrator and caricaturist

73 Famous prelude to the reading of the question posed by a TV viewer to the Cognoscenti Team in the Russian intellectual game show – *What? Where? When?* Television Club – which has successfully remained on air since September 4, 1975!

And now, as the cognoscenti go into a frenzied discussion of the division of antigravity engines and propulsion units into "genuine" and "non-genuine", let us reinforce their brainstorming efforts with a prompt: do they know a system that would be capable, after taking on a certain amount of energy, of *raising itself into space by itself* together with the cargo?

Do you maybe think that I am talking about a projectile launched "from a cannon to the Moon"[74]? No, the projectile is NOT a propulsion unit!

Naturally, a much more efficient solution is presented by *cavorite*, a material capable of "shielding off" gravitation and doubling as an antigravity engine and propulsion unit[75]!

However, after the first flight to the Moon in 1899 by a team of three Brits the secret of that material has been irretrievably lost!

So what can we do? Who knows the correct answer?

And now – attention – the correct answer (incidentally, this is another famous script line from the host of the *"What? Where? When?"* club):

The impossible is Possible!

It is the Unitsky planetary space transportation system.

Imagine a gossamer trestle with a diameter of slightly more than 40,000 kilometers circling the Earth along the equator. The trestle supports a vacuumized pipe housing a series of linear electric motors and a rotor suspended in a magnetic field.

Inside the rotor less than half a meter "thick", there are millions of tons of payload. The rotor is covered with heatproof coating.

The linear motors gradually – over the course of several days or weeks – boost the rotor to circular velocity – 10 km per second. The pipe opens up, and the rotor, now free of its magnetic field fetters, shoots up and several dozens of minutes later reaches the near space and attains the height of 300 km!

In the process, the rotor expands (fig. 3.38,a) – first due to the elastic properties of its material (phase 1), and then by operation of special telescopic systems (phase 2).

And if we define the rotor boosting system as an "antigravity" engine, the rotor itself will surprisingly turn into a centrifugal propulsion unit. The dynamized (TRIZ!) rotor launches itself into the Earth orbit together with its cargo in the same way as Baron von Münchhausen pulls himself out of the bog together with his horse!

---

74 Jules Gabriel Verne (1828-1905) – famous French writer, author or adventure and science fiction books where he predicted numerous "impossible" inventions many of which have been made by now. He wrote, in particular, the novel *De la terre à la lune*, or *From the Earth to the Moon* (1865)

75 The story of the first flight to the Moon is presented in the novel *The First Men in the Moon* (1901) written by the famous British writer and publicist H. G. Wells (1866-1946)

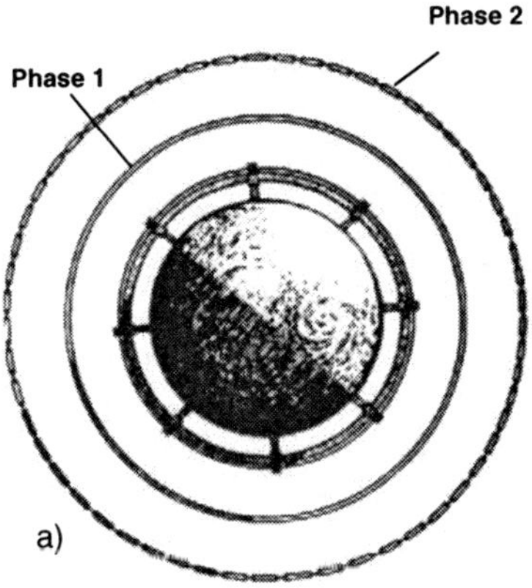

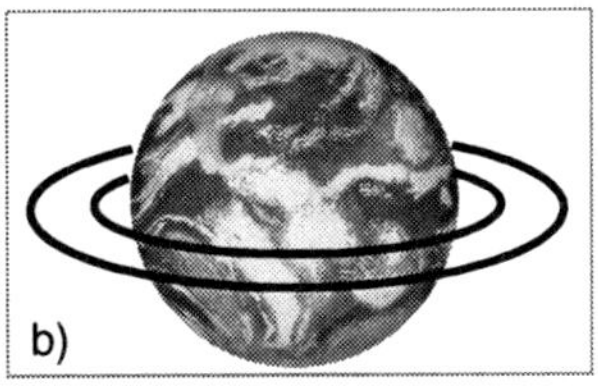

fig. 3.38. Unitsky planetary space transportation system:
a) rotor orbit launch phases – view from one of the poles;
b) rings of circumterrestrial space civilization.

The same mechanism can be used to launch into near space the capsule which houses the rotor – now rotating around the planet. Cabins holding millions of passengers and/or cargo bays carrying millions of tons of payload can be placed inside the capsule or suspended on its outer surface!

Clearly, the trestle, the rotor and the space-based industrial and residential infrastructure can only be implemented on the basis of STU that we discussed in the previous section! Incidentally, due to unique STU properties, the trestle can span three oceans!

All we have to do to make it happen is to… **disarm the Earth!** And invest the massive amounts of cash now spent on murder and wars into the future of the plant.

Example 3.9. Lengthening of Unitsky Ring

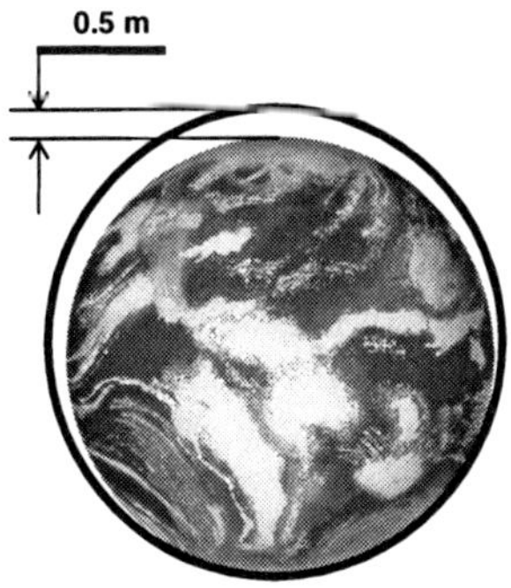

fig. 3.39

And now the problem (with a catch):

The Earth is covered with a ring (fig. 3.39). At the bottom the ring rests on the surface of the Earth, while at the top it is raised about half a meter so that a not too stout person can crawl between it and the Earth.

Question 1: What is the circumference of the ring, and by how much is it longer than the circumference of the Earth (which in this case is assumed at 40,000 kilometers)?

Question 2: By how much (in percentage terms) will Unitsky ring become longer at the height of 300 km above the Earth?

Example 3.10. Self-stabilization of the Unitsky Ring shape

fig. 3.40

Look at this merry fellow, the founder of a touring Circus Cowboys troupe, swinging an American lasso to advertise his act on his web site[76]. The lasso by itself forms a relatively regular circle, even though the momentum boosting this "ring" into motion is somewhat **ir**regular.

Self-stabilization of Unitsky ring is much more even and uniform. Why?

Example 3.11. How does one make several Unitsky Rings?

Jocular prompt: Watch the video presentation using the link to the site – there you can have a good look at how the lasso ring works.

From there it is just one step to…

Afterword[77] to the last two sections:

In 2009 we celebrated the 120th anniversary of Igor Sikorsky[78], aircraft designer who created the world's first multiple-engine bomber in Russia and the world's first helicopter in America. Soon after emigration, Sikorsky found himself with just 20 dollars in his pocket.

Being in Chicago, he "invested" that money very wisely – he bought a ticket to a concert given by Sergey Rachmaninoff[79]. After the concert the two Russian émigrés struck up a conversation. Rachmaninoff asked how much money the aircraft designer would need to start his own business. The answer was: 500 dollars. Rachmaninoff reached into his pocket, took out a bunch of banknotes – the entire fee for the concert – and gave it to Sikorsky. It was five thousand dollars – a huge sum for that time…

---

76 http://www.circuscowboy.com/en/company/artists/sherrif-scott.shtml.

77 Quoted with modifications from: *STU in Questions and Answers,* www.yunitskiy.com

78 Igor Ivanovich Sikorsky (1889-1972) – Russian and American aircraft designer, outstanding creator of airplanes and helicopters; in 1914, he set two world speed records on his biplane C-6: 111 km/h with two passengers on board, and 106 km/h with five passengers; emigrated to the USA in 1919

79 Sergey Vladimirovich Rachmaninoff (1873-1943) – outstanding Russian composer, pianist and conductor; emigrated into Europe in 1917, and into the USA in 1918

Nobody in America believed in Sikorsky's helicopter. Moreover, in the 1930-es, some 30 years after his first failed experiments with the prototype helicopter in Kiev, most engineers concurred that the solution he had chosen – with one main rotor and one steering rotor – would never work. Sikorsky managed to prove that it did work, and since the middle of the last century that solution (later dubbed "classical" throughout the world) has been used by 90% of all helicopters, including those used by US Presidents.

"Expert opinions" regarding new solutions are very often erroneous (surprise, surprise!). Well, if "experts" and "specialists" were always right, they would be the most successful and wealthy people in the world based on their knowledge of the future – and of the technologies on which they should spend their energy and money while becoming "filthy rich".

However, all human history is testimony of the reverse – the highest earners are those who invest into projects that "experts" and "specialists" would not touch with a barge pole.

To make this point, let us go through some of the most "interesting" predictions made by "experts" and "specialists" over the last 150 years. They sincerely believed they were right!

Predictions:

- I think there is a world market for maybe five computers (Thomas Watson, IBM Director, 1943);
- But what ... is it good for? (Engineer at the Advanced Computing Systems Division of IBM, commenting on the microchip, 1968);
- There is no reason anyone would want a computer in their home (Ken Olsen, Founder and President of Digital Equipment Corp., 1977);
- This 'telephone' has too many shortcomings to be seriously considered as a means of communication. The device is inherently of no value to us (Western Union internal memo, 1876);
- Who in the hell wants to hear actors talk? (reaction of H. M. Warner, co-founder of Warner Brothers, to the use of sound in cinema, 1927);
- The wireless music box has no imaginable commercial value. Who would pay for a message sent to nobody in particular? (David Sarnoff's associates in response to his urgings for investment in the radio, 1920);
- We don't like their sound, and guitar music is on the way out (Decca Recording Co. rejecting the Beatles, 1962);
- Heavier-than-air flying machines are impossible! (Lord Kelvin, President of the Royal Society, 1895);
- ... Professor Goddard ... does not know the relation of action and reaction, and of the need to have something better than a vacuum against which to react... He ... seems to lack the knowledge ladled out daily in high schools (New York Times article about Robert Goddard's[80] revolutionary work on creating rockets, 1921);

[80] Robert Goddard (1882-1945) – outstanding American scientist, one of the pioneers of modern rocket science

- Airplanes are interesting toys but of no military value (Marechal Ferdinand Foch, Professor, Ecole Supérieure de Guerre);

- Drill for oil? You mean drill into the ground to try and find oil? You are crazy! (reaction to Edwin L. Drake project, 1859);

- Everything that can be invented has been invented (Charles H. Duell, Special Commissioner, United States Patent and Trademark Office, 1899);

- Louis Pasteur's theory of germs is ridiculous fiction (Pierre Pachet, Professor, University of Toulouse, 1872);

The abdomen, the chest, and the brain will forever be shut from the intrusion of the wise and humane surgeon (Sir John Eric Ericksen, British surgeon, appointed Surgeon-Extraordinary to Queen Victoria, 1873);

- 640K ought to be enough for anybody (Bill Gates, founder of Microsoft, 1981).

fig. 3.41. First A. E. Unitsky monograph (1995) about String Systems presented by him to the author of this book. The print run was only 50 (!) copies…

Unitsky string systems are no historical exception to this rule.

Indeed, with respect to any new idea only its Creator can be a true expert and specialist. But he also has to be a Fighter[81] – so that he can fight for his understanding of the *sense of duty before mankind.*

Still, may each inventor meet a Rachmaninoff of his own…

3.6 The Invisible in the Obvious

Once a man who had grown bitter and disillusioned told me: "We are consumers. We are too used to artifacts. We do not see them as creations of talent and hard work. We see them only as things. Alas, we are like cows endlessly chewing hay or grass together with all kinds of field flowers, and feeling no urge to SEE. A cow does not need to think. Its purpose in life is to chew, because it is a representative of the suborder or ruminants (cud-chewing animals). And we, too, chew our cud, and consume things – not artifacts." This is sad – but I hope not true for all of us.

After all, our discourse is about positive things, about progress. And progress presupposes – as a minimum – separation and distancing of man from animals. We are trying to see things as artifacts so that we can discover inventions, the fruits of intelligence and skill. Incidentally, TRIZ makes use of both intelligence

[81] G. S. Altshuller, I. M. Vertkin (1994) *How to Become a Genius*. – Minsk, Belarus

(ability to think efficiently) and skill (ability to do something, as acquired by application of knowledge on the basis of experience).

And now – to bikes! For what can be more familiar (fig. 3.42) from among apparent things-artifacts around us!

http://en.wikipedia.org/wiki/Bicycle

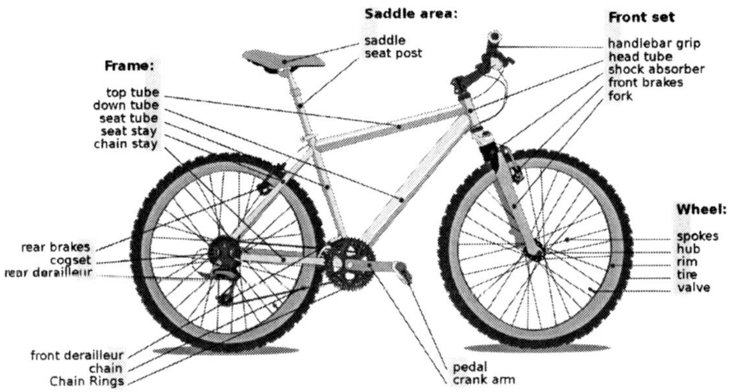

fig. 3.42. Main components of a modern bicycle[82]

We are interested in the evolution of the bicycle. This artifact – which is now considered to be the most mass-scale means of transportation in the history of civilization (despite the lack of cover or cabin of any sort!) – has been developing over a period of more than 200 years.

We will see the invisible – involvement and movement of creative thought overcoming irreconcilable contradictions and creating solutions which changed the destiny of that artifact. TRIZ will tell us about the history of, and the secrets lurking behind, this wonderful device whose developmental and evolutionary potential is still far from exhausted, at least in my humble opinion.

In the apparent[83], we will see things of which outsiders have no idea.

The history section of this account will start with two examples. In the end, they have turned out to be nothing but myths, but having lived with them for so long we are loath to part with them – as we would with an interesting fairy tale. That is why they have been recounted here (fig. 3.43) in the form of examples illustrating Problem No. 0.

Examination of all examples requires that they be read at least twice: first, in the beginning of your TRIZ learning path so that you can learn of historical facts and contradictions that were overcome in each example by inventing a new solution and, second, after you have studied this book and mastered all transformation models described in TRIZ windows.

---

[82] http://ru.wikipedia.org/wiki; file: Bicycle_diagram-ru.svg

[83] About illustrations: 1) the author thanks the owners of all illustrations used in this book; and 2) all examples have been adapted for training purposes

0 problem	**Standard Contradiction**	
	(+)-factors	(-)-factors
Accelerate the movement of man along the surface of the land by using only his muscular strength. But what should we take as the prototype: a chariot? A cart?	*Increase of speed; decrease of energy use*	*Large weight of known carriages with two, three and four wheels*
Time: 1500 ?		
Solution: This wooden da Vinci "bike" model is exhibited in the Exhibitions Museum at the Korean Invention Learning Center!		
TRIZ: *Copying, Mediator, Unite, Dynamization*		
Time: 1790 ?		
Solution: Comte de Sivrac (France) invented a two-wheeled push-bicycle with a wooden frame and without a steering mechanism (one of the side panels in the drawing is "removed")		
TRIZ: *Copying, Unite*		

fig. 3.43. Who invented this wonderful "horse" coveted by kids even today, in the 21st century?!

During the second reading you will discern – in all their fullness and pragmatic beauty – things that were invisible or not quite clear to you during the first reading.

I am sure that you will remember *Magic Vision* pictures, and when you do you will experience a powerful *déjà-vu.* Only through knowledge of transformation models – *TRIZ navigators of creative thinking* – will you be able to fully understand and appreciate the efficiency of creating methods, and the effectiveness and beauty of solutions which again and again modified this system (the bicycle) to take it, step by step, to its familiar and apparent modern form.

And to spice up examination of examples, we will tell you, by way of introduction, a truthful story (fig. 3.44) about how Comte de Sivrac could have come up with the original idea for his extraordinary invention.

If you wish, you can make up similar stories for any of the examples below. I will be happy if you send them to me.

Besides, these examples contain a method that can be used to formulate reinventing problems for any other artifacts.

May you find this reading material interesting!

And when you have reviewed the following 66 examples, go on to study Parts 2-4 of this book which provide a detailed description of Easy TRIZ instruments.

The Legend of Comte de Sivrac

1. Once upon a time the count was sitting on the bank of the river fishing. And then he sees this cart wheel rolling by. Rolling and not falling. So he thinks to himself: would be nice to take a ride on that wheel! But therein lies a dilemma – how does one ride a wheel?

So the count quits fishing and goes back home to his castle, but now he has that idée fixe about how he wants to ride the wheel. He goes to the stable, takes a wheel, straddles it with his feet planted on the shaft from left and right, and then grabs a pole and just stands there, as he is afraid to let go and start riding. Then he picks up his courage, pushes himself off away from the pole and... BANG! goes down head first! Luckily for him, there was a small haystack between the wooden floor and his forehead. Yea.

So he thinks, like, this must have been the shortest ride in history, because you cannot really ride a wheel for a long time. Besides, a cart has more wheels, and when you ride a cart you can SIT! So why not take two wheels and connect them – not on the same axis, like in a hansom, but one after the other. And there is a bonus – you can place a seat in between.

No sooner invented than done.

So the count makes himself a two-wheeled "horse" and is riding around happy as a lark. He is having the time of his life riding that thing, and people around are having the time of their lives laughing their heads off. He pushes off with his legs, you know, to have his "horse" run faster. Then he raises his legs and rolls on while the rolling is good, somehow managing not to fall. Now, if that is not a miracle, I do not know what is. A man is riding a wheel and not falling down! Soon his workers start making such "horses" for their own kids, and people start coming from other counties to look at his invention and build its likenesses for their use.

As for the count, well, he soon gets bored with rocking on and falling off his "horse", because the "horse" stubbornly refuses to turn where he wants to go. So he buries it in the attic and goes back fishing.

2. It is a pity the count did not know TRIZ system evolution laws, or he could have invented tons of interesting things, like a scooter, a skateboard or even a monocycle plus a bunch of steering gadgets!

Also, he could have written down his inventing method as a special creative technique and given it a fancy name, like "copying", "unite" or even "local property". And then he could have taught his descendants to use those.

Who knows how much farther and faster our civilization would have progressed!..

fig. 3.44. The truthful story of Comte de Sivrac as heard by Baron von Münchhausen in the late years of his life (paragraph 1) and complemented with a historical commentary by the author (M.O., paragraph 2)

I wish you success!

3.7 History of Bicycle Evolution in Pictures, Historical Facts, Contradictions and Inventive Solutions

Read, think and enjoy it!

<table>
<tr><td rowspan="2">01 problem</td><td colspan="2">Standard contradiction</td></tr>
<tr><td>(+)-factor</td><td>(-)-factor</td></tr>
<tr><td>It is difficult to turn and go around obstacles when riding Comte de Sivrac's "running machine"</td><td>Improve ease steering during turns</td><td>Rigid connection of all construction elements</td></tr>
<tr><td>Time: 1817</td><td colspan="2" rowspan="3"></td></tr>
<tr><td>Solution: Karl Drais invented a "running machine" controlled by the front steering wheel (Germany)</td></tr>
<tr><td>TRIZ: Segmentation, Dynamization, Local property</td></tr>
</table>

<table>
<tr><td rowspan="2">02 problem</td><td colspan="2">Standard contradiction</td></tr>
<tr><td>(+)-factor</td><td>(-)-factor</td></tr>
<tr><td>When the rider pushes off from the ground with his legs, movement is slow and uncomfortable, and energy expenditure is high</td><td>Increase ease of movement; reduce energy expenditure; increase speed</td><td>No transfer of energy from the rider to the wheel</td></tr>
<tr><td>Time: 1839</td><td colspan="2" rowspan="3">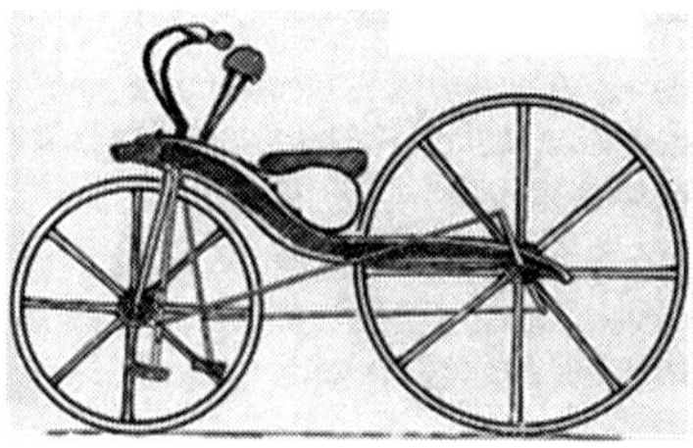</td></tr>
<tr><td>Solution: Kirkpatrick MacMillan creates a rear-wheel drive using mid-mounted treadles connected, through a system of levers, to the rear crank, which created a certain mechanical advantage (England)</td></tr>
<tr><td>TRIZ: Inverse action, Periodic action, Mediator</td></tr>
<tr><td>Time: 1850-1870</td><td colspan="2" rowspan="3"></td></tr>
<tr><td>Solution: Pierre Lallement (France, USA), Pierre Michaux (France), and Philip Fischer (Germany) independently invented pedals connected to the front wheel.
The new contraption is named vélocipède, from Latin velox – "fast" and pes – "foot".</td></tr>
<tr><td>TRIZ: Mediator, Copying, Local property, Asymmetry</td></tr>
</table>

<table>
<tr><td rowspan="2">03 problem</td><td colspan="2">Standard contradiction</td></tr>
<tr><td>(+)-factor</td><td>(-)-factor</td></tr>
<tr><td>With larger wheel diameter, it becomes increasingly difficult for the rider to brake with his legs or by back-pressing the pedals</td><td>Increase of speed</td><td>No transfer of energy from the rider to the wheel for the purpose of braking</td></tr>
<tr><td>Time: about 1845</td><td colspan="2" rowspan="3">
Vienna Technical Museum; photo by author</td></tr>
<tr><td>Solution: application of a hand brake directly to the wheel</td></tr>
<tr><td>TRIZ: Previously installed cushion, Local property, Mediator, Copying</td></tr>
</table>

<table>
<tr><td rowspan="2">04 problem</td><td colspan="2">Standard contradiction</td></tr>
<tr><td>(+)-factor</td><td>(-)-factor</td></tr>
<tr><td>With larger wheel diameter, it takes increasingly more energy to rotate the wheels</td><td>Increase of speed</td><td>Increase of friction in axles</td></tr>
<tr><td>Time: about 1845</td><td colspan="2" rowspan="3">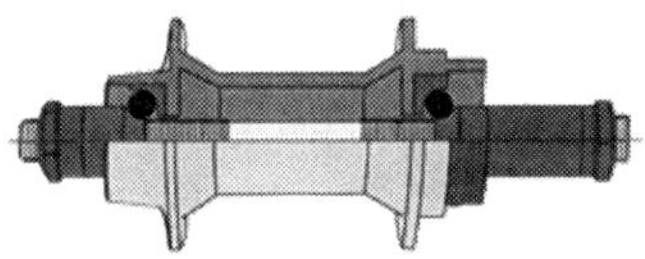
Modern assembly</td></tr>
<tr><td>Solution: wheel axles are fitted with bearings</td></tr>
<tr><td>TRIZ: Spherical-shape, Local property, Mediator</td></tr>
</table>

<table>
<tr><td rowspan="2">05 problem</td><td colspan="2">Standard contradiction</td></tr>
<tr><td>(+)-factor</td><td>(-)-factor</td></tr>
<tr><td>With higher speed, the ease and comfort of riding are increasingly impaired when the bicycle runs over small pebbles and twigs</td><td>Increase of speed</td><td>Wooden wheels, sometimes braced with metal hoops, hit obstacles very hard</td></tr>
<tr><td>Time: 1) 1845; 2) 1847</td><td colspan="2" rowspan="3">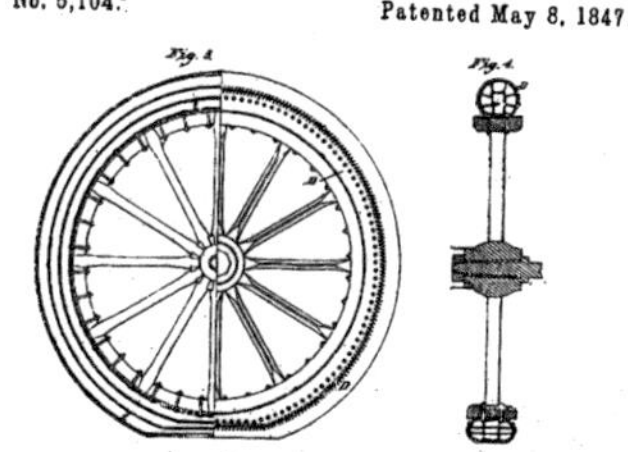</td></tr>
<tr><td>Solution: 1) Scottish inventor R. Thompson manufactures the first solid rubber tires (England); 2) Lord Lorain invents inflatable tires which were too expensive and soon were forgotten (England)</td></tr>
<tr><td>TRIZ: Mediator, Local property, Change in the aggregate state, Copying</td></tr>
</table>

06 problem	**Standard contradiction**	
	(+)-factor	(-)-factor
Large dimensions and bulky parts used to assure durability of wooden bicycles substantially increase their weight	*Increase of durability*	*Increase of weight*
Time: 1870		
Solution: James Starley working for Coventry Machine Company (England) manufactures the first bicycle with a hollow section-steel frame and wire-spoke wheels		
TRIZ: *Segmentation, Replacement of mechanical matter*		

07 problem	**Standard contradiction**	
	(+)-factor	(-)-factor
Further increase of wheel diameter becomes impossible, and speed increase stops	*Increase of speed*	*Limit on wheel size increase*
Time: about 1879		
Solution: Harry Lawson invents chain drive where the drive sprocket and the driven sprocket have different diameters		
TRIZ: *Asymmetry, Mediator, Local property*		

08 problem	**Standard contradiction**	
	(+)-factor	(-)-factor
With higher speed, the ease and comfort of riding are increasingly impaired, and more noise is produced, when the bicycle runs over small pebbles and twigs	*Increase of speed*	*Metal wheel impact*
Time: 1888, see 1847		
Solution: Veterinary John Dunlop invents inflatable bicycle tire (England)		
TRIZ: *Change in the aggregate state, Mediator, Use of pneumatic or hydraulic constructions*		

09 problem	**Standard contradiction**	
	(+)-factor	(-)-factor
When accelerating down a steep slope, the rider lost the pedals, and could easily fall down ("come a cropper") without foot props (some machines had footboards); attempts to brake by bearing down on the pedals resulted in falls and injuries	*Ability to rest without pushing the pedals; ability to brake the wheel with strong leg muscles*	*Constant rigid connection between the pedals and the wheels through the gear*
Time: 1898	1) 2) 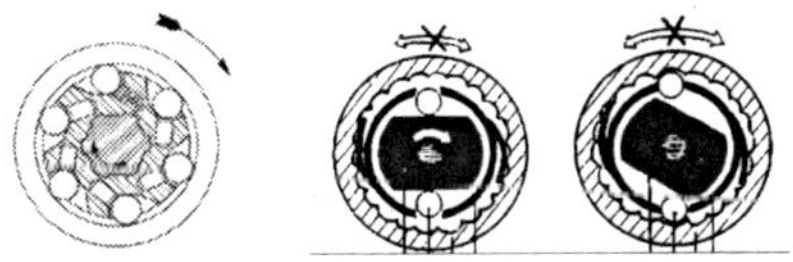	
Solution: The company Fichtel & Sachs (Germany) patented a Torpedo bicycle gear hub with a freewheel mechanism (1) and an inbuilt braking system (2)		
TRIZ: *Segmentation, Matryoshka, Uninterrupted useful function*		

10 problem	**Standard contradiction**	
	(+)-factor	(-)-factor
Army use	*Possibility to carry the bicycle*	*Uncomfortable due to the large length of the bicycle*
Time: late 19th century		
Solution: manufacture of folding bicycles (Germany)		
TRIZ: *Segmentation, Dynamization, Local property, Copying*		

11 problem	**Standard contradiction**	
	(+)-factor	(-)-factor
Army use on rough cross-country terrain	*Reduction of jogging and shaking when riding*	*Rigid connection between the wheels and the frame*
Time: early 20th century		
Solution: spring-mounted wheels		
TRIZ: *Copying, Dynamization, Local property*		

12 problem	**Standard contradiction**	
	(+)-factor	(-)-factor
Army use	*Increase of reliability and saving scarce material (rubber)*	*Inflatable tires easily damaged by sharp stones, glass, etc.*
Time: 1915		
Solution: invention of a tire-less wheel consisting of two concentric hoops linked by numerous radial springs		
TRIZ: *Matryoshka, Dynamization, Preliminary counter-action*		

13 problem	**Standard contradiction**	
	(+)-factor	(-)-factor
The chain is not reliable when dirty, and falls off the sprockets	*Protection of chain switch from dirt*	*Incomplete protection*
Time: late 19th century – early 20th century	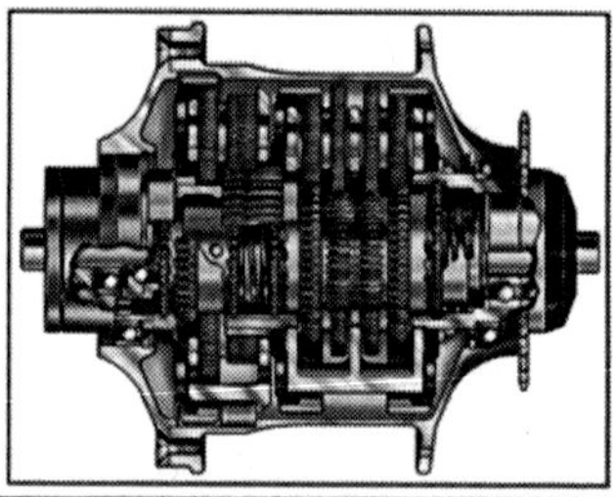 Modern assembly	
Solution: invention of a planetary gear hub		
TRIZ: *Matryoshka, Transition into another dimension*		

14 problem	**Standard contradiction**	
	(+)-factor	(-)-factor
Sports competitions require optimal use of stamina	*Variable load*	*Fixed gear transmission ratio*
Time: 1950-es	 Modern assembly	
Solution: multiple-speed shift gear		
TRIZ: *Segmentation, Copying, Local property*		

15 **problem**	**Standard contradiction**	
	(+)-factor	(-)-factor
Chain often falls off in multiple-speed gear mechanism	*Chain must be engaged only in one pair of sprockets*	*It is difficult to choose diameters to ensure that the length of the chain is fixed*
Time: 1905-es	 Modern assembly	
Solution: invention of a chain pull mechanism		
TRIZ: *Local property, Self-servicing, Mediator*		

16 **problem**	**Standard contradiction**	
	(+)-factor	(-)-factor
The chain makes noise and becomes rusty	*Lubricate the chain*	*Lubricants leave dirty stains on clothes*
Time: 1984	 Modern assembly	
Solution: The company Continental Gummi-Werke designs synthetic timing belts		
TRIZ: *Change in the aggregate state, Copying, Inverse action, Local property, Transition into another dimension*		

17 **problem**	**Standard contradiction**	
	(+) factor	(-)-factor
Inefficient utilization of leg power	*Minimal effort is generated at the highest point of pedal trajectory, maximum effort – when the pedal is in horizontal position*	*Fixed traction lever due to the constant diameter of the drive sprocket*
Time: ?	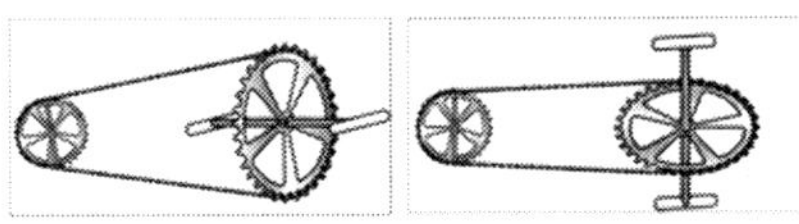	
Solution: invention of an elliptic drive sprocket		
TRIZ: *Asymmetry, Copying, Uninterrupted useful function, Local property*		

18 problem	**Standard contradiction**	
	Requirement 1	Requirement 2
Toe clips on the pedals can be dangerous when the rider needs to rapidly get off the bicycle	*Toe clips make it possible not only to push the pedals, but also to pull them*	*Toe clips can lock the foot in*
Time: 1985		
Solution: The company LOOK manufactures clipless pedals similar to downhill snow skiing bindings (France)		
TRIZ: *Universality, Previously installed cushion, Mediator*		

19 problem	**Standard contradiction**	
	(+)-factor	(-)-factor
Increased reliability and reduced weight of the wheels	*Wheels must be light*	*Spokes are difficult to adjust, heavy and easily deformed*
Time: late 1980-es	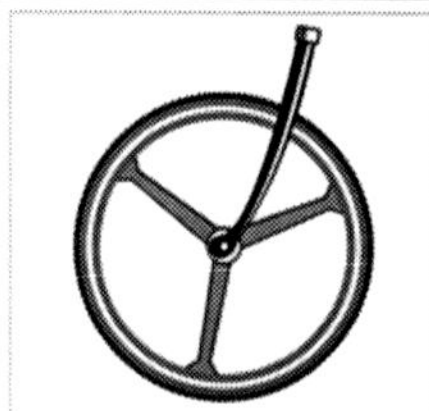	
Solution: solid disk wheels made of carbon fiber; wheels with 4, or even 3, carbonate "spokes"		
TRIZ: *Use of composite materials*		

20 problem	**Standard contradiction**	
	(+)-factor	(-)-factor
Reduction of the overall weight of the construction	*The bicycle must be light*	*There is a risk of reduced durability*
Time: 1990-es		
Solution: cast solid magnesium frames; frames made of carbon fiber and ceramic fiber; glued aluminum frames		
TRIZ: *Use of composite materials*		

<table>
<tr><td rowspan="2">21 problem</td><td colspan="2">Radical contradiction</td></tr>
<tr><td>(+)-factor</td><td>(-)-factor</td></tr>
<tr><td>Chain drawbacks: it becomes dirty, falls off, breaks; multiple-speed gear is too heavy</td><td>The chain <u>must be</u> to transfer energy to the drive sprocket</td><td>The chain <u>must not be</u> as it becomes dirty and easily falls off</td></tr>
<tr><td>Time: late 19th century</td><td colspan="2" rowspan="3">
Modern assembly</td></tr>
<tr><td>Solution: use of universal joints and bevel gears; no dirt plus solid protection</td></tr>
<tr><td>TRIZ: Separation, Copying, Local property, Inverse action, Transition into another dimension, Spherical-shape, Matryoshka</td></tr>
</table>

<table>
<tr><td rowspan="2">22 problem</td><td colspan="2">Standard contradiction</td></tr>
<tr><td>(+)-factor</td><td>(-)-factor</td></tr>
<tr><td>The chain (including the universal joint etc.) has excessive weight resulting in friction-caused power loss</td><td>It is necessary to improve gear efficiency</td><td>Transmission has too many links</td></tr>
<tr><td>Time: multiple attempts</td><td colspan="2" rowspan="3">
Joey Ruiter & StudioDesign concept</td></tr>
<tr><td>Solution: evolutional spiral! – Olympic champion suggests "inverse" Michaux-Lallement-Fischer bike! And HE knows about pedal-pushing more than anybody!</td></tr>
<tr><td>TRIZ: ideal TRIZ-Solution: transmission is maximally reduced! Local property, Inverse action</td></tr>
</table>

<table>
<tr><td rowspan="2">23 problem</td><td colspan="2">Standard contradiction</td></tr>
<tr><td>(+)-factor</td><td>(-)-factor</td></tr>
<tr><td>Many people like push-bicycles, but they are not comfortable to sit on! What can we do?</td><td>Push off with your legs while in motion</td><td>Large load on legs when standing</td></tr>
<tr><td>Time: 2008</td><td colspan="2" rowspan="3">
GlideCycle by David Vidmar. Piece of advice: take a closer look at this bike's history and evolution at www.glidecycle.com</td></tr>
<tr><td>Solution: a grimace of evolution! – this is a revenant of the "bike" invented by Comte de Sivrac and Baron Drais! And yes, you saw it right: THERE IS NO CHAIN!</td></tr>
<tr><td>TRIZ: Unite the alternative systems (scooter + bike) with partial reduction + Inverse action, Local property</td></tr>
</table>

24 **problem**	**Standard contradiction**	
	(+)-factor	(-)-factor
The bicycle is heavy and bulky, it is difficult to carry and store	*Ease of transportation and storage*	*Large weight and overall dimensions*
Time: 1978 **Solution:** continued development of folding bicycles in Germany **TRIZ:** *Segmentation, Dynamization, Local property, Copying*		
Time: 2008 **Solution:** the *Contortionist* model is a surprisingly simple realization of the Matryoshka principle – the entire construction is hidden between the wheels, and the folded bicycle is easy to roll! **TRIZ:** *Segmentation, Dynamization, Local property, Copying, Inverse action, Transition into another dimension, Matryoshka*	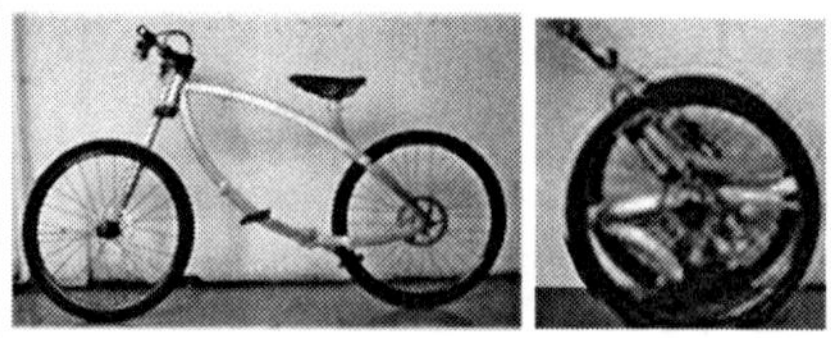 Contortionist by Dominic Hargreaves	
Time: 2008 **Solution:** a graduate of the Royal College of Arts (RCA, London) designed something which is "apparently incredible" – a collapsible wheel! **TRIZ:** *Segmentation, Dynamization*	 Duncan Fitzsimmons	
Time: 2009 **Solution:** Mexican designer Victor Aleman proposed a bicycle where all components are disassembled into small parts and then turned into a "nested doll"! **TRIZ:** *Segmentation, Dynamization, Matryoshka*	 Urban Bicycle designed by Victor M. Aleman	

25 **problem**	**Standard contradiction**	
	(+)-factor	(-)-factor
The bicycle is bulky, it is difficult to carry and store	*Ease of transportation and storage*	*Large size*
Time: 2008	 STRiDA by Mark Sanders	
Solution: smaller wheels, radical modification of the traditional frame form		
TRIZ: *Segmentation, Dynamization, Local property, Copying, Inverse action, Transition into another dimension, Matryoshka*		
Time: 2008	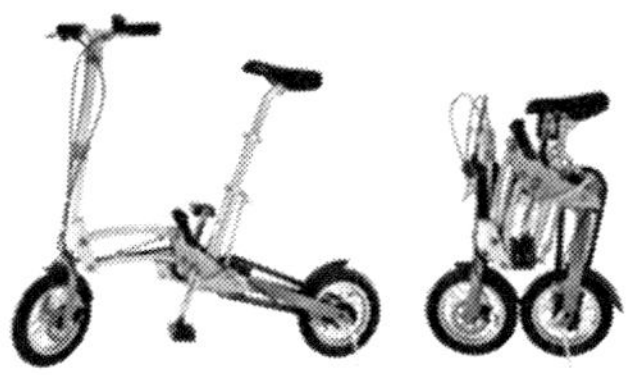 Mobiky MY16 CONFORT	
Solution: further reduction of wheel size (the wheels are now almost as small as those of a scooter) + the *Genius* model is fitted with an electric motor	 Genius Rendezvous from Mobiky & Matra MS	
TRIZ: *Segmentation, Dynamization, Local property, Copying, Inverse action, Replacement of mechanical matter*		
Time: 2009	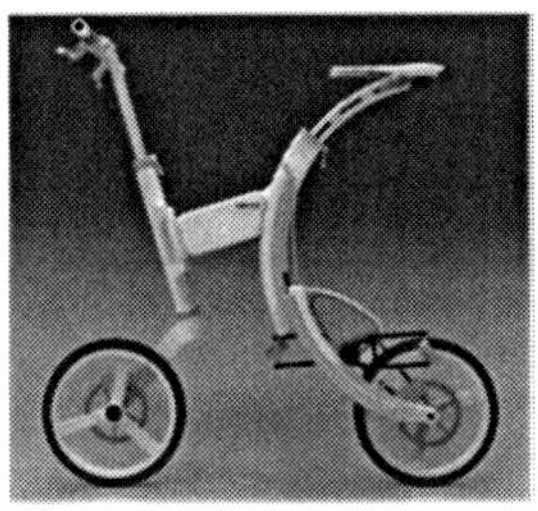  Grasshopper by David Gonçalves	
Solution: this *Grasshopper* also shatters our familiar perceptions of design and functionality, So – to paraphrase a famous maxim – the bicycle is as inexhaustible as … the atom!		
TRIZ: *Segmentation, Dynamization, Spherical-shape*		

26 **problem**	**Standard contradiction**	
	(+)-factor	(-)-factor
The chain is the most "obsolescent" system component of the bicycle!	*It is necessary to improve gear efficiency*	*Transmission has too many links*
Time: 1869		Monowheel by Rousseau of Marseilles
Solution: invention which can only be described as a stroke of genius! Who knows, maybe the future of the bicycle is its "well forgotten" past?! – see below		
TRIZ: *ideal (in a fashion) TRIZ-Solution: the system is reduced to its operating element! + Copying, Inverse action, Local property, Matryoshka*		
Time: 2008	 	Monowheel by Ben Wilson
Solution: history repeats itself at a new level of evolution! This model does have a chain – although it is quite small!		
TRIZ: *ideal TRIZ-Solution: transmission is considerably reduced! + Copying, Local property, Matryoshka*		
Time: 2008	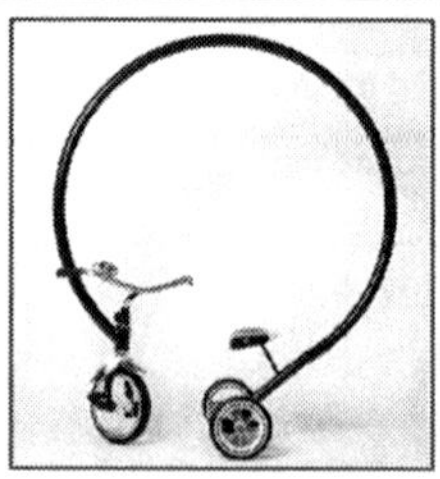	Transformer by Sergio Garcia, USA www.sergiotheg.com
Solution: this one has no chain, either! The wheels of this "bicycle" are joined by a flexible transformer frame which can assume bizarre shapes		
TRIZ: *Inverse action (very radical!), Local property, Dynamization – and simple sense of humor!*		

27 problem	**Standard contradiction**	
	(+)-factor	(-)-factor
When going uphill, it is difficult to push the pedals even in low gears; people with weak health find it difficult to ride the bicycle for a long time	*Negotiate steep hills; increase duration of the trip*	*There are no suitable gears; muscle power is insufficient*
Time: since the 1990-es	 Modern Electrobike Pi	
Solution: invention of electric bicycles with storage batteries and electric motors		
TRIZ: *Replacement of mechanical matter, Separation, Local property, Copying, Partial or excess effect, Discard and renewal of parts, Mediator, Unite*		
Time: 2009	Copenhagen Wheel by Carlo Ratti – MIT SENSEable City Lab.  	
Solution: the motor-wheel is designed by Carlo Ratti's MIT SENSEable City Lab		
TRIZ: *trend – idealization and reduction!*		
Time: 2009	NOAH by Andrei França 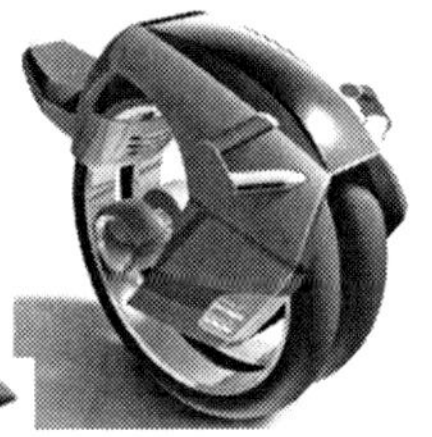	
Solution: reduce! One of the authors was apparently incapable of letting go of one "redundant" wheel. Well, at least the others did not stand on ceremony and this is what we have as a result. I do not even know what to call these machines – light motorcycles or super-heavy bicycles… As for the pedals, well, you cannot push them because they are not there. To compensate for that, there are plenty gyroscopes and a powerful electronic brain…		
TRIZ: *Reduction! + Replacement of mechanical matter, Inverse action, Matryoshka… and others*	Bombardier EMBRIO 	

28 problem	**Standard contradiction**	
	(+)-factor	(-)-factor
Use of composite materials makes the bicycle lighter – and more expensive	*Weight of the moveable object*	*Complex construction consisting of many components*
Time: 2008	Ultimate Wheel (1980-e)	Ciclotte by Luca Schieppati
Solution: in its laconic expression, this extraordinary idea is similar to the original marginal concept of a "bicycle" as a "wheel with pedals"!		
TRIZ: *Reduction + Inverse action, Matryoshka*		
Time: 2009		Wheelskates by Michael Jenkins
Solution: Australian inventor Michael Jenkins invented another "foot-wheel" concept (basically, a scooter)		
TRIZ: *Reduction + Segmentation, Separation, Dynamization, Inverse action, Local property, Copying*		
Time: 2009	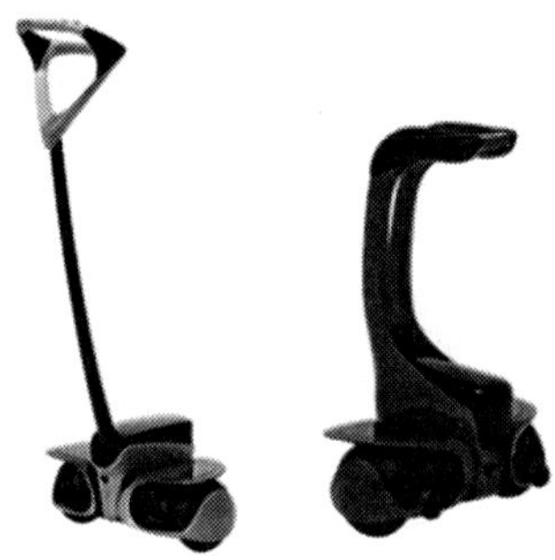	Winglet from Toyota
Solution: you can stand, and you can sit! Is this a bicycle, or is this a scooter?... Be that as it may, Dean Kamen's Segway has robbed many people of their sleep! And, naturally, there is a massive informational resource represented by multiple gyroscopes and processors – so that you don't even have to think!		
TRIZ: *reduction – to the operating elements, to the "last" wheel (well, it IS a dual wheel)!*		
Time: 2008		COOL RIDER
Solution: a "tow car"! If you complete this scooter with a saddle, it will turn into a "bike" – with the pedals cut off...		
TRIZ: *Segmentation, Separation, Local property, Inverse action, Transition into another dimension, Mediator (!), Matryoshka*		

29 problem	**Standard contradiction**	
	(+)-factor	(-)-factor
Modern bicycles are heavy and have poor power-to-weight ratios	*Increase of storage battery capacity*	*Large weight*

Time: 2006 **Solution:** and this is just a joke, not a bike but a scooter! But in this case we can say that it is a "reduced" Comte de Sivrac bike! If only it had an electric motor! **TRIZ:** *reduction – to the operating elements, to the "last" wheel!*	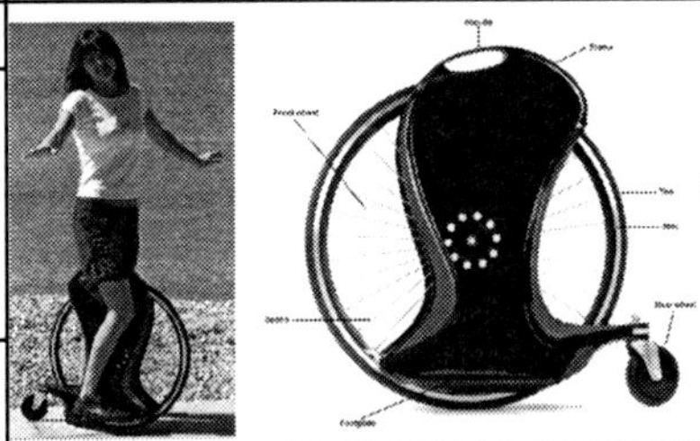 Magic Wheel
Time: since the 1990-es **Solution:** and this is a wonderful "hobby horse" of Trevor Blackwell, a very serious engineer and entrepreneur from the Silicon Valley. Only the pedals are missing! **TRIZ:** *almost total reduction if the prototype is a real bike, or expansion if the prototype is a monocycle!*	 http://focusdesigns.com
Time: 2009 **Solution:** invention of an "almost" monocyclic bike with an electric drive – what a beauty! Animations are especially impressive! **TRIZ:** *Trend – partial reduction to one operating wheel!*	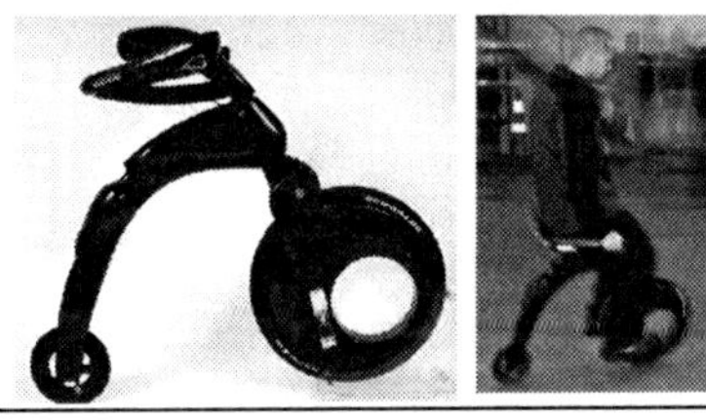 Скутер YikeBike
Time: 2008 **Solution:** something evocative of a "kamikaze in a torpedo" – this invention draws a lot of attention and causes bitter arguments **TRIZ:** *Reduction + Inverse action, Transition into another dimension + fantasy.* *My teacher Stanislaw Antonovich Baltsevich used to say: "There is no limit to both human fantasy and human stupidity!" He than went on to provide a suitable quote from the great Einstein.* (NB: the second part of my teacher's statement does not apply to this example.)	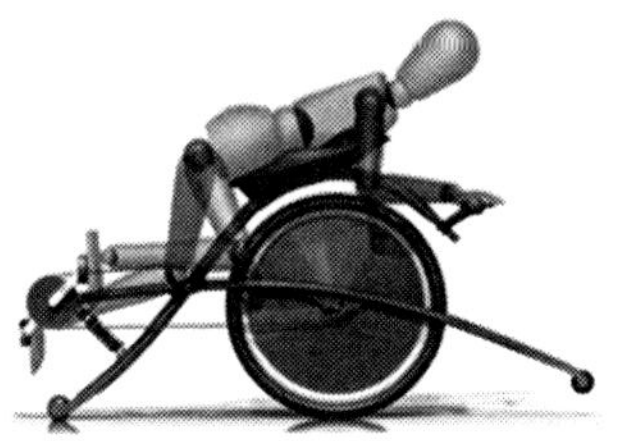 WheelU by Tomer Zer Kavod Look, standing is bad, sitting is good, and lying is even better, that's why we have recumbent models! What else would you want?! So our idea is siple – look at the alternatives and… do what you think is best!

30 problem	**Standard contradiction**	
	(+)-factor	(-)-factor
It IS TRUE – it is difficult to learn to ride the bicycle, and it is equally difficult to ride it using just one wheel…	*Comfort, safety*	*Unreliability, instability*
Time: 2009 **Solution:** This bike manufactured by Gyrobike, a company established by graduates of the Dartmouth College, USA, increases stability of the bicycle – which is quite useful when you are teaching your kid to ride it! **TRIZ:** *Replacement of mechanical matter, Separation, Copying, Matryoshka*	GYROWHEEL REINVENTING THE (TRAINING) WHEEL 	Gyrobike with Gyrowheel 
Time: 2009 **Solution:** Due to inbuilt gyroscopes, this bike/scooter can stand "by itself", while the ingenious construction of its wheel endows it with an ability to move in "all directions"! **TRIZ:** *Reduction! + Segmentation, Self-servicing, Feedback, Universality*	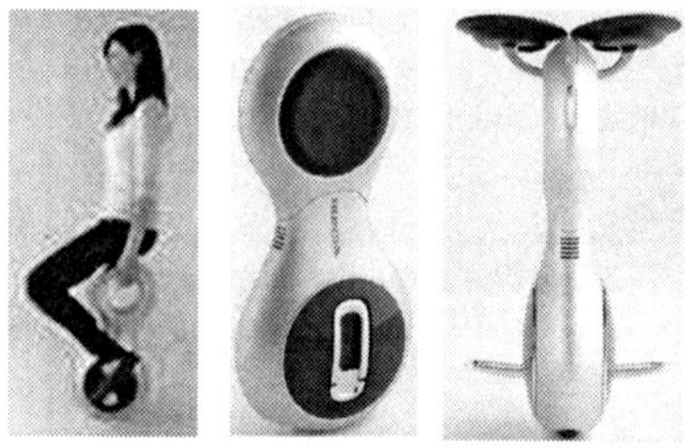	Honda U3-X
Time: since the 1990-es **Solution:** Inventor Thomas Gordon sees such robots exploring the Mars and the Moon. It rolls BY ITSELF (apparently obeying TRIZ laws). But will it be able to stand up BY ITSELF if it falls? **TRIZ:** *Replacement of mechan. matter, Separation, Local property, Copying*	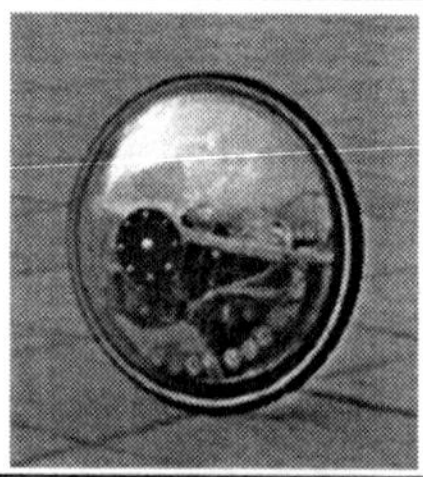	Carnegie Mellon University's Robotics Institute www.ri.cmu.edu

31 problem	**Standard contradiction**	
	(+)-factor	(-)-factor
Bicycle ergonomics, efficiency and design will continue to improve. How?	*Comfort, safety, ease of use*	*Weight, dimensions*
Time: 2009 **Solution:** design-concept of this hybrid was suggested by a 1992 Olympic champion (there is little left to add: miniature storage battery and electric motor + pedal gear) **TRIZ:** *trend – idealization!*		Chris Boardman & SkySports Design concept

Finally, to prove that the bicycle is as inexhaustible as the atom, I will give you several examples of more obscure ideas which have not gained such popularity. However, these examples have considerable cognitive and esthetic value, and to leave them out of this book would mean to detract from the knowledge and impressions of my respectable readers.

I am not commenting on these examples. Instead, I exhort you (if you like them, that is) to use them to hone your extracting and reinventing skills.

Also, I would like to draw your attention to the fact that some of these illustrations can be classified as belonging to *steampunk**, a branch of fantastic and visual arts which I believe is imbued with nostalgia for *lost generations* producing numerous engineering and social failures. Naturally, not all of these things would have been found useful, and not all of them are likeable. Still, they all leave a lot of room for the imagination…

www.buildlog.net

www.instructables.com

www.instructables.com

* http://en.wikipedia.org/wiki/Steampunk

Main problem – do not fall asleep!

Pity you cannot make it any longer – it will be too difficult to turn!

All according to TRIZ: if you cannot make it longer, make it rounder

By air!

By water!

http://www.eons.com/groups/topic
/2573825-July-28

Under water!

See Example 13.19 to compare!

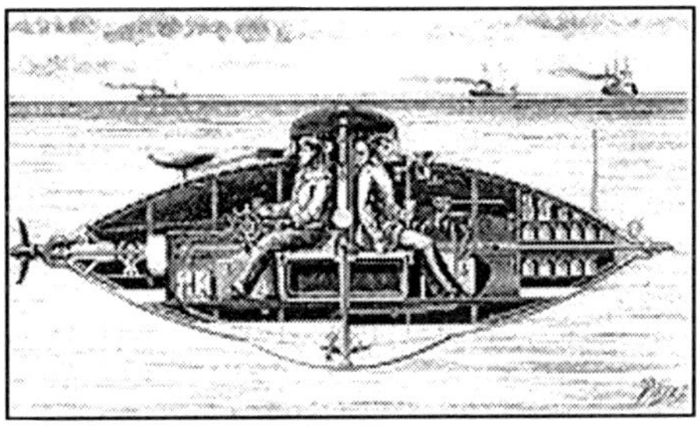

http://www.bluespace.ru/about.html

By land or by water!

www.shuttlebike.com

By land and by water!

Almost a car!

Well, if you are still not impressed, maybe this fruit of an active imagination will give you positive emotions – and you will be surprised, if not amazed?

www.gizmag.com

TRAINING EXAMPLES – FOR INTERESTING AND USEFUL PASTIME

Problem 3.2. Berlin Underground: How Do You Enter and Exit the Car?

In Moscow or Minsk, the doors in all subway cars are opened simultaneously by the engine driver. In Berlin or Paris, however, passengers who wish to enter or exit the car open the door by themselves. To do that, they turn a knob or press a button placed on the door (fig. 3.45). Why has this been done, and what creative models, including TRIZ models, are objectively present in this system compared to the "centralized door management" system?

a)  b)

fig. 3.45. Handle (a) and button (b) opening doors in subway cars

Problem 3.3. Dishwashing Fluid Bottle Cap.

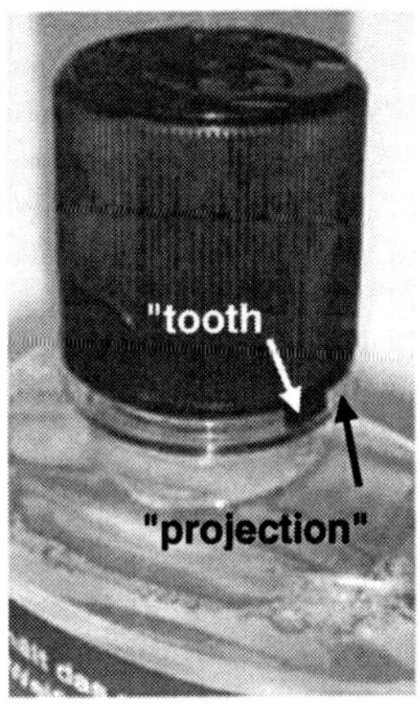

fig. 3.46. Cap with a secret

This plastic bottle contains dishwashing fluid. The bottle is normally found in the kitchen near the dishwasher. The liquid inside the bottle must not be ingested. Due to its frequent use, it is generally difficult to hide from children. Clearly, a child may get the bottle, and most certainly will try to drink from it.

To prevent this from happening, the cap of the bottle has a simple secret. As you can see in fig. 3.46, at the bottom of the cap there is a "tooth" (with another one just like that on the opposite side) which rests against a projection on the neck of the bottle making it impossible to turn the cap.

Without knowing the secret, even a grownup will find it difficult to open the bottle. How do you get the cap off?

Problem 3.4. Fan without Propeller Blades!

James Dyson (www.dyson.co.uk), the famous inventor of the cyclone vacuum cleaner where dust and small particles are separated from the air jet inside a device using centrifugal force, came up with a new gadget: a bladeless fan (fig. 3.47). The effect produced by the air jet issuing from this round frame not fitted with any "propeller" is simply astounding!

You may discern a very narrow slit along the inner surface of the fan's "intake" (which is fitted inside a ring-like structure of a smaller diameter on the farther side of the fan – which is not visible from this point). But what is the source of the air jet whooshing from the fan's "output" side which is turned our way?

fig. 3.47. This is a fan! Can you believe it?

Problem 3.5. The Candle Burned upon the Desk, the Candle Burned…

That poem by Boris Pasternak[84] suddenly came to my mind when one evening I was having a pleasant conversation in the Hannover home of my friend Heinrich, a VW employee.

Quite unexpectedly I realized that the candle (which had been burning throughout the conversation) had all the time been covered with a matte lampshade (fig. 3.48) which kept the light out of our eyes – despite the fact that over the course of the evening the candle had burnt down almost to its foundation. And the small lampshade had clearly descended, too! I looked under the lampshade and understood how this was possible.

So how did the lampshade covering the candle light move?

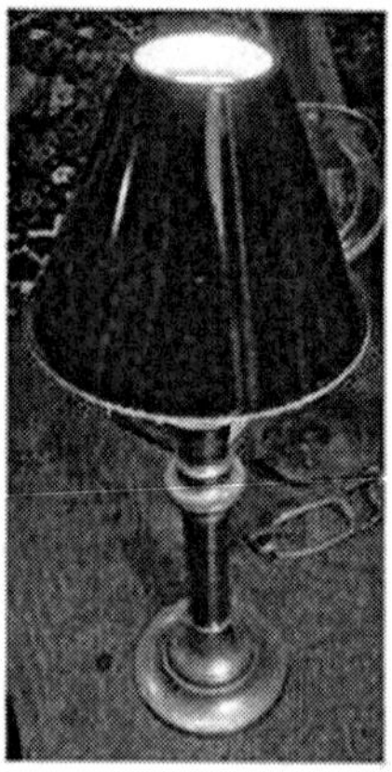

fig. 3.48. Candle with a "Self-Descending" Lampshade

Problem 3.6. Golf Ball.

On my work desk I keep a small golf ball (fig. 3.49) as a reminder that any artifact – even a small one – is a treasure trove of experience accumulated by several (sometimes numerous!) generations of people who have empirically attained a level of excellence which took millions of years to be attained by Nature.

---

84 Boris Leonidovich Pasternak (1890-1960) – famous Russian poet, writer and translator into the Russian language of works by Shakespeare, Goethe and Schiller, winner of the Nobel Prize for Literature (1958)

Somehow man was able to continuously improve that artifact, noting and gainfully employing key physical effects paving the road to perfection. By trial and error, unaware of the theory behind the effects, man managed – through thousands and millions of experiments – to find the optimal dimensions both for that ball and for many other artifacts, advancing them to the "state-of-the-art" level.

fig. 3.49

When hit by a good player, the ball flies in an almost straight line for a distance ranging from 200 m to 300 m (!), a span of flight which is two times longer than that achievable by smooth balls, and much longer than can even be imagined for archaic balls made of wood, gutta-percha (yes, you can find this word in a dictionary or on the Internet!) and other materials! Why?

Anyway, that is all fine and dandy, but... do we really have to use the "trial and error" method? Even if we rename it into the "trial and success" method, its efficiency will not be much improved!

Problem 3.7. "Magic Faucet".

This faucet[85] (fig. 3.50) is a well-loved tourist attraction in Aqua Park on Tenerife, one of the largest Spanish Canary Islands.

fig. 3.50. Water jet... with no source!

[85] You can find the answer at http://sobrecadiz.com/2008/07/11/el-parque-acuatico-aqualand-bahia-de-cadiz and some other web sites – but do not hurry to do that;
TRIZ experts say: if a phenomenon exists, all resources needed for its realization are available – you only need to figure out where they are located, and how they operate.

The stream of water issuing from the faucet which apparently hangs in midair overwhelms even those who know how this thing works.

There are many similar dynamic sculptures in various other cities and countries.

Question: how would you go about creating such an "illusion"?

Problem 3.8. BRT.

In many large cities of the world, elevated railways (fig. 3.51) historically developed parallel to underground railways, or subways.

However, potentials of both these types of metropolitan transportation systems in such cities have been fully exhausted.

During the rush hours, metro ingress and exit routes are dangerous, for example, in Moscow, as people have to jostle through unimaginable crowds. It is no longer possible to build new lines. It is very expensive, and in any event the number of such lines serving different directions is limited. At the same time, the streets of megapolises are choked with motor vehicles.

What can we do?

fig. 3.51. Station and a section of elevated railroad in Berlin
(S-Bahnhof Friedrichstrasse, Berlin)

Problem 3.9. Autobus → Trolleybus → Tramway → ?

Surface public transportation systems, such as buses, trolleys and trams (fig. 3.52) are known to be very efficient due to their high passenger capacity.

Unfortunately, city streets are congested with cars, which slow public transport vehicles down and reduce their advantage to zero. It is not possible to build dedicated lanes for buses or trams.

What can we do?

a) b)

fig. 3.52. Fourth-generation trolleybus (a) and tramway (b) featuring equipment for information exchange between the driver and the passengers (BELKOMMUNMASH, Minsk, Belarus)

RECOMMENDATION

First and foremost, think about what else you need to develop a solution in each of these examples: additional information about materials and construction, knowledge of physical or technical properties, simple acumen or, maybe, curiosity to examine how these artifacts are made.

Write your thoughts and impressions down, dividing them into the following categories:

1) specific applied knowledge;

2) creative knowledge and skills;

3) psychological aspects – wonder, interest, desire to learn more, desire to uncover the secret by yourselves, other emotions.

Primary TRIZ Models

Each tool has retroactive effect on the person using it. TRIZ is a tool designed for subtle, audacious, highly organized thinking operations. Resolving one problem does not change the thinking style. But in the course of training the student resolves dozens, hundreds of problems. Then his thinking is gradually transformed.[86]

Genrikh Altshuller

4 Extracting: Identification of TRIZ Models

4.1 Extracting the Transformation Models

Let us assume that we are studying a certain artifact[87] – the fruit (product) of an invention. In other words, that artifact *already exists and is known to us* in terms of both its properties and its construction.

Then the process of inventing that artifact is a transition from a certain previous original state "was" to a new state "is" with the same (sometimes modified, rarely completely new) Main Useful Function (MUF) which determines the purpose of the artifact.

The original *prototype artifact* is the object which existed in the state "was", and was later improved, or transformed.

The new *resultant artifact* is the object into which the prototype artifact has been "transformed", and its current state ("is") has emerged as a result of such transformation.

Clearly, both objects can coexist physically and continue to be used concurrently, for example row boats and sail boats, or sail boats and motor boats.

The state "was" always precedes the state "is", and both of them are related to past events in historical time.

We always consider two *existing* artifacts, one of which is assumed to be the *prototype artifact* because it was created *before the second artifact*, while the second artifact is assumed to be the *resultant artifact* because it was created *after the first artifact* and *on the basis of the first artifact.*

---

[86] G. S. Altshuller (1987) *Activation of the Human Factor in Education and Upbringing*. – Moscow, Znanie (Knowledge) Publishers (in Russian)

[87] The Modern TRIZ direction developed by the author of this book since the early 1990-es has expanded the empirical base for identification of efficient transformation models by postulating *objective presence of creative models in any artifacts*, in any innovative solutions and ideas; by the same token, patent descriptions are the most systematized source of identification of *technically oriented creative models* (as defined by G. S. Altshuller).

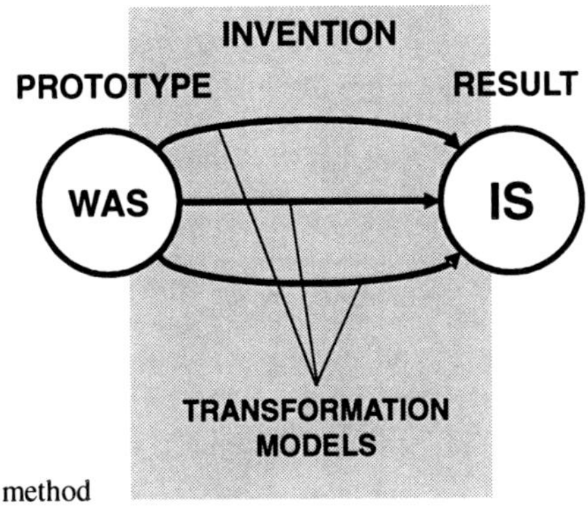

fig. 4.1. Extracting method

Extracting – fundamental teaching method in Modern TRIZ.

The purpose of extracting is to identify *creative transformation models* (fig. 4.1) which are objectively present in constructional transformations of the prototype artifact, and enable its transition to the resultant artifact (see division into model levels in section *3.2 Noospheric Modeling Concept*, fig. 3.11 and fig. 3.12).

Definition of the method **"Extracting"**	***Extracting*** – to draw out transformation models from any (given) artifact using any (given) source of information describing innovative ideas and objects.

There are two basic extracting procedures:

1) Extracting-1 – identification of all objective transformation models completely or partially present in the transition from the prototype artifact to the resultant artifact;

2) Extracting-2 – identification of dominant transformations required to assure transition from the prototype artifact to the resultant artifact, and associated with certain removed contradictions.

In this section we will focus on Extracting-1 examples, while the next section will be dedicated to Extracting-2 examples.

The Extracting-1 procedure can have two outcomes:

1) the model identified is included into the As-catalog[88]; such model is then presented as the result of the examination;

2) the model identified is not included into the As-catalog and, in the opinion of the student, is new; in such case, the student must prepare a general description of the model and send it to the author of this book at the Modern TRIZ Academy[89] for review and classification.

88 section *14.2.2 As-Catalog*

89 www.modern-triz-academy.com

First off, let us recall that to all intents and purposes we have already used the extracting procedure in section *2.2.2 Specialized Transformations*! Indeed, we extracted from the artifacts under study information not only about their technical implementation, but also – no, first and foremost! – about creative models objectively existing in such artifacts. And this IS extracting!

Let us consider several examples. First, we will present them in a special format which reflects the three artifact modeling spheres, as shown in fig. 3.12.

Then we will perform the extracting using a standard "computerless" form. Only such forms will be used in all subsequent examples. Naturally, forms with a similar standard structure are used by the EASyTRIZ™ software package at all levels.

Example 4.1. Epitome of Rational Order!

In the early 1950-es, when I was a little child, not so far from our house lived a former German POW. He could speak a little Russian, and by the time I met him he had spent many years at various Minsk construction sites. For some reason he never went back to Germany, but instead stayed on and worked as a mechanic and plumber for the local housing administration. One day he came to our apartment to fix something in the kitchen.

That is when I saw the wonderful magic box which I believed was shrouded in an inscrutable veil of mystery – because it could open and close in a blink of an eye!

When opened, it seemed huge and presented an assortment of numerous small boxes filled with shining instruments and parts. And then it all disappeared inside one big box with sides lined with some "ridges". And it was with the help of those "ridges" that the small boxes soared up and to the side when the toolbox was opened for work!

Once, after I had started my work in Germany, I saw a replica of that toolbox (fig. 4.2)! It no longer seemed as big as in my childhood, but I still recognized its convenience and its ingenious construction. So I bought it, and it has served me well for more than 15 years.

Incidentally, now such toolboxes are all but impossible to find in construction shops – their design and construction have changed, and the old tradition is vanishing!

Besides, the new metal toolboxes are not as heavy. It is hard to single out any specific model – they are all needful, and they all serve the main useful function, that of putting everything in order and making it visible and easily accessible.

The person who designed them was very imaginative.

And what would you say about the extremely efficient parallelogram-shaped lifting mechanism? Or about the handles that are made so that, when one holds the box, they prevent it from being accidentally opened?

Psychological sphere (motive, need)

Inconvenience of using multiple boxes and bags to carry around instruments and parts, striving for convenience and order. How can the situation be improved?

Creative sphere (ideal result, change)

Instinctive "models" (objective, IFR)

All critically important instruments and parts must always be within reach and in a certain order!

TRIZ models (transformations)

Segmentation, Local property, Separation, Dynamization, Copying, Transition into another dimension, Universality, Unite, generally – a Matryoshka filled with instruments and parts

Applied sphere (design, construction)

The toolbox is made in the form of a portable collapsible multi-sectional construction consisting of several small boxes easily expanded and just as easily collapsed into a solid rectangular box for subsequent transportation.

fig. 4.2. Traditional German toolbox

Example 4.2. Epitome of Multifunctional "Ideality"!

Once during a trip to China I saw a beer bottle with a truly multifunctional cap. In the evening, I came up with an example which I then demonstrated at my workshops. The students were ecstatic, as they easily managed to perform extracting and reinventing for that artifact – apparently due to close familiarity with the object (fig. 4.3)!

I chose to comment on only one transformation model – "previously installed cushion": the cap would, indeed, come very handy in a dramatic situation involving imminent death by thirst over a full bottle of beer or lemonade – because there is no bottle opener! Now, THAT is a REAL problem situation! I hope you understand the "sufferers" who find themselves in such situation because old-model bottle caps are "unopenable"!

One more thing: please write down and send to me an account of your substantiation of objectivity of presence of each model in that object!

Psychological sphere (motive, need)

Familiar "ribbed" metal caps closely hugging the neck of the bottle are hard to open without a special "opener". How can the situation be improved?

Creative sphere (ideal result, change)

Instinctive "models" (objective, IFR)

Penchant for minimalism and urge to obtain the ideal result: let the cap BY ITSELF helps the user to take it off the bottle!

TRIZ models (transformations)

Models implemented: Change in the aggregate state of the object, Dynamization, Inverse action, Local property, Universality, Use of flexible covers and thin films, Asymmetry, Matryoshka (nested doll)

Applied sphere (design, construction)

The cap is made in the form of a rolled-in aluminum lid fitted with a ring. When the user pulls the ring, the cap is torn from the top along expanding perforated tear lines. An elastic membranous gasket is glued to the bottom of the cap, so that when the bottle is recapped the gasket firmly closes the opening. The cap holds on to the bottle due to elastic properties of the material of which it is made; besides, the rolled-in hem of the cap fits into the circular groove right below the opening.

fig. 4.3. This "dragon" has in advance prepared everything those willing to partake of its "blood" may need!

Example 4.3. Epitome of "Ideal" Convenience! In the summer of 2009, when we were traveling to a workshop to be held at Vladimir State University, we were lucky to see – and use – a wonderfully inventive solution: a straw which had been placed inside a soft drink bottle in advance (fig. 4.4), and floated up high enough to jut out of the bottle neck as soon as the cap was twisted off!

The caption under the photograph says it all! Well, except that maybe we might add something for training purposes… Let us do just that!

Clearly, the prototype artifact is any bottle without the straw inside. Then a simple comparative examination of the prototype artifact and the resultant artifact easily yields the following transformation models:

1) inverse action: the user is not looking for the straw, the straw comes to the user BY ITSELF!

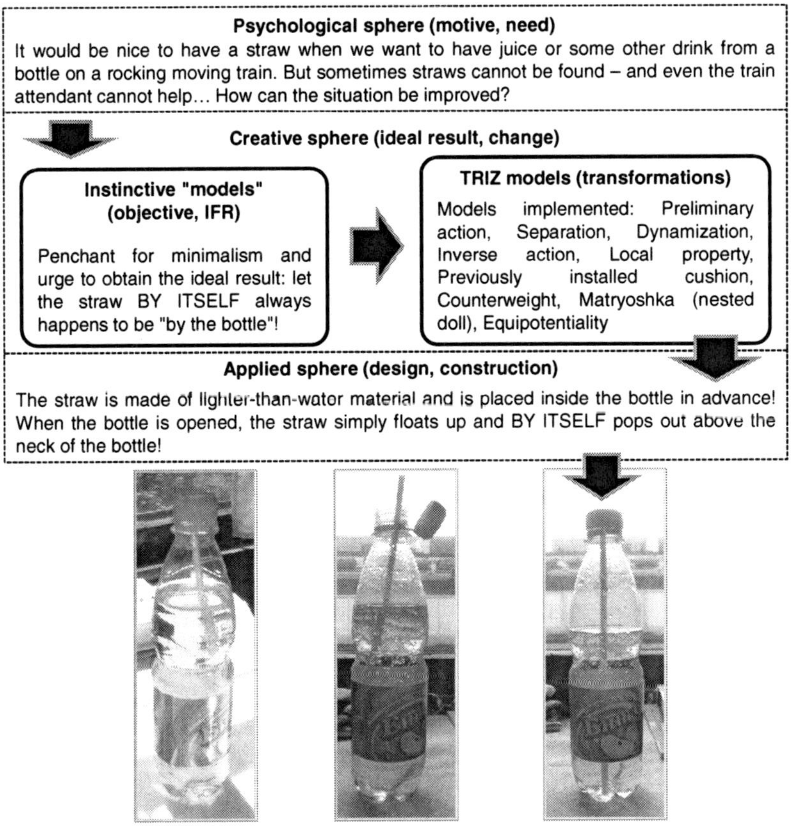

fig. 4.4. Simple, elegant, witty, efficient!

2) preliminary action: prepare objects in advance so that they can be put to work from the best position and are available without loss of time – there is no need to comment on THAT!

3) separation: the required property ("Oh come ye straw!") has clearly been realized!

4) equipotentiality: the straw not just "makes an appearance", it pops up and juts out of the bottle neck!

5) dynamization: the straw moves through space BY ITSELF!

6) local property: the straw is located in the most convenient place;

7) previously installed cushion: yes, yes, look at the previous examples!

8) counter-weight: the straw floats up BY ITSELF according to the law of Archimedes;

9) matryoshka (nested doll): the straw is inserted into the bottle in strict compliance with that model!

Essentially, we sought to demonstrate the possibility, and provide illustrative examples, of extracting transformation models from routine artifacts that we can find around us in our daily life.

And now let us have a look at standard forms and their contents so that you can use them (hopefully independently and efficiently) for many years to come – after all, you WILL be doing TRIZ etudes just as Rachmaninoff did piano etudes, or a chess master does thousands of chess etudes, or a combat art master trains his body, intuition and will.

Example 4.4. "Black Box for Each Motor Car[90].

Briefly, the MUF of the "black box" is to register objective information about the movement of a vehicle. The "black box" is fire- and water-proof, it is virtually indestructible, and it automatically transmits its location for a very long period of time.

Let us assume that each motor car is equipped with a "black box" whose MUF is to register, for a certain period of time (say, one month), the key motion metrics (speed, acceleration/deceleration, mileage) and perform video recording of the immediate environment in front of and behind the motor car while it is in motion. It is also possible to register coordinates of the motor car's location at certain time intervals. The video recording is stored for several hours.

Data cannot be modified or deleted by the user without violating the rules governing the use of the "black box". All data inside the "black box" are available only to the police (under certain conditions), some data is also available to insurance companies.

Advantages of equipping all motor cars with "black boxes" include the following: objective control and registration of motion metrics makes it impossible to distort or conceal critical information about what actually happens while the motor car is in motion, which in and of itself has efficient restraining effect on all drivers.

Let us consider creative models (fig. 4.5) implemented in such "black box". These models are objectively present in each function, in each technical solution added to the artifact as compared to its prototype.

In this case, the "prototype" is probably represented by witness accounts or measurements taken at the site of the accident *post factum*. In other words all we have – at best – is circumstantial evidence, rather than direct measurements taken while the vehicle is actually moving. Accordingly, all information we have is by definition imprecise.

So, the prototype is: non-registration of motor car motion metrics and environmental information.

In the table presented in fig. 4.5, two pluses mark models which are critically important for the implementation of the MUF of the "black box", while one plus marks models whose participation in fragmentary, or which are hard to define. Let us describe two-plus models as *dominant*, and one-plus models as *supplementary*.

---

[90] This idea has been "around" (or, more precisely, on the Internet) for a long time; I wonder when it is going to be promoted to a "standard"...

Rank	#	Navigator	Substantiation
	01	**Change in the aggregate state of an object**	
++	02	**Preliminary action**	The "black box" is installed in advance
	03	**Segmentation**	Recording is automatically divided into overlapping fragments
++	04	**Replacement of mechanical matter**	Rough measurements taken at the site of the accident are replaced with precise measurements taken by special sensors while the vehicle is in motion
++	05	**Separation**	The system is complemented with instruments assuring accurate and complete measurement and registration of motion metrics
	06	**Use of mechanical oscillations**	
	07	**Dynamization**	New recordings replace older recordings
	08	**Periodic action**	
	09	**Change in color**	
++	10	**Copying**	Motion metrics are measured objectively by accurate sensors, and copied into the memory of the "black box" Copying of the construction of, and operating procedures used by, "black boxes" used in aviation and other areas
+	11	**Inverse action**	Detailed information about the accident is registered "by the motor car", rather than by witnesses or experts
++	12	**Local property**	All measurements are taken in real time and at the actual location of the accident
	13	**Inexpensive short-life object as a replacement for expensive long-life one**	
	14	**Use of pneumatic or hydraulic constructions**	
+	15	**Discard and renewal of parts**	Older recordings are deleted to be replaced by new recordings
+	16	**Partial or excess effect**	Recordings are mostly excessive, but they become critically important in the event of an accident
	17	**Use of composite materials**	
+	18	**Mediator**	The "black box" acts as an intermediary between the motor car and its environment It also disciplines the driver
	19	**Transition into another dimension**	
+	20	**Universality**	The "black box" is, undoubtedly, a universal multifunctional recorder

fig. 4.5-beginning. Extracting-1 results for the motor car "black box" artifact

+	21	**Transform damage into use**	The complexity and the supposedly very high cost of "black boxes" – and of the new infrastructure required for their successful utilization, including development and approval of new laws and regulations – inflict infinitesimal "damage" compared to the "benefit" of saving hundreds of thousands of lives; moreover, the "damage" is further dwarfed by the huge amounts that will be saved by insurance companies and government services as accident rates go down
++	22	**Spherical-shape**	The "black box" must ensure panoramic recording of the vehicle's immediate environment
	23	**Use of inert media**	
	24	**Asymmetry**	
	25	**Use of flexible covers and thin films**	
	26	**Phase transitions**	
	27	**Full use of thermal expansion**	
++	28	**Previously installed cushion**	This model is definitely present in this solution, as conditions for the accurate description of the accident are created in advance
+	29	**Self-servicing**	The motor car "by itself" disciplines its driver and other drivers; motor cars equipped with "black boxes" can also be fitted with warning markers (stickers)
	30	**Use of strong oxidants**	
	31	**Use of porous materials**	
	32	**Counter-weight**	
	33	**Quick jump**	
	34	**Matryoshka (nested doll)**	
	35	**Unite**	
++	36	**Feedback**	If the driver violates traffic rules, the "black box" can automatically report such violations to the relevant traffic control body
	37	**Equipotentiality**	
	38	**Homogeneity**	
+	39	**Preliminary counter-action**	Motor cars which are equipped with "black boxes" and fitted with warning markers proactively counteract drivers prone to violation of traffic rules, as they create a documented record of such violation
+	40	**Uninterrupted useful function**	The "black box" remains in operation even when the car is parked; for example, it can record intervals between motion periods

fig. 4.5-end. Extracting-1 results for the motor car "black box" artifact

We have identified 7 dominant creative models and 10 supplementary models. This is the creative "potential" of this as yet non-existent artifact.

Let us look at two more examples to see various implementation options of the Extracting-1 procedure.

Example 4.5. Plastic Water Bottle

A heavy two-liter water bottle (fig. 4.6) has two deep depressions, one such depression on each side.

The MUF of the depressions is reduce the diameter of the bottle at the place where it is gripped by the user, and create obstructions (walls along the edges of the depressions) to stop the fingers in the depressions from sliding down.

Naturally, the prototype is any bottle without the depressions.

Extracting-1 yields the following models (fig. 4.7; only dominant models are listed):

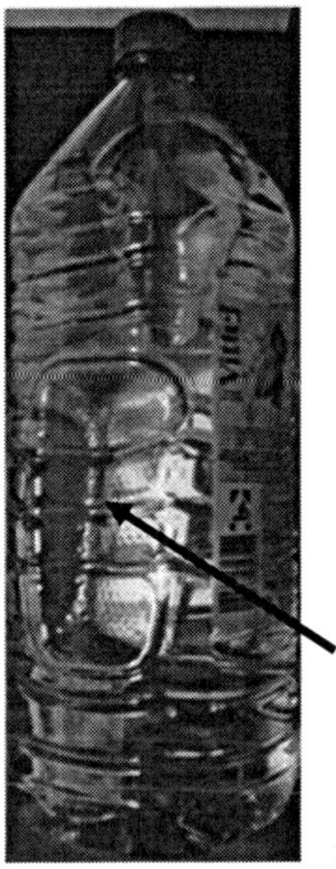

fig. 4.6. "Water bottle" artifact

++	05	**Segmentation**	The required property has been implemented by introducing a bottle "holder" (rather than by attaching a bottle handle)
++	11	**Inverse action**	Not a handle (protuberance), but a hollow (depression)
++	12	**Local property**	The depression is made approximately at the center of gravity of the water-filled bottle
++	18	**Mediator**	The depression acts as an intermediary between the fingers and the bottle
++	19	**Transition into another dimension**	The "holder" has been placed inside the body of the bottle
++	21	**Transform damage into use**	Increased complexity of construction yields a positive result: the bottle becomes more rigid – now it is not deformed when held by the user!
++	34	**Matryoshka**	See 19

fig. 4.7. Extracting-1 dominating models for the "Water bottle" artifact

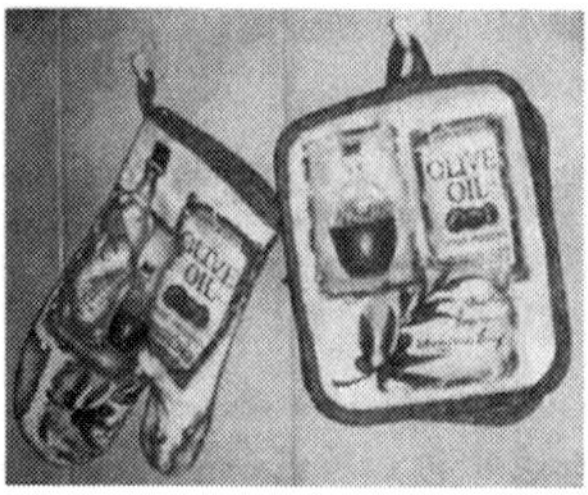

fig. 4.8. Oven mitt and pot holder

Example 4.6. Oven Mitt and Pot Holder

On the kitchen wall next to the stove (fig. 4.8) there hang an oven mitt and a pot holder – both made of thick cloth so that they can be used to grab hot sauce pans and frying pans right off the hot plate.

The prototype is any cloth which is not in such appealing "marketable" state.

Extracting-1 is shown in fig. 4.9.

++	02	**Preliminary action**	These accessories are prepared in advance
++	18	**Mediator**	All these "grabbers" act as intermediaries between the hands and the hot pans
++	34	**Matryoshka**	The oven mitt is made and operates in accordance with the "matryoshka" principle!

fig. 4.9. Dominant models yielded by Extracting-1 for the "oven mitt" and "pot holder" artifacts

Example 4.7. Mouse Hides USB-Connector in a "Pocket"

This computer mouse (fig. 4.10) has a USB receiver-transmitter (connector) hidden in a special pocket at the bottom of its body.

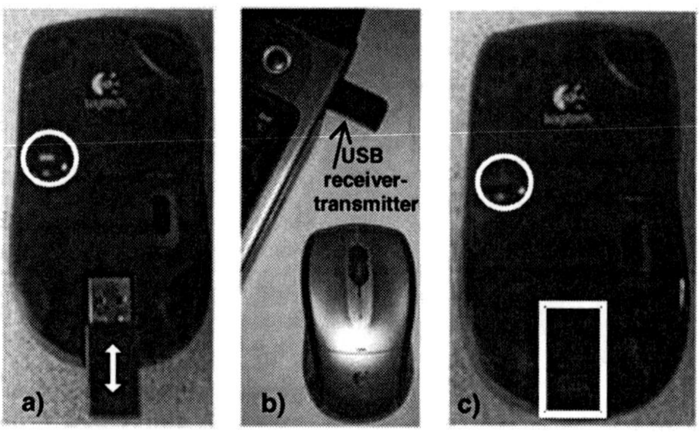

fig. 4.10. Computer mouse with retractable USB connector

The connector also doubles as a switch. When the connector is extracted from the pocket, the mouse turns on (the signal diode lights up – fig. 4.10,a), and when the connector is retracted into the pocket (marked by a white rectangle in fig. 4.10,c), the mouse turns off (the diode goes out – fig. 4.10,c).

In fig. 4.10,b the mouse is shown in operation after having established wireless connection with the notebook.

Perform Extracting-1 independently for "wired" mouse as the prototype.

4.2 Extracting the Contradictions

The Extracting-2 procedure is performed for the dominant models yielded by the Extracting-1 procedure.

For each dominant model, or group of dominant models, it is necessary to define the (standard and/or radical) contradiction which was present in the prototype artifact, and is subsequently removed as a result of implementation of that model, or group of models.

Example 4.8. Extracting-2 for Example 4.5. Plastic Water Bottle

We can take any "undepressed" bottle as the prototype artifact. If possible, the number of the dominant models should be reduced to minimum, so that there remain only one or two most important models.

For example, we can assume that the resultant artifact is free of the following contradictions that were present in the prototype artifact (fig. 4.11):

++	05	**Segmentation**	**Standard Contradiction:** the "holder" must be (exist) to hold the heavy bottle, but this may complicate the construction of the artifact.
++	12	**Local property**	**Radical Contradiction 1:** the "holder" must be (exist) to hold the heavy bottle approximately in the middle (close to its center of gravity), and the "holder" must not be (exist) as placing a handle in the middle of the bottle will impair ease of packaging and transportation.
++	18	**Mediator**	**Radical Contradiction 2:** the "holder" must be (exist) to hold the heavy bottle, and the "holder" must not be (exist) as this will complicate the construction of the bottle, and make it more difficult to design and manufacture multiple-bottle packs for easier transportation.

fig. 4.11. Results yielded by Extracting-2 for the "Water Bottle" artifact

Selection of the number, "content" and types of contradictions depends on the purpose of examination. For example, for training purposes you can practice by developing a larger number of various contradiction models. For reinventing purposes, one standard contradiction and one radical contradiction will suffice. For practical purposes (when you actually need to improve the prototype), the contradictions may have to be scrutinized more closely and in several variations.

Example 4.9. Extracting-2 for Example 4.6. Oven Mitt and Pot Holder

You can use, as the prototype, any imaginable "piece of cloth" – corner of an apron, table napkin, towel, etc. – which you instinctively grab (if it is THERE to grab, that is) when you need to take the pot off the hot plate.

++	18	**Mediator**	**Radical Contradiction:** the "grabber" must be (exist) so that you can get hold of a hot item (pan), and the "grabber" must not be (exist) because... it is "never there when you need it", not even a simple piece of cloth!

fig. 4.12. Results yielded by Extracting-2 for the "Oven Mitt" and "Pot Holder" artifacts

SELF-TRAINING

Problem 4.10. Socket Wrench with Replaceable Heads.

Perform the Extracting-1 procedure for a socket wrench with replaceable heads. Note that the heads are magnetized and, consequently, do not fall off the wrench (fig. 4.13,a) when it separates from the relevant part (nut, bolt, spark plug, etc.).

Define radical contradictions for one or two dominant transformation models.

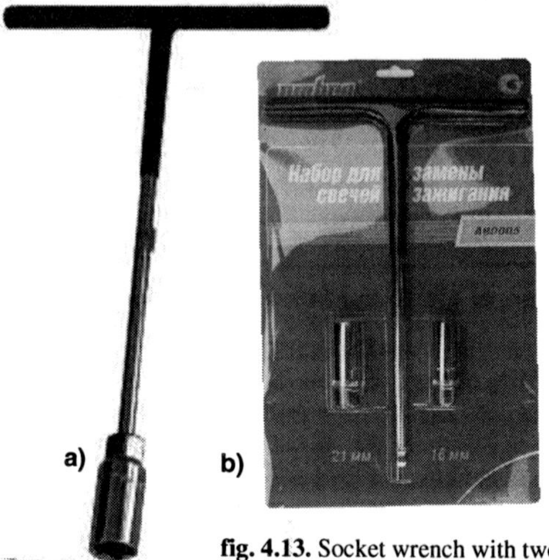

fig. 4.13. Socket wrench with two replaceable heads in working condition (a) and packaged (b)

Problem 4.11. "Articulated" Wrench.

Perform the Extracting-1 procedure for a socket wrench which consists of three moveable parts (fig. 4.14) one of which is a double two-size head.

The smaller head is inserted into the larger head and spring-loaded to ensure that it is always in working condition. When the wrench is used with large-diameter parts, the smaller head is simply pressed into the body of the larger head, and does not prevent the part from being turned.

Next, to prevent the handle from spontaneously detaching itself, it is fitted with a cone-shaped spring support which enables the worker to use the wrench at different angles and get to hard-to-reach nuts and bolts.

When you are done, perform the Extracting-2 procedure as well.

a)

b)

fig. 4.14. Socket wrench with a double head (a) and moving handle (b); inner head is pressed with a tip of ballpoint pen

Problem 4.12. Adjustable Wrench.

Perform the Extracting-1 and Extracting-2 procedures for the adjustable wrench shown in fig. 4.15.

fig. 4.15. Adjustable wrench

Problem 4.13. How Michelangelo[91] Sculpted His *David*.

It is known that in Michelangelo times (and long before that) masters made clay models of their future sculptures. The model was immersed in water (fig. 4.16) to see the contour of the sculpture in the place from which the master was going to continue his work, the place where he had left off the day before. As work went on, the level of water went down. To better see the contour, the water was tinted, or milk was used instead (however, it had to be replaced when it went sour).

This technique was used by the great Michelangelo when he was creating his 5-meter marble *King David* (it is possible also to use a model in recumbent position).

Perform the Extracting-1 and Extracting-2 procedures for this sculpting technique.

fig. 4.16. Clay model in a cup filled with milk

Problem 4.14. Invention of a Painting Joke!

Perform both extracting procedures for this merry diptych (fig. 4.18) by the Italian master Giuseppe Arcimboldo (1527-1593). Turn the book upside down and look again!

fig. 4.18. Metamorphism Diptych

Photographed by the author in Louvre, Paris, 2002

91 Michelangelo Buonarotti (1475-1564) – Italian sculptor, painter, architect, poet, thinker. One of the greatest masters of the Renaissance era. *David* – Carrara marble, 1501-1504; Accademia di Belle Arti Firenze (Academy of Fine Arts of Florence)

Problem 4.15. Glass Painting and Stained-Glass Windows.

Oil paintings made on the reverse side of a glass sheet are very bright and produce a 3D effect. This technique is also used to make stained-glass windows[92].

fig. 4.17. Painting on the glass
(oil; reverse side)

The procedure is quite complicated: first the artist makes a sketch, then an outline sketch, then the paint is laid on layer by layer, with the foreground layer applied first, and the background layer applied last. The finished painting (fig. 4.17) is viewed from the "clean" (unpainted) side of the glass!

Perform the Extracting-1 and Extracting-2 procedures for this painting technique.

92 www.icompass.ru/compas/vitrages

5 Operative Resources: Build the Solution Thus and Here!

In this section we will deal with certain notions about which it is *not easy* to say that they are *"easy"*. But their knowledge, at least at the basic level, is a must for MTRIZ Junior and MTRIZ Practitioner certification.

The new notions relate to the so-called *operative resources (OR)* of the operative zone.

Definition **Operative resources**	**Operative resources (ORs)** are actually or allegedly existing systemic-technical and physical-technical OZ components minimally necessary for solving the problem, so that if the model lacks at least one such OR, this radically reduces the probability, or completely eliminates the possibility, of finding a solution.
Addition **Occam's razor for ORs**	ORs include resources minimally necessary for OZ modeling.

The most important ORs are the key "dramatis personae" of the OZ, or *actors – inductors* and *receptors, resources* (potentially, of all types) and *functional ideal models (FIMs), including the ideal final result (IFR).*

5.1 Inductors and Receptors

First off, let us consider those OZ elements which are actively involved in the problem situation as carriers of conflicting processes and their outcomes – the ***actors***. All actors can be conventionally divided into two categories. Some of them are effect initiators, while others are effect recipients.

Definition **"Actors"**	**Actors** are core OZ elements which actively participate in the problem situation and act as carriers of the processes originating in the OZ and leading to the emergence of specific contradictive properties.
Definition **"Inductor"**	**Inductor** is an actor which creates an effect impacting another actor (receptor) through transfer of energy, information or material, and initiates a change of, or action by, the receptor.
Definition **"Receptor"**	**Receptor** is an actor which receives the effect generated by the inductor, and changes, or is spurred into action, under the influence of such effect.
Definition **"Main Receptor"**	**Main Receptor** is the product of the OZ. If the OZ represents the entire system, the main receptor is the product of such system.

This definition is important because, pursuant to TRIZ, changes normally affect those inductors which directly interact with the main receptor (see section 5.4 below).

The structure of the OZ can be illustrated by a number of graphic schemes. The simplest is the functional/structural scheme reflecting OZ elements and functional relations between such elements (informational or physical effects of diverse nature, for example, transfer of materials, mechanical or electromagnetic effect, etc.).

One and the same element can act as an inductor in one interaction (chain, contour), and as a receptor in another interaction (chain, contour).

Let us consider the functional/structural OZ model for Problem 2.3.

Example 5.1. Development of the Functional/Structural OZ Model for Problem 2.3. Clever Gate".

The qualitative (informal, "non-mathematical") description of this assignment can be presented one more time as follows:

Problem: the door of the gate can be activated in different ways, for example, by a spring.

Then the OZ can be presented as consisting of the following elements (fig. 5.1): the door; the first post to which the door is attached by hinges; the spring which is attached to the first post at one end, and to the door at the other end; and the second post which serves as a "stop-bolt" (stopper post). The two posts form the support construction of the gate.

The "roles" of the elements are designated with the initial letters of the words "inductor" (I) and "receptor" (R). If necessary, we can also introduce I and R indexing. The effect that the elements have on each other are designated with functional actions (hold, pull, etc.).

The door is the main receptor. When the door is opened, it stretches the spring, and the spring accumulates energy. When the door is closed, the spring pulls it towards the second post. When the movement is completed, the door must be firmly pressed to the second post. As a result, the second post arrests the movement of the door, and in doing so is hit by the door.

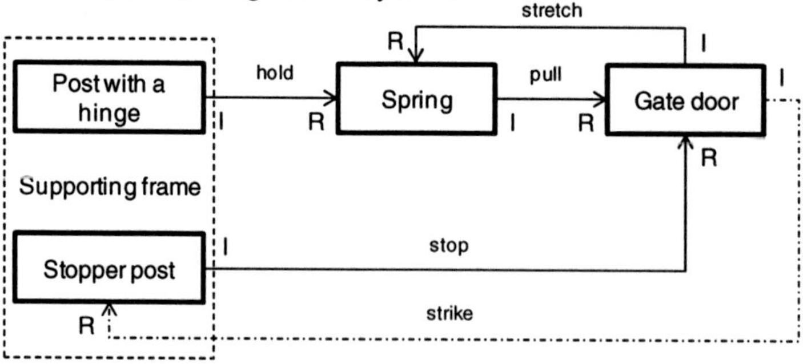

fig. 5.1. OZ structure for Problem 2.3. Clever Gate

The spring is the main inductor. Its main useful function is to pull (and, in fact, close) the door. Let us assume, however, that the "customer" demands that the "system" be minimized, and the spring be removed! Can this be done?

The general answer is: it is necessary to change the inductor... the "post" (!?)

And now let us undertake a similar review of the Functional/Structural model of the problem situation in Problem 2.7. "Short Arms".

Example 5.2. Development of the Functional/Structural OZ Model for Problem 2.7. "Short Arms"

Let us recall the nature of the problem: it is difficult to grasp a large pile of fallen leaves, and thus accelerate completion of the "cleaning session".

The main useful function of the "arm" inductor: hold the leaves.

The OZ consists of two elements (fig. 5.2): leaves, arms.

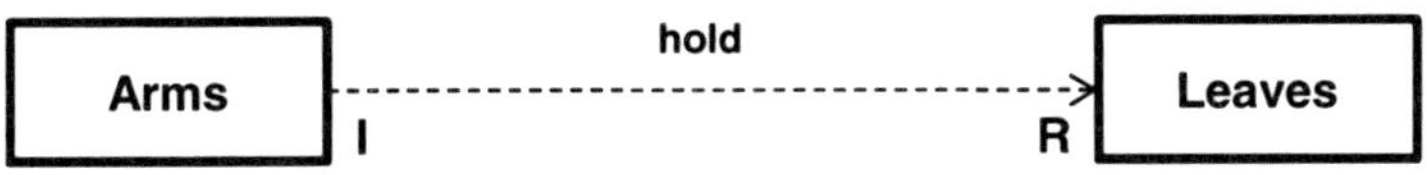

fig. 5.2. OZ structure for Problem 2.7. Short Arms

The dotted line signifies inefficient action – the cleaner fails to grasp a large pile of leaves. This is the basic problem associated with having "short" arms. Meanwhile, it is prohibited to use "complex" instruments – we need to invent something else (for future use, naturally).

The general answer is: we need to modify the "arm" inductor (!?).

Example 5.3. Development of the Functional/Structural OZ Model for Problem 2.12. "Clever" Mailbox.

Gist of the problem: how can we tell, by looking out of the window, whether there is mail in the mailbox?

The main useful function of the mailbox is to store new correspondence.

The OZ also consists of two elements (fig. 5.3): box, man.

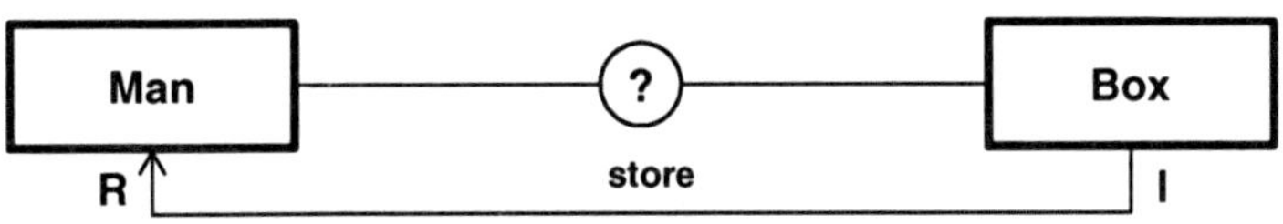

fig. 5.3. OZ structure for Problem 2.12. "Clever" Mailbox.

The man in question is looking out of the window, but cannot see whether the mailbox is full or empty. After all, the box is not transparent. So the line with the question mark signifies, basically, lack of required action, or lack of any action.

The general answer is: we need to modify the "box" inductor (!?).

Example 5.4. Development of the Functional/Structural OZ Model for Problem 2.13. Parking Lot.

Gist of the problem: what can we do to make parking lots – particularly very large parking lots, like those near exhibition halls and stadiums – look better, rather than like gray lifeless fields?

The main useful function of the parking lot is to "hold" cars. However, we demand that the MUF be supplemented with another critically important property: parking lots must "not depress" people.

The OZ consists of three elements (fig. 5.4): lot, car, man.

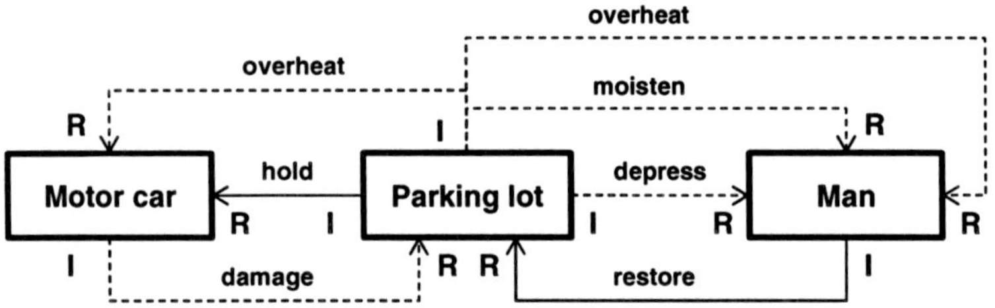

fig. 5.4. OZ structure for Problem 2.13. Parking Lot.

Clearly, the general answer is: modify the "lot" inductor (!?).

Example 5.5. Development of the Functional/Structural OZ Model for Problem 2.20. Window Washer.

Let us recall the nature of the problem: sometimes it is very difficult to wash the windowpane from both sides, because the window does not open wide enough.

The OZ consists of three elements (fig. 5.5): windowpane, arm, "washer".

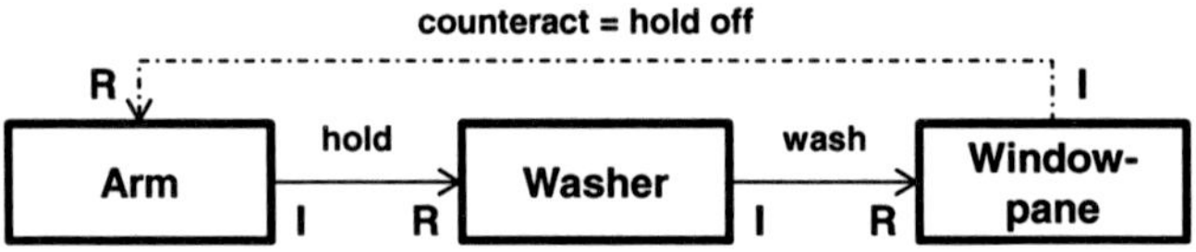

fig. 5.5. OZ structure for Problem 2.20. Window Washer.

The main receptor is the windowpane. The main inductor is the sponge or rag which is used to wash the windowpane. Let us call that instrument the "washer".

The main useful function of the "washer" is to wash the windowpane. However, the "washer" cannot reach certain parts of the windowpane because the arms are not long enough to reach the farther spots on the outside, while application of complex "rods" or "holders" is inefficient. In this case, the "windowpane" (inductor) has adverse impact on the "arms" (receptor).

The general answer is: we need to modify the "washer" inductor (!?). And this is correct because we cannot (by default) modify the "windowpane" (main receptor).

Each system, each system element and, therefore, each (supposedly) modified inductor has a certain organization, and contains all resources necessary for its continued existence.

This and similar problems are resolved by modifying (transforming) OZ resources.

Incidentally, in many cases ideas for solutions are generated while describing and analyzing such resources.

5.2 Resource Modeling

Definitions of resources and their classification are provided in the tables below.

Definition **"Resources"**	**Resources** – the totality of material and non-material (model) organizational components of the artifact[93] as an integral systemic object.
Systemic-technical resources – addition 1 to the definition of the term "Resources"	**Systemic-technical resources** – the totality of system-defining, system-forming and system-realizing components of the artifact.
System-defining components are models which determine the purpose and effectiveness of the artifact. *System-forming* components are all components comprising the systemic organization of the artifact: functional composition, process-related conceptualizations, structure-related conceptualizations and, naturally, their underpinning material and non-material constituent parts, such as construction and knowledge (information), in particular, ideas, especially conceptual ideas. *System-realizing* components include, first and foremost, the integral construction of the artifact determining all its processes and functions and shaping its actual interaction with surrounding artifacts.	
Physical-technical resources – addition 2 to the definition of the term "Resources"	**Physical-technical resources** – the totality of system-forming and system-realizing components of the artifact residing in its construction.
Construction defines the physical realization of the artifact in space-time-material and all processes involving its interaction with materials, energy and information occurring both within the construction and in the course of its exchange with surrounding artifacts.	
OZ Resources – addition 3 to the definition of the term "Resources"	**OZ resources (operative resources)** – the totality of system-defining, system-forming and system-realizing components of the artifact residing in the OZ.

The well-known classification table (fig. 5.6) divides resources into 8 types which are then lumped into two large groups – systemic-technical resources and

93 Here the term "artifact" is used in its broadest meaning, and refers to any integrated systemic formation (whether material or non-material), including any construction, information, process, phenomenon, work of art of any type, etc.

physical-technical resources. It is always preferable to select, from among the resources listed in the table (fig. 5.7), one with the leftmost value of any kind.

> One of the key TRIZ statements is the reminder (relevant for the purposes of both analysis and synthesis) that when any resource of the artifact is changed, all other resources of such artifact – and possibly resources of surrounding artifacts – are changed, too.

SYSTEMIC-TECHNICAL RESOURCES			
SYSTEMIC	**INFORMATIONAL**	**FUNCTIONAL**	**STRUCTURAL**
Related to general system properties	Related to transmission of information-bearing messages	Related to creation of functions	Related to the composition of the object
Purpose of the system, its efficiency, productivity, reliability, safety, survivability, durability, etc.	Data integrity, accuracy, validity, interference immunity, methods and efficiency of measurement, management, encoding, etc.	Main useful function compliant with the purpose of the system, auxiliary functions, negative functions, description of the operative principle (functional model)	List of components and inter-component relations, types of structures (linear, branching, parallel, closed, etc.)
PHYSICAL-TECHNICAL RESOURCES			
SPATIAL	**TEMPORAL**	**MATERIAL**	**ENERGETICAL**
Related to geometric properties	Related to temporal evaluations	Related to properties of materials	Related to properties of energy and its manifestations
Shape of the object, dimensions of the object – length, width, height, diameter, etc., shape features – presence of cavities, projections, etc.	Frequency of events, duration of time intervals, duration of time lags/leads. **Operative Time (OT):** interval of existence of the problem situation	Chemical composition, physical properties, special engineering properties	Types of applied and measured energy, including mechanical, gravitational, thermal, electromagnetic and other forces; methods of energy utilization, etc.

fig. 5.6. Classification of OZ resources

Properties of the resource	**Value:** free → inexpensive → expensive
	Quality: harmful → neutral → useful
	Quantity: unlimited → sufficient → insufficient
	Readiness for application: ready → change in progress → creation in progress

fig. 5.7. Resource value scale used to select resources which are required to solve the problem

Successful resolution of the problem relies on resources involved in analyzing and modifying the OZ and synthesizing an idea for the solution. Accordingly, the "size" of the OZ (number and composition of its elements and relations) is directly dependent on the resources which are taken into consideration.

As was noted above, sometimes to successfully solve a problem all you need to do is to undertake a detailed analysis of available resources, and properly inventory and structure such resources. In such cases, contradiction models are built only to fine-tune the original definition of the problem, to simplify (reduce) the description of the initial problem situation.

In this section we will limit ourselves to reviewing practical examples and comments to solutions from the viewpoint of transformation of resources involved.

Example 5.6. OZ Resources for Problem 2.3. "Clever" Gate".

Now, following an analysis of the initial situation (see Example 5.1) it was established that the OZ consists of three elements: support construction consisting of two posts, spring, door. Despite the simplicity of the "system", its elements possess a number of diverse properties, each such property being classified as a certain type of resources.

Let us consider the resources in the initial problem situation (fig. 5.8).

Systemic-technical resources	**Systemic**	System objective: the door must close BY ITSELF.
	Informa-tional	It is necessary to use some physical technical effect or geometrical effect.
	Functional	The door of the gate must be put into motion so that it reaches the stopper post and remains there in the appressed state (in other words, the door must close).
	Structural	The door of the gate is fixed to the first post with the help of a hinge. The second post (stopper post) is used to arrest the movement of the door as it closes. According to the terms of the assignment, the spring and other "complex" elements must be excluded!
Physical-technical resources	**Spatial**	Potentially, there is a possibility to change the shape and/or spatial location of OZ elements.
	Temporal	OT: from the moment the door is opened to the moment the door is closed, i.e. while the door is moving.
	Material	The posts and the door are made of wood.
	Energetical	It is necessary to find a source of energy to close the door of the gate. Out of all existing forces, only the weight of the door is available (gravity force).

fig. 5.8. Description of OZ resources for the Clever Gate problem

The entirety of properties and requirements presented above makes up an initial resource "composite sketch" of the problem situation.

The solution is almost uniquely dictated by the fact that we have only one available source of energy – the weight of the door. All we have to do is understand how we can use it.

In the absence of viable alternatives, we can only try to change the spatial resource: to enable the door of the gate to move towards the stopper post BY ITSELF, we need to install both posts in an inclined position, so that the door will move under its own weight (fig. 5.9,a,b).

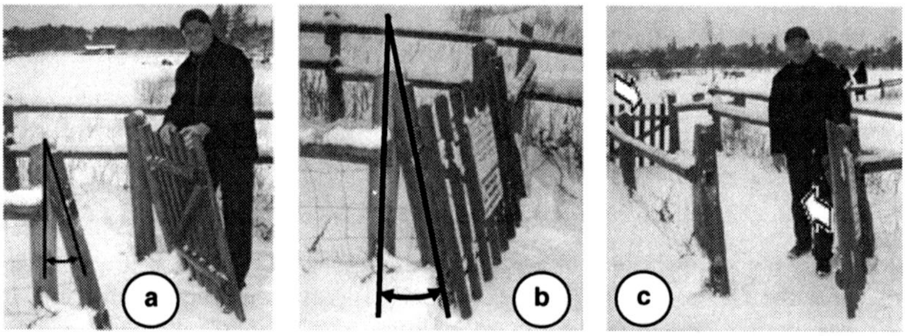

fig. 5.9. Now, that's a truly "clever" gate!

To weaken the blow, the stopper post can be fitted with a rubber plate (you can see it in the photos).

At the conclusion of the gate story, we would like to point out that there are *two* gates in the passage (fig. 5.9, c): the door of the gate in the foreground is inclined towards the plot, while the door of the gate in the background leans towards the passage (see the arrows). Now, why would we need two gates?

And one more thing: why are both doors inclined (and close) towards the passage, and not the other way round? "Fuzzy" prompt: the informational resource takes account of the interaction between the "plot" system and surrounding systems.

Example 5.7. OZ Resources for Problem 2.12. Clever Mailbox.

Of all resources (fig. 5.10 - 5.11), special attention should be paid to the following three: *functional* – it is necessary to create and transmit a "signal"; *energetical* – a source of energy is required to produce the "signal"; *spatial* – it is possible to modify the shape of the mailbox, and such modification will serve as implementation of the "signal".

Systemic-technical resources	**Systemic**	System objective: remote notification of the user regarding new mail in the mailbox.
	Informa-tional	The mailbox has two states – "without mail" and "with mail". The second state must be definitely known to the user, for example, when he looks at the mailbox out of the window.
	Functional	Transmission of a certain signal to the user. The mailbox must BY ITSELF notify the user that it contains new mail.
	Structural	It is necessary to establish a signal communication between the "mailbox" inductor and the "user" receptor.

fig. 5.10. Systemic-technical OZ resources for the Clever Mailbox Problem

Physical-technical resources		
	Spatial	1) we should take into account the fact that the distance between the mailbox and the house is rather large; 2) it is potentially possible to modify the shape and/or position of the mailbox.
	Temporal	OT: time from insertion of new mail to the moment when the user discovers the new state of the mailbox.
	Material	Some new parts may have to be made of a certain material.
	Energetical	To produce the "signal", the mailbox must receive a certain power impulse.

fig. 5.11. Physical-technical OZ resources for the Clever Mailbox Problem

Hence a very simple and still rather inventive solution (fig. 5.12): a small wooden lever in the form of a "flag" is raised when there is new mail in the mailbox – use of the spatial resource (basically, modification of the original shape of the mailbox).

The mailman puts in the new mail and raises the "flag". The user takes out the new mail and lowers the "flag". Therefore, man is the source of energy.

fig. 5.12. Signal flag of a "clever" mailbox

Example 5.8. OZ Resources for Problem 2.13. Parking Lot.

You can undertake a resource analysis by analogy with the previous examples. The dominant resource for this solution is almost certainly the spatial resource – with some input from the structural resource (fig. 5.13).

Look at the photos featuring solutions similar to the one I implemented in 2009 near one of the fountains at the All-Russian Exhibition Center in Moscow (Exhibition of Achievements of People's Economy; fig. 5.14).

fig. 5.13. Parking lots with combined "concrete + grass" paving at North-Eastern Forest University (a) and Science and Industry Technical Park (b) in Harbin, China

fig. 5.14. Path to the fountain with combined "wire + grass" paving

One part of space has one systemic property, namely, to be hard and strong enough to hold cars, and the carrier of that property is the main bearing material, namely, concrete (fig. 5.13) or wire (fig. 5.14); the remaining space has the opposing property, namely, to be Nature-friendly, esthetically pleasing, and the carrier of that property is grass growing in the cells formed by the main bearing material.

Look around, and you will find many similar examples which you can use to train your observation abilities and, naturally, your inventiveness.

Example 5.9. How Can We Reconcile the Irreconcilable?

Dad and Mom put the child to bed (fig. 5.15,a) hoping to spend some time with their parents or friends in the neighboring room (fig. 5.15,b). However, the child wakes up (or never goes to sleep). It is OK if he is resting quietly, maybe dreaming about something, like in fig. 5.15,c. It is bad if he starts crying, and it is even worse if he falls out of his crib. Familiar picture – familiar parental fears? And rest and fear do not go together very well, now, do they?

fig. 5.15. The parents want to spend some time with friends and relatives, but what does the child want? Yes, THAT child, the one who is supposed to be sleeping!

Radical Contradiction: the parents must *be close* to the child so that they can monitor him, and they must *be far* from the child so that they can meet with their friends.

The dominant negative resource is spatial – walls, or just distance. The negative outcome is interrupted visual and audial communication – the informational resource is scarce or non-existent. The systemic requirement is to implement the visual/audial monitoring function.

The solution – admittedly very simple (subject to the possibilities opened up by modern technologies) and efficient – is to use a special video camera and video monitor to create a wireless communication channel with the child (see the option suggested by the German company ENOX World GmbH from Hamburg in fig. 5.16).

The system has been complemented with an informational resource in the form of a visual/audial communication contour implementing a new function – "remote monitoring of the child". All other resources take part in system transformation, as well: structural resource (new elements and connections), spatial resource (efficient positioning), material resource (all these things need to be manufactured), and, of course, energetical resource (the new elements need reliable power supply). The system as a whole has acquired a new property: ongoing monitoring of the child wherever the parents may find themselves inside the house.

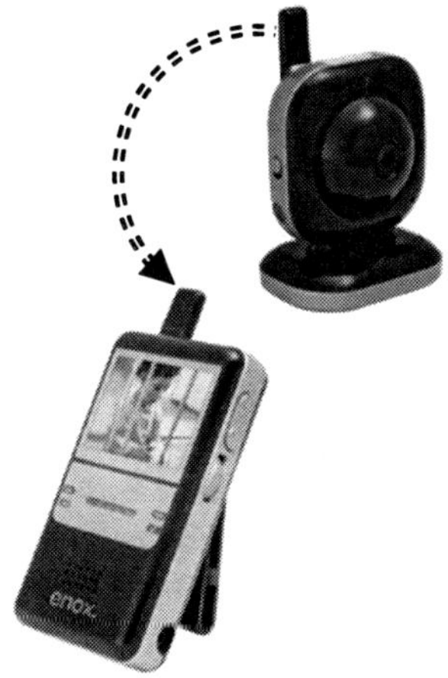

fig. 5.16. Baby Watch system elements

Additionally, installation of a humidity sensor will help identify the moment when the child needs new Pampers, while a temperature sensor will warn the parents that the child may be feeling ill.

As a result, today, at the beginning of the 21st century, young parents may have fewer troubles than their own parents did.

Example 5.10. Welcome to school – but let me check your fingerprints first!

There are identification systems using unique biometric data, such as fingerprints [or retinal scans], to grant access to certain restricted spaces (confidential data storage facilities, bank vaults, etc.) to certain people known to the system.

And now (February 2010) researchers from Vladimir State University are testing such access system (fig. 5.17) at the Lyceum of Industry and Commerce.

fig. 5.17. "Press the button and get what you want..." [94]

They are also working on tracking systems which make it possible to find any person, for example, a child, using such person's current GPS coordinates – this can be regarded as a continuation of the previous example. Besides, the system may have a "panic button" that the child can use to let his or her parents know that he or she needs help.

Such systems with their undoubtedly useful properties, as well as other similar individual and collective safety systems have nothing in common with the total control system ("Big Brother is watching you") described in George Orwell's novel "Nineteen Eighty-Four"[95].

---

94 From the lyrics of one of the 1991 hits of the Russian techno-pop band "Technology" (www.vesti.ru)

95 George Orwell (1903-1950) – famous English writer and publicist, author of outstanding novels "Animal Farm" (1945) and "Nineteen Eighty-Four" (1949)

Generally speaking, the informational resource has become the key decision-making aspect of the late 20th century – early 21st century.

5.3 Targeted Ideal Modeling

> Nusquam per novus est perficio. [96]
>
> *Cicero*

5.3.1 Ideal Final Result According to Bartini

> To solve a problem, it is necessary to outline, as compactly as possible, a group of closely correlated factors which play the decisive role in the issue under consideration, separating all secondary elements.
>
> Then you need to formulate the most contrasting **"EITHER – OR"** contradiction, opposition excluding the solution.
>
> The solution will be found in the logical combination of identify of opposites... **"BOTH – AND".**[97]
>
> *Robert Bartini*

The notion of *Ideal Final Result* (IFR) is one of the many key TRIZ concepts. As discovered by many civilizations very long ago, and as reflected in the above Cicero maxim, *nothing quite new is perfect.* And when it emerges, there begins a struggle to improve what *must become perfect.* Generally speaking, that struggle is eternal and never-ending (we will discuss it in more detail later, in section *9. System Evolution Models*).

Where does the struggle start? It starts with identification of incompatible, conflicting factors – properties and requirements. The *key* factors (well, THAT goes without saying).

Then a contradiction (for example, a standard contradiction) is formulated. It states that, if we want to resolve the conflict with *available* material resources using *traditional* methods (which basically also constitute a resource – an informational resource), we will have to choose between two factors – it is either "plus in one – minus in another", or the other way round. You may recall that we already modeled standard (and radical) contradictions in *section 2.1 Contradiction.*

---

[96] Marcus Tullius Cicero (106-43 BC) – Roman politician, philosopher and orator; the English translation of the maxim above is: *Nothing quite new is perfect.*

[97] Robert Ludvigovich Bartini (Roberto Oros di Bartini) (1897-1974) – outstanding Soviet aircraft constructor, physicist, mathematician, philosopher, writer; quoted from: I. E. Chutko (1978) *Red Aircraft.* – Moscow, Politizdat (Political Publishers; in Russian)

But what do we really want? We want both factors to be within their best ranges! And for THAT we have to invent a new constructive solution. Remember figure 2.3?

Well, sometimes the path to solution may materialize only when the contradiction has sustained changes which turn it into its antipode – the IFR. Essentially, this is the method suggested by R.L. Bartini (see the epigraph) as the starting point for finding the solution: specify the unique combination of factors and their values required to attain the ideal anticipated result.

The formula describing transition from the contradiction to the IFR according to Bartini is presented in fig. 5.18:

IS: $\uparrow F1 \text{ VS } \downarrow F2 \equiv \text{PROBLEM}$

$\Downarrow$

MUST BE: $\uparrow F1' \;\&\; \uparrow F2' \equiv \text{IFR}$

fig. 5.18. Formula of "contradiction → IFR" transition according to Bartini

Or, using the English-language model which is well-known in management theory:

WIN & WIN = IFR

This transition can also be described as the ***mandatory transition*** of system properties (fig. 3.3) towards (trend) the IFR (fig. 5.19).

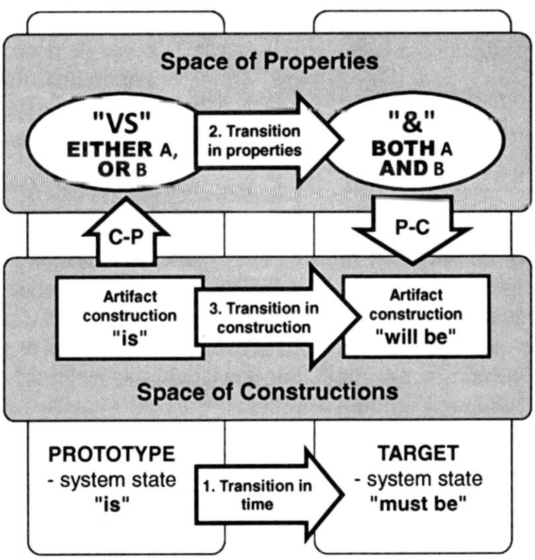

fig. 5.19. Scheme of "contradiction → IFR" transition according to Bartini

Example 5.11. Stal-6 Landing Gear.

In the early 1930-es, the world's best aircraft could fly with speeds of up to 300 km/h. [The Soviet Government] ordered to increase that maximum speed by 150%. That order was fulfilled by Bartini. One of the problems was the powerful drag due to poor wind shape of then existing aircraft. For example, wheeled landing gear consisted of several rigid props which at best were covered with cowls. Meanwhile, cowls increased (by however little) the weight of the aircraft. All designers could offer at the time was a prototype with a landing gear that clung closer to, but was not completely retracted into, the body of the aircraft.

Accordingly, the contradiction formula known for all prototypes was this:

Wheeled landing gear ► movement on the land surface
VS deceleration in the air

The Stal-6 aircraft that Robert Bartini was ordered to design in 1930 was improved based on transformation of the standard contradiction into the IFR:

Wheeled landing gear ► movement on the land surface
& absence (!) in the air

IFR: "zero deceleration" in the air! There should be no wheels in the air – there is no need for them!

fig. 5.19. Stal-6 (www.airwar.ru)

Thus was created the Stal-6 aircraft with just one retractable landing wheel (fig. 5.19) under the fuselage and skids (also retractable) at the wing-tips and under the tail! After the takeoff, the entire landing gear could be retracted into the fuselage! For the first time in the world!

And in 1933, the aircraft set the world speed record at 420 km/h!

Example 5.12. Stal-6 Steel.

Stal-6 was able to attain record speeds not only (and not primarily) due to its new landing gear construction, but also due to other revolutionary innovations.

The weight of the aircraft is one of the critically important factors in the "group of closely correlated factors". To build Stal-6, it was necessary to weld thin sheets of chrome-molybdenum steel and stainless steel to receive what today would be termed a "composite material". But how could this be done? After all, it called for "radically" different welding conditions!

The radical contradiction was this[98]: it was necessary to "merge, in one welding construction, the best properties of different steels – durability, plasticity, resistance to corrosion, etc. The best combination was that of chrome-molybdenum steel and stainless steel.

[98] I. E. Chutko (1978) *Red Aircraft*. – Moscow, Politizdat (Political Publishers; in Russian)

But they were welded in different ways. Stainless steel had to be welded by applying a short-lived powerful electric charge. If the process stretched out in time, even a little, certain substances which made the steel stainless came out of the melt drip, and in the weld seam it turned into ordinary steel.

Chrome-molybdenum steel, on the contrary, had to be welded slowly, using weak electrical charge yielding relatively low temperatures, otherwise the weld spot, rapidly cooling in the air, overheated and became fragile, and the seam broke down. These opposing requirements made the two welding modes incompatible."

The radical contradiction formula was this:

Electric current ► ↑ (strong) for stainless steel
VS ↓ (weak) for chrome-molybdenum steel

Bartini and engineer S.M. Popov found the following solution (quoted from I.E. Chutko[98]): first there was a very brief and very powerful charge; this was enough for stainless steel to retain all its components, but not enough for chrome-molybdenum steel to get overheated. Then the charge was reduced to produce the ideal temperature to complete the welding of chrome-molybdenum steel. As a result, stainless steel and chrome-molybdenum steel could be welded using existing technologies without any breakthrough inventions. Welding control functions were automated, as no man was able to promptly change welding modes (this took infinitesimal fractions of one second), so both visually and audially the new welding method was in no way different from the old methods.

The formula for the victorious IFR in this solution is this:

Electric current ► ↑ powerful and shot-lived – for stainless steel
& ↓ less powerful and longer-lived – for chrome-molybdenum steel

We believe that this example can be used to reiterate the importance of understanding, learning through training, and utilization of the noospheric creative model (see section *3.2 Noospheric Modeling Concept*).

Let us single out, in this example, various knowledge and skill levels employed to generate efficient ideas:

1) *applied level* – all requisite welding knowledge was available, but it was not efficiently used "here and now" to meet the new requirements;

2) *creative level* – it was necessary to correctly use the temporal resource, and divide separate opposing properties in time: one property is realized during one time interval, and the other property is realized during another time interval;

3) last but not least, *psychological level* – here it will be best to quote I. E. Chutko: "It goes without saying that the impasse we overcame was not technical (as all technologies used in the Stal-6 aircraft had already been used before), but psychological."

By 1934 the engineers designed an even faster aircraft, Stal-8, "with a maximum speed of 630 km/h – another huge speed jump of 200 km!" For various reasons, Stal-8 was never built, while "speeds of 650-650 km/h were attained by our serial fighters only in 1939-1940" – quoted from I. E. Chutko[98].

Example 5.13. Stal-7: The Steel Gull.

After the success of Stal-6, [the Soviet Government] ordered to design an aircraft capable of long-range (maybe even around-the-world) flights. To do that, it was necessary to install a second landing wheel in the front part of the aircraft, increase its load-lifting capacity, and improve its manageability during takeoff and landing. But all that required that the aircraft's own weight be significantly reduced – which was a tall task taking into consideration the difficulties caused by the increasing dimensions: wing mounting, additional landing gear, installation of two engines (and their placement on the wings). As a result, the weight problem soon got out of control, eliminating all benefits created by anticipated increase of speed and load-lifting capacity.

A plethora of "irreconcilable" standard and radical contradictions! And R. Bartini suggested a new aircraft which turned into a *steel "gull"* – literally (fig. 5.20 and fig. 5.21)!

The W-shaped wing – "inverted gull wing" – *"BY ITSELF"* assured convenient mounting of engines and undercarriage at the bends!

Another innovation was the "triangular" cross section of the fuselage (oval with a broad foundation) which created an ideal configuration for the fuselage-wing transition (fig. 5.21 – Er-2 bomber built on the basis of the Stal-7 aircraft).

fig. 5.20. Stal-7 (www.airwar.ru)

fig. 5.21. Er-2 bomber based on Stal-7
An asymmetric cockpit canopy can be seen at the top of the aircraft.
Source: www.airwar.ru

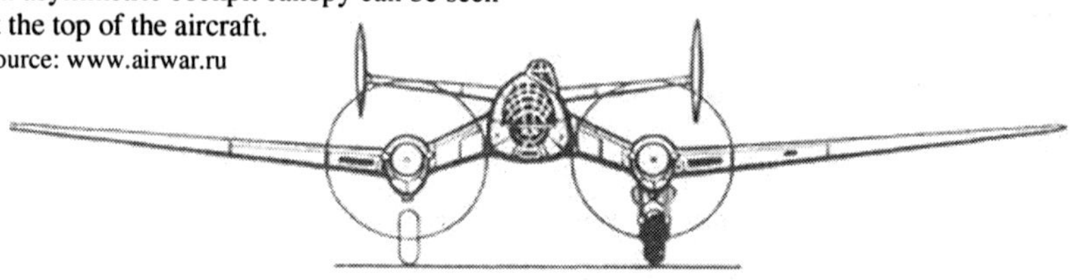

In 1936 Stal-7 made its debut appearance at the International Exhibition in Paris, and in August 1939 it set a world speed record at 450 km/h, covering the triangular Moscow-Sverdlovsk-Sevastopol-Moscow route (5,068 km) in 12.5 hours.

The planned around-the-world flight was "cancelled" by the Second World War.

Er-2 bombers participated in raids on the Nazi capital in August-September 1941, covering the distance from Moscow to Berlin and back without intermediate landings.

Example 5.14. Project R: The Ideality Curve and the "Albatross".

The world's first triangular monoplane – aircraft with extreme wing sweep – was suggested in 1933 by Voronezh designer A. S. Moskalyov (1904-1982). His projects (Sigma and Strela (Arrow), fig. 5.22) can be regarded as progenitors of all subsequent delta-wing aircraft in the world.

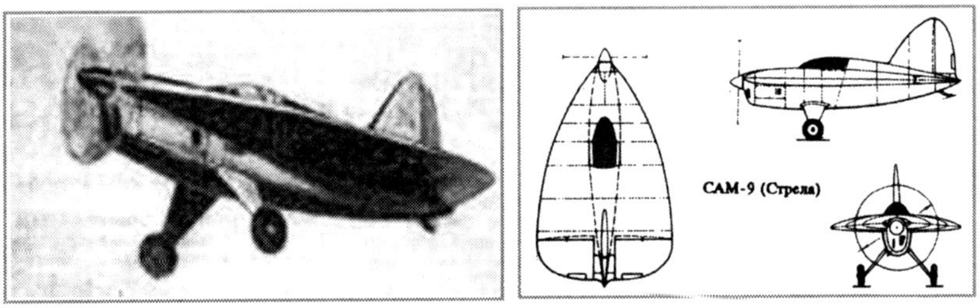

fig. 5.22. Strela monoplane (www.airwar.ru)

In 1942, when the USSR was going through one of the hardest periods in its history, [the Soviet Government] ordered development of supersonic fighters and bombers. However, by then many experts throughout the world had realized that aircraft with "ordinary" wings would not be capable of breaking the sound barrier and flying at supersonic speeds.

The increasingly portentous contradictions hindering development of future supersonic aircraft were exemplified by the ratio of landing speed to cruising speed.

For subsonic aircraft, that ratio ranged from 3 to 5. However, with maximum speeds reaching, say, 2,000 km/h, and desirable landing speeds in the vicinity of 200 km/h, the ratio went up to 10. And that was not the limit! Incidentally, for high-speed aircraft the ability to maintain relatively "low" landing speeds became the main problem.

The resultant standard contradictions can be formulated, in a simplified form, as follows:

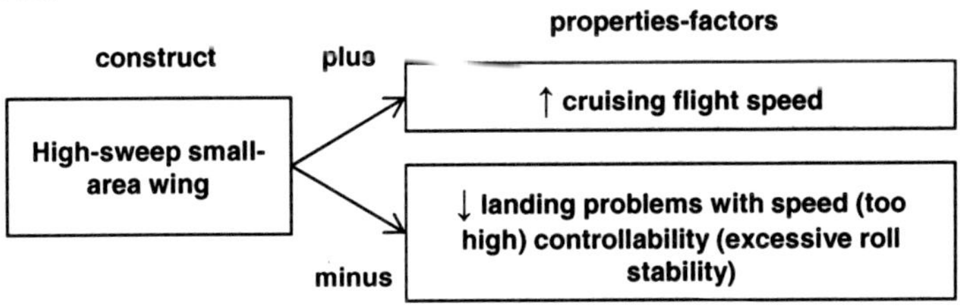

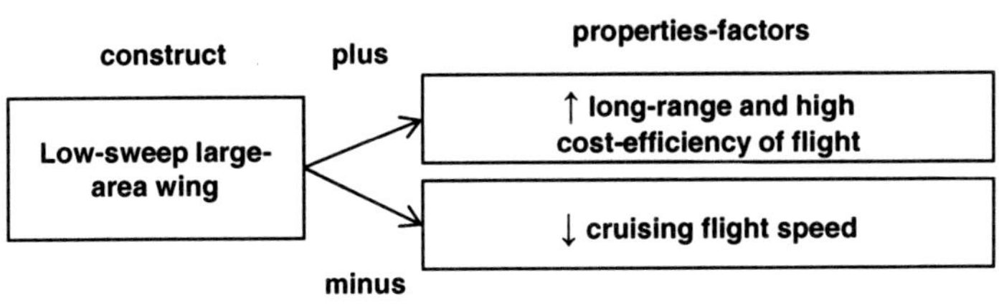

To deal with that problem, Bartini, working within the framework of the "R" project (in the year of 1942!), for the first time in the history of aviation suggested that a supersonic single-seat fighter be equipped with a variable-sweep wing (*Bartini wing*), which basically represented an ingenious application of the **"BOTH – AND"** method. It was a realization of the "ideal curve" based – literally – on the rules of aerodynamics: each part of the wing must work in the best available mode, namely, the closer to the fuselage, the larger must be the sweep!

In the 1950-es Robert Bartini used that wind in his other projects, including the supersonic long-range strategic bomber A-57 (fig. 5.23). Flight range – up to 15,000 km, flight altitude – 18-23 km, speed – up to 2,500 km/h. The aircraft was special in that it was amphibious, i.e. it could take off from, and land on, water rather than land.

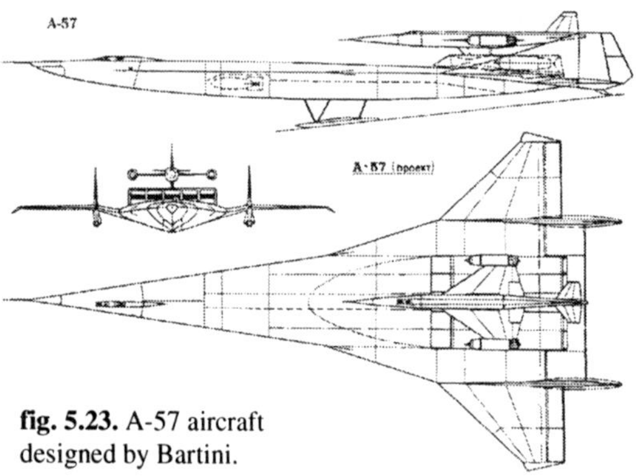

fig. 5.23. A-57 aircraft designed by Bartini.

At the same time, similar wing shapes were designed by A. S. Moskalyov. In the near future wings of that type would embellish supersonic passenger aircraft – the Soviet Tu-144 and the Anglo-French Concorde (fig. 5.24).

The shape of the aircraft resembles that of a soaring albatross.

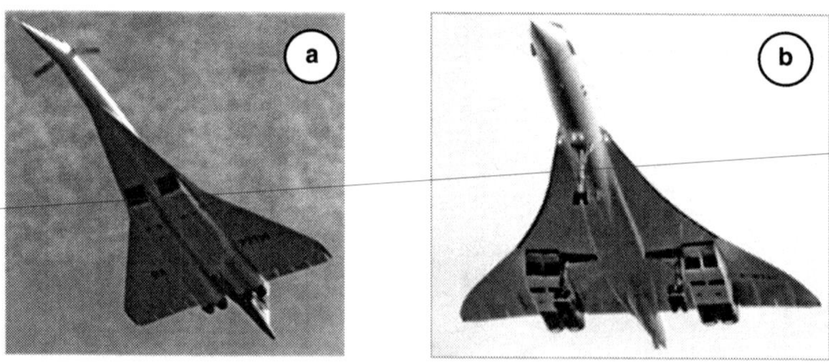

fig. 5.24. Evolution of Bartini wing: a) Tu-144 (1968); b) Concorde (1969)
Pay attention to the completely different engine configuration

Example 5.15. Aerofoil Boats and Aerofoil Aircraft (Ekranoplan).

In the 1960-es, Robert Bartini undertook an analysis of systemic limits for various types of aircraft based on the following essential factors: range, load-lifting capacity, and speed. His conclusions were fundamental and prophetic – even though they still await their implementation, as do many other of his pioneering ideas.

Relying on the I. E. Chutko[98] book, we will limit our discussion here by those facts which are necessary[99] to demonstrate the "BOTH – AND" method.

R.L. Bartini attempted to see the future of all types of transport (not just aircraft) from physical and mathematical positions. He took the key properties of a means of transportation and reduced them to three general properties. Thereby he proved – quite convincingly, too – that all existing types of transport use a ridiculously small portion of their overall potential, and that all of them are far from ideal.

Bartini said: "The airplane flies well, but its takeoff and landing are clumsy, while the helicopter takes off and lands well, but its flight is slow. Is there a way out of these contradictions? Yes. The way out is to design the body of the aircraft so as to achieve a unity of the opposites – unity of wing, fuselage and empennage functions.

I believe that with time the landing gear below the body of the aircraft will be replaced by an aerodynamic shield. The resultant air cushion will make the future aircraft – aerofoil aircraft – *"all-airfield" or, if you will, "airfield-free"*: they will be capable of taking off and landing anywhere... Airfield and VTOL aircraft will eventually perform a quantum jump in transportation.

Trains gliding along on high-pressure air cushions will be running *monorail trestle roads* at transonic or even supersonic speeds. This method will cover the bulk of transcontinental passenger and cargo transportation.

Transoceanic cargos will be carried not only by supersonic aircraft, but also by huge *catamaran aerofoil boats* capable of taking on thousands of tons of payload".

Creation of *large VTOL amphibious aircraft* would enable transport operations on most of the Earth's surface, including Arctic and Antarctic ice fields and deserts, seas and oceans.

In the 1960-es and early 1970-es, several model and sample aircraft built in accordance with Bartini's ideas were actually tested[100] (fig. 5.25 and fig. 5.26).

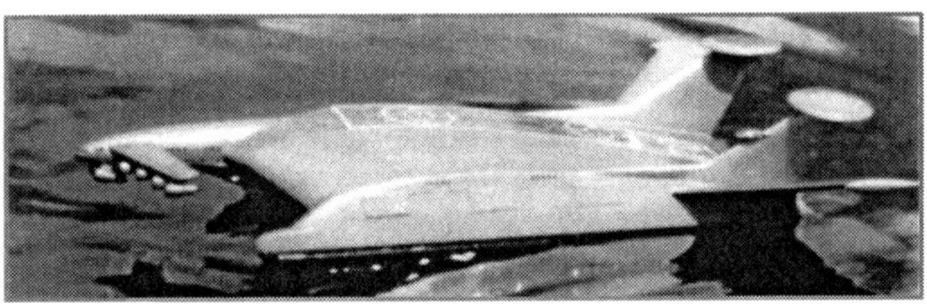

fig. 5.25. Aerofoil aircraft carrier SVVP-2500. Swimming pool testing model

[99] To keep things simple, direct quotes have been replaced by, or combined with, the author's text.

[100] Georgy Mikhailovich Beriyev (1903-1979) – famous Soviet aircraft designer, creator of several amphibious aircraft; aerofoil aircraft 14M1P was later completed by the Beriyev design bureau

fig. 5.26. Aerofoil aircraft 14M1P
Modification of VTOL amphibious aircraft VVA-14

The SVVP-2500 aircraft was supposed to be larger than a soccer field, and have a load-lifting capacity of 2,500 tons, which would enable its use as a flying aircraft carrier.

During the same time and somewhat later, other extraordinary projects were developed by the design bureau headed by R. E. Alekseyev[101] (fig. 5.27 and fig. 5.28).

fig. 5.27. Aerofoil aircraft KM
Assault landing and rescue vessel, NATO nickname: "Caspian Sea Monster". Length – 100 m, takeoff mass – 544 tons

fig. 5.28. Aerofoil aircraft Lun (Harrier)
Attack missile-carrying aerofoil aircraft, NATO nickname: "Aircraft Carrier Killer". Length – 73 m, flight speed – 500 km/h, takeoff mass – 400 tons

5.3.2. Functional Ideal Modeling

Ideal is a guiding star.
Without it, there is no firm direction…[102]

L. N. Tolstoy

The process of generation of ideas on the basis of TRIZ models and methods is an objective-directed system where IFR and FIM are defined through objectives and meta-trends (general directions and methods used to approach objectives), and the path to each objective is shaped by TRIZ transformation models. This was schematically illustrated by fig. 3.1 in section *3.1 Algorithms of Inventive Creativity.*

Here we will consider the models helping to effect transition from the IFR formulation to the future change in the construction of the system – structure of a machine, plan of a marketing campaign, functional structure of an organization, structure of interpersonal relations, etc.

---

[101] Rostislav Evgenyevich Alekseyev (1916-1980) – outstanding Soviet ship designer, creator of hydrofoils, aerofoil boats and aerofoil aircraft (ekranoplan)

[102] Lev Nikolayevich Tolstoy (1828-1910) – great Russian writer and thinker, educator, public and religious figure

Definition **"Ideal Final Result"**	**Ideal Final Result (IFR)** – the *definition of the objective* of artifact enhancement presented as the required functional state or action which best (ideally) conforms to the designated purpose of the artifact.
"Functional Ideal Model" – addition to the definition of IFR	**Functional Ideal Model (FIM)** – the formally arranged *model-metaphorical description* of the way in which the artifact must function in order to attain the IFR.

Beginners normally experience noticeable difficulties with the formulation of the IFR. The reason is psychological inertia which draws the would-be inventor back to the prototype, while the objective is to obtain a new artifact with new properties which need to be stated in the form of the IFR.

To weaken psychological inertia during IFR formulation, TRIZ practitioners have developed the following two time-proven rules:

1) At the initial stages, you must not think how and with what (with the help of what resources) you will get your solution;
2) An unknown resource or action required to obtain the desired result can be temporarily replaced with a metaphorical symbol, for example, the ***X-resource***.

Let us remind our readers of the basic FIMs[103].

First variant – Macro-FIM:

The ***X-resource*** – without making the system excessively complex or causing any negative effects – ensures, together with the other available resources, the obtaining of the

[*required functioning or IFR*].

Second variant – Micro-FIM:

The ***X-resource*** in the form of *particles of substance or energy* is located inside the operative zone and ensures, together with the other available resources, the obtaining of the

[*required functioning or IFR*].

Third variant – Maxi-FIM:

The operative zone ***by itself*** ensures the obtaining of the

[*required functioning or IFR*].

It should be noted that oftentimes once you have correctly formulated the IFR and correctly identified the relevant FIM, they will take you to the idea for the solution BY THEMSELVES. You could see this effect in action while reading the previous section.

---

103 M. Orloff (2006) *Inventive Thinking through TRIZ*, Section 8.2. Functional Ideal Modeling

According to TRIZ, FIMs become the meta-trend towards transformation of the system, as they build metaphorical bridges from the IFR directly to the desired solution (fig. 5.29).

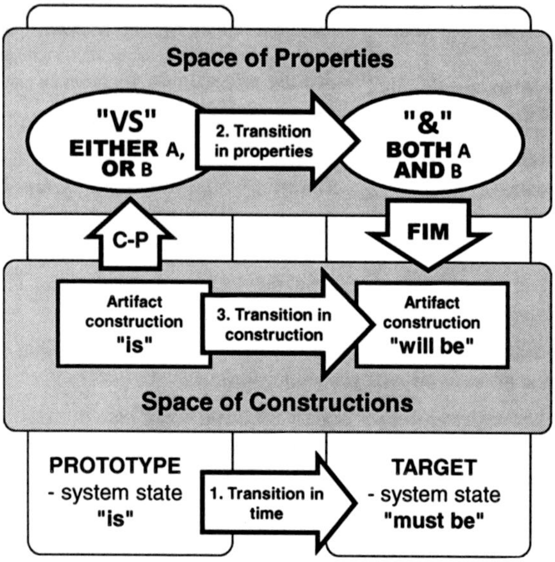

fig. 5.29. Scheme of transition from the IFR to shaping the end result – combination of the FIM and Transition in Construction (3)

According to an old oriental parable, no matter how many times you are told what is "sweetness", you will only understand it once you have tried it. You need practical experience. That is why we now proceed to practical examples.

Example 5.16. See the IFR in the Stars!

When in the early 1950-es there arose the need to build a strategic submarine fleet armed with nuclear ballistic missiles, one of the fundamental problems that had to be tackled by the newly-created Ural research institute headed by V. P. Makeyev (см. [33]) was to find an efficient way to home the missiles and warheads on designated targets.

Indeed, the targeting program – and it had to operate with 100% efficiency with respect to any target and in any direction! – had to be constantly adjusted because the location of the "launch site" (the submarine) in the world ocean changed on an ongoing basis. To perform such adjustments, it was necessary not only to precisely determine the location of the submarine, but also to recalculate the entire launch program, and then retarget *analog* missile flight control systems.

Clearly, even with state-of-the-art calculation methods, that would require time. But can you imagine just HOW hard it was in the absence of modern high-performance computers? I believe it is difficult to imagine that – just as it is difficult to understand what is "sweetness" based only on other people's accounts.

But that is only half the story... The submarine could also be pitching and rolling, which – together with other factors – could have adverse effect on the accuracy of above-water and under-water launches. Be that as it may, it was apparently difficult, if not impossible, to assure that the target was hit unless the flight was corrected en route.

The **IFR** here is clear: en route support of missile flight.

Let as also write down the **FIM** for the first variant:

> The ***X-resource*** – without making the system excessively complex or causing any negative effects – ensures, together with the other available resources, the obtaining of the
> [*en route support of missile flight*].

To realize transition to the new construction (arrow 3 in fig. 5.29), we need knowledge: taking into consideration the fact that the submarine, the sea, the land and the missile can offer no alternative flight control resources, what else can be used as an additional resource to calculate and adjust flight parameters?

There can be only one such X-resource – the STARS! And, naturally, creation and implementation of a new *stellar monitoring* flight control system.

The missile had to be taught to *find its way by the stars*, so that after leaving the atmosphere it would be able to adjust its flight by itself. There it was, the extraterrestrial X-resource that had to be activated to gain this critical advantage.

The missile had to become more "intelligent", it had to be equipped with new informational resources to detect and identify available stellar constants, use them as navigation reference points, dynamically calculate missile location and orientation parameters vis-à-vis the stars and the land, and perform the final adjustment of the flight trajectory.

Ultimately, with maximum flight range of 8,000 km, the impact error was under 1 km, while without stellar orientation, it was as high as 3-5 km.

As for the "absolute" requirement "not to make the system more complex", the structural complexity of the navigation system of the missile, as compared to its other systems and even to the original navigation system, sustained minimal changes. Naturally, the final judgment on that is reserved for the professionals.

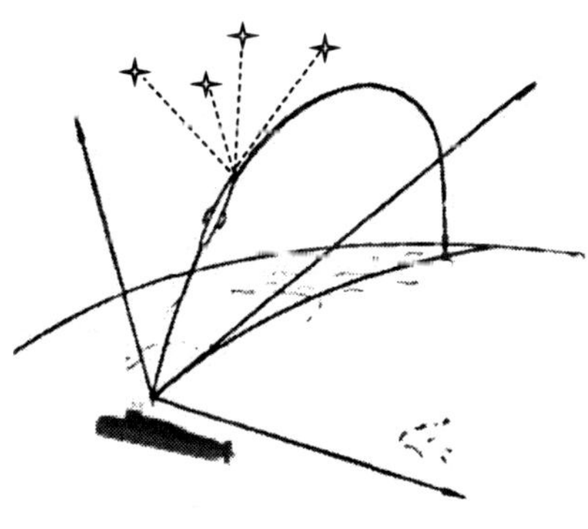

fig. 5.30. R-29 missile stellar orientation system

Our task is to identify, in this example, its creative potential and inherently existing creative models.

In other words, we need to do what designer Makeyev did: *see the IFR in the stars!*

Example 5.17. How Archimedes Helped Alexandria 2,250 Years Later.

In 280 BC, on the Island of Pharos near Alexandria in Egypt, there was built the Pharos Lighthouse, one of the 7 wonders of the ancient world, a building almost 150 meters high which projected its light 70 miles into the sea using a system of mirrors designed by Archimedes[104]. The lighthouse was topped by a 7-meter statue of Poseidon, the god of the sea.

Alexandria itself was founded by Alexander the Great[105] who had conquered Egypt some 30 years before construction of the lighthouse had begun. Later Alexandria was known not only for its lighthouse, but also for its world's largest library frequented by Archimedes, and other notable sights.

The Pharos lighthouse had stood for more than 1,600 years when it was destroyed by an earthquake in 1375. In 1993, while expanding the Alexandria Harbor, workers discovered the ruins of the lighthouse and other ancient buildings. More than 200 large and small artifacts were raised from the bottom of the sea.

Problem: how can we raise very long and very heavy column fragments and statues so as to prevent them from breaking up?

Let as write down the **IFR** and the **FIM** for the first variant:

> The ***X-resource*** – without making the system excessively complex or causing any negative effects – ensures, together with the other available resources, the obtaining of the
>
> [*uniform application of effort to raise a column from the sea bottom*].

The column needs to be fixed so that, as it is being raised, its weight is uniformly distributed along its entire length. Large cranes cannot be used because the harbor is too shallow.

We need a special resource generating the lifting force. Maybe it is time for us to recall the law of Archimedes?!

fig. 5.30. By the law of Archimedes

Scuba divers lowered to the bottom soft inflatable floats, and tied them to the artifacts they wanted to raise. Then the floats were evenly and gradually inflated with compressed air, and they slowly lifted the artifacts to the surface. Supported by the floats, the artifacts were transported to the shore where it was possible to use platforms equipped with powerful cranes.

Example 5.18. "Ideal" Bottle opener! Sometimes it is difficult to pull a cork out of the bottle because the cork starts crumbling. The corkscrew first impales the

---

[104] Archimedes (287-212 BC) – Greek mathematician, physicist, mechanic and engineer from Syracuse; made numerous geometrical discoveries; laid the foundation of mechanics and hydrostatics; created many important inventions

[105] Alexander the Great (Alexander III of Macedon, 356-323 BC) – Macedonian king who was believed to be the greatest warlord of the ancient world

cork as it drills through it, and then "adds insult to injury" by breaking it up as it is being pulled out. What can we do?

We need as bottle opener that will have no (or almost no) damaging effect on the cork.

It would be even better if (in line with the third-variant FIM) the cork would pop out BY ITSELF with minimum effort applied by the person opening the bottle.

Let us write down that **IFR** into the **FIM**: The X-resource – without making the system excessively complex or causing any negative effects – ensures, together with the other available resources, the obtaining of the

[*minimization of damage sustained by the cork and easier extraction of the cork from the bottle*].

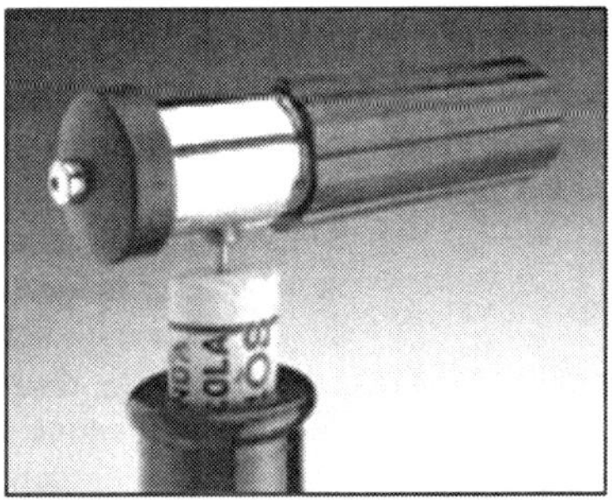

fig. 5.31. Is this really an ideal bottle opener?

What can help the cork to leave the bottle? Increased air pressure inside the bottle between the cork and the liquid, usually wine.

Indeed, there are syringe bottle openers where a thin hollow needle is inserted into the cork (or into the interstice between the cork and the glass), and then air is pumped into the bottle. The material X_7-resource in this case is air, and the energetical X_8-resource is energy expended to depress the plunger of the syringe to pump air under the cork. A strikingly elegant solution!

The somewhat provocative question in the caption under fig. 5.31 is a reminder that, in situations where such bottle opener was equipped with a small gas tank filled with compressed air, the cork was sometimes expelled too fast (together with some wine), or the bottle burst under excessive pressure.

Example 5.19. Why Are Temporary Road Markings "Glowing"? When highway lanes are temporarily realigned to accommodate road repair works, new road markings can be painted on. Their visibility, more or less adequate during the day, deteriorates at night or when it is raining. Besides, they are rapidly rubbed off by the wheels of the passing cars. But have you ever noticed the bright light cast off by a small piece of glass lying on the side of the road as it is caught in the glare of your headlights? Or the glow of various reflectors?

Let as try to apply the IFR and the FIM for the first variant:

Here goes: The ***X-resource*** – without making the system excessively complex or causing any negative effects – ensures, together with the other available resources, the obtaining of the

[*brightly lit lane markings*].

The markings must become more active inductors! For example, they must produce bright light!

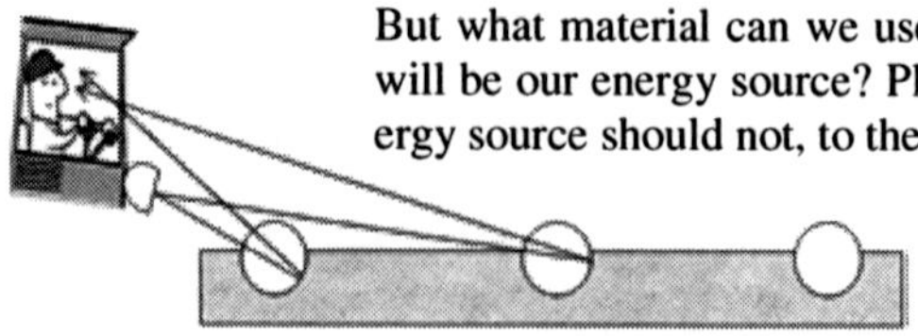

But what material can we use to obtain that effect, and what will be our energy source? Plus both the material and the energy source should not, to the extent possible, be expensive!

Here is a possible solution (Fig. 5.32): the road markings are made in the form of thin film which is glued to the surface of the road.

fig. 5.32. The film produces light "by itself"!

The film contains myriads of embedded (fused-in) glass balls with a diameter of 3-4 mm each. The balls are made from glass production waste and broken glass which was earlier simply thrown away.

The light projected by the headlights pierces the ball and bounces off the farther hemisphere producing a reflection powerful enough to be seen by the driver.

Example 5.20. House Number Plate. What can be done to make the house number plate visible even when there is no power? Let us write down the FIM for the second variant:

fig. 5.33. Self-light!

The ***X-resource*** in the form of particles of substance or energy is located inside the operative zone and ensures, together with the other available resources, attainment of the following result:

[*autonomous house number plate lighting*].

In accordance with the selected FIM, we need to create a solution where all required materials and energy would be located within the OZ of the system.

And here is the solution[106]: a panel of economic light-emitting diodes powered by an accumulator linked to a solar battery (fig. 5.33). One-day charge is sufficient for 150 hours of uninterrupted operation.

X-resources: X_7 – "intelligent" materials, and X_8 – unlimited solar energy (see fig. 5.6 and fig. 5.7 for selection of resources).

Example 5.21. Solution for Problem 2.6. Tricks and "Illusions". Let us write down the IFR and the FIM for the second variant, because in this case everything that we need is definitely located within the OZ:

The ***X-resource*** in the form of particles of substance or energy is located inside the operative zone and ensures, together with the other available resources, attainment of the following result:

[*durable connection of the knives into a stable construction resting on the glasses*].

The "algorithm" for the erection of the "bridge" (fig. 5.34) is simple: put Knife 1 into position, underpin it with Knife 2 which is then underpinned with Knife 3 which is then underpinned with Knife 1.

---

106 www.proidee.de

That's it! The circle is closed!

Naturally, you use the energy produced by your hands while making the "bridge", but when you let go of the knives, they hold on to each other by force of friction.

Here is a caveat, though: before trying to show this trick to your friends, practice it until you are positive that you can build the bridge fast and without mistakes *using any variant.*

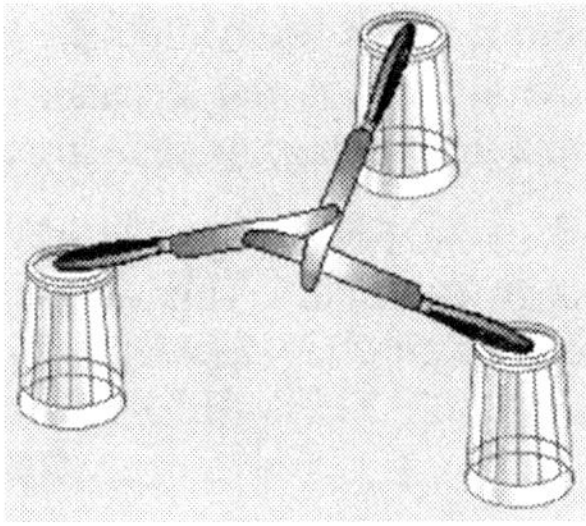

fig. 5.34. "Ant bridge"

Why am I giving you this piece of advice? Because there are several "catches" (just imagine, even this simple problem has two solutions and several dead ends!) which can rain on your parade at the most inopportune moment. Discover them by yourselves and find ways to neutralize them!

Also, to learn this trick better and improve your creative abilities, you can have a look at other problems presented by Edward de Bono[107].

Finally, a recommendation regarding the topic of this book: successful use of TRIZ requires a lot of training – it is not enough just to read another manual. Self-training will help you learn a lot of "tricks" which are difficult to explain in a book.

Example 5.22. Fighter-Bomber Fuel Tank. In the 1930-es, aircraft used special aviation fuel. However, the fumes produced by the fuel were explosive, and if a bullet hit the fuel tank, the aircraft could blow up. Attempts to protect the fuel tank with armor would increase the weight of the aircraft. What could be done?

FIM: The ***X-resource*** in the form of particles of substance or energy is located inside the operative zone and ensures, together with the other available resources, attainment of the following result:

[*prevention of fuel tank conflagration*].

Solution – based on resources "available" at the aircraft: the space vacated inside the fuel tank as fuel was being fed into the engine was filled with chilled exhaust gas.

Example 5.23. At Different Speeds These Are Different Machines! When variable-sweep wing aircraft (such as the fighter-bomber SU-24 described above) were built, it became clear that their center of gravity shifted (and their behavior in the air altered) as their wing position changed.

The problem became even more serious with emergence of the first supersonic passenger aircraft, such as Tu-144 and Concorde: at maximum speeds they became longer(!), and the center of gravity shifted by 2 meters or even more!

FIM: The ***X-resource*** in the form of particles of substance or energy is located inside the operative zone and ensures, together with the other available resources, attainment of the following result:

[*simple center-of-gravity positioning of aircraft*].

---

[107] de Bono, Edward (1968) The Five-day Course in Thinking. –The Penguin Press, London

Solution: based on availability of a massive "material" resource available on board the aircraft, it was decided to redistribute the fuel in fuel tanks inside the wings and the fuselage to maintain the "correct" position of the center of gravity.

Example 5.24. Solution for Problem 3.5. The Candle Burned upon the Desk, the Candle Burned... This simple solution is one of the best engineering discoveries where things are apparently happening without energy loss or excessive material waste, BY THEMSELVES, and with minimum change.

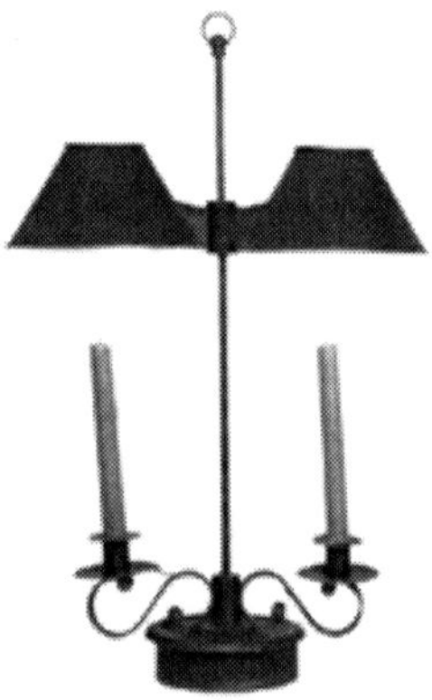

fig. 5.35. Manually Moved Lampshade

But first let us write down this "unfeasible" requirement:

> The operative zone ***by itself*** ensures the obtaining of the IFR
> [*the lampshade descends BY ITSELF as the candle is burning down*].

The operative zone in this example is the top of the candle with the burning wick. As the candle is burning down, it is becoming shorter, and soon the bright light begins to show from under the lampshade blinding the spectator.

In some lamps (fig. 5.35) you can pull the lampshade down manually to cover the light.

Now, we will show you how the "ideal" lampshade" and the "ideal" candle work together (fig. 5.36).

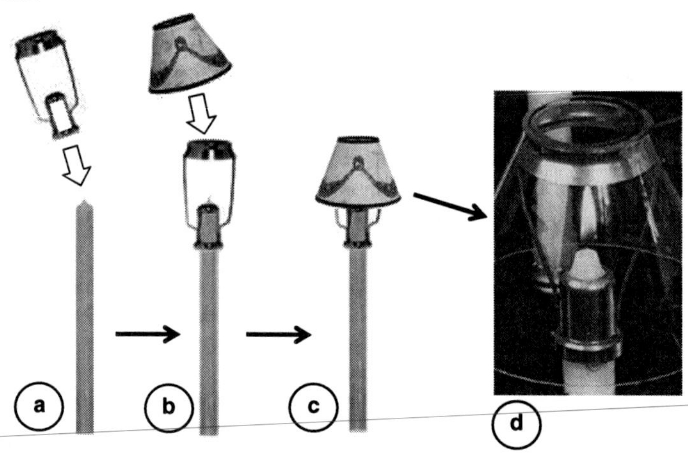

fig. 5.36. "Self-Moving" Lampshade

First, the candle is topped with a frame (fig. 5.36, a). Then the lampshade is placed on the frame (b) – and *presto chango!* – the construction is complete (c). To better show the frame topping the candle and the place where the lampshade joins the frame, we have used a glass lampshade in fig. 5.36, d.

Example 5.25. A Sample of "Ideal Minimalism"! This is the solution of the benchmark problem P3 "How can you stop drops from spilling over when pouring liquid from a bottle?"

What can you do to make some “thingamajig” fit a larger range of bottle necks, while retaining the ease of manufacture, storage and use?

Or, which is basically the same thing, according to FIM Variant 3:

The operative zone ***by itself*** ensures the obtaining of the [*universal drop stopper fitting multiple bottles*].

A gorgeous "ideal" solution is embodied in European Patent 0560777.

Many years ago I was stunned by the "minimalistic" design of the device sold under the registered trademark "DropStop". The patented device consisted of... a simple round (or, later, differently shaped) plate which BY ITSELF fixed itself within the bottle neck using elastic properties of the construction and of the material (fig. 5.37).

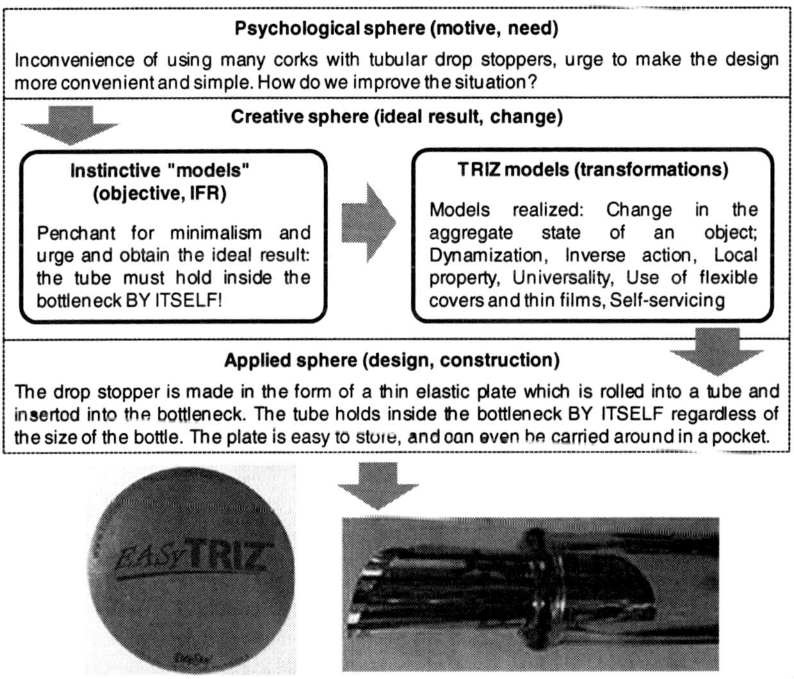

fig. 5.37. Sample of structural reduction and ideality – "DropStop"

Let us elaborate on several dominant transformation models:

- use of elasticity, elastic properties in line with the "change of the aggregate state" model;

- dynamization to change shape and size and ensure adaptation to the bottle neck: characteristics of the object (or the environment) must change so as to be optimal at each step of the work procedure;

- the static tube was transformed into a plate which becomes a "tube" when rolled – implementation of the "inverse action" principle: make a moveable part of the object fixed or a fixed part moveable;

- local property: every part of the object must exist under conditions that best correspond to its function – the plate "by itself" takes the best position inside the bottle neck.

Since I first saw it, I have been presenting this useful thing as a souvenir with our logotypes and as an "award" to students who have done well at my seminars and workshops.

But here is a question: does this "ideal" artifact have drawbacks, and can it be improved? The answer is: yes. And my students do it during our classes – after we have formulated the contradiction, the IFR and the FIM.

Example 5.26. Dumbbells. Normally, dumbbells have round disk-like or ball-like weights and a handle. A unit-cast dumbbell is shown in fig. 5.38.

fig. 5.38. Unit-Cast Dumbbell

Such dumbbell can give you an unpleasant surprise when it rolls on your toes or falls off the shelf.

Should all dumbbells be so "round"? Or is this some vestigial archaism inherited from kettlebells?

Can we make dumbbells in accordance with FIM Variant 3:

> The operative zone ***by itself*** ensures the obtaining of the
> [*non-rolling dumbbell*].

Well, somebody must have had it with round dumbbells – and that somebody realized the above FIM with the related IFR by making ribbed dumbbells (fig. 5.39).

fig. 5.39. These dumbbells will not roll along the floor or off the shelf BY THEMSELVES!

Example 5.27. Missile Hanging in the Silo... by Itself! The first submarine-based missiles where rigidly suspended inside their silos. Because of that, the silos had large diameters, leading to serious launch problems – before the missiles could go off, the silos had to be filled with a large amount of water. Incidentally, the weight of that water exceeded the weight of the missile. In the event of a salvo launch, the submarine risked to be overloaded with outside water.

This problem became increasingly acute with missile diameters becoming larger and larger.

Maybe it is possible to make the silos so that the FIM and the IFR would correspond to the third variant: the operative zone **by itself** ensures the obtaining of the
[*minimal diameter and, accordingly, volume of the silo*].

The very wording of the IFR could have prompted a review of the construction of the silo as follows: minimal diameter of the silo = diameter of the missile! But we still need to install missile suspension elements! Or do we? Can the missile "hang" inside the silo BY ITSELF? Well, then we would have to build a virtually "ideal" silo!

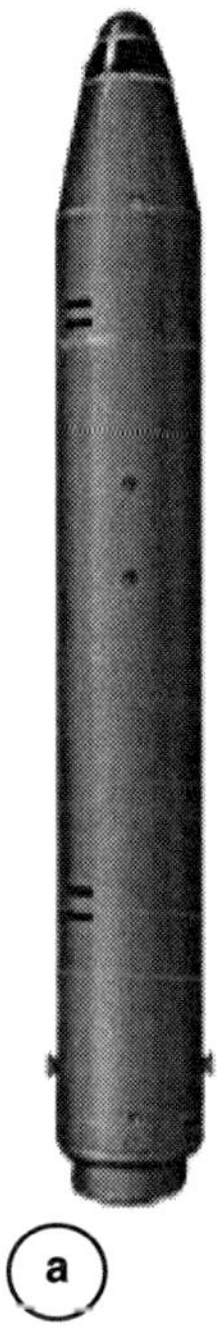

And that task was resolved in the early 1970-es for the two-stage missile R-29 (fig. 5.40,a) created under the guidance of the famous missile designer V. P. Makeyev (see [32]). For the first time, he suggested installation, on the walls of the silo, of reusable circular shock pads (fig. 5.40,b). The diameter of the silo was reduced to the minimally possible! The missile with a length of 13 m and diameter of 1.8 m hung with its lower end on the bottom of the silo. Gross launching mass – 33.3 tons, payload – 1,100 kg, warhead delivery range – 8,000 km.

Let me remind you that R-29 had the world's first stellar navigation system, while its successor R-29RM could boast the world's highest power/mass efficiency ratio! R-29RM characteristics: length – 14.8 m, diameter – 1.9 m, gross launching mass – 40.3 tons, payload – 2,800 kg, operative range – 8,300 km, impact error – less than 500 m!

Some ten years later, Makeyev design bureau created another missile – the three-stage R-39 "Sineva" ("Azure"): length - 16 m, diameter – 2.4 m, gross launching mass – 90 tons (!), payload – 2,550 kg, operative range – 8,000-1,000 km.

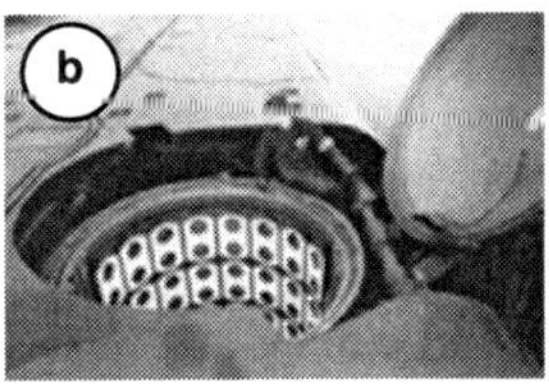

fig. 5.40. R-29

This missile was different in that it had a shock-absorber launch system where the missile was hanging inside the silo, resting with its jutting fungiform head (fig. 5.41) on the upper edge of the silo (shaded area in the drawing).

fig. 5.41. R-39

What do all these examples demonstrate? First and foremost, that all artifacts around us contain huge accumulated creative experience. These artifacts can be represented by ordinary items, such as lampshades, bottle openers or dumb-bells (what can possibly be simpler than a dumb-bell?!), or by extremely sophisticated high-tech items, such as aviation systems or submarine-based missiles. All these artifacts can yield simple (and not very simple) creative models – which can be turned into efficient instruments and used to solve new problems.

5.4 OZ Transformation Scheme

The OZ transformation process (fig. 5.42) has the following stages:

1) selection of the inductor and receptor within the OZ;

2) definition of the contradictions which prevent acquisition of required properties by the receptor, or reflect other problems existing in the OZ;

3) determination of resources, both those already existing in the OZ, and those that are required additionally;

4) formulation of the IFR and FIMs for the OZ;

5) selection of transformation models, subject to the following:

- transformations are usually applied to the inductor, its own resources and other resources available in, or introduced into, the OZ;

- realization of transformations is directed towards replacement of contradictions with the Ideal Final Result, or complete removal of contradictions.

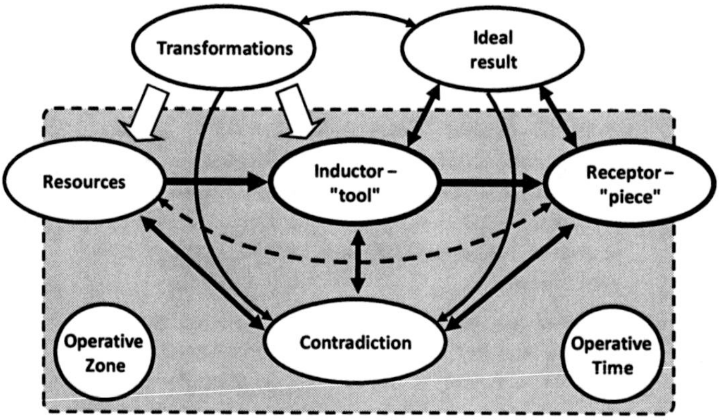

fig. 5.42. OZ transformation scheme

The method of selection of transformations is determined by the type of contradiction, or by the type of resource, or by the nature of IFR which also leads to selection of dominant resources most actively "involved" in the transformation. We have seen examples of such selections in previous sections. Other methods of selection and application of transformation models will be presented below.

At this juncture, we would like to draw the readers' attention to the very important role played by the selection of OZ actors and resources, definition of the Ideal Final Result and, naturally, selection of transformation models (fig. 5.43).

This complex multifaceted selection can generate solutions of very different levels. There are five levels of inventive solutions in TRIZ[108].

Clearly, Level III+ solutions can radically change the object[109]. This means that in strategic terms they will gravitate towards "Maxi-Solutions" which involve the change of the operative principle and, consequently, of the whole construction.

---

[108] Altshuller, G.S. (1986) *To Find an Idea. Introduction to Theory of Inventive Problem Solving.* – Novosibirsk, Nauka Publishers (in Russian; there are also later editions)

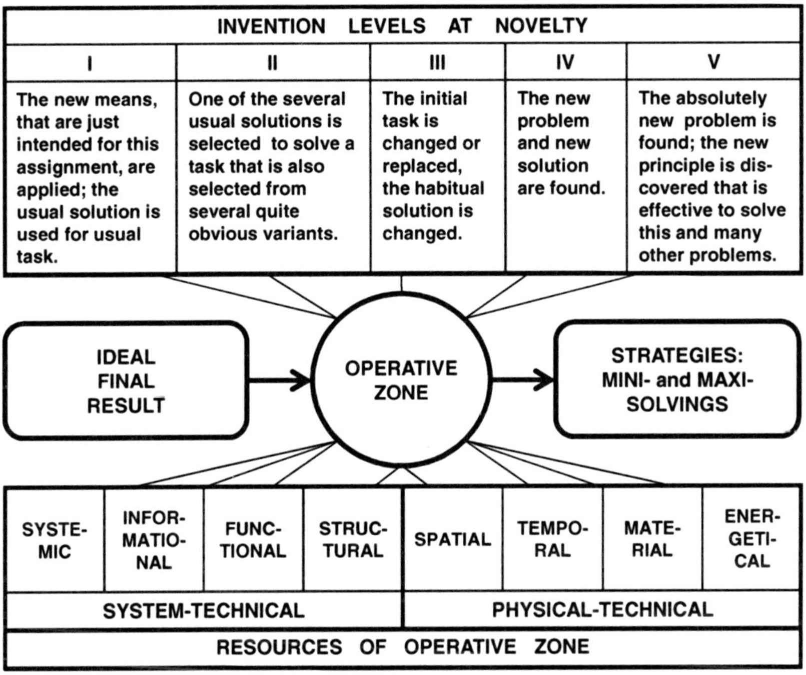

fig. 5.43. Connection between major systemic aspects of OZ transformation

However, many clients are interested in inexpensive improvements of existing technologies and products, and prefer the "Mini-Solution" strategy which tend to produce the required result with minimum changes to the system.

It should be noted that in many cases perception of the extent of "radicality" of changes depends on the overall systemic analysis of the solution and identification of new implementation options.

Thus, we saw that when solving the R-29 missile navigation problem (Example 5.16. See the IFR in the Stars!), we overcame an acute radical contradiction: to have and not to have (for various reasons) high precision of payload delivery. This provoked a modification of the flight control system and introduction of stellar navigation for the exo-atmospheric portion of the flight, i.e. implementation of a Level IV "Maxi-Solution"!

Solution of the problem dealing with installation of the R-39 missile in the silo (Example 5.27. Missile Hanging in the Silo… by Itself!) using the force of gravity and the edge of the silo, despite the apparent "unusualness" of its implementation in the then existing submarine construction and the breathtaking emotions produced by its technical sophistication, can be classified (in terms of inventive novelty and complexity) as a Level II "Mini-Solution".

[109] M. Orloff (2006) *Inventive Thinking through TRIZ: A Practical Guide*", parts *"3.2. Levels of Inventions"* and *"14.1. Development of Systems"*

6 Reinventing: TRIZ-Modeling the Invention Process

Extracting and Reinventing, based on the Meta-Algorithm of Inventing, is a modern "TRIZ-tomograph" of inventive thinking.

It works like a "time machine": reinventing allows us to explore creative thinking of any inventor from any time and era!!

Ruwim Kisselman,

PhD in engineering, inventor, head of the Inventors Club[110] *"Schöpfer" ("The Creators" – Germ.) from Bonn, Germany, organizer of the yearly seminars of the Modern TRIZ Academy in the German Museum in Bonn*[111]

6.1 Reinventing

Let us assume that we are studying a certain artifact – the product (result) of an invention. This product-artifact is in the state "is", i.e. already exists (fig. 6.1). It is matched to a certain prototype-artifact – the predecessor of the product-artifact in time. Both artifacts are vessels containing the experiences acquired in the course of their creation: technical, creative, psychological (motivational, emotional).

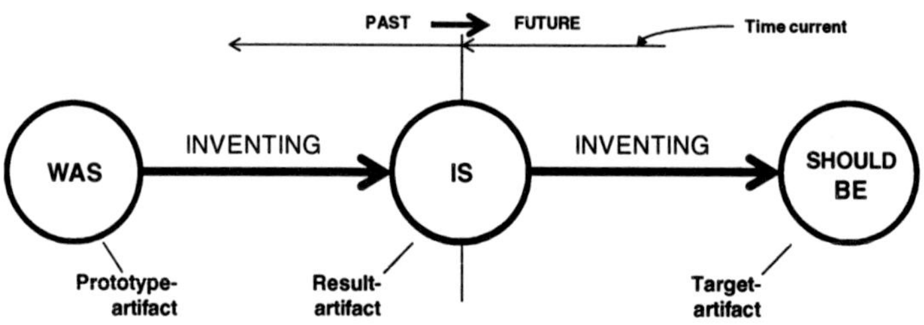

fig. 6.1. Inventions in the Course of Time

The question is this: *How do you learn to invent?*

How to you create a new target artifact? Can you use previous inventing experiences? If yes, how do you do that?

The answers to these questions are provided by the technology of reinventing developed by the author.

We need to organize – and then use on an ongoing basis – the "training–application" cycle shown in fig. 6.2. In that cycle, each new invention, together with previously created artifacts, becomes the object of analysis and the material for subsequent teaching.

---

110 Association of Inventor Clubs created in Germany earlier within the framework of the INSTI Program (*Innovationsstimulierung* – State Program for the Support of Innovative Movement in Germany; see www.insti.de and www.erfinderclubs.de); the program's successor was SIGNO – Schutz von Ideen für die gewerbliche Nutzung (Protection of Ideas for Industrial Utilization)

111 www.deutsches-museum.de

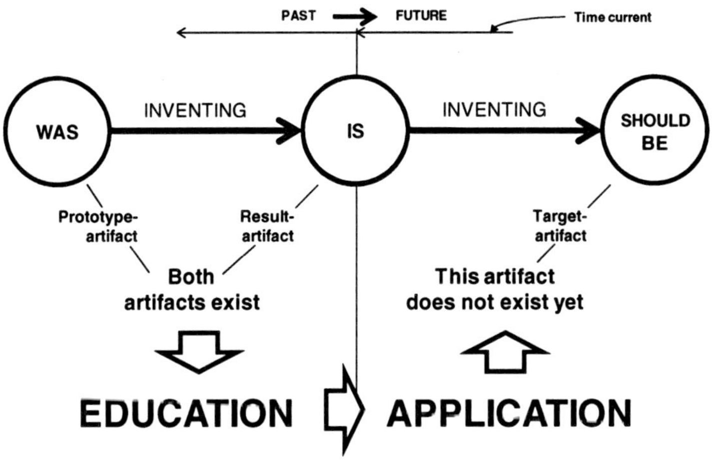

fig. 6.2. Training–Application Cycle in the Course of Time

This is where we will need the Meta-Algorithm of Invention (MAI) T-R-I-Z discussed in section *3.1 Algorithms of Inventive Creativity.*

Reinventing is the repetition (modeling, reproduction) of the process of known problems on the basis of the format and the schema of MAI by using the methods and models of TRIZ.

Reinventing – fundamental teaching method in Modern TRIZ.

When reinventing beginners are studying problems with known solutions which are illustrated in the MAI format in order to understand how these problems would have been solved on the basis of TRIZ and how in the future similar problems could be solved by using both, MAI as well as TRIZ.

Definition "Reinventing"	**Reinventing** – *modeling (reconstruction, reproduction, renewal) of the invention process.*
Addition 1	**TRIZ-Reinventing** – *modeling of the invention process* on the basis of TRIZ models.
Addition 2	The initial reinventing has the first goal to allow students to *quickly and correctly acquire* the algorithm of inventive problems solving in the format of MAI T-R-I-Z.
Addition 3	The second and supreme mission of reinventing is to reliably prepare students for an autonomous working on *any new practical problem.*

MAI is only a general frame and a general navigator for inventive problem solving and it becomes a practical instrument only in connection with models and methods of TRIZ which equip the stages of MAI and thereby turn it into a certain variant of ARIZ – Algorithm of Inventive Problem Solving.

The content of reinventing is the reconstruction and description of all stages of an invention's creation with involvement of models and recommendations of TRIZ for each stage.

The complete reinventing process, as presented in its aggregated form in fig. 6.3 and 6.5, consists of two stages: the preliminary stage – *extracting*, and the main stage – *reinventing* on the basis of MAI T-R-I-Z.

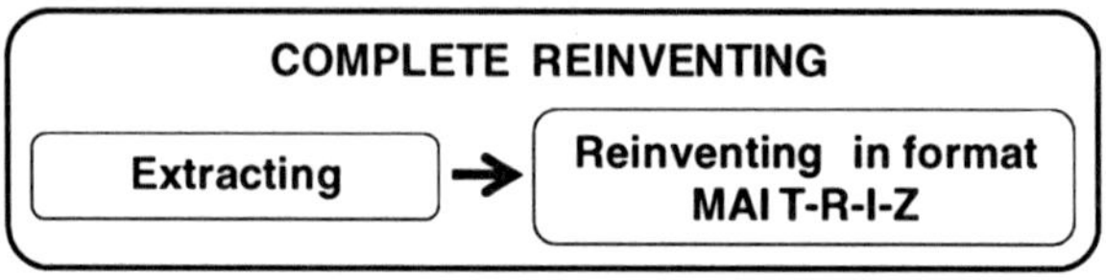

fig. 6.3. Illustration to the Definition of Reinventing

We already know that when we study a product-artifact of interest we, first and foremost, select a matching prototype-artifact, and then perform two extracting procedures:

1) Extracting-1 – to identify the transformation models objectively realized in the product-artifact vis-à-vis the prototype-artifact;

2) Extracting-2 – to identify the contradictions objectively eliminated in the product-artifact vis-à-vis the prototype-artifact. In other words, we are talking about contradictions that were PRESENT in the prototype-artifact, but are ABSENT in the product-artifact.

We need to learn to model, on the basis of MAI T-R-I-Z, the process of creating any known product-artifacts, and then use that learning to create new inventions (fig. 6.4).

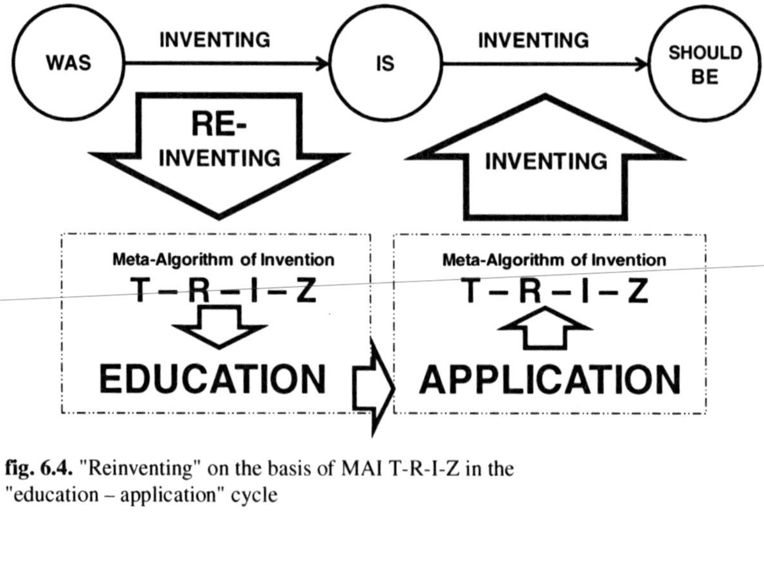

fig. 6.4. "Reinventing" on the basis of MAI T-R-I-Z in the "education – application" cycle

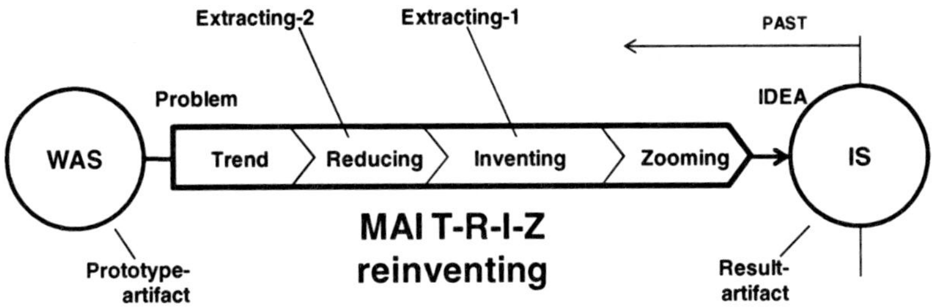

fig. 6.5. Extracting procedures for Reinventing

Example 6.1. Over-the-Table Exhaust Fan.

In South Korea there are many extremely popular restaurants where you cook yourselves using various "materials" ordered from your waiter (fig. 6.6, a). For example, you can grill meat on hot coals (fig. 6.6, b) right on your table! To remove smoke, each table is equipped with an overhanging exhaust pipe (fig. 6.7, a) which can be easily lowered down to the meat grill.

The switch and the draft control are mounted on the pipe (fig. 6.7, b, in the black oval).

fig. 6.6. The Korean "Self-Servicing" Way

fig. 6.7. An Impressive Exhaust Ventilation System!

Now, that's an interesting invention! Let us "invent" it again!

Preliminary Modeling: Extracting.

Extracting-1 yields the following models (fig. 6.8):

+	**02**	Preliminary action	The pipe intake can be fixed at any height above the table.
++	**05**	Separation	Smoke is removed from the grill zone.
+	**07**	Dynamization	The pipe can be moved up and down using a special cable (preinstalled feature). The pipe is corrugated in the middle, and when you fix it above the table its length can vary quite significantly.
++	**12**	Local property	The pipe can be fixed at the required level when in operation, and raised high above the table when idle. Switches and regulators are placed on the pipe rather conveniently.
++	**14**	Use of pneumatic and hydraulic constructions	The exhaust fan is working as a pneumatic pump extracting the air from above the grill.
++	**18**	Mediator	The pipe (and the entire system) act as the mediator between the table and the space outside the restaurant.
+	**19**	Transition into another dimension	The pipe is hanging from above – saving space.
+	**35**	Unite	All pipes are joined to larger-diameter pipes which are also connected into a single system.

fig. 6.8. Extracting-1 Results for the "Over-the-Table Exhaust Fan" Artifact

As a result of Extracting-2, let us write down one standard contradiction and one radical contradiction (fig. 6.9).

<table>
<tr><td>++</td><td>02</td><td>Preliminary action</td><td rowspan="2">Standard Contradiction: It would be nice to grill meat right on the table, but then the restaurant would be filled with smoke.</td></tr>
<tr><td>++</td><td>05</td><td>Separation</td></tr>
<tr><td>++</td><td>07</td><td>Dynamization</td><td rowspan="3">Radical Contradiction: The smoke must be "far" from the table (outside of the restaurant) to permit normal breathing, and the smoke must be "close" to the table (inside the restaurant) as it takes a long time to be cleared by natural ventilation.</td></tr>
<tr><td>++</td><td>12</td><td>Local property</td></tr>
<tr><td>++</td><td>18</td><td>Mediator</td></tr>
</table>

fig. 6.9. Extracting-2 for the "Over-the-Table Exhaust Fan" Artifact

Main Modeling: Reinventing.

Let us provide a brief description of one of the MAI T-R-I-Z reinventing variants (fig. 6.10).

TREND

If restaurant visitors grill meat at each table, the restaurant will be filled with smoke! But still, grilled meat is SO yummy! So what do we do?

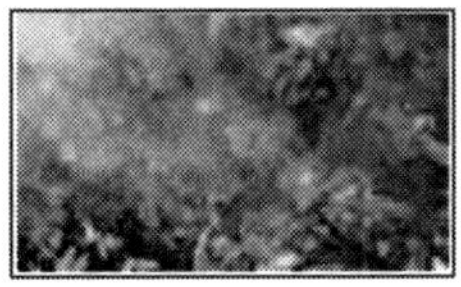

REDUCING

FIM: X-resource, together with available or modified resources, and without making the object more complex or introducing any negative properties, guarantees attainment of the following **IFR**:

[*absence of smoke in the restaurant when meat is grilled at each table*].

Standard (Technical) Contradiction

Grill → (+) *fire at each table in the restaurant* → Factors: *34 Temperature*

Grill → (-) *damaging factors – smoke, carbon monoxide* → Factors: *14 Internal damaging factors*

Navigators

01 Change in the aggregate state
05 Separation ✓
18 Mediator ✓
21 Transform damage into use

07 Dynamization ✓
12 Local property ✓

Radical (Physical) Contradiction

Grill ⇒ *must be on each table so that visitors can grill meat on hot coals* & *must not be on each table, as the restaurant will be filled with smoke*

INVENTING

***Key models – 05 Separation and 18 Mediator**: the smoke produced by the grill is removed from the restaurant with the help of an exhaust pipe.*

***Important ideas:** the smoke is removed by a moveable pipe (07 Dynamization) from right above the grill (12 Local property).*

Model 02 Preliminary action is also present: the exhaust pipe is switched on before the grill is installed. The solution also uses Model 14 Use of pneumatic constructions: the pipe is fitted with an exhaust fan.

ZOOMING

Have the contradictions been removed? – Yes. – ~~No~~.

Super-effects: *none*

Negative effects: *none*

Figure

BRIEF DESCRIPTION

A moveable exhaust pipe is installed above each table. All pipes are joined into a single exhaust ventilation system.
Dominating models: 05 Separation and 18 Mediator.
Additional models: 02 Preliminary action, 14 Use of pneumatic constructions, as well as 19 Transition into another dimension and 35 Unite.

fig. 6.10. MAI T-R-I-Z Reinventing for the "Over-the-Table Exhaust Fan" Artifact

Here we have added descriptions for the Trend and Zooming stages, and provided more detailed description of the Reducing and Inventing stages. At the Reducing stage we have illustrated formal contradiction modeling and selection of transformation models recommended by the A-matrix – or introduced by extracting (if the A-matrix offers no models to resolve the relevant formal contradiction). At the Inventing stage we have shown application (interpretation) of transformation models taken from the A_S-catalog to resolve the problem under study.

NOTE An experienced TRIZ-modeler can perform reinventing for relatively simple artifacts even easier and faster by completing the extracting exercise in the course of main reinventing. In any case, reinventing is based on extracting applied transformations from the product-artifact, and extracting contradictions from the prototype-artifact.

Example 6.2. Carousel Table.

In China tables (fig. 6.11) are often fitted with carousel-like additional "tabletops" – the may be in the form of a ring (a) or a solid disc (b, c) made of wood (a) or glass (b, c), large (a, c) or relatively small (b). On this "second floor", they place dishes with food, and anyone can turn the disc and easily take whatever they need. The disc may also hold a flower display (c) which can then be gradually examined from all sides.

fig. 6.11. Several Variants of the Chinese "Carousel Table".

When sitting at such table (well, at any other table, for that matter!), all you need to do is be attentive and polite – and:

1) before starting to rotate the disc make sure that no-one else is doing it at that moment in time – or loading food off the table;

2) considerately help others to rotate the disc which often displays noticeable inertia due to the quantity and weight of the dishes.

This reinventing looks like it was made without extracting. But take a closer look at the description (fig. 6.12), and you will see that extracting is always present in any reinventing – whether directly or indirectly (like in this example) – and that no reinventing is in fact possible without extracting!

If now you perform the extracting by yourselves, you will easily find several other useful models in this artifact.

TREND

Have you ever wanted to taste some exquisite dish on the other end of a huge table? Can we somehow remedy this familiar – and extremely uncomfortable – situation?

REDUCING

FIM: X-resource, together with available or modified resources, and without making the object more complex or introducing any negative properties, guarantees attainment of the following **IFR:**

[*easy access to dishes from any place at the table*].

Standard (Technical) Contradiction

		Factors	Navigators
Table	(+) *difficult to reach dishes – improvement required*	*10 Ease of use*	*06 Use of mechanical oscillations*
			07 Dynamization ✓✓
	(-) *large*	*18 Surface of fixed object*	*16 Partial or excess effect*
			23 Use of inert media

Radical (Physical) Contradiction

Table ⇨ *must be large to seat many guests* & *must not be large, as it becomes difficult to reach dishes*

INVENTING

IDEA: *fit the table with a rotating "carousel" which will carry the dishes **(07 Dynamization)**.*

ZOOMING

Have the contradictions been removed? – Yes. – ~~No~~.

Super-effects: *beautiful!*

Negative effects: *table construction becomes more complex.*

BRIEF DESCRIPTION

The table is fitted with a "carousel" to make any dish easily accessible to any guest. Dominating model: 07 Dynamization.

fig. 6.12. "Carousel Table" reinventing

6.2 Prototype-artifact and Product-artifact

So, let us assume that we are studying a certain artifact – the fruit (product) of an invention.

> **ATTENTION: Critically Important Methodological Recommendations**
>
> When doing extracting and reinventing, it is absolutely necessary to remember the following (fig. 6.13):
>
> 1) transformation models are formulated for the **Product-artifact** in comparison with the prototype-artifact;
>
> 2) contradictions are formulated for the **Prototype-artifact** – in the product-artifact, the initial contradictions are already removed;
>
> 3) MAI T-R-I-Z stages Trend and Reducing are "linked" to the prototype, while stages Inventing and Zooming are "linked" to the product.

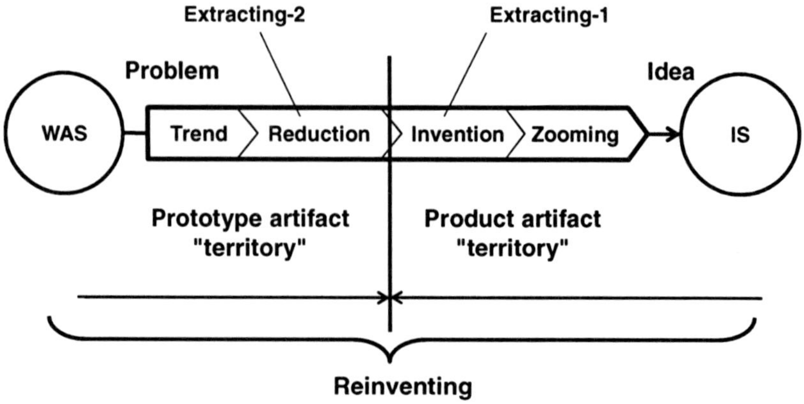

fig. 6.13. "Territories" of Correct (!) Formulation of Extracting and Reinventing Models

Additional Methodological Recommendations[112] : Reinventing Meta-Algorithm

Attentive readers must have already noticed that during the reinventing exercise artifact analysis proceeds strictly from the PRODUCT to the PROTOTYPE in accordance with the following "algorithm":

Meta-Algorithm of Reinventing Z-I-R-T

Stage 1: Review the properties and construction of the product-artifact under analysis, and compare them to the properties and construction of the selected prototype-artifact – this corresponds to the *Zooming* stage in "direct" MAI T-R-I-Z;

Stage 2: Extract transformation models objectively participating and realized in the product-artifact – this corresponds to the *Inventing* stage in "direct" MAI T-R-I-Z;

Stage 3: Extract contradictions – this corresponds to the *Reducing* stage in "direct" MAI T-R-I-Z;

Stage 4: Describe the original problem situation for the prototype-artifact – this corresponds to the *Trend* stage in "direct" MAI T-R-I-Z.

In fact, all reinventing stages are performed in reverse order compared to MAI T-R-I-Z; accordingly, we can define and call the Meta-Algorithm of Reinventing either "MARI T-R-I-Z" or "MAI Z-I-R-T".

---

[112] This material is provided for illustrative purposes only, and is not used in MTRIZ Practitioner certification.

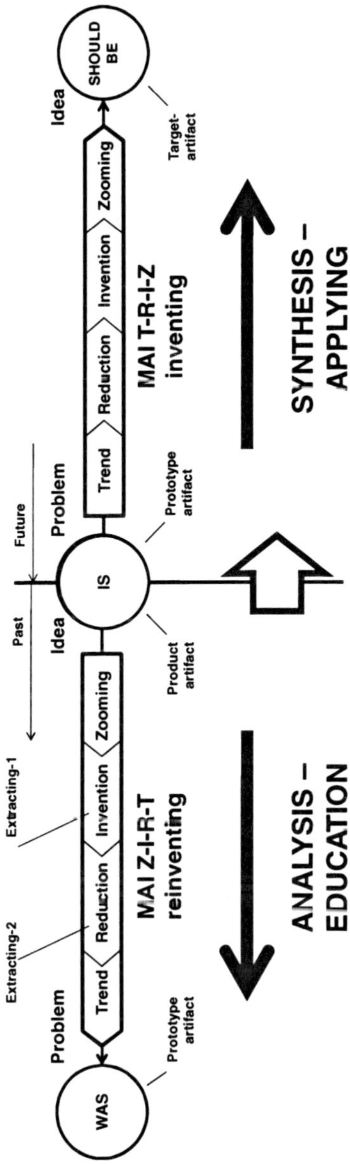

The opposition of MAI T-R-I-Z and MARI T-R-I-Z can be schematically presented as follows (fig. 6.14).

Here it is also important to note that any object that we consider a product-artifact for reinventing purposes becomes the prototype-artifact as soon as we decide to enhance it, invent a target prototype.

And in conclusion of this section we would like to share the formula, the logotype scheme which integrates in a single cycle the standard MARI T-R-I-Z reinventing-teaching and the standard MAI T-R-I-Z application-invention for creating new ideas (fig. 6.15).

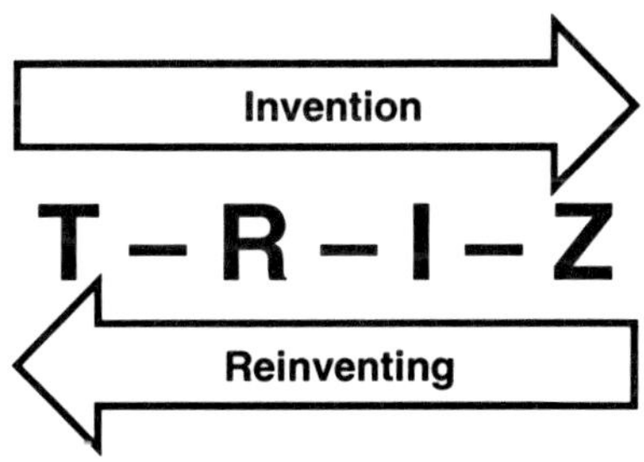

fig. 6.15. Schematic Formula of Cyclical Integration of MAI T-R-I-Z and MAI Z-I-R-T

This scheme clearly shows that in Modern TRIZ any artifacts and inventions can become the objects of study and analysis.

fig. 6.14. "Direct" MAI T-R-I-Z: Invention – Synthesis-Application; "Reverse" MAI Z-I-R-T: Reinventing – Analysis-Training

6.3 Invention Algorithm START[113]

In this textbook and course – which are basically designed for beginners – we use the start version of MAI T-R-I-Z which is restricted to the use of primary universal TRIZ models to resolve contradictions. The name **START** is an acronym (**S**implest **T**RIZ-**A**lgorithm of **R**esourceful **T**hinking, see fig. 6.16) part-timing as a mnemonic aid.

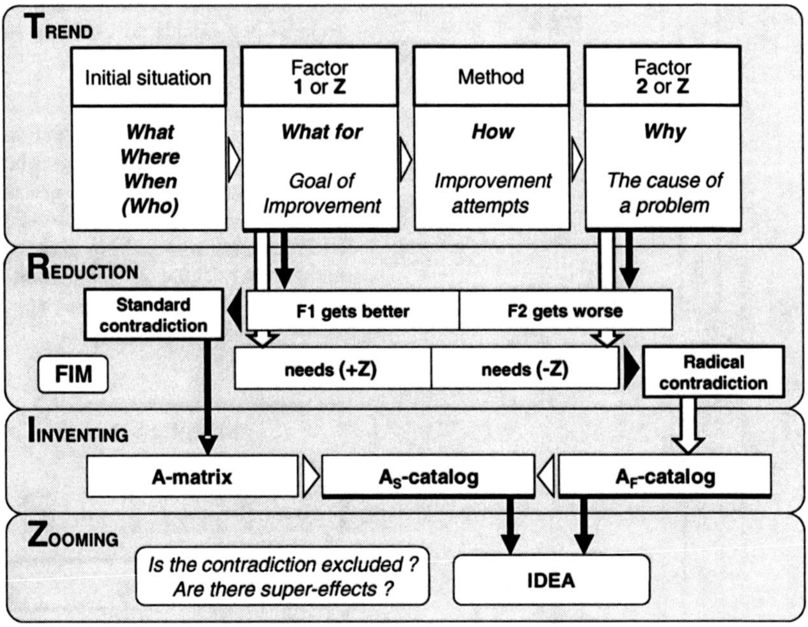

fig. 6.16. **S**implest **T**RIZ-**A**lgorithm of **R**esourceful **T**hinking (**START**)

So, START!

The main difference between START and MAI T-R-I-Z lies in the fact that each START stage is filled with specific procedures enabling development of solutions based on TRIZ models. This transforms the "meta-algorithm" (i.e. the maximally integrated, aggregated and, accordingly, abstract "path") into an executable practical algorithm which directs, organizes and disciplines the work performed to create new solutions.

113 **ATTENTION: this scheme must be committed to memory** – it is a practical working algorithm enabling resolution of MTRIZ Practitioner-level problems, and a fundamental component of MAI T-R-I-Z used at more advanced levels, such as MTRIZ Expert and MTRIZ Master; this scheme is completely integrated into the EASyTRIZ™ software package; when working without a computer, we use a slightly simplified but similarly structured scheme.

All material included into this book – from A to Z – is built on this scheme. In subsequent sections, START procedures will be reviewed in painstaking detail and, naturally, illustrated with examples.

The general comment to the scheme is as follows.

TREND

Resolution of any problem starts with an analysis of the initial situation or, if you will, with diagnosing the problem. Accordingly, in the extended Modern TRIZ version this stage is also called "Diagnostics".

The analysis naturally involves determination of "what", "where" and "when" occurs in the system so that the quality of outcomes produced by such system and, accordingly, the efficiency of its operation and makeup, become insufficient.

Answers to these questions[114] must produce a sufficiently accurate and concise description of the initial problem situation.

In any case, we then need to determine the purpose of making the situation better, the aim of improving or enhancing the system. This is done by asking "what for" (i.e. for what purpose) we need to change the existing system. The answer to that question contains, whether directly or indirectly, an indication – or at least a supposition – as to the direction in which we need the system to change.

To improve the system, we might attempt to effect changes using known methods. Essentially, these are attempts to answer the question: "How can we improve the system?"

Clearly, if none of these attempts works, we have an "inventive problem". This means that we have an obstacle which cannot be overcome with known methods and, consequently, requires an unobvious inventive idea. And this idea is what we call an "invention" – regardless of whether it is large or small, and whether it is worthy of a patent or simply enables elimination of some deficiency in the production process.

The answer to the question "why" helps to define the reasons preventing the required improvement of the system's operation.

It is the opposition, incompatibility or, sometimes, outright mutual resistance of the properties inherent in the "aim of improving" (and of these "reasons") that leads to the need to define the initial situation in terms of "contradictions" and "problems".

Thus, at the Trend stage we undertake a general review of the problem situation, informally define one or several contradictions, and determine the direction (trend) of the desired improvement, or evolution, of the system.

REDUCING

Be that as it may, the *concentrated answer* to the question "what for" – as posed at the Trend stage – is formulated in the form of the so-called Ideal Final

---

[114] The use of leading, or "controlling", questions was suggested by Quintilian, Roman rhetorician, in the first century AD – see "Inventive Thinking through TRIZ" by M.Orloff (2006), Section 4.1 "Inventing the Theories of Inventing"

Result (IFR). It is this IFR that substantially shapes the trend – purpose and pattern – of the future system transformation.

The IFR may be recorded separately, or incorporated into the formulation of the so-called Functional Ideal Model (FIM). Implementation of the FIM is aimed (trend!) at the attainment of the IFR.

Methodologically, the FIM and the IFR are fixed at the Reducing stage as teleological benchmarks for all subsequent modeling and generation of ideas, with the FIM being reduced to certain standard form.

However, the key task at this stage is to reduce the informal contradiction to some standard format.

For starters, let us recall that all contradictions are divided into two types (have a closer look at sections *2.1.3* and *2.1.4*):

1) Standard Contradiction: each attempt to improve one property of the system results in deterioration of another property of the same system;

2) Radical Contradiction: diametrically opposed requirements are imposed on the same property. This conflict is represented in the form of two opposed requirements as they apply to the same property **Z**: for one purpose, **Z** needs to be increased (trend **+Z**), for the other purpose, **Z** needs to be decreased (trend **-Z**).

Both formal contradiction models have their "own" means and "paths" of resolution.

The "path" to resolution of a standard contradiction is discussed in section *7.1 Resolution of Standard Contradictions*, while the "path" to resolution of a radical contradiction is presented in section *7.2 Resolution of Radical Contradictions.*

INVENTING

At this stage, generation of ideas is supported with transformation models borrowed from catalogs created in TRIZ. In Modern TRIZ, this stage is also called "Transformation".

To resolve standard contradictions, we use *specialized* transformation models from the so-called As-catalog (section *14.2.2 As-catalog*), while A-matrix helps to select suitable models (see section *14.1.2 A-matrix table*).

To resolve radical contradictions, we use *fundamental* transformation models selected from section *14.2.3 Af-catalog* and *14.2.4 Afs-catalog*, or from the advanced Af-catalogs in sections *14.2.5* and *14.2.6.*

At this juncture, you might want to review the scheme shown in fig. 3.12 (section *3.2. Noospheric Modeling Concept in Modern TRIZ*). According to that scheme, generation of ideas relies on the following knowledge and skills:

1) **professional knowledge** from the subject area, and skills required to resolve practical tasks related to system design and production, manage implementation and explore evolution of systems in certain areas;

2) **creative knowledge and skills** based, first and foremost, on your creative abilities and – equally important! – on your knowledge of TRIZ foundations and your skills in the application of TRIZ methods and tools;

3) **psychological knowledge and skills** – understanding of cognitive models underlying resolution of complex (and, alas, not so complex!) problems, and skills which can be used to overcome negative psychological stereotypes which hinder problem resolution, and to stimulate positive attitudes which encourage successful problem resolution and generation of efficient, and even beautiful, ideas.

On the whole, we believe that creation of an efficient idea is the fruit of fusing talent with knowledge and creative thinking skills.

ZOOMING

In Modern TRIZ, this stage is also called "Verification", which points to the forthcoming examination of the quality of newly discovered ideas. These ideas need to be "verified", checked for implementability and efficiency.

The use of the metaphorical name "Zooming" is determined by the nature of the main verification procedures, namely, by the need to verify the ideas at several levels of examination. Thus, it is necessary to assess the quality of the solution at the level of the operating zone, i.e. the area which originally hosted the cause of the problem. Then we need to "zoom up" and assess the quality of the solution, say, at the level of the entire system. Then we might want to take an even broader look at the level encompassing the interests of the user for whom the artifact (system) was created in the first place. Last but not least, in many cases it is possible (and often necessary) to measure the impact of the idea at an even higher and more general level, for example at the level of the society, or even at the level of the entire Nature.

These transitions are similar to changing the focus – *zooming!* – in a photo or video camera to change the "field of vision", the amount of space that we see through the lens. The metaphorical (but also functional) name of the fourth, final stage of MAI T-R-I-Z is originated from this stage

Example 6.3. Gabion.

Bank strengthening is a complex technical problem. One way to solve it would be to fix the bank with stones. However, the water gradually washes soil and sand from under the stones, and the manmade shoreline breaks down. How do we protect the shoreline from exposure to water – flows, waves, tides, rains?

Let us try to solve this problem with the help of TRIZ, MAI T-R-I-Z in the START format and, naturally, certain knowledge about… shore protection technologies.

Yes, of course, the author knew the answer, and somebody else, not the author, invented this solution. But it *will* be a reinventing exercise. The author will not demonstrate the solution in advance, like he did in the previous sections to illustrate the extracting and reinventing techniques. So, let us do some START T-R-I-Z inventing!

To do that, we need to review some START procedure details.

The Trend Stage

What: shoreline section. **Where:** say, a river flowing through a city park. **When:** on a permanent basis.

What for: to strengthen the shoreline section and protect it from water erosion and caving.

How: stone masonry with interstitial voids filled with sand and soil. The stones are, to the extent possible, laid as compactly as possible. Result: not always efficient.

Why: the water washes out one stone after the other until the entire construction collapses.

It is easy to see that the standard computerless form (fig. 6.16) lacks the leading ("controlling") questions, while the description of the original problem situation is quite extensive.

In most cases, this is enough to amply present the problem and hint at the possible solution. It is just that "controlling" questions help to emphasize important points while providing the information minimally (absolutely) necessary to describe the original problem situation.

The Reducing Stage.

Let us formulate only the **IFR**: lasting reliable stone structure, or reliable protection of the bank with stones.

FIM: we need to introduce some resource which is unknown (at least for the time being) – an X-resource – most probably, in the form of a new structure or material required to secure the **IFR**.

The Inventing Stage.

A solution based on Separation, Mediator and Matryoshka (nested doll) models quite logically follows from the recommendations for this set! This stage is described in more detail in sections 7 and 8. However, attentive readers must have already formed an idea about how ideas can be modeled and generated both at this stage and at previous stages.

The Zooming Stage.

The initial contradictions have been eliminated! If we look at what this construction can do from a broader perspective (change the scale, or zoom up!), we will notice a "powerful" positive super-effect: wire mesh can be shaped into arbitrary architectural and engineering forms, for example, steps, terraces, slopes, ramps, foundations, walls, etc.

TREND

To prevent water erosion of river banks, they can be strengthened with stones. However, water gradually washes sand and soil from under the stones, and the shoreline collapses. What can we do?

REDUCING

FIM: X-resource, together with available or modified resources, and without making the object more complex or introducing any negative properties, guarantees attainment of the following **IFR**:
[*reliable strengthening of the bank with stones*].

Standard (Technical) Contradiction

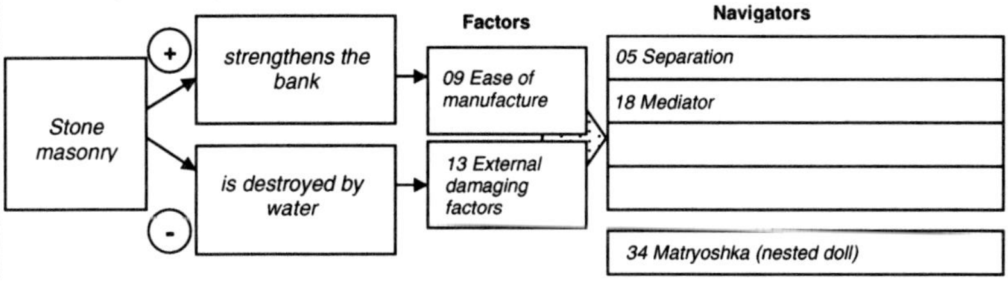

Radical (Physical) Contradiction

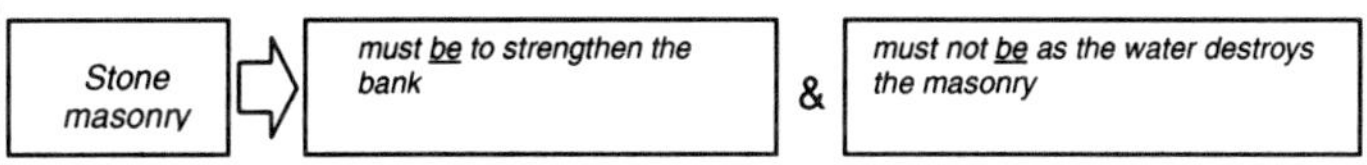

INVENTING

Key model – 18 Mediator*: the stones are fixed with durable mesh wire. Moreover, the stones can be placed inside prefabricated sections similar to mesh wire baskets – thereby giving the construction the required shape and size.*

Additional navigators *– 04 Replacement of mechanical matter: the stone masonry is given a certain structure; 13 Inexpensive short-life object: the wire is gradually eroded, but before that happens the shoreline is securely fixed with trees and bushes; 25 Use of flexible covers (mesh wire); 34 Matryoshka (nested doll): the stones are "packed" into mesh wire baskets; 35 Unite: the stones are joined together.*

ZOOMING

Have the contradictions been removed? – Yes. – ~~No~~.

Super-effects: *sustainable construction is gradually covered with soil – and later with grass and other plants – which makes the bank stronger and more pleasing to the eye.*

Negative effects: *none*

Figure

BRIEF DESCRIPTION

According to navigator 18 Mediator, stones are placed into prefabricated sections of required shape and size made of durable mesh wire.

NOTE: such constructions in the form of thick or flat baskets, called "gabions", have been known since the Middle Ages.

Military fortifications made of wickerwork cylinders with open tops and bottoms filled with soil and stones were used to protect cannoneers and gunmen.

It is believed that gabions were invented by Leonardo da Vinci to build the foundation of the San Marco Castle in Milan, Italy (http://gabiondesign.be/gabion.html)

fig. 6.17. Gabion

7 To Defeat a Contradiction: Generation of Efficient Ideas

Both the algorithm of inventive problem-solving and the standards* are based on identification and elimination of the contradiction – the main factor restraining further development of the system.[115]

Genrikh Altshuller

7.1 Resolution of Standard Contradictions: method[116] BICO – Binary In Cluster Out

And now let us consider the first "classical" method that is used to resolve standard contradictions with the help of a simple "solution prompter" – the *"technique selection matrix"*, *Altshuller matrix*, or *A-matrix* (terms introduced by the author).

Neither that matrix[117], nor the catalog of specialized transformation models (A_S-catalog[118]), is a cure-all solution. However, the matrix helps to understand the mechanisms underpinning resolution of standard contradictions for many, frequently quite intricate, situations and acquire initial practical skills in generating ideas on the basis of the simplest models so as to advance, in the foreseeable future, to more effective TRIZ tools.

Besides, both the A-matrix and the A_S-catalog are "work-in-progress" and, clearly, after you get to know these two elementary tools it will be easier for you to apply more sophisticated and efficient tools of this type.

According to the START algorithm (fig. 7.1), subsequent steps that must be taken to generate ideas involve replacement of the informal standard contradiction, as formulated at the *Trend* stage, with its standard equivalent enabling the use of the A-matrix (fig. 7.3). In fig. 7.1, this operation is shown by two arrows leading from the *Trend* stage to the *Reducing* stage.

In practical terms, this transition is made as follows:

We consider that the *plus-factor* (some line of the A-matrix) is the target, as it corresponds to the main useful function of the object and *must be improved* (strengthened, increased, etc., or in some cases decreased – see the left column in fig. 7.2).

The *minus-factor* (some column of the A-matrix) does not correspond to the main useful function of the object and *must be corrected* (neutralized, weakened, decreased, etc., or in some cases increased – see the right column in fig. 7.2).

---

* here "standard" – kind of transformation models by G. Altshuller

115 From: G.S. Altshuller, I.M. Vertkin (1994) *How to Become a Genius. Life Strategy of a Creative Personality*. – Minsk, Belarus (in Russian)

116 BICO – Binary In Cluster Out

117 Section *14.1.2 A-matrix table*

118 Section *14.2.2 As-catalog*

We know that the A-matrix has 39 inputs. Accordingly, it uses a total of 39 *formal* factors (fig. 7.2). This means that informal plus-factors and minus-factors need to be replaced with their formal counterparts.

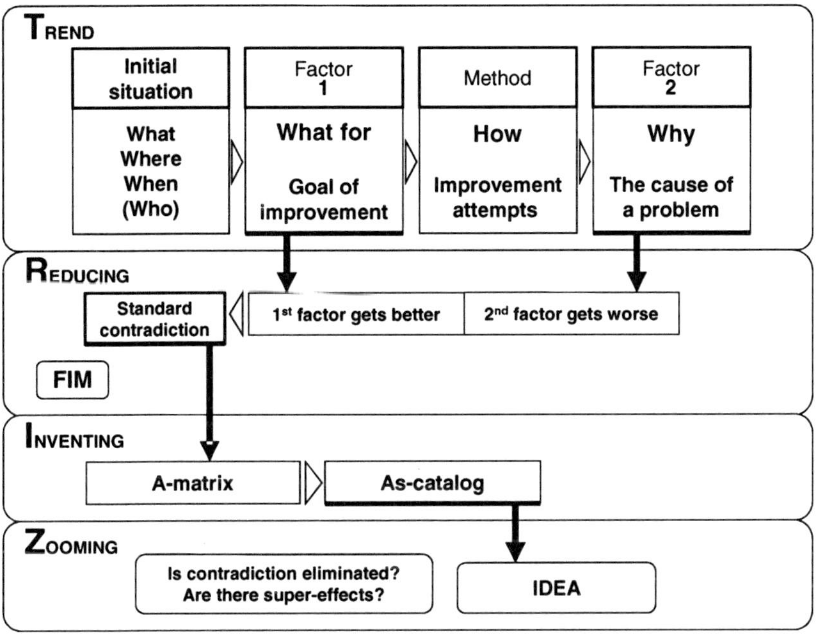

fig. 7.1. BICO method realization route in START T-R-I-Z

To do that, it is necessary to match each informal factor from the previously formulated informal (initial) contradiction with one of the 39 formal factors (section *14.1.1 Plus- and Minus-factors*). Situations where multiple interpretations are possible are considered at higher EASyTRIZ levels.

Selection of formal factors results in formulation of a *formal standard contradiction*. Plus- and minus-factors in this model are represented by numbered plus- and minus-factors of the A-matrix. Another set of terms used in MTRIZ takes into consideration the evolutionary effect of change and, accordingly, plus-factors are also referred to as *trend-factors* (as they correspond to the objective towards which the object is evolving), while minus-factors can sometimes be called *problem-factors* (as they reflect the obstacles hindering evolution of the object).

Inasmuch as the same expressions can be used to denote both plus- and minus-factors, beginners may find it difficult to correctly interpret and select the most suitable factors. To simplify interpretation pending accumulation of sufficient experience and understanding, you can use the table below (see fig. 7.2).

INTERPRETATION OF PLUS-FACTOR, OR TREND-FACTOR	#	FACTOR	#	INTERPRETATION OF MINUS-FACTOR, OR PROBLEM-FACTOR
increase required	01	**productivity**	01	decrease is bad
expansion required	02	**universality, adaptability**	02	narrowing is bad
increase usually required	03	**level of automation**	03	absence is bad, implementation difficulties are bad
increase required	04	**reliability**	04	decrease is bad
increase required	05	**precision of manufacture**	05	decrease is bad
increase required	06	**precision of measurements**	06	decrease is bad
decrease required	07	**complexity of construction**	07	increase is bad
decrease required	08	**complexity of inspection and measurements**	08	increase is bad
increase required	09	**ease of manufacture**	09	decrease is bad
increase required	10	**ease of use**	10	decrease is bad
increase required	11	**ease of repair**	11	decrease is bad
decrease required	12	**loss of information**	12	increase is bad
decrease required	13	**external damaging factors**	13	increase is bad
decrease required	14	**internal damaging factors**	14	increase is bad
factor being improved related to length	15	**length of the moveable object**	15	factor being corrected related to length
factor being improved related to length	16	**length of the fixed object**	16	factor being corrected related to length
factor being improved related to surface	17	**surface of the moveable object**	17	factor being corrected related to surface
factor being improved related to surface	18	**surface of the fixed object**	18	factor being corrected related to surface
factor being improved related to volume	19	**volume of the moveable object**	19	factor being corrected related to volume
factor being improved related to volume	20	**volume of the fixed object**	20	factor being corrected related to volume
factor being improved related to shape	21	**shape**	21	factor being corrected related to shape
increase usually required	22	**speed**	22	decrease is usually bad

fig. 7.2 - beginning. Interpretation of Plus-factors and Minus-factors

factor being improved related to time	23	**functional time of the moveable object**	23	factor being corrected related to time
factor being improved related to time	24	**functional time of the fixed object**	24	factor being corrected related to time
decrease required	25	**loss of time**	25	increase is bad
factor being improved related to quantity of material	26	**quantity of material**	26	factor being corrected related to quantity of material
decrease required	27	**loss of material**	27	increase is bad
increase required	28	**strength**	28	decrease is bad
increase required	29	**stabile structure of the object**	29	decrease is bad
increase usually required	30	**force**	30	decrease is usually bad
factor being improved related to weight	31	**tension, pressure**	31	factor being corrected related to weight
factor being improved related to weight	32	**weight of the moveable object**	32	factor being corrected related to weight
factor being improved related to weight	33	**weight of the fixed object**	33	factor being corrected related to weight
factor being improved related to temperature	34	**temperature**	34	factor being corrected related to temperature
factor being improved related to lighting	35	**brightness of the lighting**	35	factor being corrected related to lighting
increase required	36	**power**	36	decrease is bad
decrease required	37	**energy use of the moveable object**	37	increase is bad
decrease required	38	**energy use of the fixed object**	38	increase is bad
decrease required	39	**loss of energy**	39	increase is bad

fig. 7.2-end. Interpretation of Plus-factors and Minus-factors

Now let us turn to the A-matrix (fig. 7.3).

Let us assume, for example, that we have selected, from the matrix, plus-factor 15 and minus-factor 28. Then we copy, from the cell of the matrix located at the intersection of line 15 and column 28, the numbers of transformation models recommended (based on research conducted by TRIZ experts) for resolving contradictions of the type "15 Length of the Fixed Object **VS** 28 Strength": 01, 14, 15, 32.

In other words, the formula for the "general solution" is:

15 **VS** 28 = 01, 14, 15, 32.

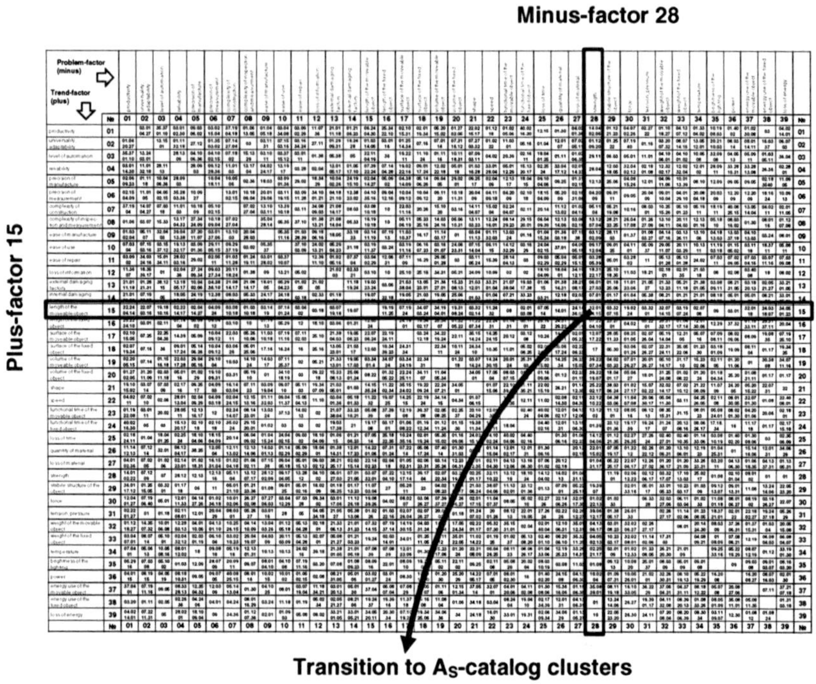

fig. 7.3. Selection of recommendations from the A-Matrix

Using these numbers, we select respective specialized transformation models from the A_S-catalog, and try to apply those models to resolve our contradiction (both formal and, accordingly, informal).

In MTRIZ, the above procedure is called the "BICO Method" (BICO = Binary In Cluster Out) based on the names of the main components of the procedure – the use of a two-factor (binary) contradiction and its input (*in*) into the A-matrix, and retrieval of a cluster (simple list, set) of numbers representing recommended transformation models as the output (*out*) of the matrix.

And now (and this is the pivotal moment of the Inventing stage) we need to interpret the transformation models we selected as they apply to the resolution of the initial standard contradiction.

Let us consider several examples.

Example 7.1. Solution for Problem 2.7. "Short Arms".

Our own arms would be ideal – only if they were somewhat "longer"! OK, let us look at what the A-matrix has to offer so that we can resolve the existing contradiction.

We suggest that the plus-factor from the informal contradiction be matched to formal plus-factor *19 Volume of the moveable object* – meaning that we want to be able to grasp more fallen leaves.

The informal minus-factor can be matched to formal minus-factor *07 Complexity of construction* – meaning the complexity of the usual devices (carts, spades, leaf blowers, leaf suckers and whatnot).

Accordingly, the formula of the contradiction takes the following shape:

19 Volume of the moveable object **VS** *07 Complexity of construction* = **03, 10.**

Let us now generate a "composite sketch" of the future solution by making an extract from the recommendations contained "within" models 03 and 10:

03 Segmentation: b) disassemble an object into individual parts. Idea for the solution: what do we "disassemble" – some known tool? Or is it going to be our *ARMS*?

10 Copying: a) use a simplified and inexpensive copy instead of an inaccessible, complicated, expensive, inappropriate, or fragile object. Idea for the solution: what if we DO make "copies" of arms?! Make "long arms"?!

The outcome is shown in the table in fig. 7.4.

Problem 2.7. "Short arms"	**Standard contradiction**	
	plus-factor	minus-factor
It is difficult to grasp a sufficiently large amount of fallen leaves when cleaning them, while almost all available tools are complex and not too convenient.	*too many fallen leaves*	*complex tools*
	19 Volume of the moveable object	*07 Complexity of construction*
Solution: the tool is made in the form of a pair of spatulas held by traverse handles; when the spatulas are used to grasp leaves, they lean against the arm above the wrist. **Dominant TRIZ models:** *copying, separation, mediator* **Additional TRIZ models:** *local property, partial or excess effect, universality, matryoshka (nested doll), unite*	 www.moderne-hausfrau.de	

fig. 7.4. Solution of the Problem in Assignment 2.7 "Short Arms"

Yes, this is reinventing! Yes, the author knew the solution! Yes, other solutions are possible!

But the path to each such solution – and this is the most important thing – WILL BE THE SAME! Regardless of whether you are doing START and BICO modeling for reinventing purposes or tackling a new problem!

Example 7.2. Solution for Problem 2.8. That's a Lotta Line!

To solve this problem, we can use a very visual, graphical representation of the standard contradiction. Accordingly, selection of formal plus- and minus-factors becomes simpler (which unfortunately is not always the case in real life!).

Let us record the model in the form of a relatively more complex graphic scheme (fig. 7.5):

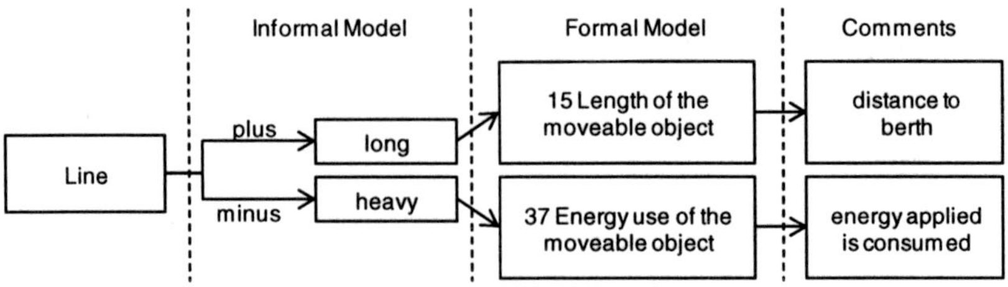

fig. 7.5. Standard Contradiction in Assignment 2.8. That's a Lotta Line!

Please pay attention to the minus-factor selection. Apparently, factor 32 or 33 (both related to weight) would be more appropriate. However, the resemblance is superficial. On the other hand, if you can envision behavior of a long and relatively thick line when you try to swing it or throw it, you will see that the real problem is "waste of effort". Because of the line's length and "non-rigidity", it will not fly, nor will it swing. All energy is consumed by its inertial mass.

Let us write the recommended models out of the A-matrix: 01, 18, 32.

Now we need to copy the most promising recommendations for the "composite sketch" of the future solution:

18 Mediator: a) use another object to transfer or transmit an action; b) temporarily connect an object with another (easily separable) object.

Idea for the solution (fig. 7.6): attach to the thick line a long light rope which can be easily – and accurately - thrown a long distance, use such rope to pull the end of the line to the berth, and then detach the mediator rope for repeated use!

Has this brought to memory the kindergarten "inventor" – see "ABC-TRIZ" – who tied a mediator thread to the rope?

Problem in Assignment 2.8 That's a Lotta Line!	**Standard contradiction**	
	plus-factor	minus-factor
It is difficult to swing or throw the mooring line, as it is heavy – and it consumes the energy applied to it.	*long line*	*energy consumed*
	15 Length of the moveable object	*37 Energy use of the moveable object*
Solution: Before the mooring, a thin weighted mediator rope is tied to the line. The rope is either thrown onto the berth, or falls into the water near the berth whereupon it is caught with a hook from a tow boat and taken to the berth. **TRIZ model:** *mediator* (a)		

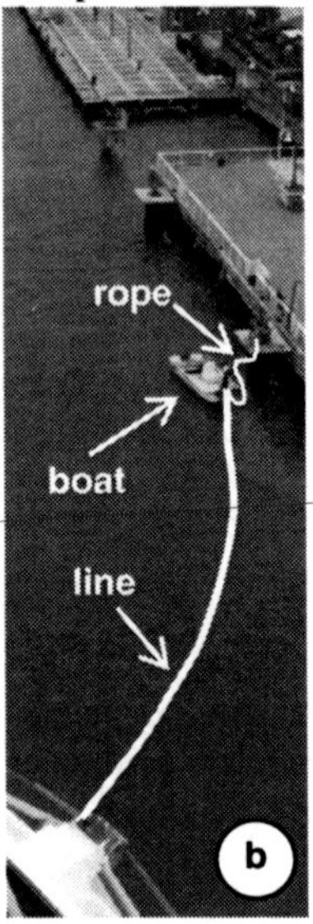

fig. 7.6. Solution of the Problem in Assignment 2.8. That's a Lotta Line!
a – Resulting table; b – Line is passed over to the berth from the boat.

Example 7.3. Solution for Problem 2.9. On the Way to a New Automotive Era: Electric Car.

Let us formulate the formal standard contradiction (fig. 7.7):

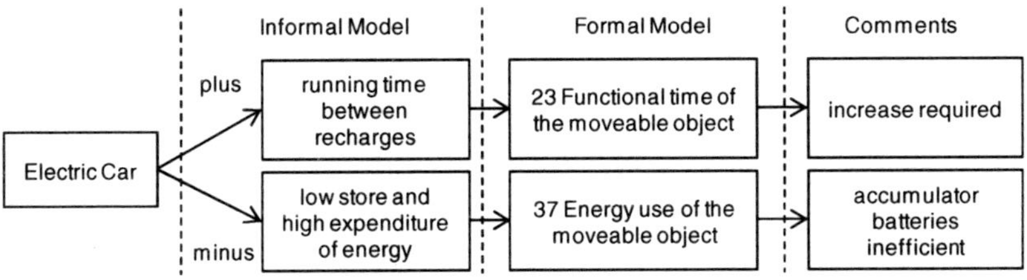

fig. 7.7. Standard Contradiction in Problem 2.9. Electric Car Recharging

Let us write the recommended models out of the A-matrix: 01, 04, 06, 20.

Now we need to copy the most promising recommendations for the "composite sketch" of the future solution:

04 Replacement of mechanical matter. b) use of electrical, magnetic, or electromagnetic fields for the interaction of objects.

Idea for the solution: Technical solutions involving the use of the roadbed as a charging device were first patented in the mid-1970-es! Transmitting electromagnetic coils are installed under the roadbed, especially before intersections where the cars often stop for prolonged periods of time, and receiving coils are installed on the cars.

A breakthrough in this area was recently made by engineers of the Korean Institute of Science and Technology (KAIST, Daejeon) who expect their charging device to have an efficiency ratio of 75%!

Problem 2.9. On the Way to a New Automotive Era: Electric Car	**Standard contradiction**	
	plus-factor	minus-factor
Electric cars would have conquered the world long ago, if they had light and compact accumulator batteries. Meanwhile, we have to look for other ways to recharge those batteries.	*increase time in motion*	*heavy and inefficient accumulator batteries*
	23 Functional time of the moveable object	*37 Energy use of the moveable object*
Solution: Engineers of the South Korean KAIST Institute install magnetic induction transmitters at intersections (and in other places along the motorway), and mount receiving devices on electric cars. **Dominant TRIZ models:** *replacement of mechanical matter, separation, matryoshka (nested doll)* **Additional TRIZ models:** *inverse action, local property, transition into another dimension*		

fig. 7.8. Solution of the problem in Assignment 2.9. On the way to a new automotive era: electric car

Example 7.4. Solution for Problem 2.10. Back to the Future! Let us formulate the formal standard contradiction (fig. 7.9):

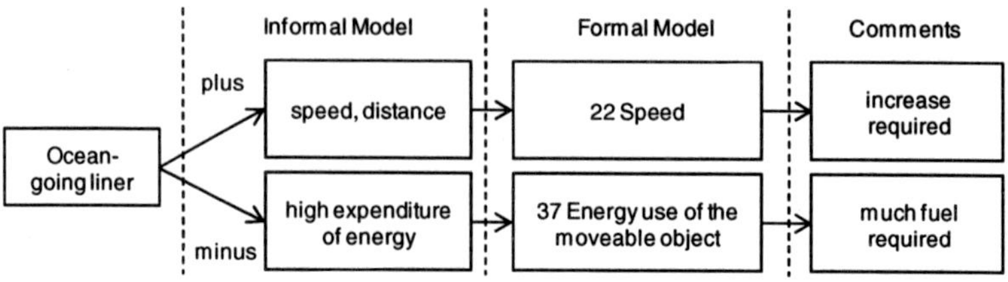

fig. 7.9. SC for the problem about ocean-going liners and "cargo-carriers"

Let us write the recommended models out of the A-matrix: 01, 07, 30, 32, then copy the most promising recommendations for the "composite sketch" of the future solution:

07 Dynamization: a) the characteristics of an object or an environment are changed to optimize every work procedure; b) disassemble an object into parts that are moveable among each other. Idea for the solution: Harness the energy of the external environment, for example, the energy of the wind, by using operator-controlled devices.

32 Counter-weight: a) compensate for the weight of an object with its connection to another object with lifting power; b) compensate for the weight of an object using interaction with the external environment (for example, with aerodynamic or hydrodynamic forces). Idea for the solution: Use favorable wind to create additional traction as additional "lifting force" to compensate for the weight of the vessel!

Idea for the solution (fig. 7.10): Go back in time – use a sail! And not just any sail, but a sail in the form of a kite! The idea was realized – and is offered to all interested vessel-owners – by SkySails GmbH from Hamburg, Germany. Fuel conservation during long voyages measured in thousands of nautical miles amounts to 10-35%, and if conditions are favorable, may be as high as 50%! In the future this idea may evolve into something even bigger!

Problem in Assignment 2.10. Back to the Future!	**Standard contradiction**	
	plus-factor	minus-factor
Ocean-going vessels consume extremely large amounts of fuel. They have to carry fuel instead of additional payload. Ocean waters are exposed to intensive contamination.	*increase voyage distance and speed*	*large consumption of energy and, consequently, fuel*
	22 Speed	*37 Energy use of the moveable object*
Solution: Engineers of SkySails GmbH from Hamburg, Germany, suggested that dirigible sails in the form of kites be installed on ocean-going vessels! **Dominant TRIZ models:** *separation, dynamization, counter-weight*		The picture (from the company's website) is adapted by author

fig. 7.10. Solution of the problem 2.10. Back to the Future!
(the picture is adapted by the author from the original published on the corporate web site)

Example 7.5. Solution for Problem 2.11. I'm the Lord of the Wheels! Let us formulate the formal standard contradiction (fig. 7.11):

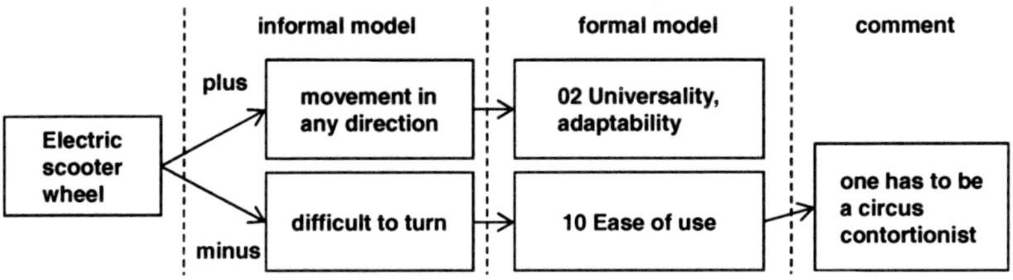

fig. 7.11. Standard contradiction for Problem 2.11. I'm the Lord of the Wheels!

Let us write the recommended models out of the A-matrix: 03, 07, 15, 16.

Then let us copy the most promising recommendations for the "composite sketch" of the future solution:

03 Segmentation: a) Divide the object into independent parts. Idea for the solution: imagine the only wheel as a set of wheels threaded as beads onto some common circular axis. The sum total of the "beads" creates a semblance of one large wheel, while each separate bead can rotate around its axis, and move the scooter at right angles to the direction of rotation of the "composite wheel".

07 Dynamization: b) disassemble an object into parts that are moveable among each other. Idea for the solution (fig. 7.12): suppose that the "tire" of the wheel is made up of many traverse rings, and that each such ring can be forced to rotate at right angles to the direction of rotation of the bearing wheel.

Naturally, to come up with such idea one has to be blessed with unbridled imagination and solid talent! Our task is simpler, though, as we are merely reinventing a well-known solution originally developed by the Japanese automobile company Honda. On the other hand, let us not forget that we learn from such examples. We are trying to understand how this invention could have been made using TRIZ methods!

Problem 2.11. I'm the Lord of the Wheels!	**Standard Contradiction**	
	plus-factor	minus-factor
Electric scooter Honda U3-X enables the rider to move in any direction without changing his or her bodily stance. To change the direction, the rider only has to slightly lean to the required side.	*Movement in any direction*	*Difficult to turn*
	02 Universality, adaptability	*10 Ease of use*
Solution: numerous small-diameter wheels have a common axis of rotation represented by a larger circle so that all such small wheels form, in their totality, a large-diameter "tire". **Dominant TRIZ Models:** *segmentation, dynamization, local property*	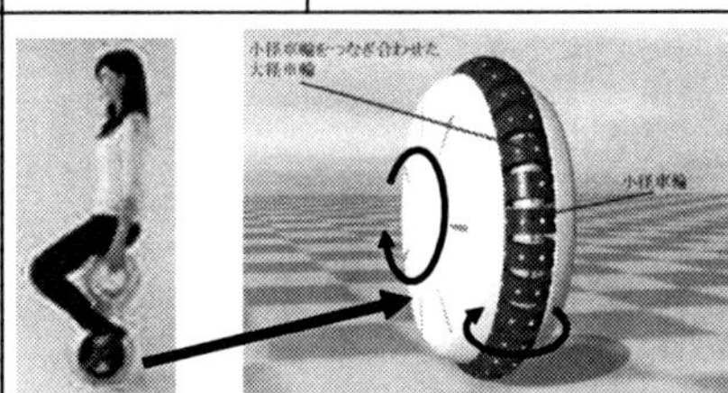	

fig. 7.12. Solution for Problem 2.11. I'm the Lord of the Wheels!

Example 7.6. Solution for Problem 4.10. Socket Wrench with Replaceable Heads

TREND

A car repair shop needs spark plug wrenches of different diameters. Naturally, it is possible to keep a separate wrench for each diameter. But there are only two or three head sizes (plug heads are made in the form of cap nuts). So car mechanics use wrenches with replaceable heads. However, heads sometimes fall off the wrench. To simplify the construction, the heads are held merely by force of friction, or by inbuilt spring-loaded balls. Unfortunately, wrenches with such balls are more difficult to manufacture. What can we do to make the wrenches convenient and easy to make?

REDUCING

FIM: X-resource, together with available or modified resources, and without making the object more complex or introducing any negative properties, guarantees attainment of the following **IFR**:

[*heads do not fall off, and the wrench is easy to make*].

Standard (Technical) Contradiction

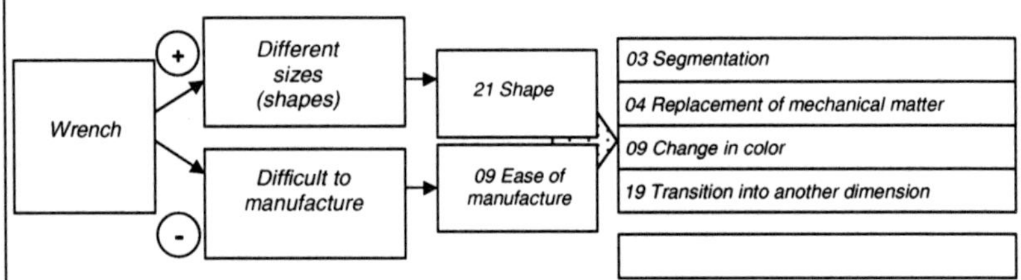

Radical (Physical) Contradiction

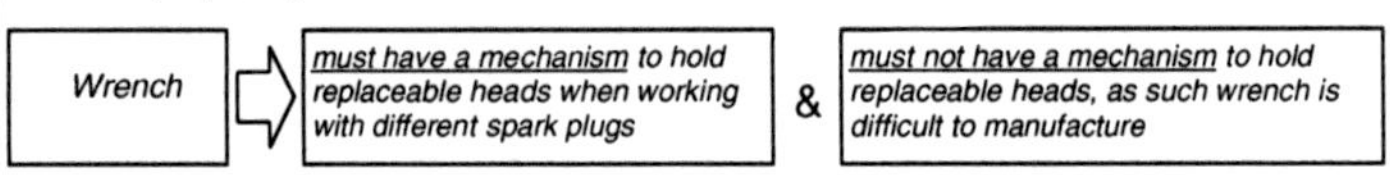

INVENTING

Key model – 04 Replacement of mechanical matter: *b) use of electrical, magnetic, or electromagnetic fields for interaction with the object.*

Key idea: *a small magnet is glued from below onto the butt of the wrench handle. The magnet then holds replaceable heads in place. The heads are easy to put on and take off, and they do not fall off the wrench.*

ZOOMING

Have the contradictions been removed? – Yes. – ~~No~~.

Super-effects: *none*

Negative effects: *none*

Picture

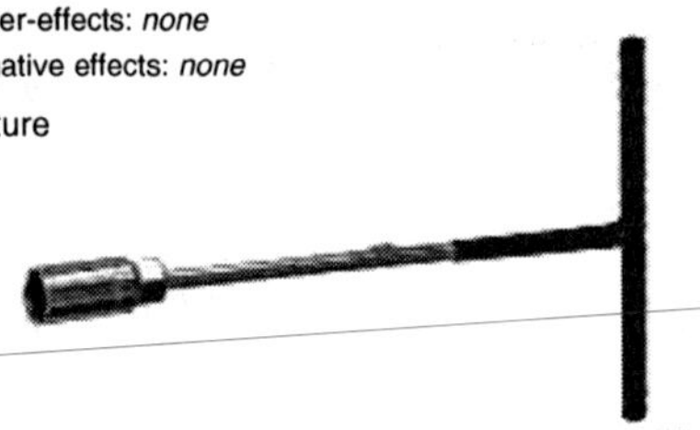

BRIEF DESCRIPTION

A small permanent magnet is installed in the butt of the wrench handle. The magnet prevents the head from falling off the handle. The solution makes use of Navigator 04 Replacement of mechanical matter.

fig. 7.13. Standard reinventing for Problem 4.10
Socket Wrench with Replaceable Heads

Example 7.7. Solution for Problem 4.11. "Articulated" Wrench

TREND

A car repair shop needs spark plug wrenches of different diameters. There exist wrenches with replaceable heads. Still, heads sometimes fall off the wrench and get lost. Besides, if the wrench has a long straight handle, it is sometimes difficult or impossible to get to a spark plug installed in a place which is difficult to access because other parts get in the way. How do we make the wrench more convenient so that it can be used in repair works more efficiently?

REDUCING

FIM: X-resource, together with available or modified resources, and without making the object more complex or introducing any negative properties, guarantees attainment of the following **IFR**:

[*heads do not get lost, and the wrench can be used in inconvenient places*].

Standard (Technical) Contradiction

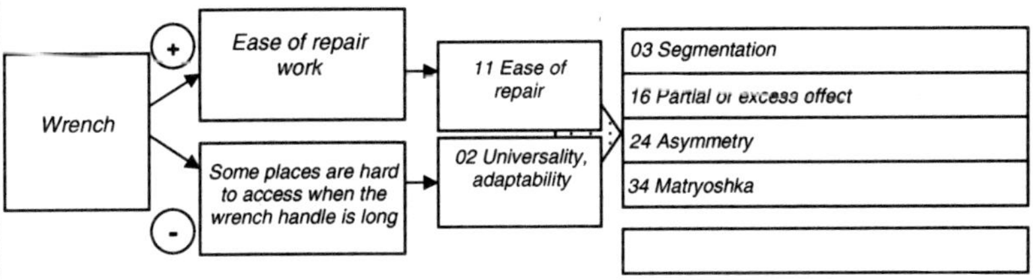

Radical (Physical) Contradiction

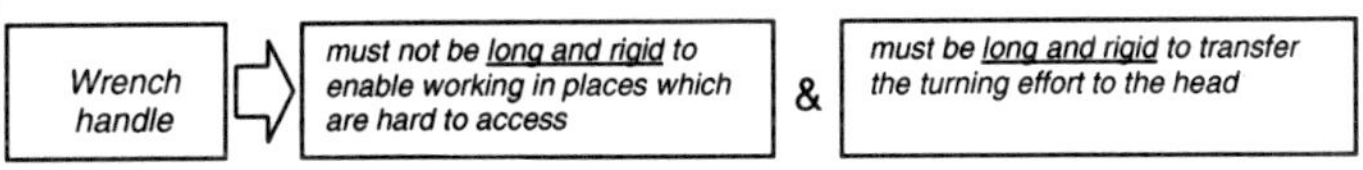

INVENTING

All models are used – 03, 24 and 34.

Key ideas:

1) according to Model 34, one head can be inserted into the other;

2) according to Models 03 and 24, the handle can be divided into several segments connected by hinge joints. As a result, the handle can be bent (and then its shape becomes asymmetrical) and used for working in places which are hard to access. The mechanic can now place the head on the spark plug and rotate the handle – regardless of where the plug is installed.

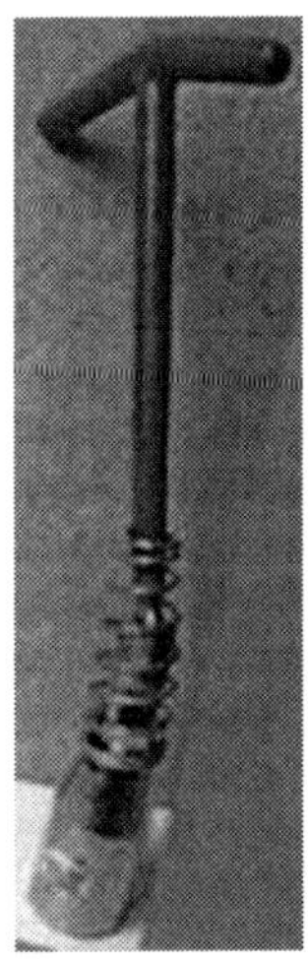

ZOOMING

Have the contradictions been removed? – Yes. – ~~No~~.

Super-effects: *none*

Negative effects: *such wrench is more difficult to make and more expensive.*

Picture

Inner head is pressed with a pen.

BRIEF DESCRIPTION

To increase ease of work in places that are hard to access, and to ensure universality of the wrench, the head with the smaller diameter is inserted inside the head with the larger diameter (Navigator 34 Matryoshka), while the handle is made of several sections connected by hinge joints (in accordance with Models 03 Segmentation and 24 Asymmetry).

fig. 7.14. Standard reinventing for Problem 4.11 *"Articulated" Wrench*

7.2 Resolution of Radical Contradictions: method[119] RICO – Radical In Cluster Out

The algorithm used to solve problems with radical contradictions is provided in fig. 7.15.

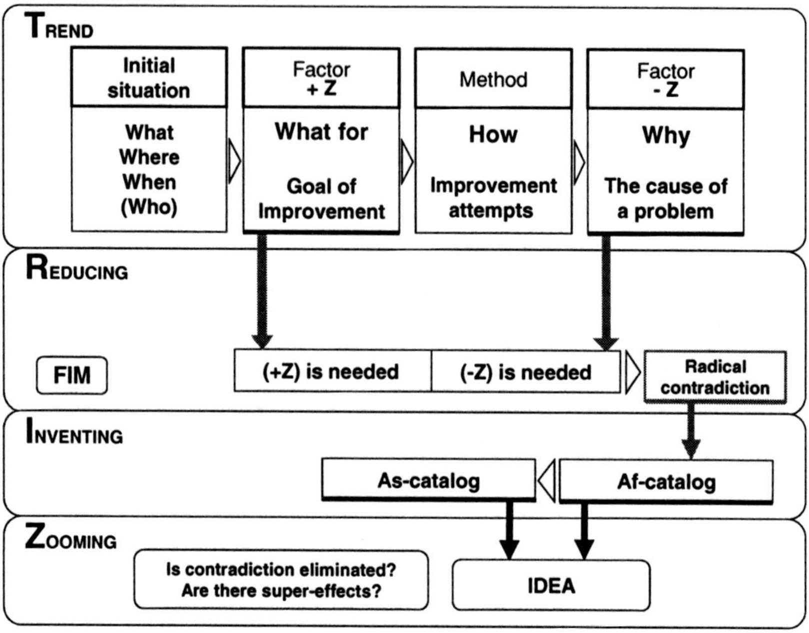

fig. 7.15. Algorithm START for resolving radical contradictions

Resolution of radical contradictions is more complex than resolution of standard contradictions. There are several reasons for that. We shall note three such reasons. First, the need to realize diametrically opposed properties in and of itself stresses out any beginner. Second, the very transition to the radical contradiction signals that relatively simpler avenues, such as solutions based on standard contradictions, may have been exhausted. Third, possibilities of effecting direct changes to certain resources are likely to have been used up (such resources are referred to as *scarce*), meaning that the required change may have to be quite drastic – and it may affect other resources.

Another difficulty lies in that it is impossible to use the A-matrix, as existence of a radical contradiction means, essentially, that the same factor must be used both as the plus-factor and as the minus-factor – with the cells lying at relevant intersections forming the main diagonal (in fig. 7.16 it is marked by a double-line border).

---

[119] RICO – Radical In Cluster Out

In reality, as suggested by the scheme in fig. 7.15, complexity of problems with radical contradictions often does not exceed that of problems routinely solved with the use of such instruments as specialized A_S-catalog navigators, precise OZ structure analysis, detailed examination of operating resources and, in many cases, fortunate wording of the IFR and the FIM.

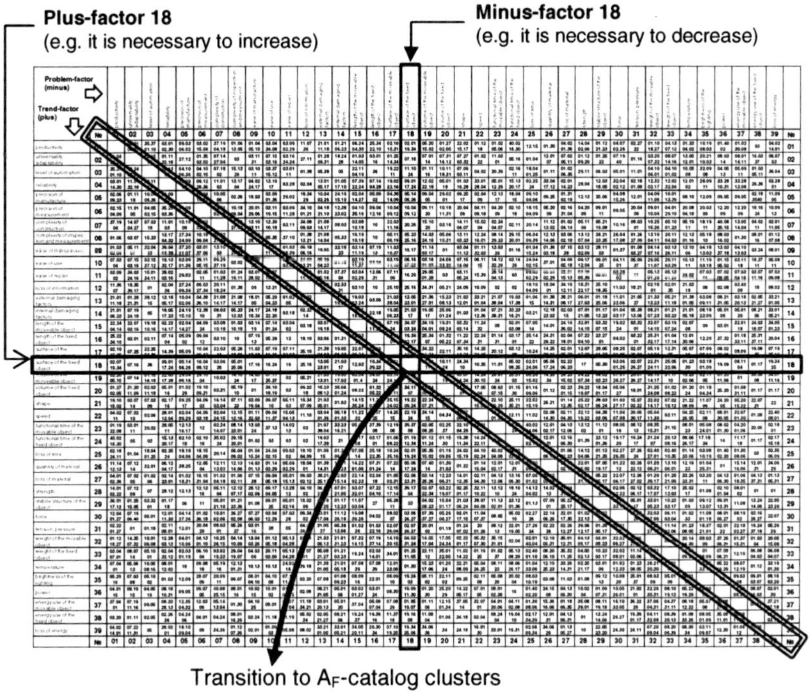

fig. 7.16. "Position" of Radical Contradictions in the A-matrix

Accordingly, for "mass-scale" cases – and for express analysis purposes – solutions can be found based on the following two recommendations:

1) formulation of the IFR in the form "BOTH – AND": whatever the initial radical contradiction, it must be transformed into the IFR according to Bartini (see section *5.3.1. Ideal Final Result According to Bartini*), i.e. represented in the form of the conjunction "BOTH – AND"; normally, only one of the two properties causes problems, and it is that problematic property that finds itself in the focus of attention when a new idea is being synthesized.

Thus, in Example 7.6 the wrench **BOTH** *must have a head locking mechanism* for one function **AND** *must not have that mechanism* as it makes construction of the wrench more complex, and its manufacture more difficult. Naturally, the wrench must have the mechanism in any event, but in accordance with the second requirement it is assumed that the mechanism is so simple that it is "absent" and can be "disregarded". It is an "ideal mechanism": the function is performed, while the mechanism is "not there".

In Example 7.7 the wrench **BOTH** must be *long and rigid* to transfer force to the part being turned **AND** it must be *short and flexible* to be used in locations that are difficult to access. Both these properties-states are required.

Hence it must be clear that, to build an idea for the solution, one of those two states should be perceived not literally – as a metaphor:

- in the first case, it is the state *"simple mechanism"*, because the state *"some other (rather complex) mechanism"* is ALREADY there;

- in the second case, the target state is *"short and flexible"*, because the state *"long and rigid"* is ALREADY there.

2) analysis and transformation of "non-scarce" resources: our attention must be focused (like the sights of a gun) on the IFR (fig. 7.17), while future changes must rely on resource transformation models.

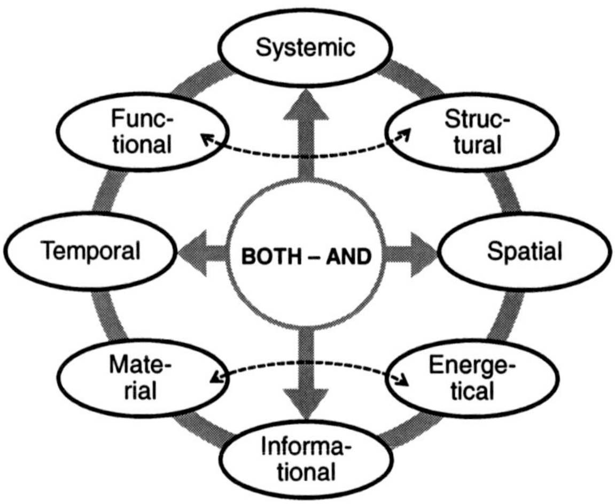

fig. 7.17. IFR "surrounded" by resources that can be used to transform the artifact

Here bidirectional (thin dashed and thick solid) arrows connect resources which frequently change together.

To resolve radical contradictions, we use the Af-catalogs[120] and the Afs-catalog[121] which, in fact, is a directory of resources complete with a description of principles governing their modification. These "principles" are used as fundamental transformation models.

Experience accumulated in the course of reinventing thousands and solving hundreds of problems serves as the basis for the following general conclusions:

1. All solutions rely on modification of resources in the operating zone.

2. The operating zone must be regarded as a dynamized model whose "size" depends on resources used, and can range from a small assembly or even one separate component to the entire system or even super-system.

---

[120] Sections *14.2.3 Af-catalog*, *14.2.5* and *14.2.6* for advanced *Af-catalogs*

[121] Section *14.2.4 Afs-catalog*

The main fundamental principles governing resolution of radical (and, through them, all other) contradictions can be reduced to the following:

1. The main principle governing resolution of radical contradictions is: one part of the dominant (key, decisive) resource has property **A**, while another part of such resource has property **non-A** (or **Ā**); accordingly, the system as a whole realizes the ideal requirement "**A**-and-**Ā**".

2. There may be situations where the dominant resource realizes property **Ā**, while the system as a whole realizes property **A**.

Finally, in most general terms, the main modification methods are the following:

1. Introduction of the "positive" dominant resource into the operating zone, with subsequent joint modification of the dominant and other available resources.

2. Withdrawal of the "negative" dominant resource from the operating zone, with subsequent modification of other available resources.

3. Introduction of the positive resource and withdrawal of the negative resource with concurrent modification of the positive resource and other available resources.

Some practical aspects of resource transformation are also discussed in section *9.3. DISK Meta-Model.*

Inasmuch as a more detailed review of fundamental transformations goes beyond the scope of this book, let us limit ourselves to examination of several illustrative examples.

Example 7.8. Road Intersection.

Let us consider possible solutions (fig. 7.18) of the traffic safety problem at a road intersection using different dominant resources. Even though these solutions have a common theme ("road intersection"), they deal with different types of intersections and problems arising in different specific situations.

The Main System Contradiction (MSC) for any intersection of two (or more) roads can be formulated as follows:

MSC: higher speed **VS** lower safety

Indeed, the higher the total number of motor cars, the more important it is to boost the throughput capacity of road intersections, *inter alia*, by increasing the speed at which motor cars cross the intersection. On the other hand, higher speeds at intersections have adverse impact on safety and increase the number of accidents.

The appropriate radical contradiction can be formulated as follows:

RC: speed of motor cars at the intersection must increase
VS must not increase (to prevent non-compliance with safety standards)

Let us consider the possibility of using, and the role that can be played by, various resources in solving this problem in different situations.

ROAD INTERSECTION	EXAMPLES OF THE RADICAL		
Type of intersection, or nature of the problem ⇨ **MSC:** higher speed **VS** inferior safety **RC:** speed must grow (supra-system objective) **VS** speed must not grow (because of inferior safety)	Intersection overload	Unregulated intersection of equally ranked roads	Unregulated intersection of unequally ranked roads
EXAMPLE OF SOLUTION PRINCIPLE	Construction of a detour (promotion to supra-system)	Formulation of rules governing the crossing of intersections of equally ranked roads	Clearly marked priority of the main road
EXAMPLE OF SOLUTION SCHEMA	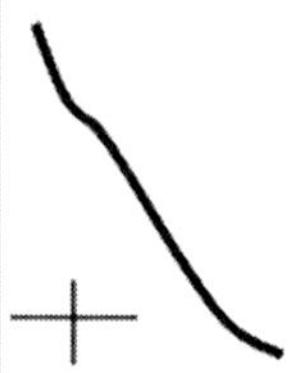	**a) Car on the right has priority (right-hand traffic)** **b)** 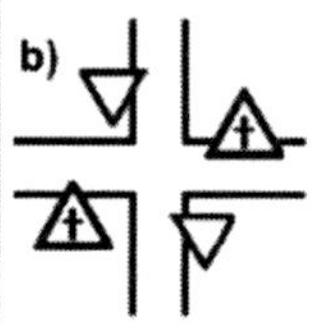	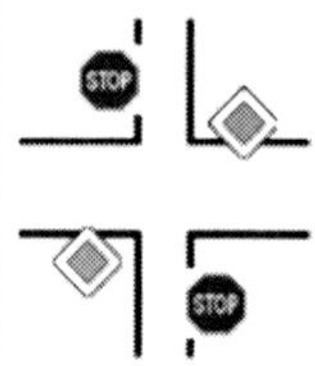
RESOURCE dominant ⇨ contributory ⇩	**01. SYSTEMIC**	**02. INFORMA-TIONAL**	**03. FUNCTIONAL**
01. SYSTEMIC	█		
02. INFORMATIONAL		█	▨
03. FUNCTIONAL		▨	█
04. STRUCTURAL		▨	▨
05. SPATIAL	░		░
06. TEMPORAL		░	░
07. MATERIAL	░		
08. ENERGETICAL			

█ **- unconditionally dominant resource** ▨ **- additional dominant resource**

fig. 7.18. Resource-oriented fundamental transformations

CONTRADICTION SOLUTIONS FOR 8 DOMINANT RESOURCES				
Unregulated intersection of equally ranked roads	Intersection with extremely heavy traffic	Intersection with heavy traffic	Unregulated intersection of unequally ranked roads	Intersection of main and subsidiary roads
Division of an X-junction into two T-junctions	Separation of roads into different levels	Formulation of crossing priority rules	Road with better surface has priority	Acceleration when entering the main road; deceleration when exiting the main road
a) 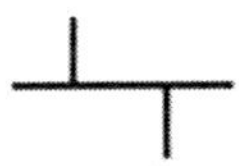 b) 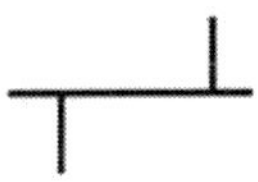	a)  b) 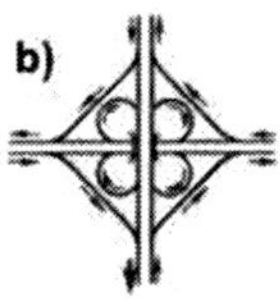		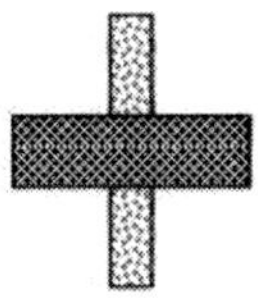 Usually complemented with priority signs	
04. STRUC-TURAL	**05. SPATIAL**	**06. TEMPORAL**	**07. MATERIAL**	**08. ENERGETICAL**

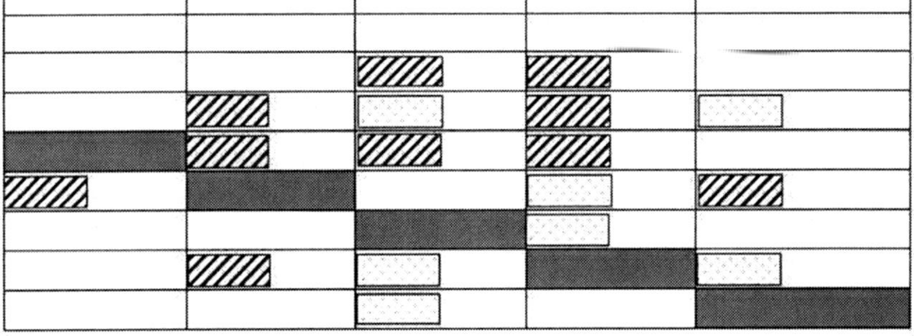

 - **auxiliary resource**

01. Systemic Dominant Resource. The intersection is located in a small town which lies between, and is the crossing point for the traffic generated by, two or more large cities. Construction of a highway going around the town, or of a detour road, would radically reduce the volume of transit traffic. The main resource here lies beyond and outside of the intersection *per se*. To be more exact, it is located within the **road system** to which the intersection belongs. Reduction of the volume of transit traffic does not require any significant modification of the intersection itself. Instead, a sign pointing the way to the highway (detour) (informational resource) may be installed at a relatively large distance from the intersection. Clearly, construction of the highway will require spatial and material resources.

02. Informational Dominant Resource. Installation at the intersection of equally ranked roads of certain signs (*Yield* and *Intersection with Secondary Road* – introduction of a new informational resource) will stipulate the rules for the crossing of the intersection, thereby increasing the level of safety. Important additional resources involved in modification of the standard operating procedure previously used at the intersection include the following: 1) functional resource – the solution regarding the assignment of the "main road" and "secondary road" status is made subject to the actual volume of traffic; preference is given to the road with heavier traffic; 2) structural resource – a certain structure is implemented governing relations between road users, namely, priority of one direction over the other.

03. Functional Dominant Resource. In this solution, unequal ranking of the roads is reinforced by giving priority to the road with heavier traffic (impact of the functional dominant resource), the decision being supported by the width of that road (see the drawing). Additional informational resources are reinforced accordingly: the main road is marked with the sign *Main Road*, the secondary road – with the sign *Stop*. The structural resource is also formulated in a more radical fashion in line with the desired functional differentiation.

04. Structural Dominant Resource. It is known that the degree of severity of accidents which occur at road intersections is higher in the event of a head-on or side collision, rather than a "grazing" collision. Crossing of an ordinary intersection gives rise to many situations which potentially may end with a head-on or side collision. Such intersection can be (subject to availability of sufficient spatial resource) transformed into two T-junctions which are much easier to cross (as supported by available statistical evidence).

05. Spatial Dominant Resource. The roads are placed at different levels, which represents a radical (in practice) and "ideal" (according to TRIZ) solution of the intersection crossing safety problem, because the intersection "exists" geographically and functionally as a crossing of two roads, and it "does not exist" because motor cars are physically located at different levels.

The role played by other resources can be reviewed by the readers independently.

May it only be noted that in option b) the illustration for the spatial resource shows numerous T-junctures (structural resource) participating in realization of a multiple-level bypass.

06. Temporal Dominant Resource. This is a realization of the dynamic version of traffic signs *Main Road* and *Stop* with periodic inversion of priorities.

Here we also suggest that the readers undertake an independent analysis of the role played by other resources.

07. Material Dominant Resource. In this solution, unequal road ranking is realized by granting priority to the road with better surface. This by definition increases the role of the functional resource, as the road with better surface is apparently built to accommodate relatively heavier traffic.

08. Energetical Dominant Resource. In this solution, the dominance of the energetical resource is not as apparent as in the other examples. However, under certain circumstances, the energetical resource plays an important role in securing the safety of entry to, and exit from, the speed highway.

The thing is, before entering the highway the car has to rapidly accelerate to the same speed as the cars already there. The acceleration lane is not always long enough to make that possible. If the entry lane slopes down and towards the highway, the car's kinetic energy is boosted by the gravitational energy generated by the car's own weight. Direction of that energy coincides with the direction of the car's movement, which facilitates its acceleration.

Conversely, exit from the highway sometimes requires rapid deceleration. If the exit lane slopes up and away from the highway, additional gravitational energy working against the direction of the car's movement helps it to decelerate.

For training purposes, the readers may also think through option b) for the 5th (spatial) resource, and determine in which situations the role played by the energetical resource is positive and negative.

These examples will help the readers to find out, in each practical situation, which resources can be used, what role is played by each resource (dominant, supplementary, auxiliary), and how each resource (and other available resources) can be changed to improve the operation of the system under review and resolve systemic contradictions.

Example 7.9. Solution for Problem 2.15. Ads on the Bus.

Let us write down the radical contradiction in the "BOTH – AND" form according to Bartini with respect to the main inductor which has both positive and negative impact:

The ad ► *must be (in the opinion of the advertiser)*
AND *must not be (as it reduces visibility for the passengers)*

Clearly, the first factor *has to be realized.* All we have to do it find a way to realize the second factor ("must not be"), as well.

The problem is aggravated by the fact that the film must stay on the bus at all times, i.e. it is impossible (is it?) to resort to separation of the conflicting properties in time (see the following example).

The requirement "must not be" can be interpreted as follows: the passengers may not notice the advertisement placed on the side of the bus (because it does not prevent them from looking out the windows), or the advertisement may even be useful by transforming "damage into use", for example, by keeping the interior of the bus shady on a sunny day (which, incidentally, does happen).

In this case, the obstacle is created by the material resource – if the film of which the advertising poster is made is not transparent. But what if it IS transparent?

To make the film transparent, we can make it thin (spatial resource) or porous, with openings (spatial and structural resources).

Known solution: the film has a porous structure so that the view out of the window appears only slightly shaded, as if the passengers were wearing sun glasses.

Solution underlying principle according to TRIZ: part of space possesses one property (bears the paint for the advertising), while the other part possesses another property (ensures transparency of the film).

The same goes for the structure: some (solid) elements of the film are bearing paint and realizing property **A**, while other (transparent) elements are not bearing paint and create property **Ā**, while the system as a whole (advertising poster film) is realizing the solution which resolves the contradiction in the form of the following combination: "**A**-and-**Ā**".

Example 7.10. Solution for Problem 2.16. Comfortable Bus Step.

Another "bus example" with the following general "solution" according to Bartini:

The step is ► *high* (when the bus is moving)

AND *low* (when the passengers are boarding or alighting)

fig. 7.19. Bus tilting at a stop

The dominant creative resource in this example is time: in all assignments with the *conflicting properties related to different time intervals* the solution is built on the temporal resource by relegating incompatible properties to different time intervals. In other words, when the bus is static, the step must be "not high", while when it is mobile it must be "high".

Known creative solution: dynamization – the tilting bus. It is also possible to make extending and descending steps, etc.

Additional creative and technical resources: functional – function of an adaptive step (or, to be more exact, of an "adaptive" tilting bus); structural – introduction of new elements.

Example 7.11. Solution for Problem 2.17. Like a Fairy Tale!

Here we will, in all seriousness, model this joking assignment using TRIZ methods.

"Clothing" solution: The girl wrapped herself up in a fishing net. Dominant creative resources: spatial and structural (see Example 7.9).

"Transportation" solution: The girl moved in the same fishing net that she had on, having tied the net to the tail (!) of a donkey. Let us define the net as the "transportation platform", even though in reality is but a simple "drag".

Dominant creative resources: functional – she was "dragged" along, i.e. she was not using known modes of transportation; spatial – the conflicting property was placed outside known systems ("not riding, not walking"); and structural – introduction of new elements and links which are not used in known systems (the drag was tied to the tail of an animal).

"Road" solution: The girl was touching the road through the fishing net with her great toe. Dominant creative resources: functional and structural - the girl was on some "transportation platform", but she was being dragged in the rut ("not in the road") – that is, she was not "walking" along the road, but she was "walking" in the sense that her toe was touching the road.

You can suggest your own interpretations to model the solutions found by this inventive girl. Incidentally, do not forget that failure to meet the conditions formulated by the "assignment originator" would have grievous consequences both for the girl and for her father[122]. This means that you should take modeling – and analysis of the girl's solutions – very seriously!

As they say, a tale may be a lie, but those who understand (*Sapienti sat est*[123]) may still learn something from it!

Example 7.12. Solution for Problem 2.18. Memory Eternal.

Let us complement the process of solving the problem with a model of the standard contradiction:

SC: 02 Universality, adaptability (to time, outside of time)
VS 21 Shape (changed, lost)

A-matrix cluster: 03, 07, 27, 32.

According to the A_{FS}-catalog (see *Primary TRIZ Instruments*), recommendation *03 Segmentation* relates to the structural resource, while recommendation *07 Dynamization* relates to the temporal resource.

On the one hand, "segmentation" means division of the whole into parts, down to minute particles, say, those existing in the form of dust, smoke or light. On the other hand, a monument is traditionally perceived as something massive made of granite or marble. Therefore, the *"segmentation"* transformation model from the very start points to the need to *forego the use of massive objects.*

"Dynamization" means creation of mobile elements in construction, or mobility of the entire object. However, if we discard the "massive monument" idea, and pursue the "segmented construction" idea to the particle level, *"dynamization"* may only mean *mobility of those particles.*

And now it becomes apparent that "technocratic" recommendations according to TRIZ do not carry the day. Here we find ourselves at one of the TRIZ boundaries, and we have to make use of the two other noospheres inherent in any solution.

---

122 Fairy tale *The Peasant's Wise Daughter* by world-renowned German writers, philologists and folklorists brothers Grimm, Jacob (1785-1863) and Wilhelm (1786-1859)

123 A word is enough to the wise (Lat.)

We need fundamental professional knowledge: monumental architecture, modern technical capabilities (up to and including light video installations), cultural and ethical traditions and trends, political realities and trends, etc.

There is also the psychological sphere which in the end includes, among other things, the so-called *insights* – instantaneous ideas which emerge as manifestations of the subconscious, "instinctive" intellect, and the immense universe of preferences and creative (even fantastical, mystical etc.) images.

One of such highly creative ideas deeply steeped in historical and archetypal memory was the idea of "eternal flame" suggested in 1921 by the journalist and writer Gabriel Boissy (1879-1949) to commemorate the French soldiers who died during WWI. That eternal flame was started at the foundation of the *Arc de Triomphe* (Arc of Triumph) in the *Avenue des Champs-Élysées* in 1923.

Over the time that has passed since then, the "eternal flame" idea has gained popularity in many other countries.

The Eternal Flame near the Tomb of the Unknown Soldier in the Alexander Orchard near the walls of the Moscow Kremlin was started on May 8, 1967, in memory of the Soviet soldiers who died in WWII. It was ignited from the flame which was started on November 6, 1957, on the Mars Field in Saint-Petersburg.

Problem in Assignment 2.18 Memory Eternal	**Radical Contradiction: the monument must be**	
	plus-factor	minus-factor
With time, monuments decay and fall apart. So what is the most "adequate" symbol of memory eternal?	*eternal*	*not eternal*
Solution: create a monument in the form of a flame which is technically maintained by continuously feeding gas into burners installed inside symbolic sculptural and architectural elements. **Dominant TRIZ Models:** *change in the aggregate state of the object, segmentation, dynamization* **Additional TRIZ Models:** *inverse action, local property, replacement of mechanical matter, separation, change in color, copying, transition into another dimension, uninterrupted useful function.*		

Example 7.13. Solution for Problem 2.20. Window Washer. Let us formulate the IFR:

The "washer" ► must be ***inside*** (because you have to operate from inside the building!)
AND must be ***outside*** (because you need to wash the outer side of the window pane)

What can we say regarding the necessary resources which are accessible for transformation?

Let us turn to the Afs-catalog and look at the fundamental model called *"Separation of the conflicting properties in structure"* as it is worded in the catalog: *one part of the system possesses one property, while the system as a whole possesses the other (opposing) property.*

Now let us write out supportive specialized navigators:

03 Segmentation: disassemble an object into individual parts, raise the degree of disassembly (reduction into parts) of an object;

04c Replacement of mechanical matter: transition from static to dynamic fields, from temporally fixed to flexible fields, from unstructured fields to fields with a specific structure;

12 Local property: transition from homogenous to heterogeneous structure so that each part performs its functions in the best possible conditions;

18a Mediator: use another object to transfer or transmit an action.

The top part of the rod is the inductor. Accordingly, we need to segment it using models 03 and 12 (with one part outside and the other part inside), and then join the segments using some "mediator"; to do that, we need to use model 04 to replace the structure of the previous "mechanical field" with some other structure.

It should be borne in mind that the role of the mediator can be played not only by a material object, but also by a field, i.e. by energy. Let us turn to the fundamental model called *"Separation of the conflicting properties in material (energy)"* and to subsequent sections of description of specialized model *04b. Replacement of mechanical matter*: use of electrical, magnetic, or electromagnetic fields for the interaction of objects.

Eventually we would be able to invent a "washer" consisting of two magnetically-connected parts! But… such devices are already on sale. For example, some of them are used to clean fish tanks.

<table>
<tr><td rowspan="2">Problem 2.20. Window Washer</td><td colspan="2">Radical contradiction: the washer must be</td></tr>
<tr><td>plus-factor</td><td>minus-factor</td></tr>
<tr><td>The window cannot be opened wide. How do we wash the outer side of the window pane?</td><td>outside</td><td>inside</td></tr>
<tr><td>Solution: The instrument for washing window panes from the outside is made in the form of two plates which contain magnets and have specialized washing surfaces. To prevent accidental falls, the outer and inner plates are joined with a cable.
Dominant TRIZ models: fundamental transformation in structure
Additional TRIZ models: fundamental transformation in energy.</td><td></td><td>
</td></tr>
</table>

Example 7.14. Solution for Problem 2.21. Makeyev Rockets

IFR: the missile *must be long* (because it must have two stages)
AND the missile *must be short* (to fit inside the submarine)

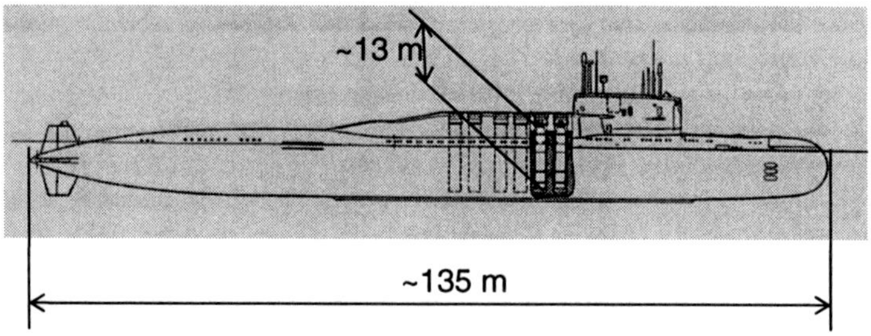

fig. 7.20. Height limit for nuclear submarine-carried missiles

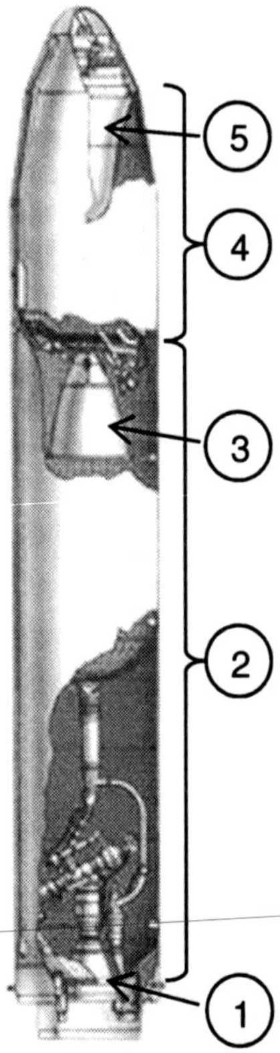

fig. 7.21. "Matryoshka" of Missile R-29 Vysota (http://makeyev.ru)

The most scarce resource is spatial. Meanwhile, workable solutions may be provided only by two resources – the *spatial resource* and the *structural resource.* Let us turn again to the A_{FS}-catalog and write out the models which appear to be the most promising as they deal with structural and spatial "contraction" of objects:

19 Transition into another dimension: ... use a many-storied composition;

34 Matryoshka (nested doll): an object is inside another object that is in turn inside yet another object, an object runs through a hollow space in another object;

11 Inverse action: instead of an action prescribed by the conditions of an assignment, complete a reverse action;

12 Local property: transition from homogenous to heterogeneous structure so that each part performs its functions in the best possible conditions;

16 Partial or excess effect: when it is difficult to achieve the desired effect completely, ...try to achieve a bit less or a bit more.

Here is the famous solution developed by designer V.P.Makeyev: the missile construction is based on the *matryoshka* principle. For the first time in the world, stage 1 engines (1) and stage 2 engines (3) are "embedded" in stage 1 fuel tank (2), while the head (5) is "embedded" in stage 2 fuel tank (4)!

Besides, part (5) is turned upside down in accordance with the *inverse action* principle! The *inverse action* principle is also applied to the entire missile – it is shortened instead of lengthened!

Principle 16 is used because the fuel load is relatively smaller as the fuel tank volume has decreased.

Now it becomes clear why designers jokingly referred to the engines and head of this missile as "drowned men".

Example 7.15. Solution for Problem 3.7. "MAGIC FAUCET".

Naturally, this assignment must be treated as a joke. And we will refrain from demonstrating the "real" invention – with or without TRIZ. But we will perform the familiar reinventing procedure according to TRIZ.

For that purpose, we will assume that the operating principle of this "Magic Faucet" is known, and that required models and contradictions have already been obtained by extracting.

TREND

Here is the assignment: a "water jet" is topped by a huge faucet producing a non-stop water flow, and there is no feeding pipe – the faucet apparently hangs in midair! We know from TRIZ that if a phenomenon exists, all requisite resources are available! QUESTION: how such construction could have been invented using TRIZ methods?

REDUCING

FIM: X-resource, together with available or modified resources, and without making the object more complex or introducing any negative properties, guarantees attainment of the following **IFR**:

[*water is continuously fed to the faucet to obtain a non-stop water jet*].

Standard (Technical) Contradiction

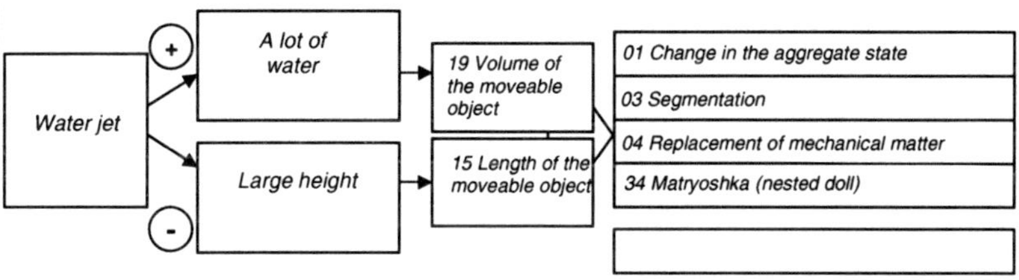

Radical (Physical) Contradiction

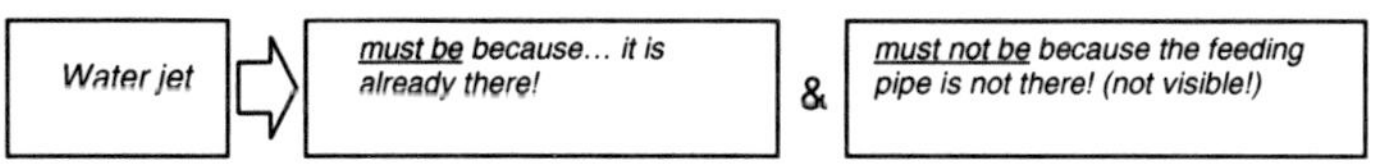

INVENTING

Key model – 34 Matryoshka (nested doll): *the general idea is simple the pipe is hidden inside the water jet!*

ZOOMING

Have the contradictions been removed? – Yes. – ~~No~~.

Super-effects: *an excellent aesthetically pleasing attraction!*

BRIEF DESCRIPTION

The faucet rests on a solid glass pipe. The water passes through the pipe (inside it, naturally!) and gets to the faucet. Then it flows out of the faucet and falls down on the outer side of the pipe so that the air jet "hides" the pipe inside it. Principle implemented: 34 Matryoshka (nested doll).

Send us reinventings of various objects – both funny and "serious", as in the previous example!

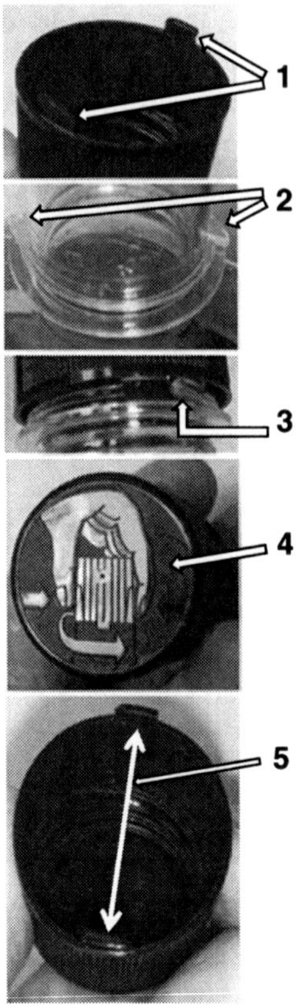

fig. 7.22. The secret of the cap is in its elasticity!

Example 7.16. Solution for Problem 3.3. Dishwashing Fluid Bottle Cap.

To model the process of inventing this solution, let us formulate the **IFR**:

the cap must ▶ *easily rotate* when being closed **AND** *not rotate* when being opened by an "unauthorized" person.

Maxi-FIM: the cap **BY ITSELF** prevents the opening of the bottle (by those who do not know the "secret").

Almost all resources are involved in this simple solution.

Structural resource: there are additional elements ("teeth" (1) on the cap, and "clasps" (2) on the bottle neck) creating a "catch" (3) which prevents the cap from being rotated counterclockwise (when the bottle is being opened).

Informational resource: 1) when the cap is compressed in a certain direction, the "teeth" unlock from the "clasps, and the bottle can be opened; 2) the cap carries a drawing (4) disclosing the "secret" and explaining how to open the bottle.

Functional resource: the cap has new functional properties which are not present in ordinary caps.

Spatial resource: the "teeth" and the "clasps" change the shape of the cap and the bottle in the direction required by the IFR and the FIM.

Material resource: the cap is made to be elastic (5).

Energetical resource: compression of an elastic cap creates an energy store sufficient for the cap to restore its circular shape BY ITSELF.

Problem 3.3. Dishwashing Fluid Bottle Cap	**Radical Contradiction: the cap must**	
	plus-factor	minus-factor
What can be done to prevent children from opening bottles containing dish washing fluid?	*rotate*	*not rotate*
Solution: the cap has two "teeth" jutting out at its lower edge; the "teeth" get caught inside special "clasps" on the neck of the bottle, and cap rotation is arrested. To open the bottle, the cap needs to be compressed in certain spots. The cap "clicks closed" by itself when rotated clockwise (when the bottle is being closed). **Dominant TRIZ Models:** *fundamental transformation in structure and in space*		

Example 7.17. Solution for Problem 2.22. Time is Powerless!

This happened in the 6th century BC. Pharaoh worshippers had to act in accordance with the famous Francis of Assisi maxim (see [55]): they chose to bow before the insuperable force of Christian conquerors, and wisely decided that time itself would restore things to their original state.

And that meant that when the temple at Philae was being turned into a church, and the frescoes were being plastered over… they had to do nothing at all! On the contrary, if they could, they would encourage the conquerors to make the plaster layer as thick as possible, and cover the newly-created white "canvas" with drawings of crosses and Christian holy texts. Ages would pass, and the world would maybe see again the wonderful paintings and colors of the ancient Egyptian culture!

And it did happen as they foresaw (fig. 7.23). Naturally, it took a very long time, and by then the ancient Egyptian religion had ceased to exist. Discovery of Egypt by archeological expeditions started in the 21st century, and soon scholars found incredible artistic riches of the ancient Egyptian culture. The frescoes in many temples had survived for millennia!

In 1980 the Isle of Philae was submerged by the waters of the Nile upon completion of the second stage of the Aswan Dam, but its copy was "reconstructed" (here is a good chance for you to make a reinventing according to TRIZ!) on the Isle of Aghilkia whose shape was modified to imitate that of the Isle of Philae. All major monuments were separated in 40,000 blocks and transferred to the new site!

Coming back to the reinventing of these ancient-history events and their later outcomes, we would like t o note that preservation of the frescoes (see also the example on the next page) was objectively supported by the following resources:

spatial resource – the entire surface of the frescoes was covered with plaster;

material resource – the plaster preserved, for many centuries, the frescoes and bas-reliefs of ancient temples of Egypt and, in particular, of the Isle of Philae;

temporal resource – the delayed event occurred, while if things had been different this extraordinary cultural heritage could have been destroyed.

fig. 7.23. a) Restored interior of the Isis Temple in Philae, Egypt (http://architecture.artyx.ru);
b) view of the new Isle of Philae (photo by author, January 2011)

Icon of Christ the Savior

WITH FAITH IN FATHERLAND!

Two icons which are sacred to the Russian history and culture – those depicting Christ the Savior and Saint Nicholas the Miracle Worker – were found in 2010 on the Spasskaya Tower (Savior Tower) and the Nikolskaya Tower (Saint Nicholas Tower), and subsequently restored at the initiative of the Foundation of Saint Andrew the First-Called.

Icon of Saint Nicholas the Miracle Worker

The Icon of the Savior first appeared above the gate of the Spasskaya tower in 1521 to commemorate salvation of Moscow from the invasion of the Crimean Khan Magmet-Girey. In 1737 the renewed Icon of the Savior survived a fire, and in 1812 – the invasion of the Napoleonic army. The latest renovation was made in 1896. Creation of the Icon of Saint Nicholas the Miracle Worker dates back to the late 15th or early 16th century. When the French were retreating from Moscow in 1812, they blew up the Nikolskaya Tower, but the icon remained intact! It also survived in 1917 despite the massive damage inflicted by bullets and shrapnel (see below the picture of the thanksgiving prayer held in 1918). The order to paint the icons over was issued in 1937, on the 20th anniversary of the October Revolution. The painters, however, must have been real masters of their craft – even though they could have paid with their lives for what they did: they hid the icons under mesh wire which was covered with a thick layer of plaster and only then painted over. This is how the icons were preserved almost unscathed. Rediscovered and rededicated, these Over-the-Gate Holy Icons of the Moscow Kremlin will continue to feed the faith of new generations in the Fatherland, in restoration of Russia, in prosperity of its long-suffering peoples and lands, and in its future worthy of its great history. As for us, we will continue to learn from history – reacquire our knowledge and revive our ability to create, understand beauty, defend the truth, respect and increase our heritage.

Work is begun: look back upon the centuries long passed...

The Spasskaya Tower, 2010

And there emerges the face of Christ!

The Nikolskaya Tower, 1918

www: fap.ru; patriarchia.ru; kp.ru

Example 7.18. Solution for Problem 3.8. BRT. BRT, or Bus Rapid Transit, are dedicated city street lanes reserved for high-speed bus traffic.

However, in many streets, even those that are very broad, there is already not enough space to create such lanes. That is why the BRT concept is best implemented using the "third dimension" – the height. It is suggested that BRT lanes be built in the "second story". This is 2-3 times cheaper than using the same space to build railway lines.

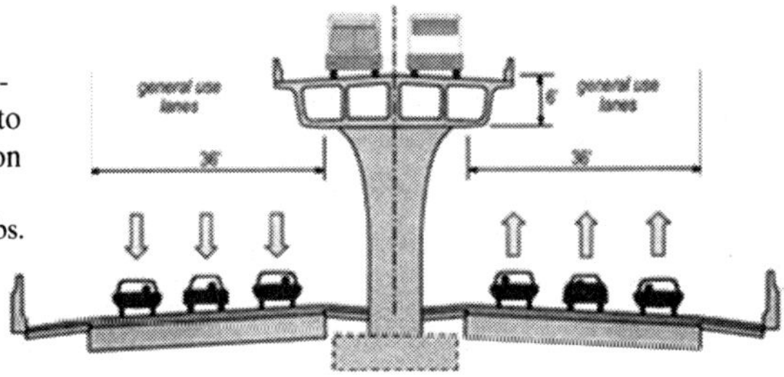

fig. 7.24. Transition into another dimension according to TRIZ – possible construction of bus trestles (http://lightrailjacksonville.webs.com/brtvslrtii.htm)

Example 7.19. Solution for Problem 3.9. Autobus – Trolleybus – Tramway – ?

An extraordinary and surprisingly efficient idea has been proposed by the Chinese company *Hashi Future Parking Equipment Co., Ltd.* – a very high and wide "hollow bus *cum* tramway" moving above two motor car lanes located between its wheels! The vehicle can be recharged at or between stops. The four sections of the vehicle can seat 1,200 passengers. Perform extracting and reinventing for this wonderful invention.

fig. 7.25. Ingenious invention: Transition into another dimension + Matryoshka + many other things (http://hsfuture.com/ehbus.html)

Example 7.20. Solution for Problem 2.23. Art and Craft of Ice.

For the IFR, the original radical contradiction can be rewritten as follows: ice bricks must be ▶*many* in the sculptures **AND** *few* in the territory of the exhibition (as they cannot be made by the artists).

The solution uses virtually all types of resources. Let us mention the following:

systemic resource – ice is cut with tractor-mounted band-saws not far away, on the large nearby Sungari river, and carried by trucks (fig. 7.24,a), i.e. the main operation is withdrawn out of the system (the site of the exhibition);

temporal resource – the sculptures are erected by special teams (fig. 7.24,b) over the course of almost two months (this process resembles assembly of large Lego erector set parts), and finished by the artists in the last two or three weeks.

fig. 7.26. This is how they built an ice city in Harbin, China

Examples of solutions considered in this and the previous sections are, undoubtedly, inventive – even though there is a temptation to say that they are "based on routine common sense". Still, they represent acts of creation[124]. Ideas for these solutions only seem to be simple. However, *they were not apparent until someone explained them and showed us how to make this or that artifact* – see a small insert about the phenomenon called *"disappearance of the miracle"* at the end of section *3.2 Noospheric Modeling Concept in Modern TRIZ.*

Be that as it may, if we want to solve more complex – and especially new, unfamiliar – tasks, we will have to go through one of the mandatory steps in the modeling of creative solutions: we will have to analyze contradictions. And we will continue to learn on simple examples! Because it IS simpler, and therefore easier to understand and more efficient. And the really important thing is not what the problem is about, or how efficient the solution is, or whether we like it or not. The really important thing is that we can demonstrate the existence of certain models – and do it in a way that will help you remember those models and use them to solve both training and real-life problems.

---

[124] Innumerable magnificent examples of such work are provided by the *Crafty Hands (Oshumeliye Ruchki)* show created by the First Channel of the Russian Television, similar columns in numerous magazines, such as *Science and Life*, and thousands of web sites.

Advanced Primary TRIZ Models

Everything should be made as simple as possible, but not simpler[125].

Albert Einstein

8 Integrated TRIZ Models

8.1 Extended Classification of Contradictions

In the problems analyzed in the previous sections, we focused on standard and radical contradictions. However, in practice problem-solving also involves other types of systemic contradictions. A brief description of such contradictions, including those already familiar to the readers, is provided below.

General contradiction	**General contradiction** (in classical TRIZ, *administrative* contradiction) – systemic requirement which reflects the general need to attain a certain property or state, or remove an obstacle preventing required operation of the system.
Ordinary contradiction	**Ordinary contradiction** – systemic requirement represented by a certain target property that needs to be either (1) increased, strengthened (positive type: only plus-factor, or trend-factor, is specified) or (2) corrected or, *vice versa*, weakened, removed (negative type: only minus-factor, or problem-factor, is specified).
Radical Contradiction	**Radical contradiction** – unity of two diametrically opposed systemic requirements as they apply to the same property.
Standard Contradiction	**Standard contradiction** – unity of two incompatible requirements as they apply to different properties.
Compositional contradiction	**Compositional contradiction** – unity of two or more standard contradictions.

Analysis of compositional contradictions made up of radical contradictions lies beyond the scope of this book.

Fig. 8.1 presents a classification of the main systemic contradictions. Let us look at several examples tailored for beginners.

We will start with a general contradiction whose understanding requires some experience. We keep encountering this type of contradiction when we set general objectives. Now, the examples.

[125] Albert Einstein (1879-1955) – outstanding German physicist, one of the founders of modern physical theory, winner of Nobel Prize in physics (1921)

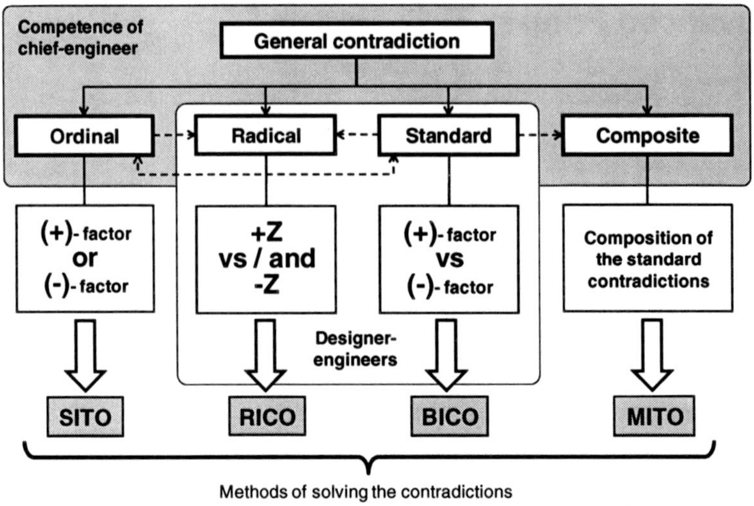

fig. 8.1. Classification of contradictions and contradiction resolution methods

Example 8.1. Training of marathon swimmers. We need to organize year-round training of swimmers who will be participating in long-distance and super-long-distance swimming competitions in open water bodies.

Problem: year-round training in open water bodies is impossible because of unsuitable weather conditions. Training in ordinary swimming pools involves the need for the swimmer to turn around each time he reaches the edge of the pool, which impairs his technical skills and reduces his speed.

Original idea (variant): construction of a specialized pool.

What can be done? First off, let us briefly formulate the gist of the problem.

General contradiction: *it is necessary to increase efficiency of pool training of marathon swimmers.*

This wording does not disclose either the nature or the cause of the problem. But in many cases this is the way problems are presented by their originators. The originators themselves may, or may not, understand what it is exactly that they mean. For example, in this case it would be expedient to transform the overly general presentation of the problem into a listing of more specific requirements: eliminate the need for turnarounds at pool edges, make the water track "endless", or build a specialized pool with a certain degree of complexity, etc.

Example 8.2. Training of divers. Training of divers* – people who jump into water from special platforms or towers – involves hundreds, if not thousands of jumps. Many such jumps are required to master certain complex movements. There are no guarantees that entry of the diver into the water will be smooth. If things go wrong, collision with the water may prove to be rather painful, or the diver may be injured.

* Scuba diving aficionados are also called "divers"

Problem: possible failure of the diver to enter the water in a technically correct manner during a training session increases the risk of injury.

General contradiction: *it is necessary to improve the safety of diver training.*

Again, the formulation of the general contradiction is deficient: the task appears to be more or less clear, but this clarity is deceptive. For example, what exactly is the meaning of "improve the safety"? Do we have to just mitigate the risk of injury or completely eliminate that risk? Also, it is not clear which resources are available or, *vice versa*, not available and cannot be used to solve the problem. For example, does the diver have to enter the water, or is it possible for him to merely imitate entry without any actual (physical) contact with water? Finally, there is general confusion as to the criteria to be used to evaluate possible solutions.

This is the main drawback of excessively general problem formulation – which in most cases looks like this:

obtain some general result
(without specifying what can helpful in finding a solution,
or what can prevent that from happening)

Incidentally, specific formulations can also have the same drawback – lack of apparent instruments and resources that can be used to solve the problem. For example, recall the orders issued to designers of new Soviet fighter planes in the 1930-es (Examples 5.11-5.15): the current world speed record is, say, 300 km/h, therefore you must, as soon as possible, assure speeds of up to 450 km/h! That's it! Full stop.

You have your orders, you are Problem Solvers, so go and solve this one! We don't care how you do it!

Similar general contradictions are present in many other problem situation. Reconsider from these positions descriptions of problem situations in Problems P4 and P5 and in Examples 2.3-2.6, 2.12-2.23, 3.2 (problem within the problem), 5.16-5.18, 5.20, 5.22-5.27, and 6.3.

Let us consider two more simple and apparently clear examples.

Example 8.3. Glass of tea. General contradiction: *it is necessary to get hold of a hot glass to drink tea from it.*

This only describes the existence of a general need to perform the action without specifying the problem. We can only guess that, inasmuch as the glass is hot, the one who is holding it can burn his fingers.

Example 8.4. Organizer page corner. General contradiction: How do we make sure that the user can quickly and neatly tear off the corner of the page?

This is a general formulation of the problem without specifying what prevents the user from quickly tearing off the corner of the page, or what is considered "neat" and "not neat".

Such statements (formulations) can be encountered in virtually all problem situations. They mark the beginning of explanation of almost any problem. And that is why they are so ***general*** – they give only an ***initial general idea*** of the existence of a certain problem *which needs to be solved.*

The structure of the problem situation is not disclosed – not at this stage. It is not clear what objective is being set, and what prevents its attainment.

Still, despite its "fuzziness", the general contradiction does delineate the area – and sometimes the direction (both purpose and trend) – in which an attempt to solve the existing problem will be made. This direction is expressed from the point of view of the user, consumer, client (superior system) who does not elaborate on the essence of the problem, at least at the beginning, but wishes to obtain a certain result.

For example, for the situations described above:

(1) In Example 8.1, we should assume that there are certain properties typical for a training session conducted in an ordinary swimming pool which are not consistent with the properties typical for an open water marathon competition, the net result being that the swimming pool training session is not adequate for the objectives set. Even before the client's claims are described in more detail, we can imagine that pool training is characterized by higher water temperature, the need to expend energy on multiple turnarounds each time the swimmer reaches the edge of the pool (because the pool is too short), etc.

(2) In Example 8.2, it is clear that the main safety threat associated with diving is heavy impact at the time the diver enters the water – and this threat must be disclosed and removed!

(3) In Example 8.3, our guess as to what prevents the user from holding the hot glass needs to be verified, and then the problem must be eliminated!

(4) In Example 8.4, we need to clarify what exactly prevents the user from quickly and neatly tearing off the corner of the page, and then help him or her overcome the existing problem!

General contradictions are reflections of the problem situation, even though they express systemic contradictions only implicitly. In other words, in a general contradiction the existence of an objective which is hard to attain, or of an obstacle which must be removed, is implied – and can be clarified.

ATTENTION: When solving real-life problems, general contradictions always must be transformed into contradictions of any other type (see fig. 8.1).

Naturally, *the path to the required outcome is not known.* And to make progress in solving the problem, it will be necessary to structure its description, define outcome evaluation criteria, identify existing obstacles, available resources and restrictions. To do that, TRIZ, systemic analysis and quality analysis offer special instruments, such as function/cost analysis.

Classification of the main contradiction resolution methods is presented in fig. 8.1. The abbreviations used are deciphered below:

SITO – Single In Tuple* Out: only one A-matrix entry point is used – either through the plus-factor (for positive ordinary contradictions) or through the minus-factor (for negative ordinary contradictions);

---

* Tuple – sequence, ordered list of elements

RICO – Radical In Cluster Out: this method uses two A-matrix entry points, and both factors have the same name; in practice, to find a solution, it is necessary to use additional tables to select auxiliary specialized navigators;

BICO – Binary In Cluster Out: this method uses two A-matrix entry points, while the plus-factor and the minus-factor have different names;

MITO – Multiple In Tuple Out: combination of several standard contradictions requires the use of multiple plus- and minus- A-matrix entry points.

More detailed descriptions of these methods are provided in the appropriate sections of this book.

In conclusion of this section, we will note that structurally contradictions form a "mono-bi-poly" system which is well known in TRIZ (fig. 8.2).

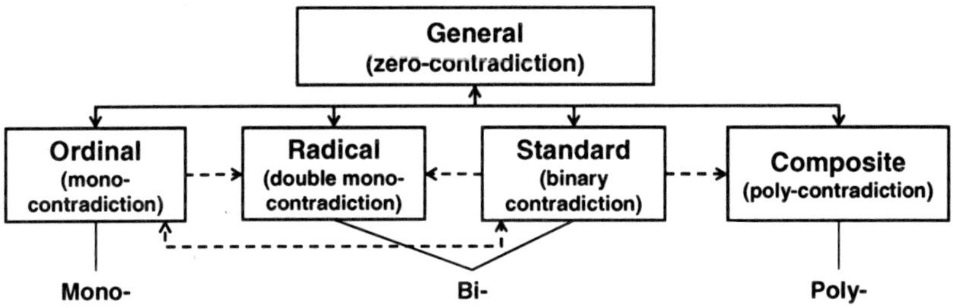

fig. 8.2. Classification of contradictions on the basis of systemic TRIZ model "mono – bi – poly – mono"

The system "starts" with a general contradiction which can be designated as a "zero contradiction", as it does not contain any expressly formulated contradiction, but rather some indistinct general requirement or wish.

The general contradiction can first be transformed into an ordinary contradiction which can be designated as a "mono contradiction", as it operates with an expressly specified positive (*plus-factor*) or negative (*minus-factor*) aspect.

The ordinary contradiction is adjacent to the radical contradiction: as soon as we pose two mutually exclusive requirements to any ordinary property (i.e. demand that the value of the main parameter of such property be increased and decreased at the same time), such ordinary contradiction is formally "radicalized".

Other relations indicative of mutual transformations of contradictions are shown by dashed arrows, and discussed in higher-level Modern TRIZ Academy programs.

Finally, we would like to draw your attention to the fact that fig. 8.2 also illustrates the possibility of any contradiction (including a compositional contradiction) reverting to the general contradiction (in which case there occurs a "poly-zero" systemic transition).

In this context, the "zero contradiction" is eminently meaningful from the viewpoint of TRIZ ideology, as it may signify that the problem has ceased to exist at the given level, and that both the problem and the process of solving the problem have been promoted to a higher system level than, say, the level of the operating zone being modified.

Then the existing contradictions are, indeed, "reset to zero" in the initial operating zone, and the current problem is replaced with a new problem existing at a higher level and possibly formulated in the form of a new general contradiction.

8.2 Resolution of Ordinary Contradictions Using the SITO (Single In Tuple Out) Method

Ordinary contradictions are the closest to general contradictions. Virtually in every case there is an opportunity to transform a general contradiction into an ordinary contradiction. To do that, it is necessary to specify either the ultimate objective, or the obstacle preventing satisfaction of a certain need and, accordingly, formulation of an ordinary contradiction of the positive or negative type.

Example 8.5. Training of marathon swimmers (continued from Example 8.1). The following statement can serve as the first approximation making our general contradiction "closer" to an ordinary contradiction: *How do we improve the quality of swimmer training by excluding the need for them to perform turnarounds as they reach the edge of the pool?*

This statement can be replaced with an ordinary contradiction of the negative type which specifies the functional drawback that needs to be removed: *How do we eliminate the need for the swimmers to perform turnarounds as they reach the edge of the pool?*

From this initial statement, we can proceed to an ordinary contradiction of the positive type which specifies the property that needs to be reinforced: *How do we create a "very long" (ideally, "endless") water track?*

Example 8.6. Training of divers (continued from Example 8.2). The following statement can serve as the first approximation making our general contradiction "closer" to an ordinary contradiction: *How can divers train safely so as to exclude the possibility of injuries when they fail to enter the water in a technically correct manner?*

Here an ordinary contradiction of the negative type is more appropriate, as the drawback that needs to be eliminated has been named expressly: possible injury when the diver enters the water. Consequently, the contradiction can be formulated as follows: *How can we exclude the possibility of injuries when the divers fail to enter the water in a technically correct manner?*

Example 8.7. Glass of tea (continued from Example 8.3). First approximation to an ordinary contradiction: *How can one take a hot glass to drink tea from it, if the glass burns the fingers?*

Ordinary contradiction of the negative type: *How do we eliminate the scalding effect that the hot surface of the glass has on the fingers?*

Example 8.8. Organizer page corner (continued from Example 8.4). First approximation to an ordinary contradiction: *How do we help the user to quickly and neatly tear off the corner of the page?*

Ordinary contradiction of the positive type specifying the property that needs to be reinforced: *How does one tear the corner of the page off quickly?*

Second variant of the positive ordinary contradiction: *How does one tear the corner of the page off neatly?*

Ordinary contradiction of the negative type specifying the functional drawback that needs to be eliminated: *How does one avoid random deviations in the shape and size of the piece which is quickly torn off the corner of the page, resulting in the organizer acquiring a shabby, tattered look?*

In this case specific qualifiers, such as "quickly torn off" and "tattered look" (or, conversely, "neat look") act as both restricting conditions and evaluation criteria of the future solution.

Let us consider application of the SITO method to resolution of ordinary contradictions.

The SITO method is most frequently used to obtain express solutions, i.e. tentative/trial solutions of a certain problem where not all plus- and/or minus-factors have been identified, but there is a possibility to test the problem (and the relevant system) for "resistance to change".

There are two variations of the SITO method (SITO+ and SITO-) depending on whether entry to the A-matrix is gained via a plus-factor or a minus-factor.

The essence of the method is very simple: an appropriate plus- or minus-factor is selected for a certain informally described property as the point of entry to the A-matrix. Then it is necessary to select, from the appropriate line (if it is a plus-factor) or column (if it is a minus-factor) of the A-matrix, the navigators which are mentioned in such line (column) most frequently. This yields a ranked list of navigators, with top positions occupied by those with the highest frequency (fig. 8.3).

It is recommended to first try the highest-ranked navigators (i.e. those closer to the top of the list).

Clearly, there is no need to make such selection each time you build an ordinary contradiction, as you can always use tables prepared in advance (see sections *14.2.7 Table for the SITO-plus method* and *14.2.8 Table for the SITO-minus method*).

This approach is based on the assumption that, somehow or other, all navigators located in the same line "participate" in changing the appropriate plus-factor, and all navigators located in the same column "participate" in changing the appropriate minus-factor. All you have to do is select the navigators that are used more frequently than others.

Examples illustrating application of TRIZ models are presented here as descriptions and outcomes of joint work done by two characters – the "Inventor" and the "Client" – seeking to resolve the problems posed by the Client.

The examples are described in a simplified form, and such descriptions may leave out certain details, sometimes rather important details.

On the other hand, solutions provided are not subject to publication restrictions stipulated by non-disclosure agreements protecting the interests of the Client as may have been signed by the Inventor. Our sole task is to demonstrate the immense diversity of problem situations, and illustrate efficient application of TRIZ instruments.

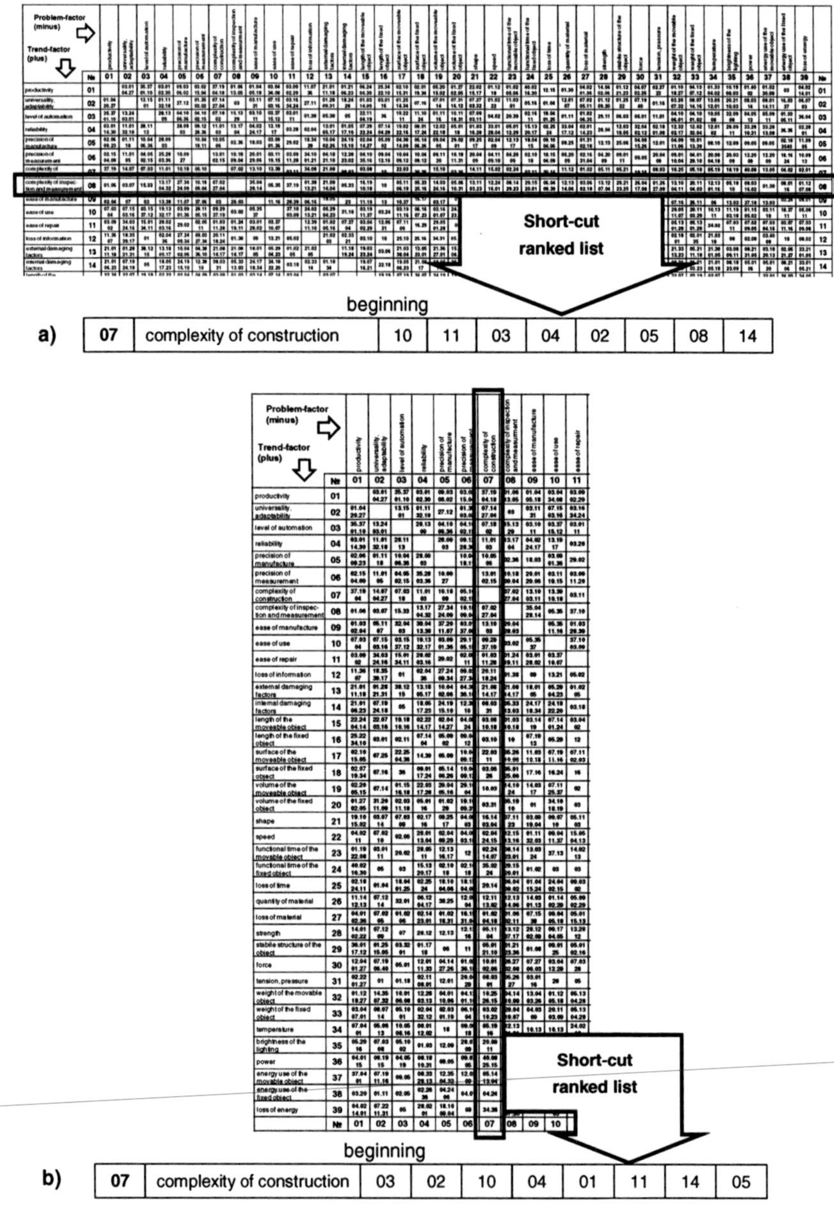

fig. 8.3. Creation of ranked lists for the SITO+ (a) and SITO- (b) methods

Series of examples: improvement of the labeling machine

To send a letter to a new address (if the intended recipient has changed his or her current address), a label with such new address is glued onto the envelope. In large postal centers, this operation is performed by special labeling machines.

The conventional component scheme of the machine (one of many supplied by the Client to postal centers) is shown in fig. 8.4.

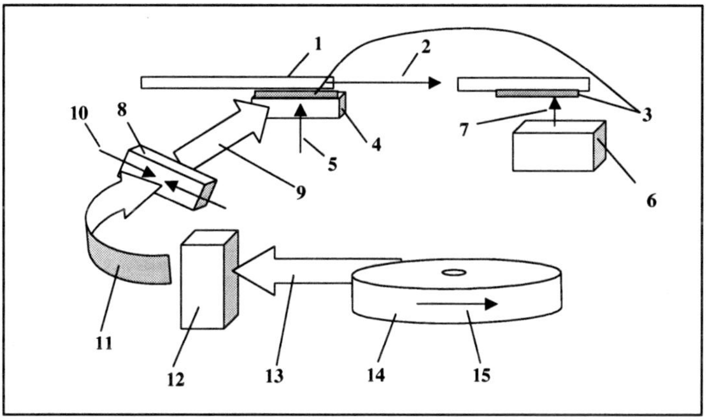

fig. 8.4. Component scheme of the labeling machine

The envelope 1 moves in direction 2, while the label 3, with glue applied to one side, is pressed to the envelope by a pneumatic clamp 4 which feeds a burst of compressed air in direction 5. Then a printer 6 prints the new address on the label with the help of a paint jet 7.

The shears 8 move in direction 10 and cut a piece of certain length (the label) from the tape fed by a dispenser 12 in direction 11. The cut-off label is then moved to the clamp 4 in direction 9.

The dispenser 12 pushes through shears 8 a neatly measured piece of tape wound off the roll 14 in direction 13. When the tape is being wound off, the roll rotates in direction 15.

Envelopes have different sizes, but the machine automatically determines the moment when the leading edge of the envelope enters the operating zone, and instructs the clamp to attach the label prepared for that envelope.

Letters need to be processed faster and faster – and this affects operating standards governing the use of labeling machines. Meanwhile, despite the increase of the processing speed (main parameter), labels must still be placed exactly in the designated spot.

Let us review (in this and following sections) several examples illustrating ongoing modernization and improvement of the labeling machine. The relevant tasks have been formulated by the Client and resolved together with the Inventor over the course of several years.

Example 8.9. Two-times increase of the main parameter. When the Inventor tried to increase the envelope travel speed 2 times, it turned out that the label often missed the required spot or was attached askew. This happened both because the impact of the air burst on the surface of the label was not always uniform, and because the thin label was easily deformed.

The Inventor came up with a solution using the Meta-Algorithm of Invention (MAI) as follows:

Step 1: Trend (diagnosis). Things will get harder!

The label attachment operation is shown in fig. 8.5.

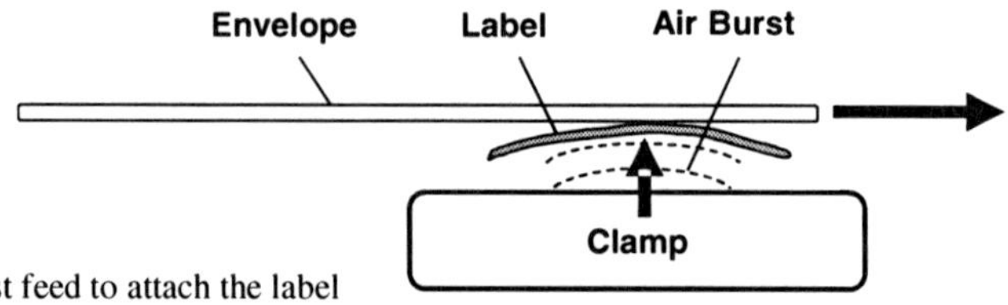

fig. 8.5. Air burst feed to attach the label

The main causes of the problem are as follows:

First, the label becomes deformed, and its gluey side touches the surface of the envelope too late (or, sometimes, too early).

Second, there emerge, in the air adjacent to the surface of the moving envelope, aerodynamic effects preventing normal convergence of the label and the envelope.

Third, sometimes the air burst bends the label in the middle, and when such label, attached only at the center, starts moving together with the envelope, its leading edge gets hit by the oncoming rush of air. As a result, the label curves up, shifts sideways or even comes unstuck, and the entire operation ends in a failure.

Step 2: Reducing (reforming). What cannot be changed?!

Two models give a relatively good idea of the roots of the problem.

One of the possible standard contradictions:

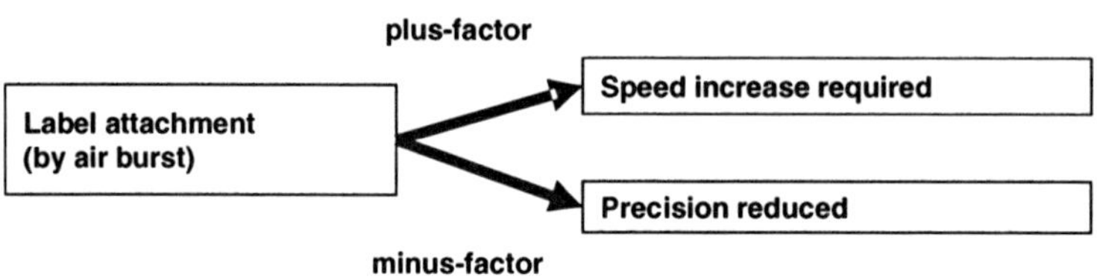

fig. 8.6. Possible wording of the standard contradiction for the labeling problem

A radical contradiction variant:

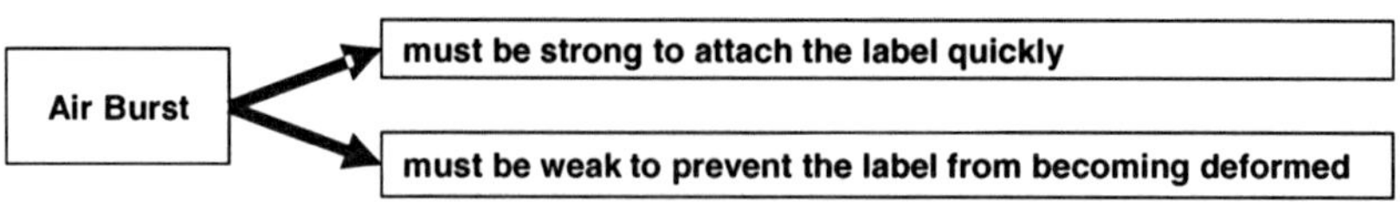

fig. 8.7. Possible wording of the radical contradiction for the labeling problem

Required "ideal result": the label is attached "**quickly-AND-precisely**" at any speed, however high!

Let us consider the resources available in the *operating zone* adjacent to the surface of the envelope, including the label and the portion of the envelope surface where the label is to be attached.

Object (systemic) resource: required productivity increase is achieved primarily by increasing the moving speed of the item (envelope), which gives rise to the problem (poor precision). There is no "room for maneuvering" here, as even a 2-times speed increase has proven problematic.

Informational resource: it may be possible to improve precision of measurements to optimize the air burst timing.

Functional resource: air burst parameters can be fine-tuned, but this looks too complicated. It is difficult to generate an air burst of the "right shape" and release it exactly at the "right moment".

Structural resource: the directions in which the envelope and the label are moving are mutually orthogonal, i.e. they move at right angles with respect to each other (see bold arrows in fig. 8.5). When the envelope begins to move faster, it becomes increasingly difficult for the label to "hit the target". Changes in this area are quite possible.

Spatial resource: it is possible to manage the "shape" of the air burst, but this solution does not seem viable (see *Functional resource*).

Temporal resource: there are virtually no "additional" possibilities (see *Informational resource*).

Material resource: the Client has forbidden to change the material of which the label is made.

Energetical resource: increasing the power of the air burst will only exacerbate the problem. However, it is possible to align the directions of energy flows, which is close to the results of structural resource analysis.

Bottom line: within the framework of the original operating principle of the machine, only one resource – *structure of movement* – shows any promise at all in terms of finding a suitable solution.

To formulate an express solution, we can apply the SITO+ method subject to the following wording of the positive ordinary contradiction:

"it is necessary to increase the speed of movement of the item (envelope)".

This corresponds to plus-factor *22. Speed* in A-matrix.

Step 3: Inventing (transformation). Creating conditions for the "ideal result"!

From *SITO-plus table*, we select line 22 which contains the following set of navigators: 04, 01, 11, 02, 08, 15, 30, 05.

Our attention is drawn to navigator 04,c *Replacement of mechanical matter* which can be interpreted as follows: *replace... unstructured fields with fields with a specific structure.* In our interpretation: *replace "badly" structured and "uncoordinated" movements with "well" structured and "coordinated" movements.*

Pursuant to model 02,b *Preliminary action*, the following recommendation is quite promising: *prepare objects in advance so that they can be put to work from the best position and are available without loss of time.*

There is a nascent trend to modify and coordinate the trajectories of movement of the envelope and the label. The trajectory of movement of the envelope cannot be changed. All we can do is change the trajectory of movement of the label.

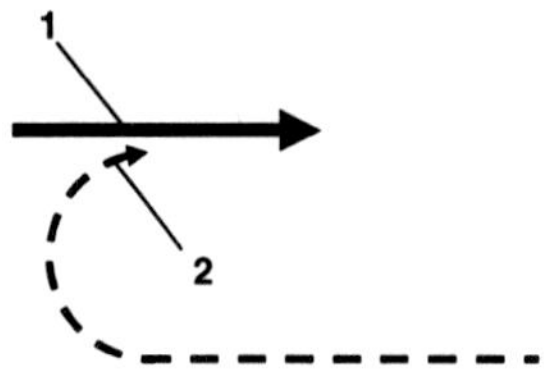

fig. 8.8

Here is an idea for coordinating the trajectories – the trajectory of movement of the envelope is straight (1), and the trajectory of movement of the label is "rotational" (see fig. 8.8).

During such coordinated movement, the label is "rolled onto" the envelope (getting glued to it in the process), and then the envelope and the newly-attached label move together to the site of the address printing operation.
However, implementation of these trajectories in the "old" machine proves impossible. Moreover, subject to the overall trend (further increase of the speed at which the envelope moves along the process line), we come to the conclusion that the air burst label attachment operating principle has exhausted itself.

We need a new assembly. A possible new technical solution is presented in fig. 8.9.

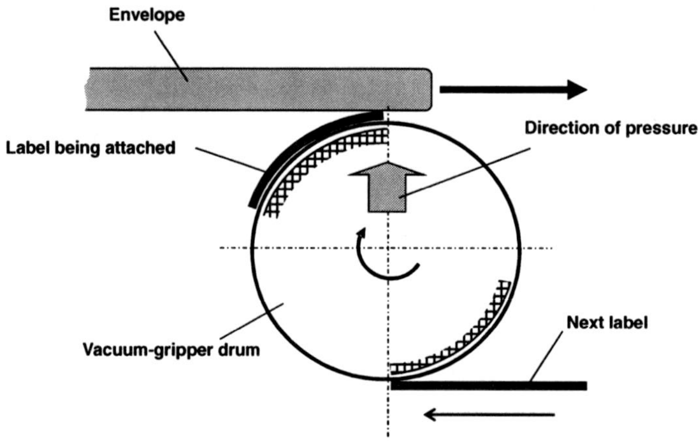

fig. 8.9. Label attachment mechanism in the form of a vacuum-gripper drum

It is proposed that the label attachment device be made in the form of a rotating drum with one, two or more "windows" in its surface (cross-hatched areas in the drawing). The labels are pressed to the surface of the drum by atmospheric pressure, as air from the inside of the drum is sucked out by an inbuilt pump.

Step 4: Zooming (verification). Is this a good solution?

Analysis reveals the following: First, the original contradictions have been completely eliminated.

Second, the new principle is very promising in terms of further increase of the speed of operation! The Client conceded that the idea underlying this solution was efficient, even though it required a radical modification of the operating principle used by the label attachment assembly.

Third, the Client greeted the emergence of a "powerful" positive "super-effect": the letter processing shop would now become much quieter, because there would be no more incessant "rub-a-dub-dub" noise produced by the air burst generator of the old machine!

Incidentally, the Inventor did not know about the noise problem produced by the "old" machine, and the "quiet" nature of the new machine was truly an unexpected and pleasant "surprise" side effect. This became one of the most important advantages of the new machine!

Example 8.10. Four-times increase of the main parameter. Some time later, it became necessary to double the envelope movement speed again. This revealed a very unpleasant drawback which proved to be very hard to get rid of, at least in the beginning, despite the apparent and simple cause of the problem.

The problem was the erratic tape feeding behavior of the dispenser (position 12 in fig. 8.4) due to the slipping of the pulling rollers on the gluey side of the tape.

The Inventor designed a solution as follows:

Step 1: Trend (diagnosis). The problem won't give up!

The situation is illustrated by the drawing in fig. 8.10.

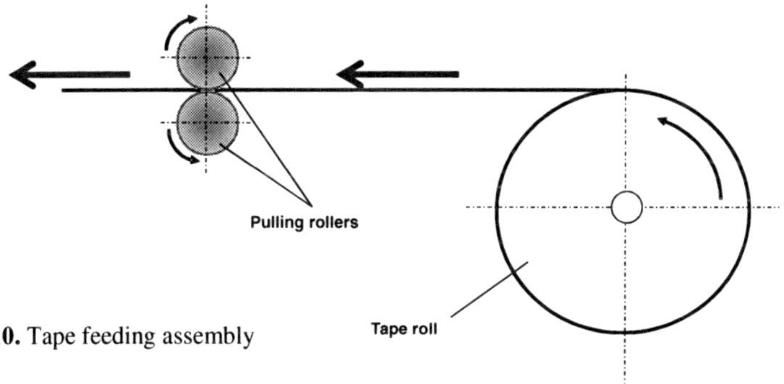

fig. 8.10. Tape feeding assembly

When the pulling roller on the gluey side of the tape slipped, the leading edge of the tape failed to reach the cutting position (fig. 8.11,a).

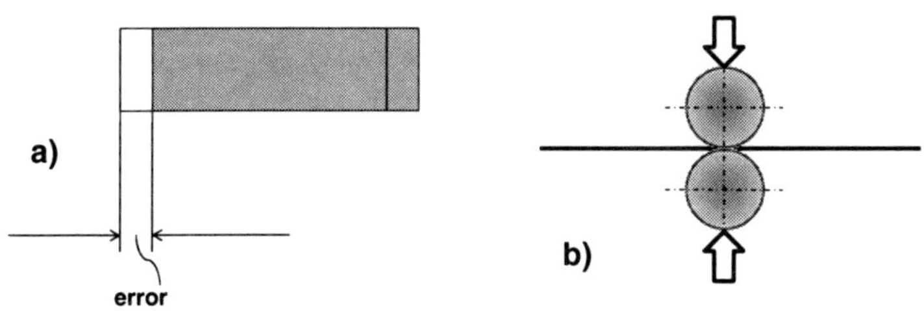

fig. 8.11. Attempt to correct the error (a) by increasing the roller pressure (b)

When the rollers were pressed together a little harder to prevent the slipping (fig. 8.11,b), the tape became deformed. Naturally, fitting the rollers with "teeth" was out of the question!

Step 2: Reducing (reforming). Again – look at the root of the problem!

The following two models give a relatively complete idea of the problem.

Original standard contradiction:

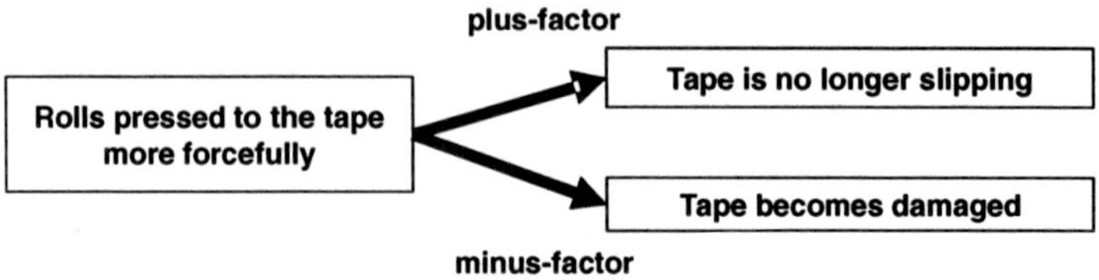

fig. 8.12. Possible wording of the standard contradiction for the slipping roller problem

"Irreconcilable" radical contradiction:

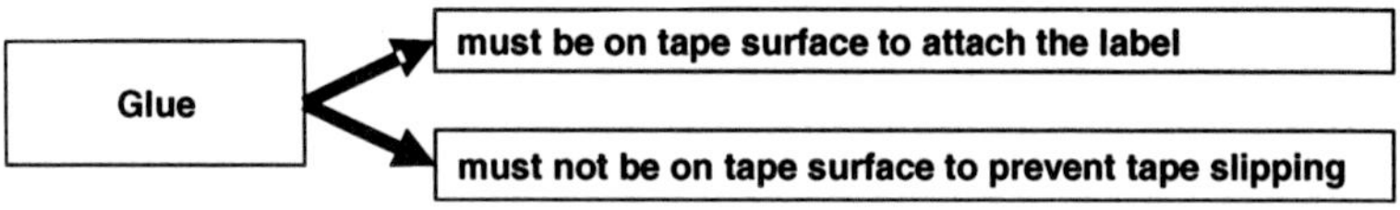

fig. 8.13. Possible wording of the radical contradiction for the slipping roller problem

Required "ideal result": pull the tape "**precisely-AND-without deformations**" at any envelope movement speed!

However, the nature of the radical contradiction excludes the possibility of solving the problem without radical changes.

To formulate an express solution, we can apply the SITO- method subject to the following wording of the negative ordinary contradiction: **it is necessary to remove the negative impact of the forceful pressure exerted on the tape by the rollers.**

This requirement corresponds to minus-factor *13 External damaging factors* from the *SITO-minus table*. The table offers the following list of navigators: 21, 01, 05, 03, 04, 38, 06, 08.

Step 3: Inventing (transformation). Resource is exhausted – reinforcements on the way!

Unfortunately, we can hardly use navigator *21 Transform damage into use*, as strong pressure is needed to assure that the tape is fed continuously despite the growing speed and the presence of glue.

There is another promising navigator, though: *01,b Change in the aggregate state of the object* which can be interpreted as follows: *change concentration or consistency (of the glue)*. Let us note *in passim* that to reduce consistency to "zero" is to... remove the glue!

Pursuant to model *05 Separation*, the following recommendation is quite promising: *separate the "incompatible part" ("incompatible property") from the object or include the only really necessary part (necessary property) into the object.*

Apparently, we simply need to remove the glue from at least one section of the tape to stop the rollers from slipping!

Navigator *03,a Segmentation* prompts: *disassemble the object into individual parts.*

Well, we can separate the rollers into parts which will be touching only those sections of the tape which are free of the glue. Alright, that DOES look like a viable solution!

The same solution can be obtained by using a fundamental transformation in space: *one part of the operating space possesses property A, while another part of the operating space possesses property not-A.* The entire system has both required properties – provided this does not impair any other properties.

OK, here is the tentative solution of this problem: some parts of the tape are not covered with glue, and the points of contact between the pulling rollers and the tape are located at such "glueless" spots (fig. 8.14).

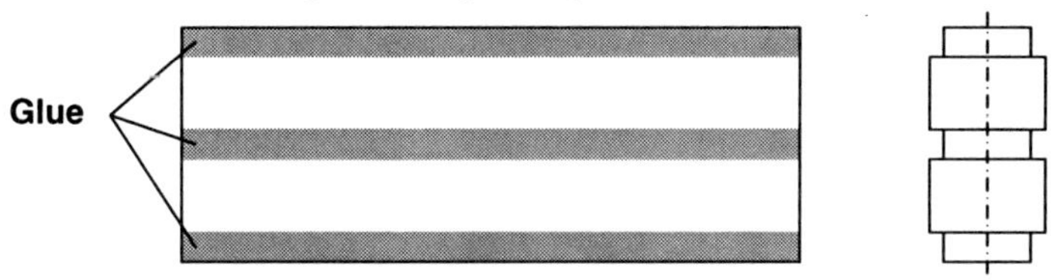

fig. 8.14. Possible solution: glue stripes

Step 4: Zooming (verification). Have all problems been resolved?

The Client agreed that the idea suggested was quite efficient and not too complex in terms of implementation.

There was also a very useful super-effect – glue consumption decreased, and so did tape production costs. It would be easy to renegotiate delivery conditions with the supplier. With numerous postal centers using A LOT of tape, the overall economic effect would be quite impressive.

However, the Client pointed out to a very serious negative effect: it would be necessary to change the existing label "specifications", and that might take years… Still, those who pay attention to such things may have noticed that "striped-glue" address labels on magazines and envelopes have already been around for years…

8.3 Resolution of Compositional Contradictions Using the MITO (Multiple In Tuple Out) Method

According to the MITO method, the number of standard contradictions defined is determined by the terms of the problem. Navigators from all clusters are written out for each contradiction. Then a list is created, where the navigator most frequently encountered in all selected clusters is placed at the top of the list followed by the navigator with the second-highest frequency, etc.

Similarly to the SITO method, it is recommended that the navigators be used in the order of precedence starting with the one at the top of the ranked list.

Clearly, it is difficult to do all that work manually. However, in the EASyTRIZ software package this procedure is completely automated.

Let us illustrate the application of the MITO method using examples further improving the labeling machine described in the previous sections.

Example 8.11. Ten-times increase of the main parameter. Some time later, the Client needed to double, or even triple, the envelope movement speed one more time. The problem was that at higher speeds the shears did not have enough time to snip off labels, while the use of the start-stop tape feeding mode led to the tape bending under the influence of air, which gave rise to various cutting defects.

As a result, the Inventor wound up with at least two operating zones, each with its own problems, and both of them together sporting about a dozen contradictions.

And only the label attachment assembly (Example 8.9) continued to successfully operate despite the increase of the envelope movement speed! This is what you get when you find a felicitous solution with a broad margin of safety covering substantial future growth of the main parameter.

The Inventor and the Client's experts started modernization of the machine with elimination of the shears problem.

Step 1: Trend (Diagnosis). This should have been expected!

The situation is illustrated by the drawing in fig. 8.15.

The mobile knife performs reciprocal movements at right angles to the tape movement direction (shown by the double arrow). For a short time interval, the tape stops to be cut off; this, however, also becomes a drawback, as it leads to the tape bending when it is jerked back into motion.

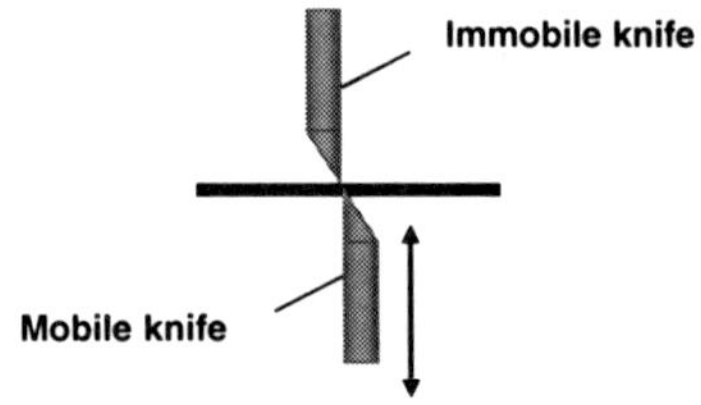

fig. 8.15. Shears in operation

However, the main problem is that the mobile knife is delayed because of the need for it to withdraw to its starting position while it lets the tape pass on. Besides, the knife movement speed is adversely affected by the force of inertia.

Step 2: Reducing (reforming). The resource is exhausted!

The following two models specify the nature of the problem in sufficient detail:
Standard contradiction:

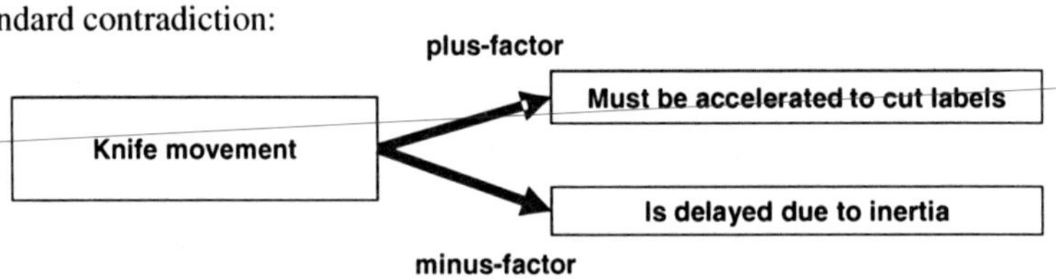

fig. 8.16. Possible wording of the standard contradiction for the delayed shears movement problem

Another "irreconcilable" radical contradiction:

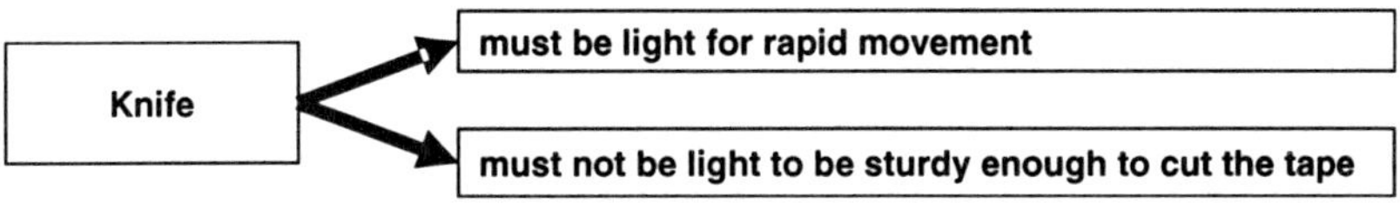

fig. 8.17. Possible wording of the radical contradiction for the delayed shears movement problem

Required "ideal result": cut the tape off "**precisely-AND-quickly**" while maintaining the required tape movement speed!

Unfortunately, the force of inertia cannot be eliminated, as the knife has a certain mass.

As concerns tape deformation, we have reached the limit of "resistance" of our material to the bending forces generated by rapid movement through the air. Therefore, the radical contradiction can be resolved only by modifying other resources – those that still retain the ability to be modified!

In this case, there are clearly two distinct causes which jointly produce tape cutting defects.

Modeling of the first cause – knife inertia – yields the following results (the entries below indicate contradictions and clusters – simple lists of navigators – selected from the relevant A-matrix cells):

+22 Speed **VS** –32 Weight of the moveable object = 04, 05, 11, 30;

+22 Speed **VS** 10 Ease of use = 04, 09, 11, 37.

Ranked list: 04^2, 11^2, 05, 09, 30, 37. In other words, we have two dominant navigators (04 and 11) both having the rank of 2 (shown in superscript).

Let us write down a suitable recommendation from *04,c Replacement of mechanical matter: transition from static fields to dynamic ones, from temporally fixed fields to flexible fields, from unstructured fields to fields with a specific structure.*

Modeling of the second cause – "non-rigidity of the tape" – yields the following results:

+22 Speed **VS** –28 Strength = 10, 12, 22, 32;

+22 Speed **VS** –15 Length of the moveable object = 11, 22, 32.

Ranked list: 22^2, 32^2, 10, 11, 12. There are two dominant navigators – 22 and 32.

According to *22 Spherical-shape*, the possible recommendations are: *a) change from linear parts of the object to curved ones, from flat surfaces to spherical ones…; c) change to turning movements.*

Step 3: Inventing (transformation). The material resource has been exhausted – the solution must rely on other resources!

There emerges an idea that instead of the reciprocal movement we should resort to continuous rotational movement; also, pursuant to model *10 Copying* (together with the previous recommendations), we need to use a circular knife as a copy of some circular saw. Interestingly, recommendation *32 Counterweight* is apparently automatically implemented if we choose to use a circular knife (fig. 8.18).

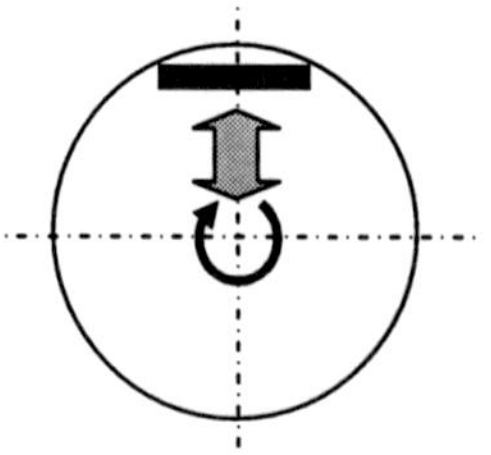

fig. 8.18. On the way to a new "shears" construction

Indeed, inertia disappears, because the circular knife can – and must – continuously rotate in the same direction.

Step 4: Zooming (verification). But the contradiction is still there!

Unfortunately, the circular knife also needs to be moved towards and away from the tape (shown as a black strip), and we are thrown back to square one (reciprocal movement, shown with a thick double arrow)!

Even worse, to cut the wide tape the reciprocal movement of the knife will have to be "extended"! What solution would you suggest, esteemed readers?
This example is completed in section *9.4 DISC Meta-Model*, Example 9.8. Improvement of the Labeling Machine.

Some time later the Inventor undertook, at the request of the Client, a detailed analytical review to identify perspective ways of improving the labeling machine. The task was to find a way to increase the speed of movement of the envelope more than 10 times compared to the original speed.

However, the key operating principle of the machine had to remain unchanged: it still had to attach labels and print a new address on each such label!

The hypothetical improvement analysis was also performed by the Inventor using the four-step MAI T-R-I-Z scheme.

Example 8.12. Improvement of the label attachment assembly.

Step 1: Trend (diagnosis). Technical assessment and identification of problems

The review immediately revived the "old" problems – delayed shears movement and imprecise tape positioning at the cutting point. Besides, tape roll inertia prevented the rollers from maintaining the required tape feed rate, while addition of a roll-spinning engine would make the machine prohibitively expensive.

Step 2: Reducing (reforming). The resource is exhausted – again!

The following models provide a detailed description of the root causes of the problem. Standard contradictions:

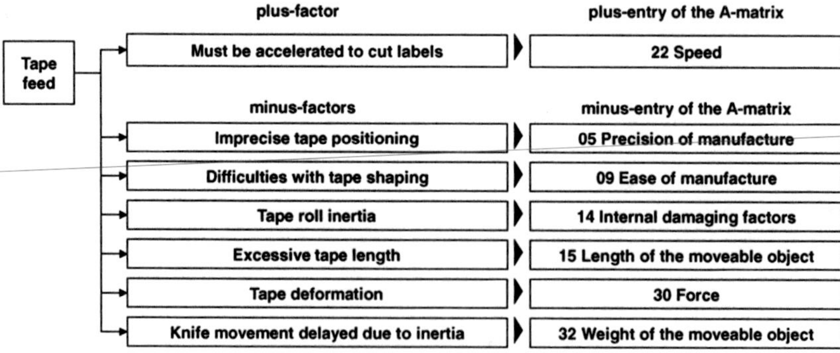

fig. 8.19. Main "standard" conflict models for the original problem situation

Step 3: Inventing (transformation). The operating principle is exhausted – we need to take up a maxi-task!

Analysis shows that the use of the old operating principle (where labels are created by cutting off sections of tape from a roll) cannot be used any longer, as it causes too many problems. To form a new strategy, we need to write out the sets (clusters) of recommendations (models) from the A-matrix for all contradiction pairs:

22 **VS*** 05 = 02 04 09 29

22 **VS** 09 = 01 03 11 32

22 **VS** 14 = 01 05 18 33

22 **VS** 15 = 11 22 32

22 **VS** 30 = 04 07 08 11

22 **VS** 32 = 04 05 11 30

Having processed this set of clusters, we receive an extended cluster with ranked models listed in the order from the highest-ranked to the lowest-ranked. The rank (superscript index) is determined by the aggregate incidence of each model in all clusters:

11^4, 04^3, 01^2, 05^2 и 32^2, 02, 03, 07, 08, 09, 18, 22, 29, 30 and 33.

The required "ideal result" in this case can be formulated only as a "maximal" ideal model: *the labels are present, while the label-cutting machine is... absent!*

Is this really so fantastic? Maybe, everything is much simpler?

So, here comes: *11 Inverse action* recommends: *a) instead of the action prescribed by the conditions of the assignment, complete a reverse action.*

Another model, the "good old" *04,c*, says: *replace... unstructured fields with fields with a specific structure.*

Model *01 Change in the aggregate state of the object* is hardly applicable to this situation.

Model *05 Separation*: *separate the "incompatible part" ("incompatible property") from the object or include the only really necessary part (necessary property) into the object.*

Here is an interesting one: *32 Counterweight*: *a) compensate for the weight of the object by connecting it to another object; b) compensate for the weight of the object using interaction with the external environment.*

Pursuant to model *02,b*, the following recommendation is quite promising: *prepare objects in advance so that they can be put to work from the best position and are available without loss of time.*

Model *03 Segmentation* offers this recommendation: *a) Divide the object into independent parts.*

---

* Here and below, the "plus" and "minus" signs in the standard contradiction formula may be omitted, as it is assumed throughout that the plus-factor comes first, and is followed, after the VS sign, by the minus-factor.

Enough! Now we can make a "composite sketch" of the solution where labels are made using some other method which is consistent with the "ideal result". Labels are made from the tape, and:

1. the main "bad" (inefficient) operations have to go – do NOT cut off labels, do NOT "pull" the tape (model 11),
2. ready-made labels must be "introduced" to the machine (model 05),
3. the ready-made labels must be delivered directly to the envelope (model 32),
4. the labels must be prepared in advance and placed in a position from which they can be rapidly attached to the envelope (model 02), and
5. the tape must already be cut into labels (model 03).

The solution directly follows from this set of properties: the tape roll must be made of labels cut-off in advance! This makes it possible to get rid of the shears, while labels can be "fed" by taking them off the roll one by one (fig. 8.20).

Note that the gluey side of the labels is turned towards the inside of the roll which pretty much determines the method to be used to form the roll, but once they are detached from the roll, their gluey side is turned outside towards the envelope which makes it possible to use the available vacuum-gripper drum. The drum "grips" each label by its dry, "glueless" back, and delivers it to the envelope "gluey-side-down".

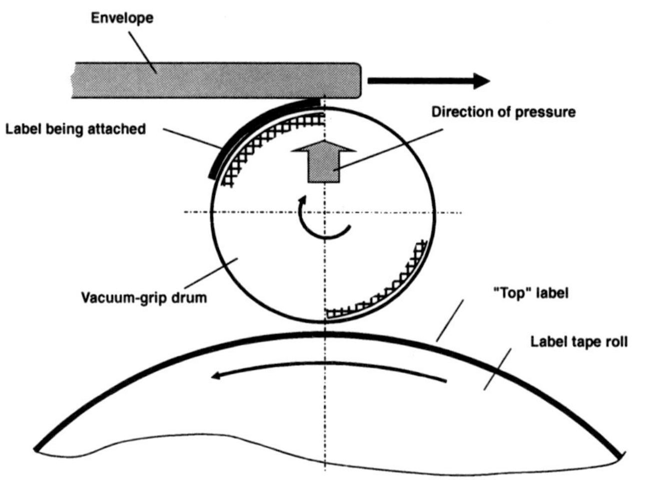

fig. 8.20. An almost "ideal" labeling machine

Step 4: Zooming (verification). What, no more problems?!

The Inventor did what he was supposed to do: if "zooming" is restricted to the machine, it will become apparent that the machine has been reduced to one main assembly – its "working element" – the device which delivers and presses the label to the envelope!

On the other hand, the Client and machine users (postal centers) have encountered a completely different – but equally serious – problem: nobody makes such label tape rolls! If we do the "zooming" at the level of cooperation with suppliers and users, we will see that sometimes solutions create new problem situations outside of their immediate operating zones.

Still, the solution is realistic and eminently patentable!

Finally, in line with TRIZ recommendations for this stage, we will now "zoom" into the future and see whether the solution remains "improvable".

The result – now truly "ideal" from the viewpoint of "reduction" of the machine – is shown in fig. 8.21. Here labels get to the envelope directly from the roll. When the time comes, the roll is pressed to the envelope which "captures" the label by itself. In reality, the technical solution for the roll-envelope joint is rather more sophisticated.

Note that in this case the gluey side of the labels making up the roll is facing outwards.

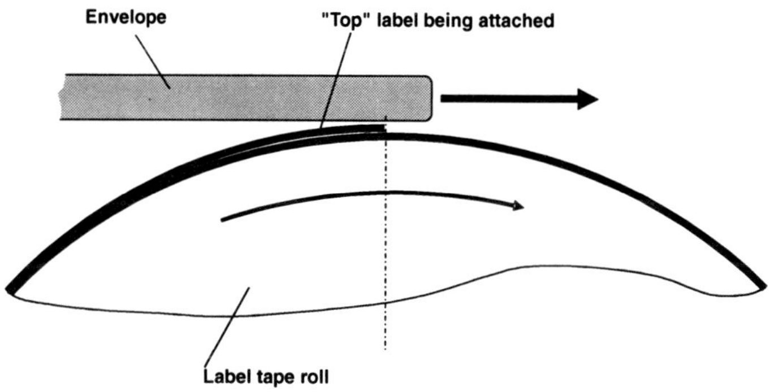

fig. 8.21. "Ideal" labeling machine

In conclusion of this series of examples, we would like to mention the fact that most solutions developed by the Inventor together with the Client's experts have been patented[126]. Equipment offered by the Client won many tenders, and its machines are supplied to many countries.

Generally speaking, the use of such methods as SITO and MITO is justified mostly for express analysis purposes when it is necessary to find a "quick fix" for the problem.

Clearly, such analysis is best performed in the instrumental environment provided by the EASyTRIZ software package, where computation of ranks in the MITO method is automated, and the entire process can be easily documented.

And, naturally, not every problem can be solved from the first attempt. When that happens, proceed to the second cycle (see the next section) supported by EASyTRIZ.

---

[126] All solutions and schemes are shown in a simplified way. They differ from their actual engineering implementations, and are used here exclusively for training purposes to explain TRIZ and MTRIZ methods.

8.4 Cycles and Branches in Problem Solving

When we are looking for solutions of complex problems, there may be variants – branches and cycles.

Branch – different ideas may be created for the solution of one and the same original problem. Branching is often the result of formulation and analysis of different contradiction models.
To eliminate different contradictions, the A-matrix recommends different sets of transformation models, hence different ideas for the solution of one and the same original problem. Such solutions are defined as *branches*. Sometimes they may be regarded as separate, virtually independent problem-solving projects which have identical starting data at the Trend stage.

Cycle – a chain of successively improved solutions. Here each link of that chain is also called a *cycle*. Cycles are based on the results of Zooming performed upon completion of each stage.
A new cycle starts if Zooming identifies deficiencies in the proposed solution. Such solution then becomes non-final, intermediate, and deficiencies discovered in it have to be eliminated, or the relevant property has to be improved – i.e. a new problem-solving cycle is initiated. In the chain of solutions, the data of each completed cycle become the starting data for the next cycle.

Branches may give rise to cycles and *vice versa*, and some projects have very complex problem-solving structure (fig. 8.22).

Let us consider an example illustrating emergence of cycles during a reinventing exercise.

Example 8.13. Man walks through the "wall".

Ten years ago, at the Industrial Exhibition in Hannover, Germany, I saw a wonderful wall made of… water! Thousands of water jets created an impression of a solid waterfall around one of the exhibition pavilions, and served as screens for projecting video clips. All entrances to the pavilion were quite ordinary, without any waterfalls. I performed reinventing for that wall in Example 49 in my first book[127].

And now – an even greater wonder of architecture and technology!

In the summer of 2008, Saragossa, Spain, became the site of the World Exhibition *Water and Sustainable Development.*

In a special article[128] and on their web site[129] the creators – architects and engineers – describe a new inventive solution used to build the information center of the exhibition.

[127] M. Orloff (2006) *Inventive Thinking through TRIZ.*

[128] http://web.mit.edu/newsoffice/2007/waterbuilding-0711.html

[129] www.digitalwaterpavilion.com

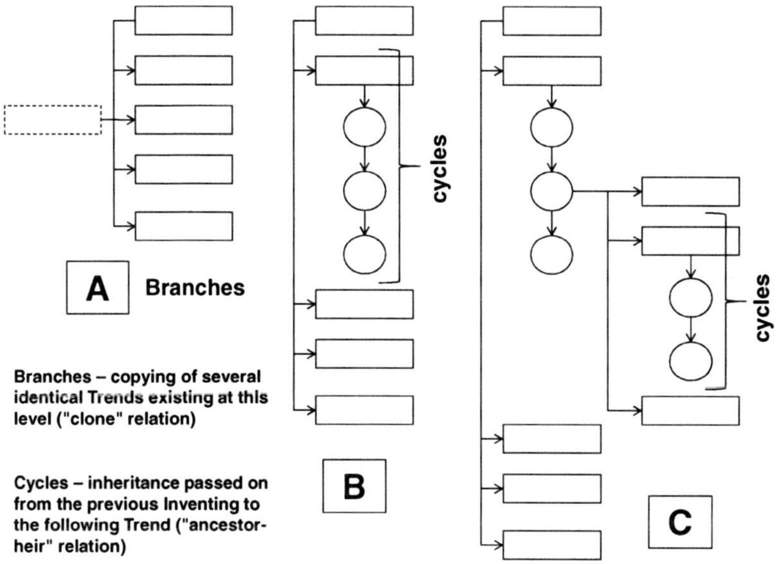

fig. 8.22. Project structures: A – branching,
B – emergence of a cyclic sequence, C – branches and cycles

Cycle 1

Trend 1

It is possible to build a water wall and use it to project video clips or static visual data.

Such water wall has no "doors" – just open "passages" in places where water jets are absent on a permanent basis.

Question: can the water wall be modified so as to enable the visitors to walk "through" it in any place?!

If we stick to the "old" idea, then to create such opportunity we need to… turn off the water (?!), but then the "wall" will disappear.

Such solution is not acceptable.

Reducing 1

So, we are dealing with a radical contradiction underlying the original problem situation: the (water) wall *must exist* around the exhibition pavilion, and such wall… *must not exist* as visitors must be able to freely walk in and out of the pavilion at any place.

Inventing 1

On the one hand, the solution is quite evident and "brainstormable".

Such solution could be based on the same principle as that described in the fairy tale about Ali Baba and the forty thieves. When Ali Baba came to the mountain, he enunciated the magic words: "Open Sesame", and the entry to the treasure cave opened. When leaving, he said: "Close Sesame", and the cave closed, with the entrance becoming part of the mountain invisible to the uninitiated.

To implement such solution, it is enough to install motion sensors. Then water feed can be stopped (without any magic words!) in any place once a visitor approaches, enabling him to enter or exit the pavilion. Once the visitor is "safely" away from the pavilion, the sensors resume the water feed.

On the other hand, we are interested in whether this solution contains any objective creative models, and whether an idea for the solution can be found using TRIZ instruments.

This solution corresponds to the models describing fundamental transformations in time and space:

- *solution in time:* the "wall" exists during the period of time when the visitor is some distance away from it, and it does not exist when the visitor is walking "through" it;

- *solution in space:* the "wall" is absent at the point of visitor entry/exit, and present in all other places.

Zooming 1

Well, this solution is quite satisfactory, as it removes the original "irreconcilable" contradiction.

It may be referred to as "dynamic water stream management".

The solution makes the project more complicated, as it requires installation of control valves and a management system using numerous sensors.

Now let us think whether this disadvantage – excessive system complexity – can be turned into an advantage...

Cycle 2

Trend 2

The proposed water "wall" has a complex construction. Can this solution be improved or developed?

Reducing 2

To better present the problem, let us build a standard contradiction:

The water on/off function has been introduced to assure "universality, adaptability" by creating entry/exit points at any location, but this increases "complexity of construction", i.e. complexity of the entire system.

Formal standard contradiction:

02 Universality, adaptability **VS** 07 Complexity of construction = 04, 07, 14, 27.

Another contradiction formula:

If we decrease – *improve* – "complexity of construction", it will increase – *impair* – the impact of "external damaging factors" affecting the visitor.

Second formal standard contradiction:

07 Complexity of construction **VS** 13 External damaging factors = 08, 14, 17, 21.

Inventing 2

Let us make a "composite sketch" of the future solution.

The following transformations can be used in the solution:

04 Replacement of mechanical matter: c) replace… unstructured fields with fields with a specific structure;

07 Dynamization: a) characteristics of the object are changed so as to optimize each step of the operating procedure;

21 Transform damage into use: c) reinforce the damaging factor until it ceases to inflict damage.

Model *14 Use of pneumatic and hydraulic constructions* is already present in the system under analysis.

Starting from the end, we now build an ideal final "image" of the idea for this solution: water jets are turned on and off with increasing frequency, and it becomes possible to divide the jets into fragments as small as single drops; as a result, the structure of the "wall" can be managed – for example, it can be shaped into letters or pictures.

Idea: expand functionality of the "wall" by introducing a programmable water jet (water drop) management function capable of "painting" graphic images[130] (fig. 8.23).

Zooming 2

The original standard contradiction has been eliminated.

The radical contradiction (similar to the one we reviewed during the first cycle) has also been eliminated, but in this case by means of a structural transformation: the jet (as a line, as a system) on the whole exists, but its parts do not.

We have also obtained a wonderful systemic super-effect: ability to use water jets (and even single water drops) to create dynamic "paintings". This solution represents an outstanding architectural and technical invention.

You say it is too complex? But this complexity is counterbalanced by creative and technical perfection of the idea! Look at the animation published on the web site of the authors of this idea, and you will not remain indifferent!

---

130 Reproduced from www.digitalwaterpavilion.com

fig. 8.23. Generation of graphic images by water jets

You might say that this solution is basically a remake of the inkjet printer operating principle. Well, it is – and it isn't! The output produced by an inkjet printer would be similar or identical to what we saw, if the pavilion could move as a tape – then this method could be used to cover its surface with water paintings.

Here the images are "flying" within the water jet plane BY THEMSELVES! This solution is like an implementation of the navigator called *Inverse action*: the "sheet" (black background behind the moving water jets) is fixed, while the totality of intermittently flying drops forms a dynamic painting covering that "sheet".

The description of this reinventing is both simplified in terms of the models analyzed, and excessive in terms of the text used – due to reasons which are both of methodological (training) and stylistic (literary) nature. For remote education or professional reinventing purposes, such description, as generated by the EASy-TRIZ software package, is normally more concise and rigorous.

9 System Evolution Models

9.1 Four Attributes of a Problem

Transformation of a task into a problem is forced by its increasing complexity (fig. 9.1).

Complexity increase, in turn, is defined by four main attributes. Two of those attributes are related to the formulation of the task, while the other two directly bear on its solution. In line with this differentiation, we recognize two levels of problem situations that we will be calling "manager problem" and "developer problem".

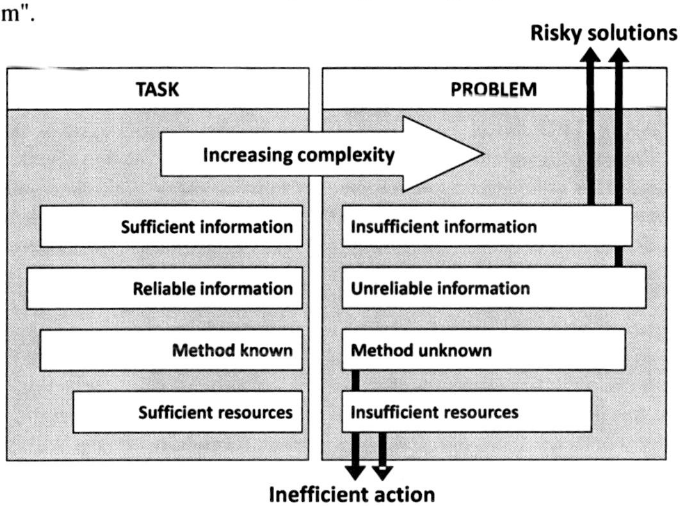

fig. 9.1. Causes and effects of increasing task complexity

Manager problem	If the source data related to the task are incomplete or unreliable, the consequences may include the following: indistinct formulation of the purpose of development, imprecise determination of the causes and sources of the problem, insufficient or incorrect listing of existing and available resources that can be used to solve the problem, etc. These deficiencies in the formulation of the problem create, from the very beginning of the work aimed at improving the object, a risk that the desired objective will never be attained.

This situation can be called a "manager problem". Indeed, it is the managers, including the chief designer, that are responsible for the formulation of the problem and definition of the ultimate objective. Errors in this area are extremely costly. For example, wrong interpretation of competitor actions can directly lead to wasteful use of corporate resources to create products that *will not be competitive a priori.* The only solace is that eventually such products may find their niche in the market – but then again, this may never happen.

Developer problem	If developers do not know what method can be used to solve the problem – which often happens in stress situations, when the solution is required "here and now" and, naturally, no apparent resources are available – their actions may prove to be inefficient.

This situation can be called a "developer problem". Indeed, it is the developers that are responsible for specific modifications of the system under development. The cost of errors in this area falls within a broad range, but is normally lower than the cost of management errors.

An ideal situation is one where managers are capable of providing well-substantiated and essentially correct system evolution projections, including those which apply to competing systems, while developers are sufficiently proactive and creative to design efficient conceptual frameworks and solutions giving the company a competitive edge in the market.

Let us briefly note the importance of continuous system evolution forecasting on the basis of TRIZ laws. The absence of such ongoing modeling is the main cause of time shortages and stress situations experience by design teams. It also creates conditions conducive to development of erroneous or inefficient solutions.

This chapter is a brief introduction to the methods of improving solution efficiency and reducing concomitant risks on the basis of available TRIZ instruments.

9.2 Negative and Positive Systems[131]

In this section, we will discuss the main sources of problematic interactions occurring over the life cycle of the system. In that context, the outside world can be conveniently presented as a unity of two systemic formations[132] (fig. 9.2): positive systems, or *con-systems*, and negative systems, or *counter-systems.*

Positive systems (con-systems)	Unity of systems which, being extraneous to artifacts of a given species, have a **positive effect** on the rate of evolution of such artifacts, enhance their functionality and increase their efficiency. Con-systems act as generators of positive impacts affecting all or any types of artifacts belonging to that species, or even individual specimens of such artifacts. There are three main types of con-systems: - developer/producer systems joined into a single creative system; - modernization systems; and - maintenance systems.

[131] This section reproduces the title and contents (with extensive unmarked quotes) of a chapter from the book written by my Teacher, a well-known military system engineer, Professor A.M. Shirokov (1924-2003): *Quality Evaluation of Radio-Technical Systems at Early Stages of Development* – Minsk, Minsk Higher Air Defense Engineering School (MVIZRU; in Russian), 1970.

[132] M.A. Orloff *Objectives and Structure of Constructive Integrated Modeling Theory in Complex Computer Design Automation Systems.* – Anthology: *Radio Electronics Problems*, Computer Science Series, Issue 15, 1991. – p.67-87 (in Russian)

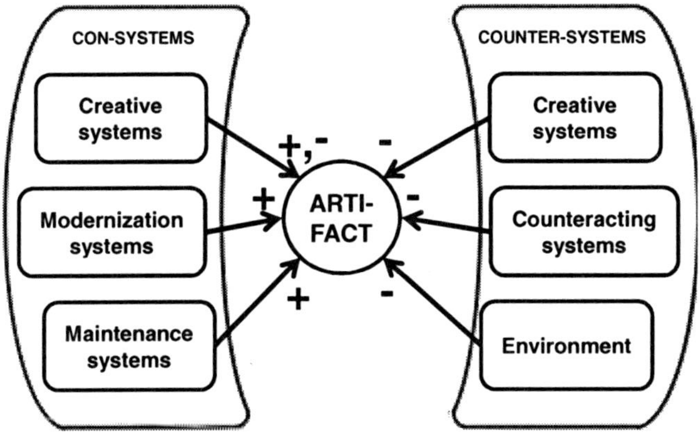

fig. 9.2. Life cycle interactions between an evolving system (artifact) and generative/ambient systems

Negative systems (counter-systems)	Unity of systems which, being extraneous to artifacts of a given species, have a **negative effect** on the rate of evolution of such artifacts, impair their functionality and decrease their efficiency. Counter-systems act as generators of negative impacts affecting all or any types of artifacts belonging to that species, or even individual specimens of such artifacts. There are three main types of counter-systems: creative counter-systems; counteracting systems; and the environment (perceived as a system).

The **creative system** is a social and technical system which can generate artifacts (in particular, technical systems). Inasmuch as it is an open system, it continuously receives inputs of information in all its manifestations, i.e. in the form of knowledge, new ideas, skills, materials, etc. Transformation of (primarily) information produces new artifact types belonging to the given species. Time intervals between emergence of new artifact types are random, but efficiency of such new types cannot be lower than that of their predecessors.

Clearly, therein lies a paradox: each new artifact produced by the creative system has a negative "impact" on its earlier "edition"; as a result, the positive creative system is also a generator of negative impacts from the viewpoint of previous artifacts.

Modernization systems are designed to improve efficiency of objects in operation by implementing, in the structure of such objects, changes which are relatively insignificant compared to such objects in their entirety. As a rule, modernization systems are organized for expensive and sophisticated long-life objects. One of the useful effects of operation of modernization systems is neutralization of negative system impacts. The overall number of, and the intervals between, modernizations depend on the value of the object and the cost of updates.

Maintenance systems are designed to decelerate object efficiency reduction; they are created together with their objects, and remain in operation throughout their useful lives. The main impact generated by maintenance systems is the fact that the operating quality of their objects is maintained at an acceptable level.

Environment as a system affects any object (technical system) throughout the period of its operation. Environmental impacts may not always be noticeable over short periods of time. Long-term exposure, however, results in physical ageing of objects under the influence of aggressive environmental factors.

Such ageing is caused by changes in the physical and chemical structure of materials comprising object parts, and in turn changes operating parameters of functional assemblies and reduces the overall efficiency of affected objects.

Counteracting systems are usually represented by various military systems. They are created with the express purpose of suppressing other artifacts. Counteracting system impacts can have different manifestations, including physical destruction of artifact elements, disruption of management chains, generation of false signals, distortion of transmitted signals, etc.

As a rule, exposure to such negative systems is very short, but the impact of such exposure can be devastating, with a very high probability of artifact efficiency being reduced to zero. Such examples are numerous:

- antitank shells as anti-systems for tanks;
- jamming systems as anti-systems for radio locators;
- fighters as anti-systems for bombers, etc.

Creative counter-systems generate new systems of the same type as the system being researched (improved). Such new systems are more efficient that the system being researched. As a result, the relative efficiency of the research target is reduced.

This phenomenon is very complex. It resides both in psychological and economic domains, its impact may depend on whether the system in question is household or industrial, etc.

In the case of weapon systems, emergence of new counteracting systems has stronger effect on reduction of efficiency of research targets than does emergence of more efficient same-type systems, but the psychological devaluation effect has adverse impact on maintenance personnel.

Still, the integrated effect of new systems is most efficient – in the end old systems are completely negated, and resources invested into their creation prove to have been wasted (except for the skills acquired while developing such systems).

It should be noted that all types of positive and negative systems are present in any operating zone, either nominally or factually, making up the environment of such zone. This results from the fact that any system element, even the smallest one, was developed and produced and is potentially being improved by a creative system, and can be regarded as a candidate for modernization and maintenance or a target of possible counteraction or replacement by a new element.

To top it off, here is a methodologically pertinent definition[133] conceptually related to negative systems:

Harmful machine	Unity of external and internal forces (and systems used to reproduce such forces) whose impact on the given artifact has adverse effect on its quality and operating efficiency and/or leads to its physical ageing and termination of its operation and existence.

This metaphor-model puts development of instruments counteracting negative systems in the context of a well-known TRIZ model called "abstract (useful) machine"[134].
According to that model, any "harmful machine" and any "useful machine" can be presented as a composition of the following parts: working element, transmission, engine, source of energy, steering system, frame.
To protect the "useful machine", it is necessary to identify both the most vulnerable and fragile spots of the "useful machine", and the most vigorous and dangerous deleterious impacts that can be produced by the "harmful machine".
Conversely, when organizing counteraction on behalf of the "useful machine", it is necessary to find the most vulnerable and fragile spots of the "harmful machine".

9.3 Evolution of System Classes[135]

Evolution of technical system classes is an objective law of their existence.

Aleksandr Shirokov

Any class of technical systems evolves as a unity of similar system classes created by teams of developers and industrialists at various times. Differences between system types emphasize the changes that occur in systems.

The creative system (fig. 9.2) is a generator of technical systems belonging to a particular types and classes. The system creation process can be interpreted as the process of amalgamation of information owned by the creative system (ideas, knowledge, experience, materials, etc.) and materialization (representation, recordation) of that information in a newly created construction.

---

[133] The idea for this metaphor belongs to V.A. Lenyashin, a TRIZ expert and the "patriarch" of TRIZ implementation at SAMSUNG.

[134] M. Orloff (2006) *Inventive Thinking through TRIZ*, Section 8.2. *Resources* and Section 14.2. *Ideal Machine*

[135] In memory of the Teacher and his lessons which nourished my interest in system analysis, I am using the title of, and the key notions (quotes not marked) described in, another book written by A.M. Shirokov (1970) *Quality Evaluation of Radio-Technical Systems at Early Stages of Development* – Minsk, MVIZRU (in Russian)

The need to accumulate new information with a view to create another system type results in each technical system class being "expanded" in time. Inasmuch as each subsequent system type must be better than its predecessor, the class as a whole evolves with the course of time.

The rate of such evolution depends on the amount of information inputted into the system when each new type is being created, i.e. ultimately on the rate of accumulation of amalgamated information related to the system belonging to a particular class, to wit: ideas, knowledge, experience, materials, etc.

Any specific sample or type can be evaluated based on its efficiency. We can assume that there is some correlation between efficiency **E** and informational content **Q** of any system:

$$\mathbf{E} \leftrightarrow \mathbf{Q}.$$

The level of efficiency of a given system at a given point in time depends on its technical and operational parameters. The general assumption is that in the course of evolution the level of efficiency of the system increases.

Among technical characteristics, there is always one that dominates the system and has the most effect on efficiency. Sometimes that characteristic is assumed to be the equivalent of efficiency, which is not always true. An analysis of changes of the dominant characteristic yields several interesting conclusions regarding development of system parameters.

The first conclusion deals with the shape of evolution curves. As a rule, growth of each characteristic is constrained by the physical properties of the system, or by the limited implementation range of ideas embedded in the system. As a result, fully represented evolutional relationships have the form of a logistical curve (fig. 9.3).

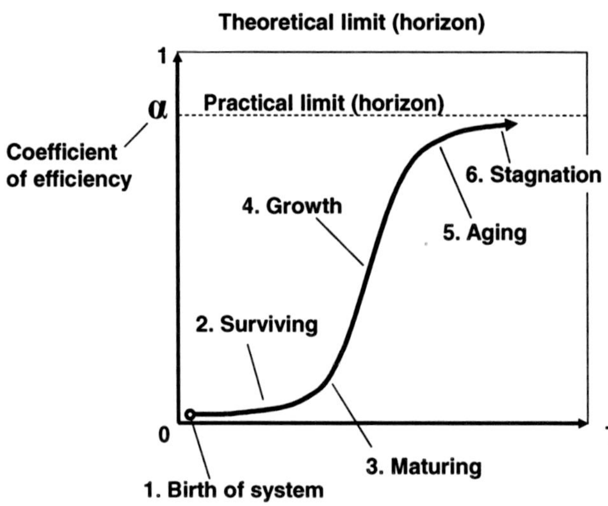

fig. 9.3. Logistical curve (S-curve) featuring evolution of the main parameter of the system class

The second conclusion is that specialization of individual system types leads to the emergence of two or more classes within one primary class. As a result, such emergent classes first form a supra-class, and then a family, while further evolution gives rise to a new order and new class of systems. The values of dominant specific indicators are grouped around the evolution curve; upon emergence of supra-class, they form a band which is further subdivided into several bands as new families come into existence (fig. 9.4). Here we need to differentiate between intra-specific development (fig. 9.4,a) and evolution as change of system types within the given class (fig. 9.4,b).

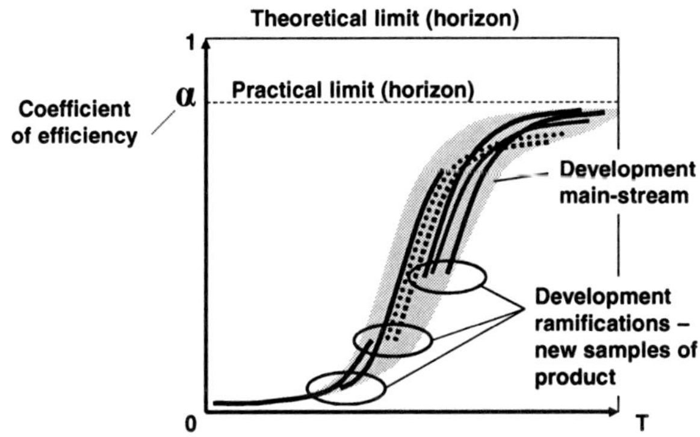

a) development as a series of intra-specific ramifications within the class

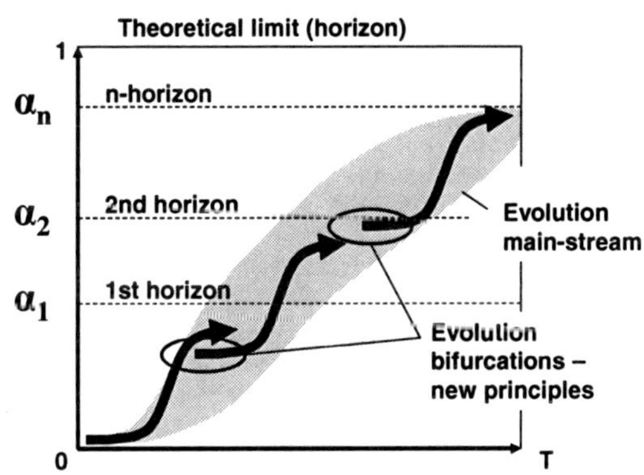

b) evolution as a series of bifurcations into classes

fig. 9.4. Families of developmental and evolutionary S-curves

Example 9.1. Ramification. When in 2006 I was using the schemes discussed in this section to complete one of my projects commissioned by Samsung headquarters in Suwon[136], I learnt that in 2005 the company had made 164 types of mobile phones.

---

[136] Samsung Advanced Institute of Technology, Suwon, Republic of Korea

Evolution of technical system classes is an objective law of their existence. The task of the creative system is to approximate the actuality and potentiality of the evolution curve, and minimize the systems implementation period. In most cases, there is no need to overhaul the entire system: it is expedient to make only those changes which secure maximum efficiency increase at the given level of investment.

Duration of the systems implementation period is largely determined by the fact that system-related complaints can be received only upon expiry of a certain period of practical operation of specific system samples. If we examine progression of efficiency hikes occurring after each practice-driven requirements upgrade, the mutual influence of this and other important aspects of system evolution can be presented in the form of a well-known spiral model (fig. 9.5).

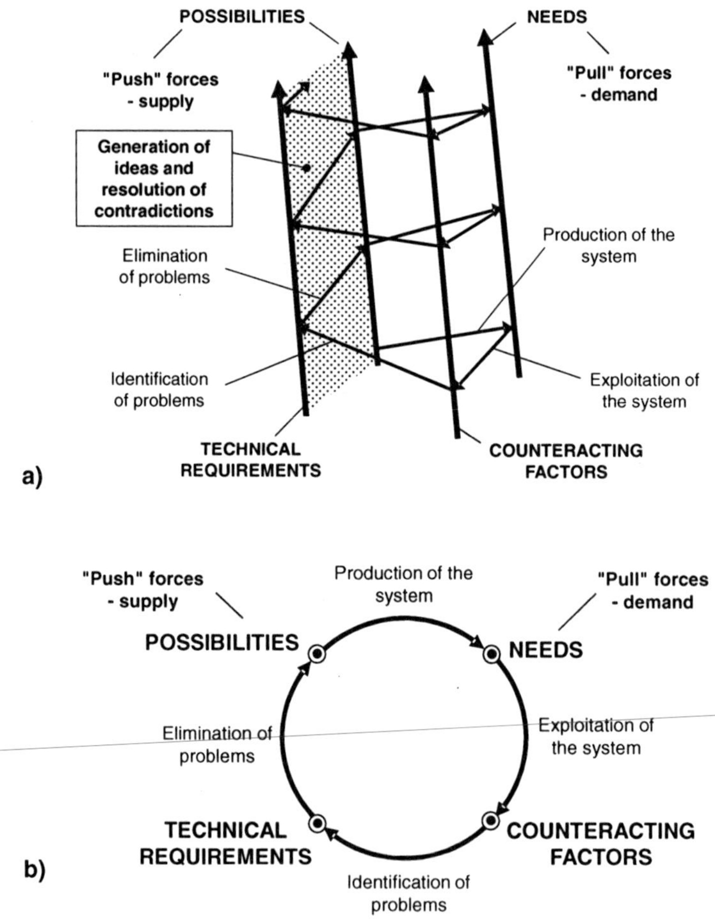

fig. 9.5. Spiral system evolution model

In some cases it is convenient to present the spiral scheme as a single cycle (fig. 9.5,b), subject to the understanding that each segment is repeated at a higher level upon completion of the full cycle in the direction indicated by arrows.

In the course of its use (operation), each artifact is exposed to harmful (counteracting) influences generated by interacting systems and the ambient environment.

Examples:

9.2. The motorway physically connects spatial points, and serves to move people and cargoes; however, the same motorway also wears out (and sometimes physically destroys) vehicles and people and/or cargo carried by such vehicles;

9.3. The Internet and mobile phones have created huge opportunities for communication and transfer of learning; on the other hand, they have also been used to commit new crimes (material and moral), and organize activities of various terrorists, including the defectives who disseminate junk (spam).

Identification of problems leads to formulation of new technical requirements, and ultimately encourages improvement of existing systems. However, exclusion of problems is possible only by creating new ideas about how to resolve the contradictions underlying such problems.

Example 9.4: the intersection of motor roads remains an extremely dangerous place due to the ever-present risk of motor car collisions. The radical contradiction of the intersection is that two motor cars moving from different directions can reach it at the same time. Two known solutions of this problem involve elimination of the underlying contradiction in time (installation of traffic lights) or in space (construction of roads at different levels).

Implementation of any of these solutions may be impeded by limited restructuring capability of existing systems. The main restricting factor is the lack of financial and material resources.

Example 9.5: the use of traffic lights does not guarantee complete absence of accidents at road intersections. On the other hand, it does not require substantial financial or material investments compared to dividing the roads in space by building tunnels or bridges.

If sufficient resources are available, it becomes possible to change the existing systems or create completely new systems.

Inasmuch as basic needs keep growing, all processes discussed above are "raised" to a higher level to meet new demand. However, in many cases changes are "prompted" by new possibilities which generate new "supply" for the benefit of consumers, or open new opportunities for artifact manufacturers.

Examples:

9.6. Road safety can be radically improved by development of traffic infrastructure; this will lend additional significance to information support of automated traffic management systems preventing emergence of dangerous situations, *inter alia*, by forced deceleration and emergency immobilization of transportation vehicles;

9.7. Safety of digital communication networks can be drastically improved by development of identification message markers (pointing to the owner and source of information) and creation of the harmful message tracking infrastructure; it is assumed that unmarked messages will be prevented from being transmitted inside the network or stored on individual computers.

Let us imagine an artifact as a continuously evolving system[137] (fig. 9.6) which consists of sub-systems, and is in turn included, as one of the sub-systems, into a higher-ranked system (super-system). The joint evolution of all scheme components can be interpreted as a change of the picture shown by all 9 "screens" following a change of point **T** – the "historical" moment in time. G. Altshuller held this model in high esteem: "Purpose of TRIZ: relying on the study of objective laws of evolution of technical systems, formulate thinking organization rules within the poly-screen scheme."

Such "thinking organization rules" include the following:
- use of the "idealization" notion, and
- construction and resolution of contradictions.

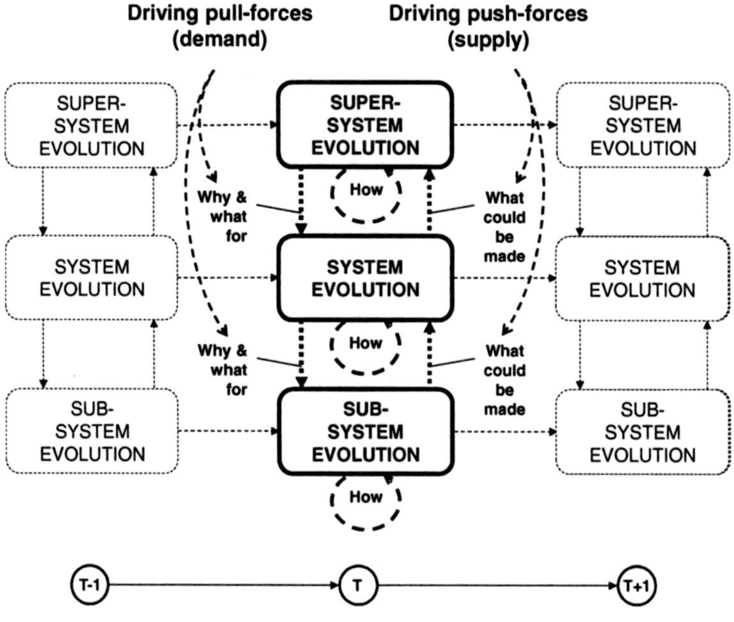

fig. 9.6. 9-screen system evolution model

To illustrate the "idealization" notion, we have provided fig. 9.7 according to which any artifact can be presented as a "formula of a complex system" which reflects evolution of a set of models[138] related to artifact functioning {**Z, E, F**} and organization {**M, P**}.

---

137 G.S. Altshuller (1986) *To Find an Idea. Introduction to the Theory of Inventive Problem Solving.* – Novosibirsk, Science Publishers (in Russian)

138 M.A. Orloff *Systemic Approach to Development of Technologies for Modeling Electronic and Computing Devices.* – Anthology: *Radio Electronics Problems*, Computer Science Series, Issue 10, 1987 (in Russian)

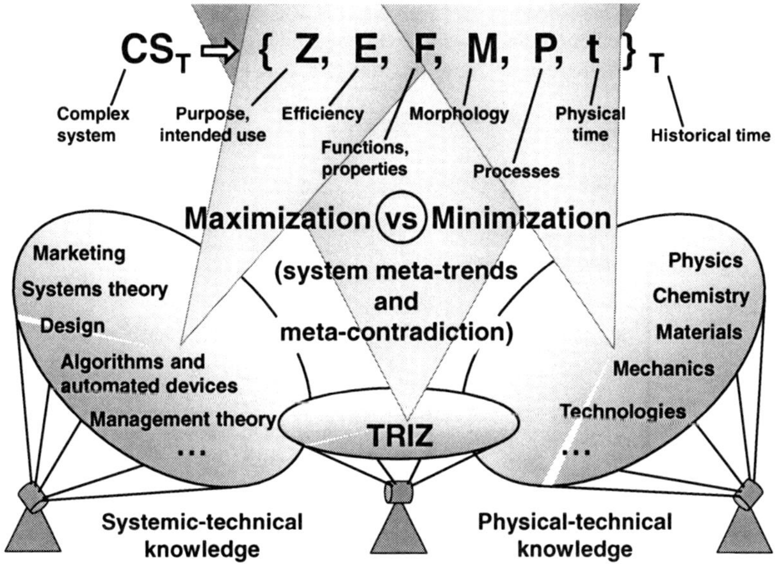

fig. 9.7. "System formula", system meta-trends and meta-contradiction

Aspects {**Z, E, F**} represent properties associated with the "functionality" of the system, including its purpose, efficiency, and functions.

Aspects {**M, P**} represent properties associated with the "physicality", construction of the system, including its dimensions, weight, energy consumption, waste products, and contaminants.

"Idealization" of the system involves increasing its "functionality" as the sum total of its positive properties, and decreasing its "physicality" as the sum total of system implementation costs. These two meta-trends can be joined (fig. 9.8) in the "idealization formula" as a ratio of the system's "functionality" to its "physicality":

$$E_{ideal} = \frac{\text{Functionality}}{\text{Physicality}}$$

fig. 9.8. System "idealization formula"

Directions of the meta-trends are shown in fig. 9.9.

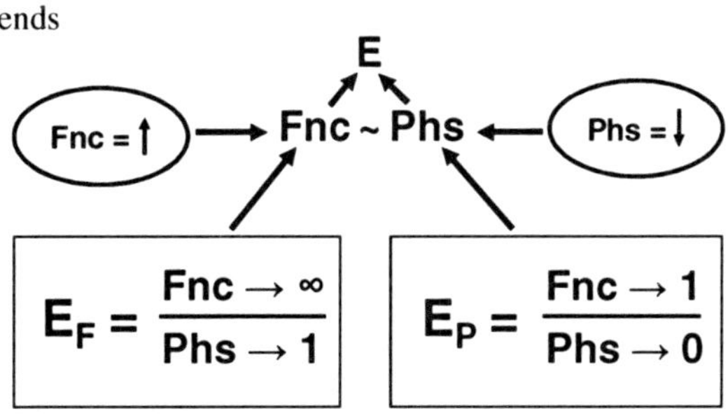

fig. 9.9. Directions of meta-trends during system idealization

Each "point" of ramification or bifurcation is a conflict of two main driving forces: growing needs (demand) and growing abilities (supply). If abilities cannot meet growing needs, there arise contradictions (fig. 9.10).

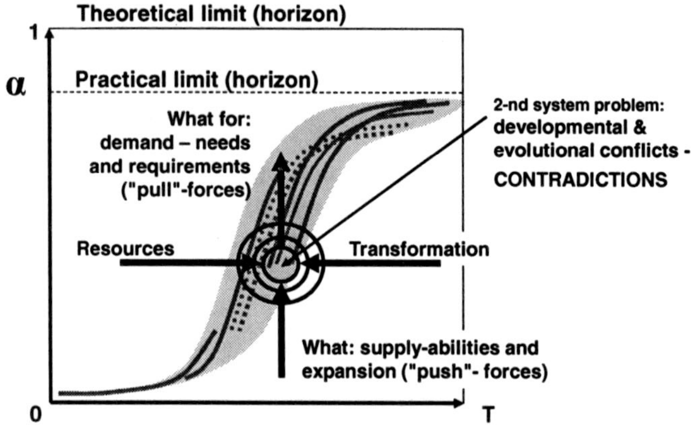

fig. 9.10. OZ representation on the S-curve

Those contradictions are intensified when OZ resources prove to be deficient or depleted to the point of exhaustion, and when transformation potential proves to be limited.

In fact, development and evolution can also be presented as a trajectory depicting emergence and elimination of contradictions and associated operating zones (fig. 9.11).

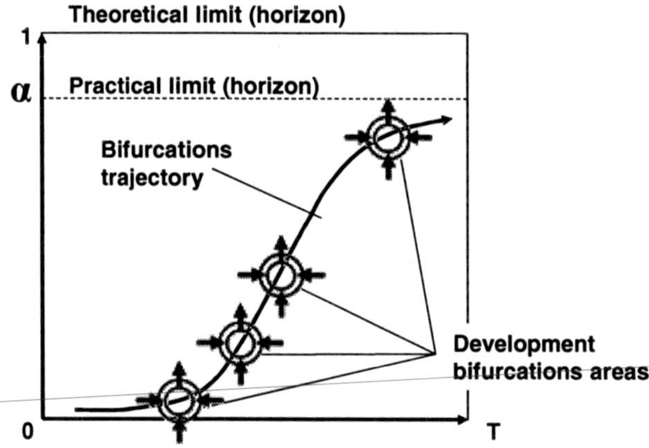

fig. 9.11. S-curve as the motion trajectory of operating zones and ramifications

Subject to the foregoing, it is possible to devise a procedure (fig. 9.12) for tentative evaluation of the level of complexity of system modification tasks. Clearly, the level of complexity increases from the level of Mini-tasks to the level of Maxi-tasks.

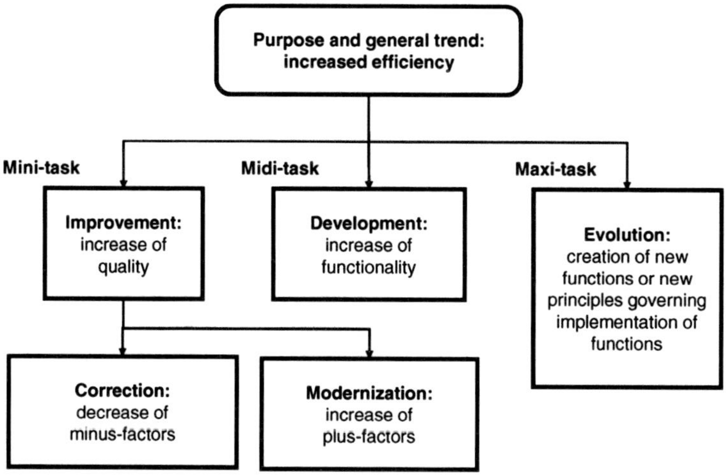

fig. 9.12. Tentative evaluation of the level of complexity of system evolution tasks

9.4 DISC Meta-Model

> To overcome technical contradictions, you must rely on your knowledge of the laws governing evolution of technical systems.[139]
>
> *Genrikh Altshuller, 1970*

So, the main "influencing forces" leading to change can be generally presented as follows (fig. 9.13):

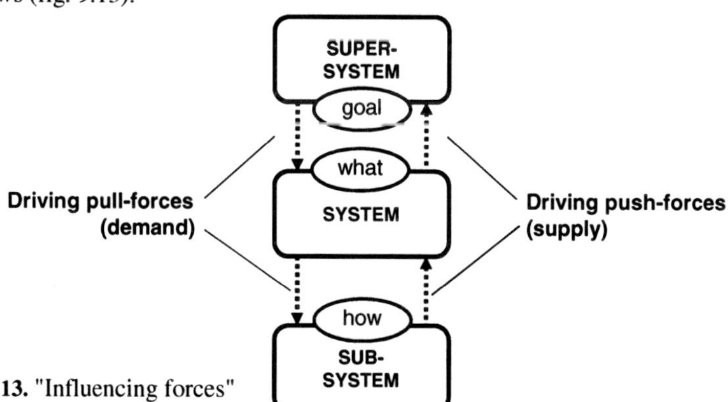

fig. 9.13. "Influencing forces" of change

[139] According to G.S. Altshuller (1986) *To Find an Idea. Introduction to the Theory of Inventive Problem Solving.* – Novosibirsk, Science Publishers (in Russian)

System change usually starts with sub-system change – it looks simpler. It is necessary to decide **"how"** to change sub-systems so as to ensure that the system meets the new requirements.

This process yields a renovated system which offers its super-system some new functionality – **"what"**.

The super-system establishes the "purpose" of system evolution – **"what for"**.

Interaction of these forces gives rise to numerous development conflicts and contradictions, and changes numerous operating zones which are "moving" along the S-curve or, to be more exact, are plotting that S-curve with new solutions.

And the main "mechanism" of constructive change (i.e. evolution) is selection and implementation of efficient transformations producing solutions which answer the question **"how"**. These aspects are presented in their totality in fig. 9.14.

Elimination of contradictions is achieved through transformation of resources. Consequently, it is necessary to understand, at least approximately, "how" various transformations affect specific resources, and "which" operations can be conducted with specific resources within the framework of specific transformations.

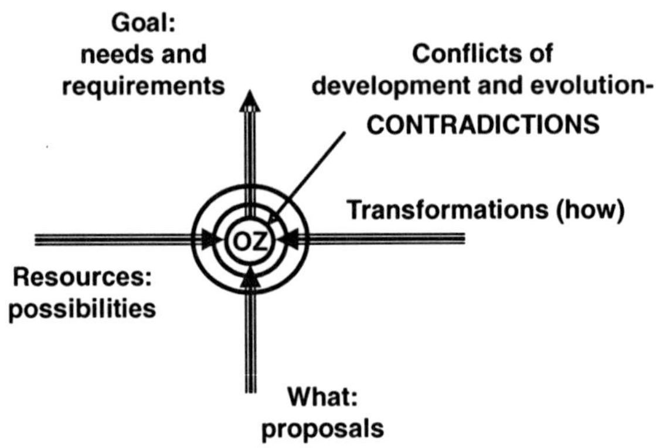

fig. 9.14. "Influencing forces" in the OZ

To this end, we must first discuss what "meta-operations" (most extensive and general transformations) occur with resources and objects at various stages of the object's life cycle and, consequently, at various "points" and stages of the S-curve (fig. 9.15).

First, there are four constructive meta-transformations:

D – *differentiation*;

I – *integration*;

S – *separation*, or segregation; and

C – *combination*, or composition.

To remember these operations, we can designate them by a mnemonic term "**DISC** Meta-Model".

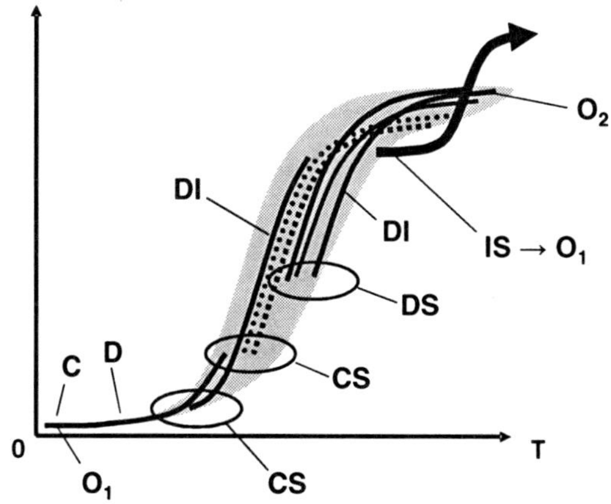

fig. 9.15. "Meta-operations" on the S-curve

Combination - an umbrella term to describe many widespread, relatively "simpler", "instinctive" operations, including the following: group, join, use an intermediary, transpose, recompose, place inside, take outside, replace all or part, copy, make by analogy, etc.

Combination is one of the creative operations most often "instinctively" used by people when they are engaged in brainstorming. We can assume that combination is one of the fundamental[140], "axiomatic operations" of Nature – assemble, create new compounds to test their properties and preserve the most efficient variants (genetic mutation, development and mutation of the protozoa, development of populations, Darwinian evolution).

This is basically enumeration of possibilities, the "trial-end-error" method. By the way, it would be just as reasonable to define it as the "trial-and-success" method!

The point is the price of victory, the duration and cost of enumeration.

Can enumeration of possibilities be directed and minimized?

Yes, and the instrument for that is TRIZ.

Differentiation - undoubtedly, the second most popular (in terms of intensity of use) "instinctive" operation which includes the following transforming actions: increase-decrease, add-subtract, adapt, fine-tune, and – again – copy, make by analogy, even make the reverse, etc.

Without alignment of qualities and properties, it would be impossible to create systems simply by combining elements. There is a need for adaptation, accommodation, mutual adjustment of elements, "wearing in" (this is an engineering term

---

[140] The high degree of similarity of such models is illustrated by numerous examples provided, in particular, in the following book: Bryan P. Bergson (1995) *On the Combination-Separation Principle: A Metaphor for Evolving System Processes.* – Elsevier Science Inc.: Technological Forecasting and Social Change, 50, p. 171-183

describing a situation where two surfaces are adjusted to one another by continuous mutual friction). The same goes for gear teeth in a new gear train. The same goes for two people who have just met – they need to "get used to one another" before they can do something together.

There is no doubt that this is another fundamental, "axiomatic operation" of Nature.

Integration - a relatively more complex operation. Essentially it means that after several elements have come together (combination) and mutually adapted (differentiation), there arises a new system with new properties which could not have been attained by individual elements if their realization had been separate or incomplete.

Integration is, basically, emergence of a system defined as an object with new systemic properties which were not inherent in system elements prior to their merger.

Thus, motor car components lying scattered around a warehouse – all of which are required to assemble a working motor car – are NOT a motor car. They will become a motor car, *make up* a motor car, only after they have been assembled, adjusted and integrated in a single construction. Clearly, there is an abyss between the properties of the system we call a "motor car" and the sum total of properties of all its components existing separately.

Separation - another relatively more complex operation, even though Nature abounds in examples of separation. Separation means the emergence of a new entity, such as a branch or a tree, or a leaf on a branch, or a new tree growing out of a seed.

As concerns the S-curve, we can propose an interpretation of its creation on the basis of the "meta-operations" discussed above (fig. 9.15).

Almost any future system starts with a combination **C** of certain new original components. For example, a motor car was born when they took a tramway carriage or a coach, and topped it with a steam (gasoline, electric) engine.

Differentiation **D** is required virtually at all stages and points of the S-curve, and always involves, as a minimum, adjustment of parameters of individual components to ensure the very possibility of their interaction, and then optimization and enhancement of that interaction.

Separation **S** can occur in two ways – by segregation of a new system type of the given species upon completion of a rather radical differentiation (**DS** type), or by combination of the system with some new property, or component possessing such property (**CS** type), so that such combination also creates a new system type (in terms of its property and construction, naturally). In either case, the net result is emergence of a new branch on the S-curve.

Integration **I** occurs at later stages of the S-curve evolution or, in any event, after the differentiation. That is why integration is often specified on the S-curve as an operation of the **DI** type. Integration signifies ongoing improvement of the system in accordance with the meta-trends of idealization (see fig. 9.9).

After that, it becomes possible to effect a transition of the **IS** type – transition to a new species of systems in the same class, to a new generation system.

In fact, all such transitions can be interpreted as a cyclical process with a quantum jump from generation to generation (fig. 9.16).

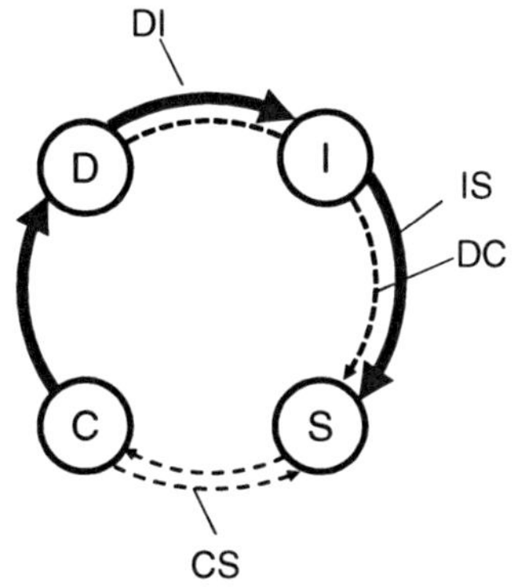

fig. 9.16. DISC Meta-Model as a cycle

Our discussion of events occurring along the S-curve would be incomplete without a description of the following two special situations:

Origination - emergence of a new idea, new solution, new system.

This event can be used to designate two situations:

1) birth $\mathbf{O_1}$ of a new system, a pioneer invention or scientific discovery, the fruit of a successful combination **C** at the beginning of the S-curve;

2) transition of the **IS** type, at the boundary of a family of S-curves, to a new S-curve (bold line above), emergence $\mathbf{O_1}$ of a new species of same-purpose systems replacing a well-integrated predecessor system which has reached the limit of its development.

Obliteration - completion of the life cycle of the system, termination of production, termination of operation, disposal.

This event marks primarily displacement of previous-generation systems with new systems – $\mathbf{O_2}$ point on the S-curve.

The full **DISCOO** Meta-Model is presented in the form of a cycle in fig. 9.17, and in the form of a spiral in fig. 9.18.

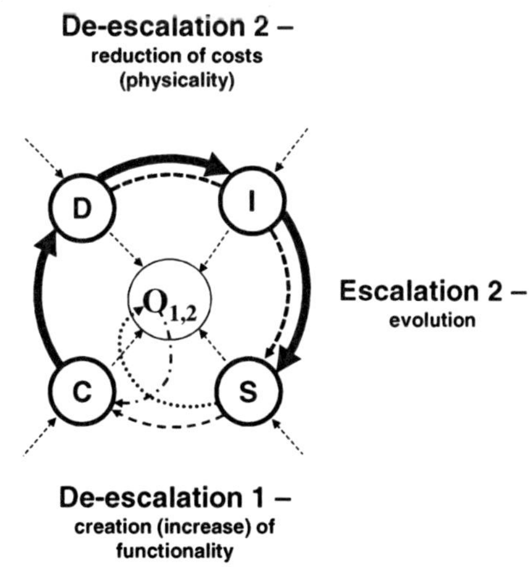

fig. 9.17. DISCOO Meta-Model

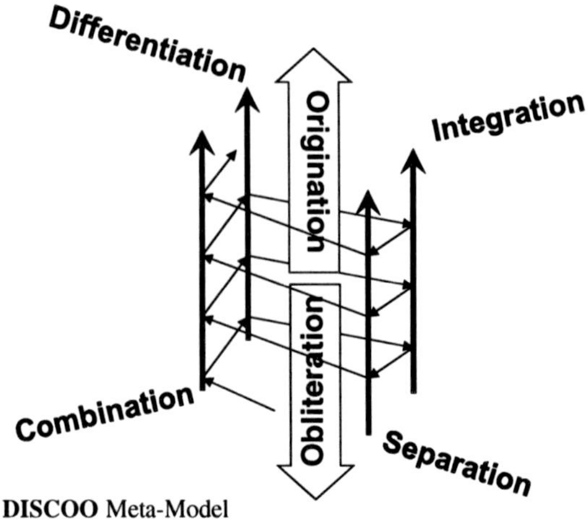

fig. 9.18. Spiral **DISCOO** Meta-Model

De-escalation 1 - Process and result of organization of components into a potentially operable system (formally, transition of the **SC** type from discrete, separated components to a functionally oriented composition), and capture of additional functions in the course of ramification (creation of new types).

This process can be used to designate primarily formation of completely new systems – according to Prigozhin[141]: *from chaos – to organization and order.*

Escalation 1 - Process and result of organization of components into a fully operable system (formally, transition of the **CD** type from a functionally oriented composition to a workable system).

De-escalation 2 - Process and result of system efficiency increase through integration, predominantly through coordination of components and reduction of all types of costs (formally, transition of the **DI** type from a workable system to a highly efficient system of the given species).

Escalation 2 - Process and result of evolutionary system efficiency increase through development of a new operating principle and formation of a new species of same-purpose systems (formally, transition of the **IS** type from a highly efficient system of the given species to a new species of systems of the given class).

And in conclusion of this section, let us discuss the connection between primary transformation models and the newly introduced DISC meta-operations.

The DISC-Matrix (fig. 9.19) establishes empirical connections between specialized navigators from the A_S-catalog.

---

141 Ilya Prigozhin (1917-2003) – Belgian biophysicist, Nobel Prize winner

#	Protomodel	RESOURCE							
		Sys	**Inf**	**Fun**	**Str**	**Spat**	**Tem**	**Mat**	**En**
01	change in the aggregate state of an object			DS				**dDI**	
02	preliminary action		CS	CS	**dCS**		**dCS**		
03	segmentation		D	DS	**dDS**	DS	DS	DS	
04	replacement of mechanical matter			DS	**dDS**			**dDS**	DS
05	separation	CS	CS	CS	**dCS**				
06	use of mechanical oscillations		D				DI	**dDI**	**dDI**
07	dynamization			D	D		**dD**		
08	periodic action		CD	C			**dCD**		
09	change in color			D				**d**	
10	copying		**dDI**	**dCD**	**dCD**	CD		CD	
11	inverse action		D	**dD**					
12	local property			**dD**	**dD**	D			
13	inexpensive short-life object as a replacement for expensive long-life one	**dD**		D	**dD**		D	D	
14	use of pneumatic or hydraulic constructions				IS				IS
15	discard and renewal of parts				**dCS**			**dCS**	CS
16	partial or excess effect	**dD**	**dD**	D					
17	use of composite materials				CI			**dCI**	CI
18	mediator		C	C	**dC**				
19	transition into another dimension		D			**dD**			
20	universality			**dI**					
21	transform damage into use			**dI**					
22	spherical-shape		I			**dI**			I
23	use of inert media			C				**dC**	
24	asymmetry		D	D	**D**	**dD**			
25	use of flexible covers and thin films			CS	C	**dCS**		CS	
26	phase transitions							**dI**	**dI**
27	full use of thormal expansion							**dI**	**dI**
28	previously installed cushions	C		**dC**	**dC**				
29	self-servicing	I		**dI**	**dI**				
30	use of strong oxidants							**dI**	**dI**
31	use of porous materials				I	**dI**		**dI**	
32	counter-weight								**dC**
33	quick jump						**dD**		
34	matryoshka (nested doll)				**dC**	**dCI**			
35	unite			CI	**dCI**	CI	CI		
36	feedback		**dCI**	CI	**dCI**				
37	equipotentiality					**dD**			**dD**
38	homogeneity							**dD**	
39	preliminary counter-action		CS	CS	**dCS**		**dCS**		
40	uninterrupted useful function						**dI**		

fig. 9.19. DISC Matrix (d – dominant resource)

Generally speaking, each navigator can be applied to almost any resource. However, the DISC-Matrix shows the resources to which the navigators are applied most frequently – this is indicated by the DISC symbol at the intersection of the line corresponding to the selected navigator, and the column corresponding to the selected resource.

Thus, in fig. 9.20 showing a fragment of the DISC-Matrix, you can see that navigator 01 is most frequently used to modify the functional resource and the material resource.

#	**Protomodel**	**RESOURCE**							
		Sys	**Inf**	**Fun**	**Str**	**Spat**	**Tem**	**Mat**	**En**
01	change in the aggregate state of an object			DS				**dDI**	

fig. 9.20. DISC Matrix fragment

At the same time, navigator 01 is used primarily to transform the material resource (state of the substance) which is indicated by the symbol "d" (for "dominant").

In addition to that, navigator 01 modifies the functional resources mostly through its differentiating impact **D**; for example it can increase flexibility or elasticity to an extent where there emerges a new function replacing the previous function (separation **S**, at least for the duration of the operating time), for example, "soft water".

The impact of navigator 01 on the material resource can be either differentiating or integrating.

For example, the use of magnet-rheological liquid creates certain new possibilities in terms of modifying the density of the "liquid" within a broad range, i.e. the system is clearly being subjected to a resource modification of the differentiating type **D**.

In implementation of the "sun protection" function in sunglasses with chromatic glass[142], navigator 01 modifies the material of the system (glass) using the integrating model **I**.

This is accompanied by De-escalation 2 (an entire mechanical system is compressed into material!) and separation **S** giving rise to a new system species (event $\mathbf{O_1}$), which results in termination (event $\mathbf{O_2}$) of the previous system with mechanical sun-protection shutters.

However, in real life both types of sunglasses coexist (as is often the case), as many consumers like retro-style sunglasses; besides, the "old" construction with

[142] M. Orloff (2006) *Inventive Thinking through TRIZ*, Sec. *12.2. Fundamental Navigator Catalog*

raised/lowered shutters realizes a positive function – rapid change of the level of transparency at the discretion of the user.

Finally, let us note that the DISC Catalog (fig. 9.21) establishes a link of the "reverse" type – from resources to specialized navigators – while specifying the nature of the impact each navigator has on the given resource within the framework of the DISC model.

RESURCE	Supporting specialized navigators
01 Systemic	05 13 16 28 29 CS D D I I
02 Informational	02 05 06 08 10 11 16 18 19 22 24 36 39 CS CS D CD CD D D C D I D CI CS
03 Functional	02 05 10 07 08 09 11 12 16 18 20 21 23 24 25 28 29 35 36 39 CS CS CD D C D D D D C I DI C D CS C I CI CI CS
04 Structural	02 03 04 05 07 12 14 15 17 18 24 25 28 31 34 35 36 39 SD SI DS CS D D IS CS CI C D C C I C CI CI CS
05 Spatial	03 10 12 19 22 24 25 31 34 35 37 DS CD D D I D CS I CI CI D
06 Temporal	02 03 06 07 08 33 35 39 39 40 CS DS DI D CD SD CI SD CS I
07 Material	01 03 04 06 09 13 15 17 23 25 26 27 30 31 38 DS DS DS DI **d** D CS CI C CS I I I I D
08 Energetical	04 06 14 15 17 22 26 27 30 32 37 DS DI IS CS CI I I I I C D

fig. 9.21. DISC Catalog

Example 9.8. Improvement of the labeling machine – continued from Example 8.11.

Cycle 2.

Trend 2. The original data for this cycle are the results of Zooming of the solution in Example 8.11 (see fig. 8.18): the circular knife must be moved towards the tape to do the cutting, and then away from the tape to enable the tape to move forward by the length of a new label. How can we get rid of the reciprocal movement of the knife?

Reducing 2. **Standard contradiction:** the knife must rotate, but this hampers the feeding of the tape.

Let us write down the formula of the contradiction:

01 Productivity **VS** 15 Length of the moveable object = 04, 06, 24, 30.

Radical contradiction: the knife must cross the entire width of the tape to cut it, and it must not cross the tape to enable it to move forward past the knife by a predetermined distance.

IFR: the knife rotates **AND** does not prevent tape movement.

Inventing 2. The first three navigators are similar in that they recommend to modify the structural resource; moreover, in all cases the preferable modification is that of the D type, i.e. one which involves creation of a range of parameter change:

04 Replacement of mechanical matter: c) transition from static to dynamic fields, from temporally fixed to flexible fields, from unstructured fields to fields with a specific structure;

06 Use of mechanical oscillators: a) cause the object to oscillate;

24 Asymmetry: a) move from a symmetrical shape of the object to an asymmetrical one.

We need to make the cutting edge of the knife "oscillate" transversely using the principle of asymmetry – which will set the new trajectory (structure) of the knife's movement.

Idea for the solution: to get rid of reciprocal movement of the circular knife, its axis of rotation was displaced – the "old" axis is marked with figure 1, the new one – with figure 2! The Client's experts suggested an eccentric knife (fig. 9.22)! This solution uses TRIZ model *24. Asymmetry.* And it makes use of the *spatial resource*!

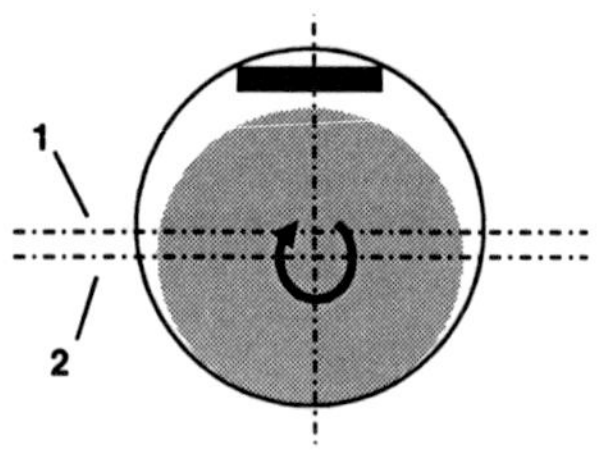

fig. 9.22. Asymmetrical construction of the shears

Also, rotation of the knife never stops! The tape moves by the length of a new label when the knife below is turned towards it with the smaller-radius part! And the tape moves on through space which is "knife-free", even though the knife keeps moving!

Other TRIZ models used include *33 Quick jump* and *08 Periodic action.* The trajectory of the smaller-radius part of the knife is shown as a dark gray circle. An important role is also played by the *temporal resource.*

Zooming 2. It turns out that the knife must be big to accommodate progressive movement of a relatively broad tape.

Cycle 3.

Trend 3. The original data for this cycle are the results of Zooming of the solution in cycle 2, to wit: we need to reduce the size of the knife!

Reducing 3. **Radical contradiction:** the knife must be big to quickly cut the tape across its entire width, and it must not be big because there is not enough free space inside the machine to accommodate a big knife.

IFR: the knife must be small, but at the same time it must be big enough to quickly cut the tape across its entire width.

Inventing 3. To solve this problem, we need first to consider possible operations with the spatial resource.

Let us write out of the DISC Catalog (section 22) Line 05 *Space* (fig. 9.23).

05 Spatial	**03**	**10**	**12**	**19**	**22**	**24**	**25**	**31**	**34**	**35**	**37**
	DS	CD	D	D	I	D	CS	I	CI	CI	D

fig. 9.23. DISC Catalog fragment describing Resource 05 *Space*

A look at the first several navigators is enough:

12 Local property: b) different parts of an object must have different functions; c) every part of the object must exist under conditions that correspond best to its functions;

19 Transition into another dimension: a) the object acquires the ability to move only along a line, but also in two dimensions (on a plane); an improvement is also possible upon transition form movement on a plane to movement in a 3D space;

22 Spherical-shape: a) change from linear parts of the object to curved ones, from flat surfaces to spherical ones; from parts shaped like cubes or parallelepipeds to spherical structures.

Idea for the solution: the tape must be shaped like a pipe segment (fig. 9.24) – change shape from 2D to 3D!

Figure 1 indicates the "old" diameter of the knife and the position of the "previous" axis of rotation of the knife, figure 2 – the "new diameter of the knife and the position of the "current" axis of rotation.

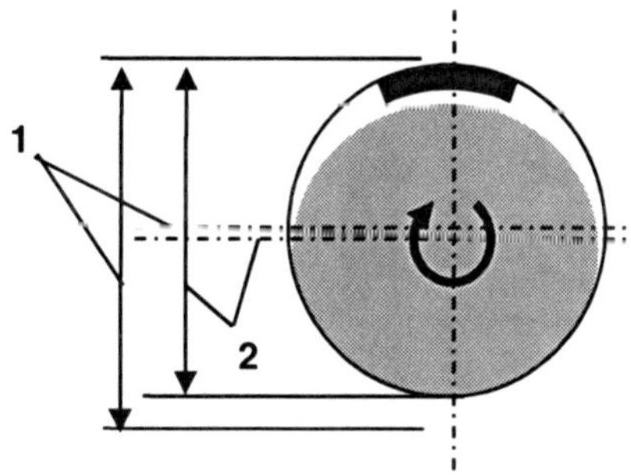

fig. 9.24. Shape of the tape is aligned to the trajectory of movement of the cutting edge of the knife

The solution is essentially based on alignment (De-escalation 2: differentiation **D** of parameters for the purpose of integration **I**) of the shape of the tape and the trajectory of knife movement when it is cutting the tape.

Zooming 3. Three super-effects have been identified:

1) the tape has ceased to bend (which previously led to production of deficient labels);

2) the speed of movement of the tape can be increased even more due to the higher rigidity of the "pipe-tape";

3) the speed of cutting of the tape has increased due to the reduction of free movement of the cutting edge of the knife.

The Client's experts independently developed the ideas suggested by the Inventor, using also other TRIZ models, and designed efficient and rather simple assemblies. The key solutions were patented.

Based on the models discussed in this section, the author implemented a series of highly efficient predictive projects for a number of companies. For example, creation of artificial "windows" or "transparent walls" with renewable "landscapes", dynamic imaging and gaze direction monitoring had been predicted by the author in the *Wall- & Window-Screen Technology* project completed for another Client several years before the same solution was marketed by the world-renowned Philips (and then other companies). Also, the *Mobile Phone FLAT* project predicting certain functions and trends leading up to the creation of a thin integrated tablet (including screens on both sides) had been implemented long before Apple iPads and Samsung Galaxy Tabs.

This provides yet another confirmation of the well-known fact: good ideas are suggested by different inventors – sooner or later, and sometimes simultaneously. It just means that their time has come! And TRIZ can bring that time closer.

Practitioner's Mastery

...To make so that
the complex becomes simple,
the simple becomes routine,
and the routine becomes pleasant.

Konstantin Stanislavsky[143]

10 EASyTRIZ Training Baseline

10.1 EASyTRIZ Instrumental Software

Over the course of many years that I have spent teaching and consulting, I have come to the conclusion that primary TRIZ instruments presented in this book simply MUST become the object of mass-scale training and practical application, as they give the user an accurate and sustainable understanding of fundamental TRIZ models – contradictions and simplest methods of their resolution.

The EASyTRIZ software package (fig. 10.1) contains everything that is necessary to organize user activities during training and subsequent practical work, and to support resolution of new problems, future self-training with the pool (depositary) of examples accumulated in the joint databank, exchange of information with other members of the association of MTRIZ Academy users, and for other purposes.

PROJECT MANAGEMENT: disclosure of previously implemented projects and creation of new projects. The term "project" is applied to any completed work, for example, extracting of models from a selected artifact, reinventing of an artifact, or resolution of a new problem. Accordingly, it is assumed that each project has a name (fig. 10.2), and can have several reruns (branches and cycles).

TRAINING CENTER: the software package offers a convenient embedded interface with the MTRIZ Academy – there is no need to type in addresses, and there is functionality supporting direct access to the archive section storing previously completed assignments, direct correspondence with the tutor (trainer), etc. The

143 Konstantin Sergeyevich Stanislavsky (1863-1938) – famous Russian actor, theater director and teacher, co-founder of the Moscow Art Theater, creator of the theatrical method known as the "Stanislavsky's System"

software package also has a number of additional features specifically designed to assist tutors, translators and administrators of the MTRIZ Academy.

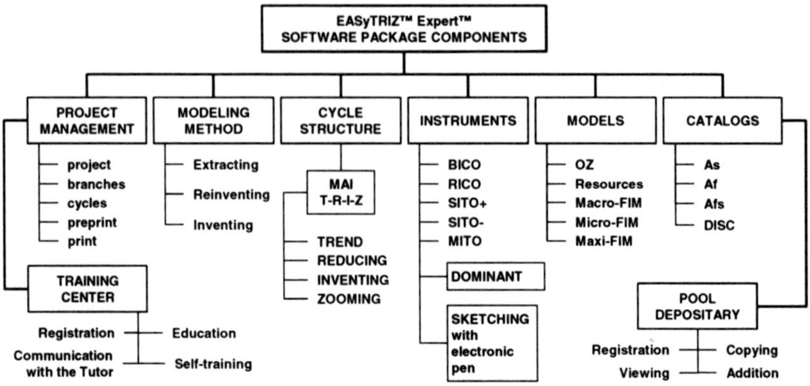

fig. 10.1. EASyTRIZ-Expert software package structure and functions

MODELING METHOD: type of work and, accordingly, nature of the project selected from one of the following three options: extracting, reinventing, and resolution of a new real-life problem (inventing).

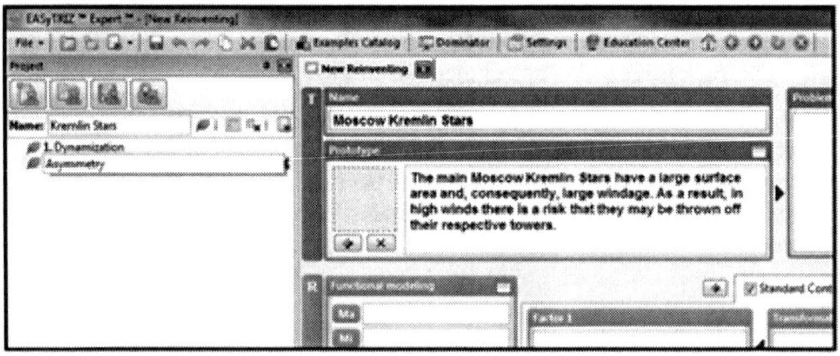

fig. 10.2. Project management window (left) in the EASyTRIZ-Expert software package

CYCLE STRUCTURE: default value – Meta-Algorithm of Invention in its expanded START T-R-I-Z version. This means that each of the following items has a standardized structure:

- examples in the books published by the MTRIZ Academy;
- examples in the catalogs of the EASyTRIZ software package;
- examples in the pool depositary;
- training reinventing;
- all training assignments stored at the Training Center, etc.;
- real-life inventing.

This helps to implement and maintain standardization of training, as well as acquire and use correct thinking patterns while solving TRIZ problems.

INSTRUMENTS: the EASyTRIZ-Expert software package uses all instruments described in this book. In the EASyTRIZ-Practitioner software package, the list of instruments is limited to two baseline methods – BICO and RICO.

DOMINANT: a special user-customizable window enabling the use of color stimulation to encourage creative activity.

MODELS: in the course of work with artifacts under review, it is possible to additionally provide more detailed information on known models describing the OZ (operating zone), various resources and FIMs (functionally ideal models).

CATALOGS: the software package has an inbuilt baseline standard catalog of examples for specialized (As-catalog) and fundamental (Af-catalog) transformations. These models are marked with labels linking them to the DISC Meta-Model.

POOL DEPOSITARY: the software package has an embedded interface for accessing the examples databank of the MTRIZ Pool.

Example 10.1. Working with EASyTRIZ-Expert windows for the simple one-run reinventing operation (fig. 10.3).

SKETCHING: this option has been included into a software package of this type for the first time in the world! This warrants a more detailed description which is provided below.

When the user opens any window to input non-standard additional information, the software package offers to use a digital (electronic) ball-point pen to draw sketches on ordinary paper.

A sketch is a combination of graphic objects and hand-written text.

There are two operating modes: online and offline. In the online mode, while drawing a sketch in one color and using one line type preset in the pen, the user can change the color and type of lines displayed on the screen and modify other functional settings.

In the offline mode, the drawing can be made while the computer is off. In such case, the drawing is saved in the pen. The user can then transfer the sketch into any window and edit graphic and text objects.

NB

This gives the user an instrument with which, instead of thinking about his drawing technique (as he has to, for example, in PowerPoint), he can focus on the subject matter of the sketch and on the problem under review!

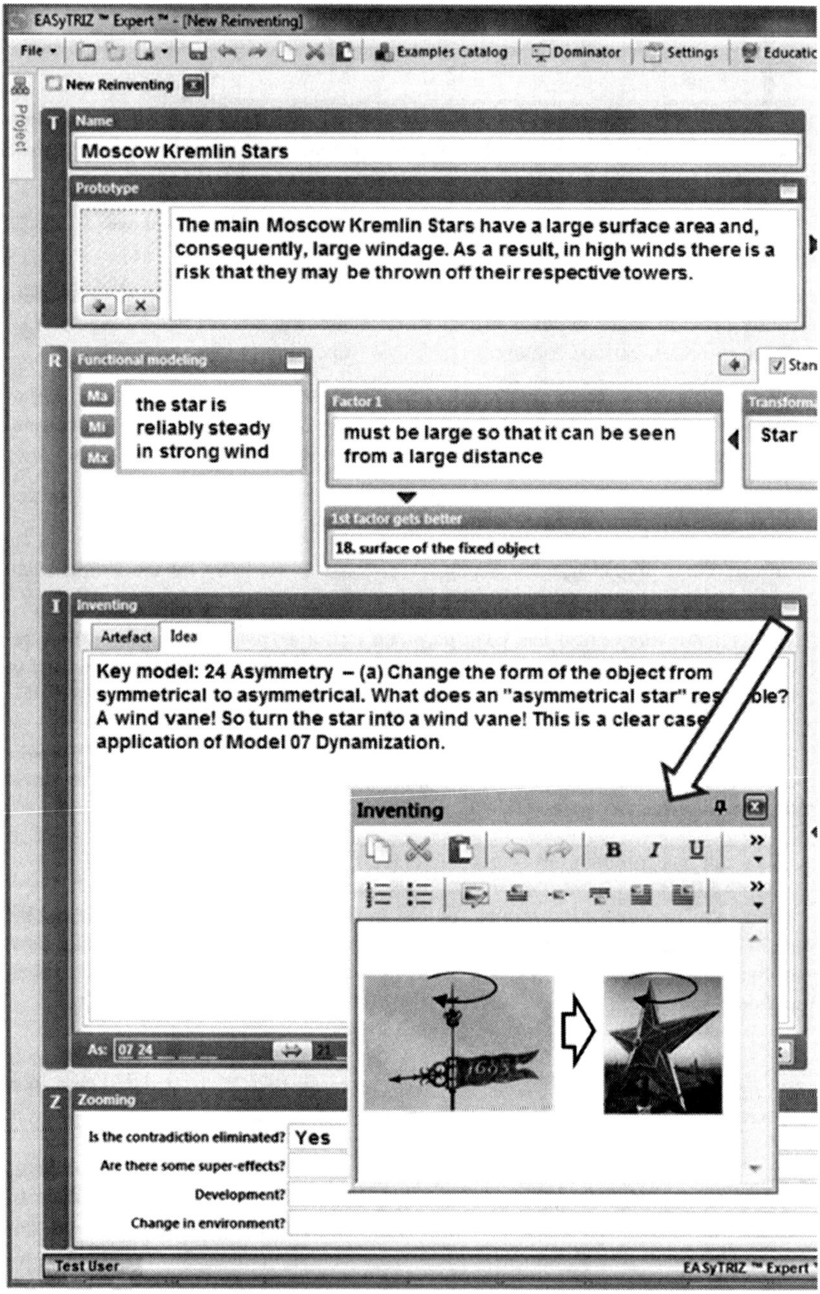

fig. 10.3. Solving Problem P1 in the main work window of the EASyTRIZ-Expert

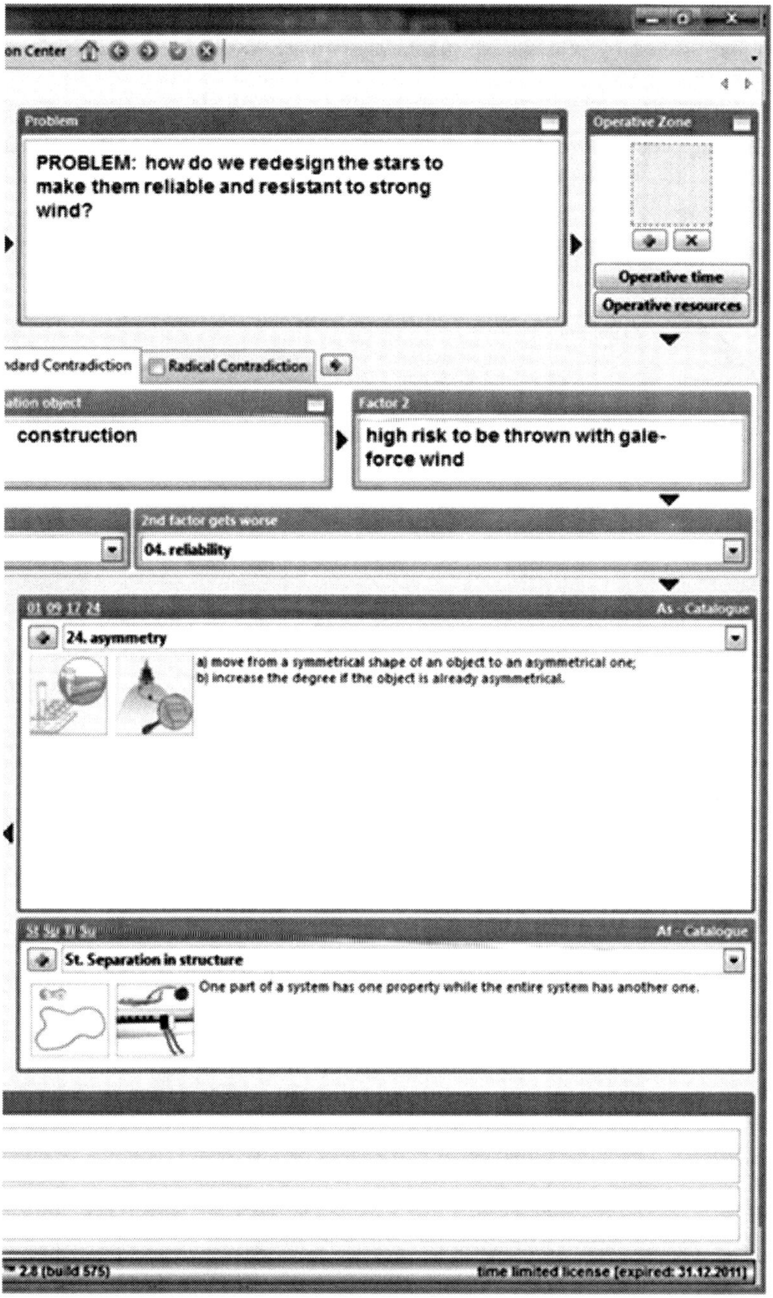
Problem
PROBLEM: how do we redesign the stars to make them reliable and resistant to strong wind?
Operative Zone
Operative time
Operative resources
Radical Contradiction
construction
Factor 2
high risk to be thrown with gale-force wind
2nd factor gets worse
04. reliability
01 09 17 24
As - Catalogue
24. asymmetry
a) move from a symmetrical shape of an object to an asymmetrical one;
b) increase the degree if the object is already asymmetrical.
A1 - Catalogue
St. Separation in structure
One part of a system has one property while the entire system has another one.
time limited license [expired: 31.12.2011]

The quality of sketches (in terms of resolution and line smoothing) is on a par with that offered by professional graphical stations! Moreover, manually drawn objects are automatically replaced (if necessary!) with "correct" geometrical figures – rectangles, circles, text boxes, etc.!

Naturally, the system also stores the original handwritten sketch. An example of what a sketch looks like before and after the user inputs it into the system is shown in fig. 10.4.

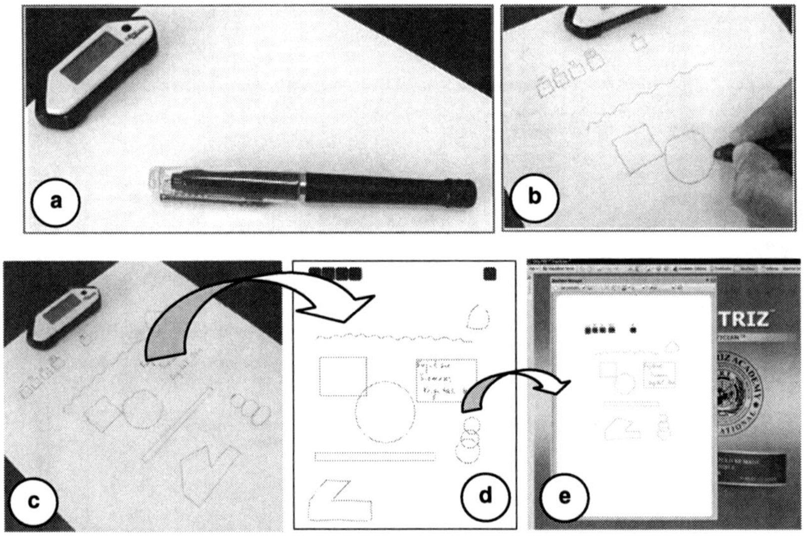

fig. 10.4. Transfer of the manual sketch into the EASyTRIZ-Expert software package

10.2 Distance Training Technology

The totality of training aids – textbooks, software, methodological materials, remote training course, etc. – forms the instrumental foundation of the training technology used at the Modern TRIZ Academy. The content of the matters under review is described in this book. Methodological peculiarities constitute the "know-how" of the Academy. At this point it will suffice to briefly present the structural organization (fig. 10.5) of the training program and process designed, first and foremost, for remote implementation (e-Learning or Distance Training). The main steps towards the mastery is shown in fig. 10.6. Certification of MTRIZ Junior is oriented to school teenagers and to the students of early semesters. All practical issues are dealt with at the MTRIZ Academy web sites.

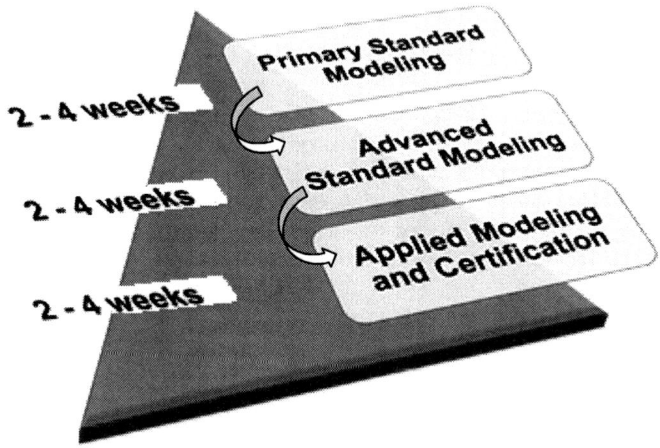

fig. 10.5. Three MTRIZ Practitioner training blocks

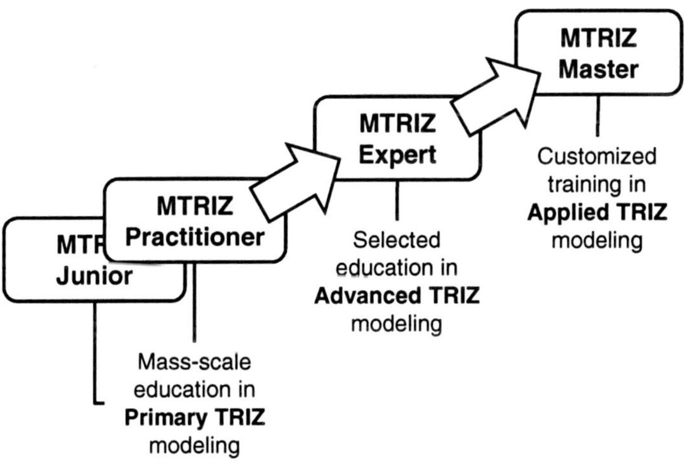

fig. 10.6. MTRIZ training: main line towards the mastery

11 Energy of Progress

11.1 On the Way to a New Innovation and Invention Theory

The future of any state depends on its intellectual potential, ability to produce innovative ideas and resolve constantly emerging contradictions of growth, and ability to project the course of events and, based on such projections, mitigate risks and maintain sustainable progress. This thesis can be illustrated by the following description of strategic platforms which generate positive driving forces and define the general vector of development at various stages of human history (fig. 11.1).

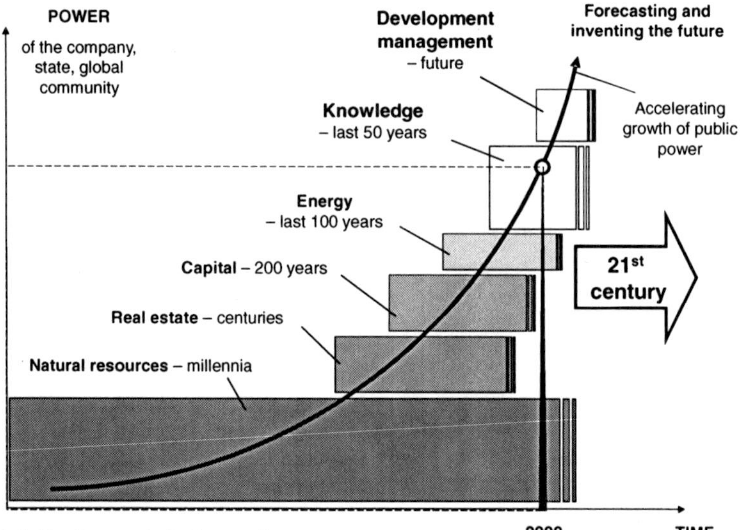

fig. 11.1. Fundamental platforms of progress

Particular strategic importance is ascribed to the intellectual potential of a special type – that related to inventions and innovations. It is the inventive creative energy that puts society into motion, promotes technical and then social progress, creates leadership through inventing large- and small-scale solutions for production, education, politics, and all other areas of human activity.

It is this aggregate inventive potential that encourages global economic change, sustainable increase of the quality of life, reproduction – generation after generation – of talented people motivated to creatively fulfill their potential for their own benefit and for the benefit of their families, their collectives, their people and the state.

Repetition and borrowing does not create leadership by definition. Leadership requires pioneering innovations and inventions. And that calls for theories and practical methods that inherently have everything that it takes to assure subsequent development and mutual enrichment through joint utilization of intellectual treasures amassed in the fields of engineering, philosophy, sociology, politics, and virtually every other area of general theoretical and applied knowledge (fig. 11.2).

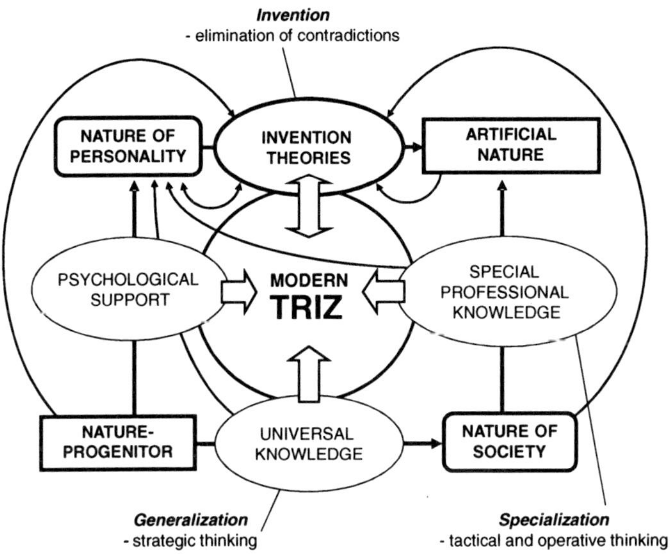

fig. 11.2. Formulation and application of invention theories

We need to critically review the principles and mechanisms used to accumulate and propagate the intellectual potential of the state.

Paradoxically, until now there has been no place in the world where creative invention methods and models would be collected, reorganized and reinjected into everyday practical work.

Experience accumulated by outstanding innovators and inventors is not studied or generalized, it is not transformed into new knowledge, and it is not taught to innovators and inventors of new generations. Each innovator has to walk by himself the same "path of errors" as his predecessors. How preposterously and wastefully archaic!

Mankind loses – almost irretrievably – gigantic intellectual riches embodied in the knowledge and skills which reside in innovators and inventors! These riches are not accumulated, not studied, not transformed into new intellectual resources that can be used to encourage further public progress!

The only trail-blazing example of systemic scientific study of inventions is presented by TRIZ. However, TRIZ is not spread throughout the world to the extent that it definitely deserves. Besides, TRIZ itself is in need of radical renewal and transformation.

And the first thing that has to be done in this respect is to create TRIZ presentation models and methods that would assure simple and rapid internalization and correct utilization of TRIZ fundamentals. This is the starting point of Modern TRIZ.

To implement this new stage of development, it is necessary to retrace – at this new turn of the spiral of history – the path walked by G. Altshuller and his successors at the time when TRIZ emerged as an independent discipline. This means that it is necessary to perform – again! – selection and analysis of empirical material.

There is no other reliable way. Only from practice to theory. Naturally, subject the fact that study of new empirical material can and should be approached from a modern theoretical platform.

However, before we undertake such full-scale study, we need to outline, at least briefly, the current state of innovative analysis. And when we do that, there will emerge an amazing picture reflective of the real understanding (or, rather, "non-understanding"!) of the real attitude of managers operating at all levels of corporate organization both to TRIZ and to teaching innovative thinking.

11.2 Opel, VW and Others: "Masses VS Elite"

First let us analyze economic efficiency of innovations created at one or two large industrial enterprises in Germany. This will give us an idea of the state of affairs in a large industrially developed country, and help us decide whether there is any sense at all in theorizing about innovations and inventions. Then, based on the results of that analysis, any concerned reader will be able to come to his own conclusions subject to his preferences and interests.

Evaluation of economic efficiency is performed by extrapolation and multiplication of real-life data from two large German carmakers – Open and Volkswagen (VW) – which we believe to be reliable benchmarks for our purposes.

In our evaluation, we will use the following method:

- first, we research the volume and economic efficiency of innovative proposals filed by employees of the two companies;

- taking into consideration the mass-scale nature of innovative proposals filed at the benchmark companies (tens of thousands per year with the participation of tens of thousands of innovators), we then assume that the ratio of intensity of innovations and their average economic efficiency will be approximately the same at all automotive industry enterprises;

- finally, we perform proportional linear extrapolation of economic efficiency figures posted by the benchmark companies to the entire industry based on some common indicator, such as the number of employees.

Example 11.1. Economic efficiency of innovations.

The table in fig. 11.3 contains public information about innovative activities at Opel and VW enterprises. Asterisks (*) mark confirmed reliable data obtained from open sources. Other data are approximations and derivations based on certain assumptions.

No.	Indicator ↓ Company →	Opel	Volkswagen
01	Number of innovative proposals per year	72,791* (2001)	150,000* (2006)
02	Annual economic impact, € million	75*	168*
03	Average economic impact per proposal, €	1,000	1,120
04	Fees paid to innovators, € million	11*	23*
05	Average annual fee per proposal, €	350	153
06	Total number of employees	33,000*	151,000*
07	Number of innovators	8,250	37,750
08	Average number of proposals per innovator	9	4
09	Number of outstanding innovators	28*	49*
10	Fee per outstanding innovator, €	51,000*	51,000*
11	Efficiency of outstanding innovations (at least), €	400,000*	400,000**
12	Contribution of outstanding innovations to annual economic impact, € million	11.2	19.6
13	Contribution of outstanding innovations to annual economic impact, %	15	12
14	Contribution of "mass-scale" innovations to annual economic impact, € million	63.8	130.4
15	Contribution of "mass-scale" innovations to annual economic impact, %	85	88
16	Economic efficiency ratio per "outstanding" idea	400 : 1	400 : 1
17	Ratio of "outstanding" ideas to total ideas	1 : 2,600	1 : 3,000
18	Ratio of "outstanding" innovators to total employees	1 : 1,200	1 : 3,000
19	Ratio of "outstanding" innovators to total innovators	1 : 300	1 : 770

** Efficiency figure is extrapolated from the Opel indicator due to the lack of reliable data on VW.

fig. 11.4. Reinventing as an MTRIZ-based reconstruction of the inventing process

Taking into consideration the fact that, on the average, only 25% of employees at both companies file innovative proposals, the average number of proposals per one active innovator ranges from 4 to 9 (line 08 in table 1).

Line 15 confirms that the "bulkiest" contribution to total economic efficiency is made by "mass-scale" proposals. Their share amounts to at least 80%, while "elite" proposals account for about 20%.

Line 16, however, shows that "outstanding" innovative proposals boast much higher "individual" efficiency, beating the averages by hundreds of times!

It should be noted that "outstanding" innovations are mostly represented by solutions of particularly complex technical and technological problems resulting in revolutionary breakthroughs in the ongoing development of products and technologies. Without such innovations, progress becomes impossible.

That is why "outstanding" innovations cannot be replaced with "mass-scale" innovations – these phenomena have different nature and significance. However, in terms of aggregate economic efficiency, contribution of "outstanding" innovations to the annual bottom line can be less weighty than that of "mass-scale" innovations.

Here is some food for thought: on the one hand, not even a million high jumpers showing average results can replace one record-breaking high jumper; on the other hand, no outstanding soccer player can play alone against a full team. Both examples are important, because they are about attaining different objectives.

The method demonstrated above, and the results obtained, are sufficiently reliable to perform approximate calculations. A similar method is used by the German Economic Institute to evaluate efficiency of innovative proposals filed by employees of German industrial enterprises.

The following data published by that institute are important for our subsequent calculations:

a) in 2007 the aggregate economic impact of innovative proposals filed at 290 medium-sized and large German enterprises (including Opel and VW enterprises in Germany) amounted to € 1.49 billion;

b) the total number of innovative proposals filed reached 1.4 million;

c) the average economic impact per proposal was € 1,064;

d) the total number of employees at the enterprises which participated in the review was about 2 million people;

e) the average economic impact per employee was € 684;

f) on the average, innovators who filed at least one proposal accounted for 25% of the total number of employees;

g) the average number of proposals per employee came close to 0.72;

h) the average number of proposals per innovator per year reached, at least at some enterprises, 17(!);

i) proportional segmented extrapolation to 5,000 similar German enterprises shows that the overall economic impact of innovative proposals filed in 2007 may have been as high as € 27 billion;

j) industrial and non-industrial innovations accounted for 71% and 29% of the aggregate economic impact, respectively.

Conclusions

1. Due to the high aggregate economic impact of "mass-scale" innovations, the first thing that we need to do is implement *mass-scale training in MTRIZ foundations*, and thereby assure direct and rapid application of MTRIZ to increase the quality and efficiency of such "mass-scale" innovations.

Indeed, MTRIZ methods help to increase the efficiency of mass-scale innovative activity at individual enterprises *without long-term special training.*

2. MTRIZ technologies enable a detailed scrutiny of outstanding innovations and their creation stories, which in turn makes it possible to develop, based on the information obtained during such scrutiny, highly efficient training examples, and incorporate such examples into the training curricula for mass-scale education of industrial workers.

Naturally, selection and education of elite innovators is a critically important separate task. However, the elite will emerge as a result of such mass-scale education giving rise to a mass-scale innovative movement. To that end, we will be reviewing technologies for ongoing training of innovators on the basis of application of MTRIZ models to assure effective accumulation and transfer of experience generated by outstanding innovators.

11.3 Inventive Ideas Pool

This is a good place to explain *what exactly*, out of the entire body of expertise created by innovators, is popularized at large enterprises, *how* it is done, and *what* is the scope of managerial activity in terms of dissemination of best practices and training of innovators.

The answer will not be very long: information circulated about new achievements is restricted to descriptions of the *technical nature of proposals.* And that is all.

Knowledge about objectively existing models which were used to create an outstanding creative solution remains undisclosed!

Have a look at fig. 11.4 (this is another version of the familiar fig. 3.12).

We postulate – and there are sufficient grounds for that – that the process of creating a major idea is the unity of three "sub-processes" related simultaneously to different "creative" (and different "systemic") levels of thinking, and to different "noospheres" of the *object of thinking* (artifact) and the *subject (organ) of thinking* (brain, personality).

That is why if we restrict our review to technical description of the innovation, we objectively remain at the level of applied knowledge. And that says it all. Let us recall Goethe[144]:

to place oneself at the level of the objects is to learn;
to take objects in their depth is to invent.

[144] Johann Wolfgang von Goethe (1749-1832) – outstanding German thinker, poet, philosopher and naturalist

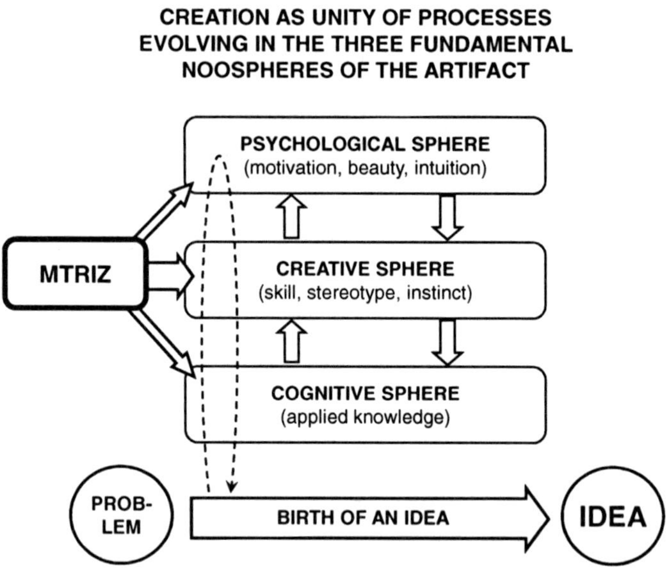

fig. 11.4. Reinventing as an MTRIZ-based reconstruction of the inventing process

Suffice it to say that the managers' "non-understanding" is based on the absence of knowledge about the structure of the creative process, as shown in fig. 11.4, and, naturally, on the absence of knowledge about TRIZ.

Accordingly, in addition to technical description of the artifact (the product of innovation), it is necessary (to the extent appropriate) to provide a description of reinventing according to TRIZ, and subsequently reinventing at the highest level of creativity – modeling of beauty, harmony, and intuitive insight. But that still resides in the realm of the future.

Today it is possible to create, for any artifact, a description of creative models objectively existing in such artifact. Then (fig. 11.5) the usual contribution to the pool of creative samples (technical description of the artifact) can be supplemented with a description of reinventing according to MAI T-R-I-Z. This will create a universe of empirical examples that can be used both for training purposes and for practical application while solving new problems.

And now imagine that over the course of one year all you have contributed to the pool is one reinventing. Imagine also that the pool has 1,000 users, and that each such user has contributed one reinventing.

This means that you – and all your colleagues, all pool members – have become co-owners of 1,000 examples! And this means that the "rate of return" on your "investment" – your contribution – is almost 100,000%!

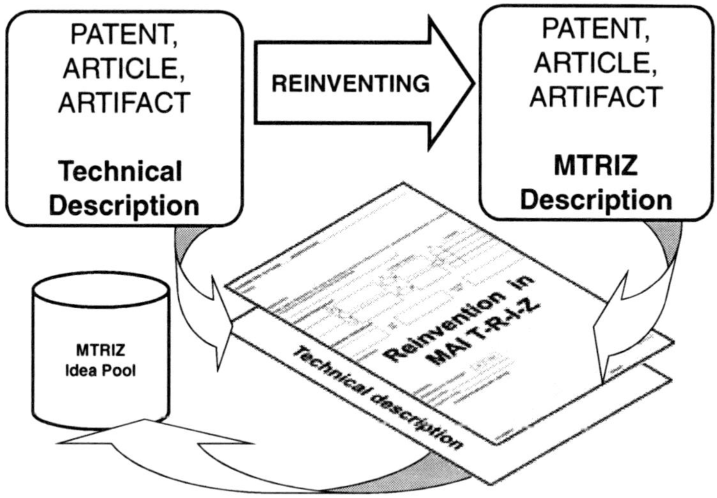

fig. 11.5. Creation of a pool of MTRIZ reinventing examples

What other area of knowledge can offer such profitable investment opportunities? This is the colossal potential of the MTRIZ Idea Pool. This situation was very aptly described by the famous playwright Bernard Shaw[145] (fig. 11.6):

"f you have ***an apple*** *and I have* ***an apple*** *and*
we exchange these apples,
then you and I will still each have ***one apple****.*

But if you have ***an idea*** *and I have* ***an idea*** *and*
we exchange these ideas,
then each of us will have ***two ideas****".*

Bernard Shaw

If thousand partners put each only ***one idea***
into TRIZ Idea Pool,
then each of partners will have ***thousand ideas****.*

M.O.

fig. 11.6. Intellectual and pragmatic efficiency of a collective pool of creative samples

Our goal is to transform this idea into a modern "creative machine".

A schematic representation of this "Intelligence Amplifier" (see reference [26]) is shown in fig. 11.7.

[145] George Bernard Shaw (1856-1950) – world-famous British (Irish and English) playwright, Noble Prize winner, author of numerous plays, including *Pygmalion*

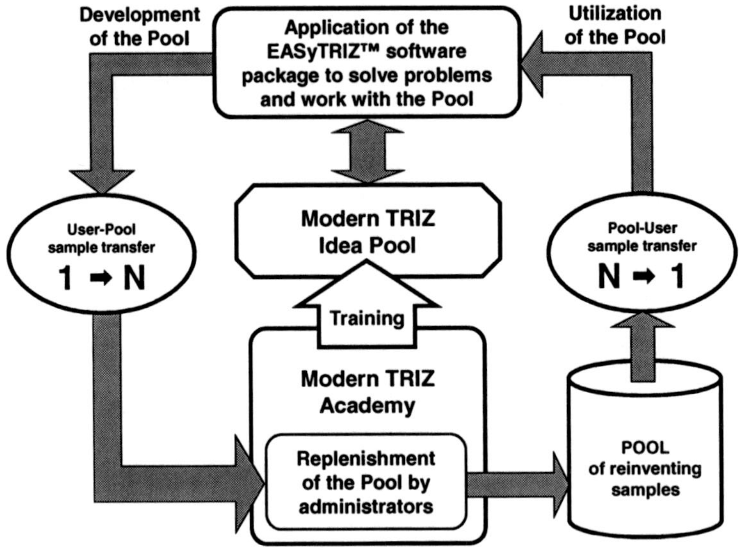

fig. 11.7. Information flows for MTRIZ Idea Pool users

Certified Modern MTRIZ users create an association whose members are granted access to an idea pool, a collection of reinventing samples for highly efficient – and simply interesting – examples of innovations and inventions. The main component of the pool is the depositary of artifact reinventing samples (examples databank). Pool members can perform a wide gamut of operations with the depositary.

First and foremost, they can view examples while solving new problems (transfer **N** → **1**). The second most important function is the transfer of examples by administrators for incorporation into the pool (transfer **1** → **N**).

Work on creating new ideas is performed, and access to all pool functions is granted, through the common interface of the EASyTRIZ™ software package.

Pool participants can be physically located in any place in the world, or they can all work in some corporate intranet. For this purpose, it is possible to create specialized databank versions and customized corporate pools.

The Modern TRIZ Academy publishes digests. It also consults and trains pool members.

Users can periodically review new examples by themselves – essentially engaging in ongoing self-training.

Functionality of the pool will be continuously expanded.

Technically, the pool can exist as a network structure (fig. 11.8) joining individual users, enterprises, corporate networks etc. into a common environment specializing in the transfer of examples, i.e. of accumulated expertise presented in standard format. It is also very important that access to innovative knowledge and examples stored in such network can be granted to colleges and universities and to schools. Examples can be based on reinventing of numerous great inventions kept in museums. *And it is possible because creation never grows old!* Ageing can affect only the physical artifacts and the *technical solutions* on which they are based, but the artifacts themselves remain interesting and useful for studying *creative solutions*, for extracting *transformation models.*

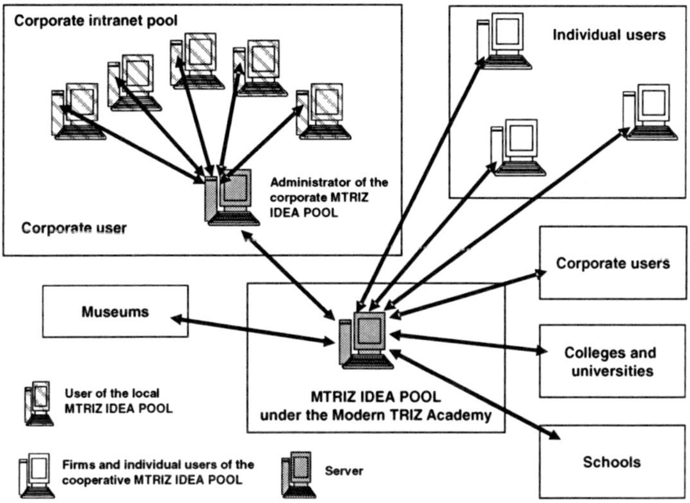

fig. 11.8. Hierarchical network structure of the open MTRIZ Idea Pool

Gradually the pool can branch out into specialized depositaries focused on individual industries or areas of knowledge (fig. 11.9).

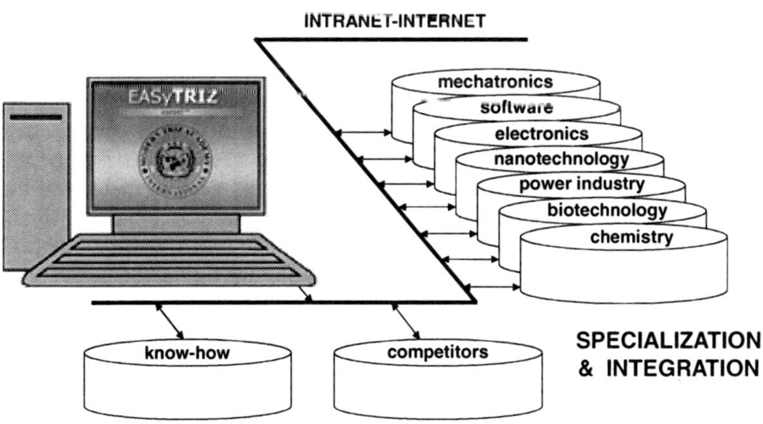

fig. 11.9. Specialization and integration of knowledge in the MTRIZ Idea Pool

12 Reinventing of Automobile Production Systems Evolution[146]

12.1 From Craft Production to Mass Production System

In this section, the case studies will focus on the innovations occurred in the evolution of Craft Production to Mass Production System.

12.1.1 Interchangeability of Parts

For centuries, after the Industrial Revolution in England in the late 18th century, economic production was based on the notion of craftsmen. Everything was crafted by skilled workers who had the requisite materials, tools, and most important, skills.

Although in the late of 1800s – when the first manufacturers of automobiles flourished – machinery and mechanization had been used as primary instruments of production, the skills of the craftsmen still played an important role. The skills and methods was the source of pride that made the craftsman was exclusive to each other.

The Prototype – Independent manufacturing stations with individual gauging system

In a Craft Production System, no two automobiles were built exactly the same. This was due to the parts manufactured had high inconsistency of dimensions. This made the assembly process could only be performed by skilled craftsmen, who still needed long time to fix and fit the parts each other. Abovementioned problems were due to the parts were manufactured in several machine shops that have their own gauging systems.

The first company formed exclusively to build automobiles was Panhard et Levassor (P&L) in France[147]. It was a very famous manufacturer at that time and its Système Panhard, a breakthrough design with front-mounted engine and rear wheel drive, became standard for all other automobiles. Although using the same blueprint, the company could not produce even two identical cars. It was because the parts were supplied by different machine shops, and each of them used their own gauging system and specifications. This condition was exacerbated by limitation of the technology in processing the raw material.

When these parts came to the automobile company for fitting, the parts specifications could be described as "approximate". The first things need to do by the workers were try to fit the parts to each other until the whole vehicle was com-

[146] This part is developed together with my student Sugeng Wahyudi (Indonesia) graduated from my Modern TRIZ class by Master of Science Program at Berlin University of Technology with Thesis "Modeling the Production System Innovations with Modern TRIZ"

[147] Gorgano, G. N. (1985) *Cars: Early and Vintage, 1886-1930*. – Grange-Universal, London

plete. The job required a lot of skill, time and effort since each part did not suit to each other. This caused a "dimensional creep" phenomenon, since by the time the workers reached the last part, the whole vehicle could differ significantly in dimension from the other car that was built by using the same blueprints[148].

Those were the reasons why automobile company at that time could not produced cars in large quantity with the same dimension and quality.

fig. 12.1. Fixing and fitting of car body in a craft workshop

fig. 12.2. Machine shop of Auto Hacket Motorcar Co., Michigan, in early 1900s

The Result of Contradiction Analysis

The "plus-state" of the "independent manufacturing stations with individual gauging system" can be determined as in A-Matrix as **number 22 (speed)**, because the parts could be produced in several parallel machine shops that increase rate of availability of parts. The "minus-state" of the "independent manufacturing stations with individual gauging system" can be determined as in A-Matrix as **number 05 (precision of manufacture)**, because those parts became not fit to each other due to different gauging systems.

Inventing Idea

Combining the plus and minus state factors above, the results will show several points in As-navigators. The target is to keep separate manufacturing stations so that the rate of parts production will still be maintained while they are able to produce parts with consistent dimensions that are fit to each other. After combining the A-Matrix 22 and 05, the idea navigator resulted :

- **Navigator 02 (preliminary action)** – according to this navigator, the manufacturing machine shops can prepare the parts in advanced by using standard gauging system so that the assembly process can be processed without loss of time for fixing and fitting the parts.
- **Navigator 18 (mediator)** – according to this navigator, specific gauging devices were used as mediators between the company and the suppliers in order to have standardized measurement of parts.

---

148 Womack, J. P., Jones, D. T. and Roos, D. (1990) *The Machine that Changed the World.* – Rawson Associates, New York

The Product-artifact (Solution) – Interchangeability of Parts

> The way to make automobiles is to make one automobile just like another automobile.[149]
>
> *Henry Ford*

Started in 1903, Ford Motor Company adopted the Craft Production during its early years as the common production system available at that time. When it introduced its famous Model T in 1908, the production volume rate was very small, while the demand was increasing. When Ford discovered the disadvantages of this production system, he began to improve it one by one.

Many people think that the key to Mass Production was the continuous assembly line. This opinion is not quite correct because the continuous assembly line could not work if the parts could not be attached to each other in a fast and simple way. To achieve this, the assembly line should be supplied by standard parts so that all of the parts could be fitted without any difficulties. This was the concept that called "interchangeability of parts".

To achieve this condition, Ford insisted that all the machine shops that supplied parts to his company needed to use the same gauging system in their manufacturing process. This was a simple idea, but a breakthrough at that time, since there were no companies had thought about that.

By the standardization of the gauging system, all the parts delivered to Ford's assembly line could be easily fitted each other, and reduce the assembly process time.

fig. 12.3. Identical Body Parts produced in Fisher Body Company

fig. 12.4. Chevrolet workers produced stocks of identical front fender with a double-action press machine (1959)

[149] ibid.

With this condition, Ford could also reduce the number of highly-skilled fitters in his company and saved labor cost. In the end, the company could produce more cars in lower price. Simultaneously, the quality of the products could be relatively uniform.

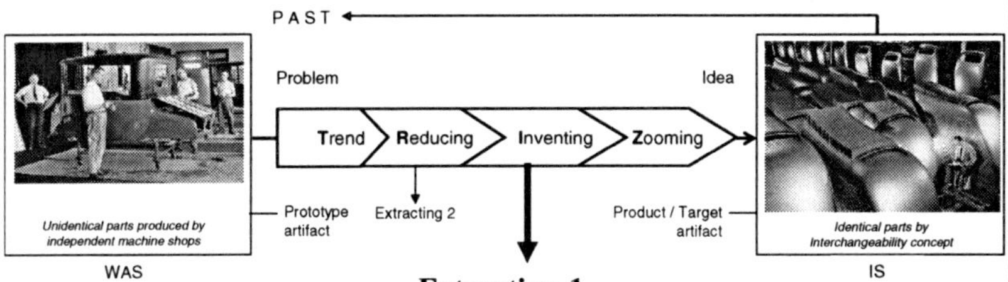

Extracting-1

LC*	No.	Navigator	Substantiation for the Extracting
++	02	**Preliminary action**	Prepare parts in advanced by using standard gauging system so that the assembly process can be done without loss of time.
+	10	**Copying**	Standard gauging system are copied and used in every machine shop that supplies the parts to the company.
+	11	**Inverse action**	Instead of letting the machine shops use their own gauging system, a standard gauging system needs to be enforced to them.
++	18	**Mediator**	A specific gauging device is used as mediator between the company and the suppliers in order to have standardized parts measurement.
+	38	**Homogeneity**	The gauging devices that used by machine shops shall have similar structure and made of similar material in order to give standard measurement result.

Extracting-2

Standard Contradiction: Independent manufacturing stations with their own gauging system would increase the delivery of the parts, but the parts produced would not fit to each other so that it prolong assembly time due to fixing and fitting activities.

Radical Contradiction: The manufacturing stations should be unified so that they could produced parts that fit to each other, but they should also be separated, so that the manufacturing speed could increase.

FIM: X-resource, together with available or modified resources, and without making the object more complex or introducing any negative properties, guarantees attainment of the following **IFR**:
[*automobiles with consistent similarity*].

fig. 12.5. Extracting: Interchangeability of Parts

TREND

In a Craft Production System, no two cars were built exactly alike. This product variation issue was due to the parts manufactured had high inconsistency of dimensions. This also made the assembly process could only be performed by skilled craftsmen, who needed long time to fix and fit the parts each other. These problems were caused due to the parts were manufactured in several independent machine shops that have their own gauging systems.

Fixing and fitting activities in Craft Production

PROBLEM: How can we build automobiles with consistent similarity ?

REDUCING

FIM: X-resource, together with available or modified resources, and without making the object more complex or introducing any negative properties, guarantees attainment of the following **IFR**: [*automobiles with consistent similarity*].

Standard (Technical) Contradiction

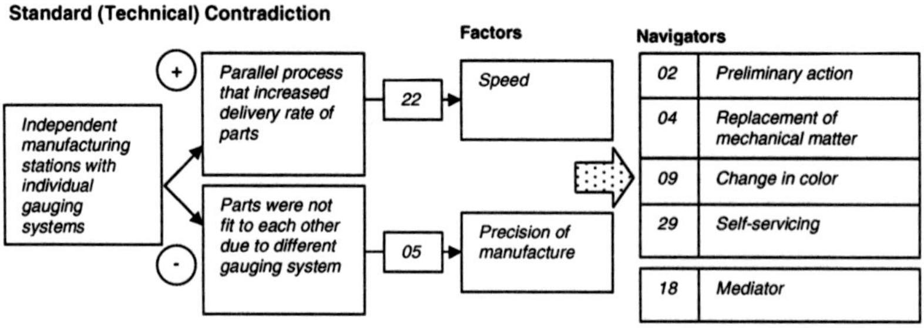

Radical (Physical) Contradiction

The manufacturing stations	⇨	***should be unified**, so that they could produce parts that were fit to each other*	&	***should be separated**, so that manufacturing speed could increase*

INVENTING

Group of navigators to model the idea for the 'interchangeability of parts' concept:

02. Preliminary action:
Prepare parts in advanced by using standard gauging system so that the assembly process can be done without loss of time.

18. Mediator:
Specific gauging devices were used as mediators between the company and the suppliers in order to have standardized measurement of parts.

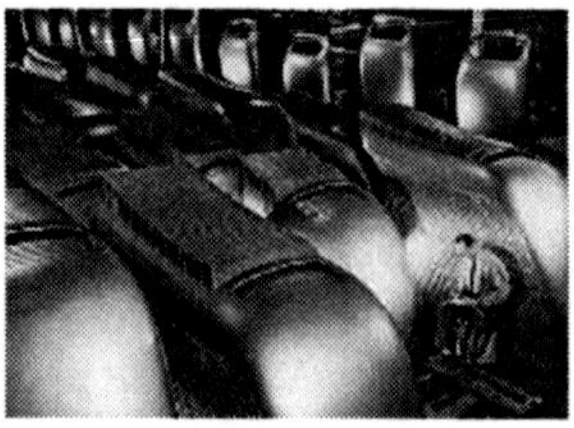

Identical parts produced in Mass Production with interchangeability concept

ZOOMING

Have the contradictions been removed? – Yes. – ~~No~~.

Super-effects: Yes. It opened possibility to have identical automobiles. It also reduced time of assembly.

Negative effects: Needed to coordinate all the parts supplier to use standard gauging system.

BRIEF DESCRIPTION

The "Interchangeability of Parts": This was the fundamental concept of the Mass Production System developed by Henry Ford. The idea was to ensure the parts were fit to each other so that ordinary workers – instead of skilled craftsmen - could assemble automobile in a short period of time. In 1908, with the introduction of interchangeability, Ford assembler's average cycle time for Model T could be reduced from days to 514 minutes (8.56 hours). This huge reduction was because the workers did not need to fix and fit the parts during the assembly process.

fig. 12.6. Reinventing: Interchangeability of Parts

12.1.2 Dedicated Machine Tools

In the era of Craft Production System, general purpose machines or tools were used to all sorts of processes. Most of the manufacturing activities by using these machines were still performed manually by craftsmen.

The Prototype – General Purpose Tools

In a craft automobile industry, the company just acted as designer and assembler. Major parts were made and supplied by many independent machine shops run by individual craftsmen. These craftsmen had their own machineries and tools that used for making all kind of parts ordered by the automobile company. Most of the tools were general ones, with some of them – by their own innovations – have been adapted to some specific functions.

In 1890s, Panhard et Levassor workforce was mainly composed of skilled craftsmen who hand-built only several hundred automobiles per years. The craftsmen were using general machines or tools that were used to all sort of processes: drilling, grinding, cutting, stamping, forming, and other operations on metal and wood.[150]

These general machines have flexibility in processing many kinds of parts. But they required set-up and adjustment by highly skilled machinist. The product result also had poor accuracy that required a lot of fixing.

With the raising demand of cars in the early of 1900s, it was impossible to rely on those ordinary machines, since the parts were not produced in correct and consistent dimension and shape. This obstructed productivity because the fitters needed long time to fit the parts during assembly. The production cost became high because it required highly skilled workers to do the jobs.

There were also not many skillful workers available at that time that caused high competition among the car manufacturers to employ them. This competition drove up wages and put the craftsmen in higher bargain position to get best paid job. In the end, it would increase the labor cost and the companies need to sell the products in a higher price.

fig. 12.7. Car body assembly with general tools

fig. 12.8. Finishing activity in craft manufacturer

150 ibid.

The Result of Contradiction Analysis

The "plus-state" of the "general purpose tools" can be determined as in A_S-Matrix as **number 02 (universality, adaptability)**, because the craftsman needed to provide and use only few tools to perform many kind of jobs. The "minus-state" of the "general purpose tools" can be determined as in A_S-Matrix as **number 05 (precision of manufacture)**, the tools had low precision in producing the parts that caused the parts did not have correct and consistent dimension and shape.

Inventing Idea

Combining the plus and minus state factors above, the results will show several points in A_S-navigators. The target is to increase the precision characteristic of the tools so that in the end the workers do not require long time to fit the parts each other during assembly process. After combining the A_S-Matrix 02 and 05, the idea navigators resulted :

- **Navigator 12 (local property)** – according to this navigator, the transition from a similar to a different structure of machine tools is required so that every tool can complete its function under the best condition.
- **Navigator 02 (preliminary action)** – according to this navigator, it is necessary to make partial or complete change of the tools so that they can be put to work from the best position and are available without loss of time.
- **Navigator 05 (separation)** – according to this navigator, the incompatible property of the tools shall be separated so that each tool can have the best property to complete its work.

The Product-artifact (Solution) – Dedicated Machine Tools

> Therefore in 1909, I announced one morning without any previous warning, that in the future we were going to build only one model, that the model was going to be "Model T", and that the chassis would be exactly the same for all cars, and I remarked: Any customer can have a car painted in any colour that he wants, so long as it is black. [151]
>
> *Henry Ford*

To achieve standardization with high quality of precision of parts, high volume of products, and low cost, dedicated machines for special purpose of tasks were required.

As has been mentioned earlier, the fundamental key of the Mass Production System was the interchangeability of parts that laid in standardization of gauging system. But this was not enough. Ford required this interchangeable parts need to

[151] McNairn, W. and McNairn, M. (1978) *Quotations from the Unusual Henry Ford.* – Quotamus Press, Redono Beach, CA

be produced in lower cost. In a Craft Production System – where a single machine could do many tasks – many set-ups and adjustment needed to be performed by skilled machinist that require long time and high labor cost. So, the key of producing inexpensive interchangeable parts laid in machines that could perform jobs in high volume with low set-up costs between pieces.

At his factory in Highland Park, Ford provided machines dedicated to produce a single item only and also nearly automated, so they could produce high volume of highly accurate parts with tight tolerances required to eliminate hand fitting[148].

For example, Ford introduced two types of dedicated machines, fifteen machines for milling blocks and thirty machines just for milling heads.

Ford also purchased stamping presses, used to make sheet-steel parts, with die spaces large enough to handle only a specific part.

By making machines that could only do single task, Ford could dramatically reduced set-up time. His engineers also perfected simple jigs and fixtures for holding the work piece in this dedicated machine that allowed unskilled workers – with a short period of training – to operate them. This meant the company could reduce its dependency on special machinist and in the end it could reduce the labor cost.

This dedicated machines strategy was running well because Ford only produced one model of car in a long period of time[152] (Model T was produced from 1908 – 1927, followed by Model A from 1927 – 1931). This was made possible by the homogeneity of the customer demand at that time that could be satisfied only by low variants of products.

fig. 12.9. Car body assembly with dedicated press machine

fig. 12.10. Dedicated machine tools in a mass production manufacturer

[152] HFHA – Henry Ford Heritage Association (2009): The Ford Story. hfha.org/HenryFord.htm#Ford-Motor-Co

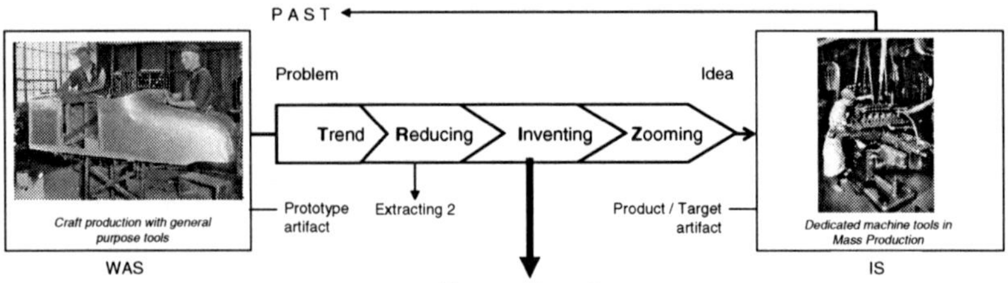

Extracting-1

LC*	No.	Navigator	Substantiation for the Extracting
++	02	**Preliminary action**	Make necessary (partial or complete) change of the tools so that they can be put to work from the best position and are available without loss of time.
+	03	**Segmentation**	"Disassemble" the universal tool into "individual" tools with their own specific task.
++	05	**Separation**	Separate the incompatible property from the tools so that each tool can have the best property to complete its work.
+	11	**Inverse action**	Instead of using universal tool for performing many tasks, dedicated tools can be used for each specific task.
++	12	**Local property**	Transition from a similar to a different structure of machine tools so that every tool can complete its function under the best condition.

Extracting-2

Standard Contradiction: Universal tools could be used in making many parts, but they had low precision that could not produced parts with correct and consistent dimension and shape.

Radical Contradiction: The machine and tools should be common so that the worker could easily do the job, but they also should be unique so that the parts produced had high precision.

FIM: X-resource, together with available or modified resources, and without making the object more complex or introducing any negative properties, guarantees attainment of the following **IFR**:
[*high precision of parts*].

fig. 12.11. Extracting: Dedicated Machine Tools

TREND

In an automobile company that use Craft Production System, the craftsmen used general purpose machine & tools to perform drilling, grinding, cutting, stamping, forming and other operations on metal and wood. This obstructed productivity because the parts were not produced in the correct and consistent dimension and shape, so in the end the craftsmen needed long time to fit the parts during assembly.

Craft production with universal tools

PROBLEM : How could we increase the precision of the parts ?

REDUCING

FIM: X-resource, together with available or modified resources, and without making the object more complex or introducing any negative properties, guarantees attainment of the following **IFR**: [*high precision of parts*].

Standard (Technical) Contradiction

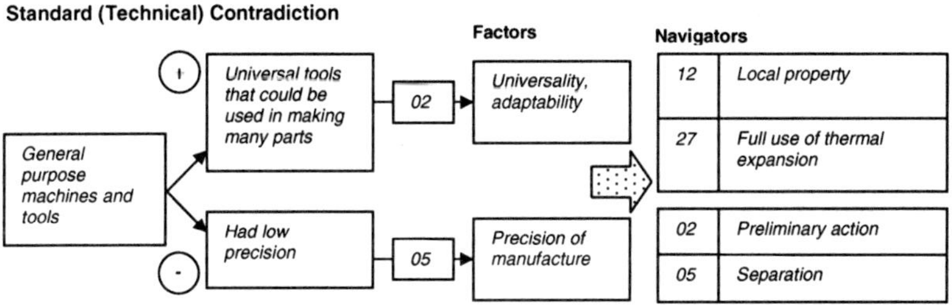

Radical (Physical) Contradiction

The machines and tools ⇒ ***should be common**, so the worker could easily do the job* & ***should be unique**, so that the parts produced had high precision*

INVENTING

Group of navigators to model the idea for the 'dedicated machine and tools' concept :

12. Local property :
Transition from a similar to a different structure of machine tools so that every tool can complete its function under the best condition.

02. Preliminary action :
Make necessary (partial or complete) change of the tools so that they can be put to work from the best position and are available without loss of time

05. Separation :
Separate the incompatible property from the tools so that each tool can have the best property to complete its work.

Dedicated machine tools in Mass Production

ZOOMING

Have the contradictions been removed? – Yes. – ~~No~~.

Super-effects: Yes. It increased the precision of parts manufactured

Negative effects: Additional cost for providing dedicated machine tools to manufacture the parts

BRIEF DESCRIPTION

Dedicated machine tools : This concept was introduced by Henry Ford to overcome the weakness of Craft Production System at that time. The use of dedicated machine tools became one pillar of the success of Mass Production System since the factory could manufacture large number of parts in high precision in a short period. The workers also did not have to spend long time to fix and fit the parts during assembly process.

fig. 12.12. Reinventing: Dedicated Machine Tools

12.1.3 Design Simplification

The Prototype – Engine with Fragmented Parts

The four-stroke internal combustion engine is the type that commonly used for automobile and industrial purposes today (cars, trucks, generators, etc.). This concept was developed in 1877 by Nikolaus Otto[153]. Later on it was leveraged by Wilhelm Maybach, who built the first four-cylinder, four-stroke internal combustion engine for Daimler Motoren Gesellschaft in 1890. The engine[154] has an output of five hp (3.7 kW) at 620 rpm. This was an epoch-making invention that replaced the two-stroke engine concept and later on used widely by many automobile companies.

The four-stroke cycle engine is more efficient, last longer and less polluting than the two-stroke cycle one. The disadvantages of this concept are it requires more moving parts and manufacturing expertise.

In the early 1900s, for the automobile industries that still adapted craft production, the production of four-stroke cycle engine involved many parts need to be manufactured. The cylinders need to be casted separately and the assembly required several highly skilled fitters. This contributed to the long production time and high labor cost in producing cars.

The design of the automobile at that time was also very complex that an owner would need to employ mechanic as his/her personal staff. This caused the vehicle could only be owned by wealthy people, and became one prohibiting factor to make automobiles as common consumer goods.

fig. 12.13. The first 4-stroke, 4-cylinder automobile engine invented by Wilhelm Maybach in 1890

fig. 12.14. Robert Peugeot in a Peugeot car in 1897. Only wealthy person could afford this kind of vehicle

---

[153] Nikolaus August Otto (1832 – 1891) – a German inventor who made the first internal-combustion engine with four-stroke concept

[154] Hodzic, M. (2008) *Gottlieb Daimler, Wilhelm Maybach and the "Grandfather Clock";* www.benzinsider.com/2008/06/gottlieb-daimler-wilhelm-maybach-and-the-grandfather-clock

The Result of Contradiction Analysis

The "plus-state" of the "engine with fragmented parts" can be determined as in A_S-Matrix as **number 09 (ease of manufacture)**, because it was easier for the manufacturer to produce the parts since they had simple shapes. The "minus-state" of the "engine with fragmented parts" can be determined as in A_S-Matrix as **number 10 (ease of use)**, because it required long time and skillful workers to assemble the moveable parts into final product.

Inventing Idea

Combining the plus and minus state factors above, the results will show several points in A_S-navigators. The target is to reduce the number of the parts need to be assembled so that ordinary workers can perform the activity in short period of time. After combining the A_S-Matrix 09 and 10, the idea navigators resulted :

- **Navigator 11 (inverse action)** – according to this navigator, it is advised to make the "moveable" parts of the engine fixed, or the opposite.
- **Navigator 35 (unite)** – according to this navigator, similar parts or parts for neighboring operations can be united.

The Product-artifact (Solution) – Design Simplification

> I will build a car for the great multitude. It will be large enough for the family, but small enough for the individual to run and care for. It will be constructed of the best materials, by the best men to be hired, after the simplest designs that modern engineering can devise.[155]
>
> *Henry Ford*

Since Model T were using the 4-cylinder, 4-stroke engine, Ford[148] also found similar problem faced by other automobile companies: long assembly time and high labor cost. This obstructed Ford's ambition to have a cheap, massive production of automobiles. He believed that one of the important keys to achieve that goal was high production rate supported by simplicity of design and ease of attachment.

To overcome those problems, Ford's idea was to develop an engine design that reduced the number of parts needed and made these parts easy to attach. It was C. Harold Wills[156], one of Ford's senior engineer who finally made the new four-cylinder engine block casted in one piece, with a detachable cylinder head, which was very daring at that time (Tebo, 2010).

This design decreased the production time since it reduced the number of moving parts in engines and simplified the assembly process. It also omitted the requirement for skilled fitters that in the end could save the labor cost.

---

155 Ford, H. and Crowther, S. (1922) *My Life and Work*. – Nevins and Hill, Ford TMC

156 Childe Harold Wills (1878 – 1940) – an American engineer, one of the first employees of the Ford Motor Co. and contributed to the design of the Model T. Founder of Wills Sainte Claire automobile company

To attract consumers from the average level, Ford had designed cars with unprecedented ease of operations and maintainability. He wanted that ordinary users with low ability in mechanical skills would be able to fix the cars by themselves with modest tool kits.

For example, the owners[148] could remove carbon deposits from the engine valves with the Ford Valve Grinding Tool, which came with the vehicle. Owners could also buy spare parts at Ford dealer and replaced the broken parts by themselves. The procedures of solving 140 common problems of the car were provided in a simple question-and-answer form of Model T's owner's manual.

The simplicity of design and ease of attachment gave Ford tremendous advantages over his competitors that catapulted Ford to the head of the world's automobile industry.

fig. 12.15. Ford's 4-cylinder, 4-stroke engine that casted in one piece

fig. 12.16. A Ford Model T that had been modified as farm truck. Simplicity of design made this model of vehicle favored by farmers for modifications

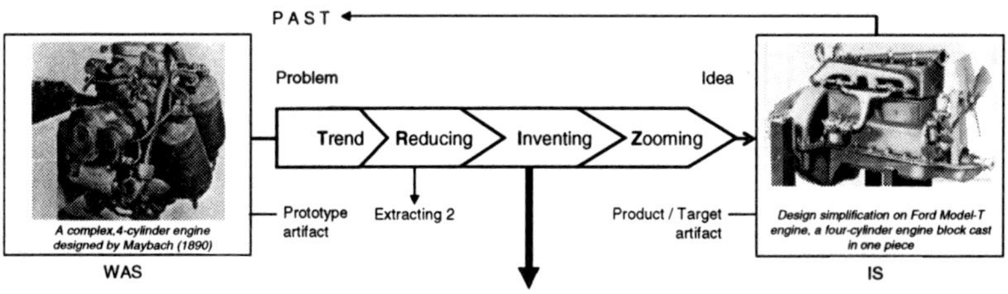

Extracting-1

LC*	No.	Navigator	Substantiation for the Extracting
+	02	**Preliminary action**	Partial change on the design of the machine need to be done.
+	05	**Separation**	Separate the parts only if necessary.
++	11	**Inverse action**	Try to make the moveable parts of the engine fixed, or the opposite.
+	12	**Local property**	Every part should exist under condition that it should be able to fit easily during assembly process.
+	16	**Partial or excess effect**	If it is difficult to assemble the engine from completely separated parts, then some parts of the engine can be unified first during manufacturing.
++	35	**Unite**	Unite similar parts or parts for neighboring operation.

Extracting-2

Standard Contradiction: Engine with fragmented parts will give simpler process during manufacturing process, but the company need to hire specialists to assemble the engine in a long time.

Radical Contradiction: The engine construction should be complex so that it could carry out its function, but it also should be simple so that ordinary workers could assemble it fast.

FIM: X-resource, together with available or modified resources, and without making the object more complex or introducing any negative properties, guarantees attainment of the following **IFR**:

[*fast assembly process of engine*].

fig. 12.17. Extracting: Design Simplification

TREND

In engine type used during the Craft Production System was comprised of large number of individual parts and cylinders that need to be joined and bolted together by skilled specialist. This caused the assembly process for the engine required a long time. This also caused high cost due to the process could only be performed by specialists.

PROBLEM : How could we reduce the time required to assemble the engine ?

The first 4-cylinder engine designed by Wilhelm Maybach (1890)

REDUCING

FIM: X-resource, together with available or modified resources, and without making the object more complex or introducing any negative properties, guarantees attainment of the following **IFR**: [*fast assembly process of engine*].

Standard (Technical) Contradiction

Engine with fragmented parts

(+) *Simpler process to manufacture the individual parts* → 09 → Factors: *Ease of manufacture*

(-) *Needed a specialist to assemble the engine parts in a long time* → 10 → *Ease of use*

Navigators

05	*Separation*
11	*Inverse action*
16	*Partial or excess effect*
35	*Unite*

Radical (Physical) Contradiction

The engine construction ⇒ ***should be complex**, so that it could carrv out its function* & ***should be simple**, so that ordinary workers could assemble it fast*

INVENTING

Group of navigators to model the idea for the 'design simplification' concept :
11. Inverse action :
Try to make the moveable parts of the engine fixed, or the opposite.
35. Unite :
Unite similar parts or parts for neighboring operation

Ford Model-T Engine, a four-cylinder engine block cast in one piece

ZOOMING

Have the contradictions been removed? – Yes. – ~~No~~.

Super-effects: Yes. It decreased the complexity of the parts needed to be assembled. Ordinary workers could replace the specialists.

Negative effects: Needed to change the engine design and manufacturing

BRIEF DESCRIPTION

The "Design Simplification": This was an important concept that contributed to rise of the Mass Production System. With this concept, less time were required to assemble automobile's engine. Ford reduced the number of moving parts in the engine, i.e. the engine casting comprised of single complex block. By contrast, competitors casted each cylinder individually and bolted them together during assembly. With this innovation, Ford company could reduce the labour cost due to less specialists were required to do the work.

fig. 12.18. Reinventing: Design Simplification

12.1.4 Delivery of Parts

The static workstation that had been adopted by Craft Production System in automobile company brought consequence that the vehicles sat in a place while the craftsmen need to bring all the tools and materials into the location.

The Prototype – Self-picking Parts

Since a lot of parts were required to build a vehicle, the job of assembling the car usually could be considered as a simpler task compared with handling the materials that had to be brought in. This was complicated by the location of parts that stored quite far from the assembly stations. The craftsmen also needed to find them first in the stock piles before could get what they were looking for.

In the early years of Ford's Model T production, a craftsman was given a full responsibility in making the whole vehicle, including picking-up all the parts and semi-finished products required.

Since hundreds of parts were required, it would take a long time for the craftsman to go to each material store, picked the parts, and put it in his workstation before he could continue his work.

These repetitive activities were not only time-consuming – it also decreased the productivity of the craftsman and obstructed the production output.

fig. 12.19. Ford's early assembly line. All required parts were brought near the workstation

The Result of Contradiction Analysis

The "plus-state" of the "self-picking parts" can be determined as in A_S-Matrix as **number 07 (complexity of construction)**, because it did not require complex layout arrangement to introduce the parts storage into the production line. The "minus-state" of the "self-picking parts" can be determined as in A_S-Matrix as **number 03 (level of automation)**, because the workers needed to leave their workstation to pick the required parts. This activity also required long time and delay the assembly process.

Inventing Idea

Combining the plus and minus state factors above, the results will show several points in A_S-navigators. The target is to reduce the waiting and transportation time due to the workers need to pick the required parts by themselves. The new alternative will make the workers to be easier to get the parts but without putting all the parts in the production line that will make the line crowded. After combining the A_S-Matrix 07 and 03, the idea navigators resulted :

- **Navigator 07 (dynamization)** – according to this navigator, it is advised to make the required parts 'moveable' to the location of the workers.
- **Navigator 18 (mediator)** – according to this navigator, it is possible to use another object to transfer all the required parts from the store to the workstations.

The Product-artifact (Solution) – Delivery of Parts

> Use work slides or some other form of carrier so that when a workman completes his operation, he drops the part always in the same place – which place must always be the most convenient place to his hand – and if possible have gravity carry the part to the next workman for his operation ... Use sliding assembly lines by which the parts to be assembled are delivered at convenient distances. [157]
>
> *Henry Ford*

As has been discussed earlier, the problem of Craft Production was that the craftsmen wasted their time to pick-up the parts every time they needed them.

The step Ford took to make his plant more efficient was to deliver the parts to each workstation so that the assemblers could remain at the same spot all day.

The improvement project was handled by Charles Sorensen[158], Ford's engineer, and Charlie Lewis, Ford's foreman. They arranged storage rooms for light-handling stocks and bulky parts, like engines and axles, closer to the assembly stations. They then arranged with the stock department to bring up batch of materials and parts required and delivered them to each workstation regularly.

With this arrangement, the craftsmen just needed to wait in their workstations and could focus with their activities. No more time wasted for craftsmen to pick-up the parts and this means higher productivity and output.

This innovation would become an important supporting strategy for the implementation of Ford's Mass Production – the parts that came to the worker, and no the opposite concept like had been implemented for many years in Craft Production.

fig. 12.20. Ford's assembly line with its improved parts delivery system in 1910s. Some of the parts were put in a sloped-rack that supplied the parts by gravity force

fig. 12.21. Parts delivery system in Austin automobile company in 1950s. The workers had lists of major items needed for each car, and they had to ensure correct items (engine, gearbox, etc.) were put in the correct boxes before sending them to the workstations by conveyor

---

157 McNairn, W. and McNairn, M. (1978) *Quotations from the Unusual Henry Ford.* – Quotamus Press, Redono Beach, CA

158 Sorensen, S. E. (1956) *My Forty Years with Ford.* – Norton, New York

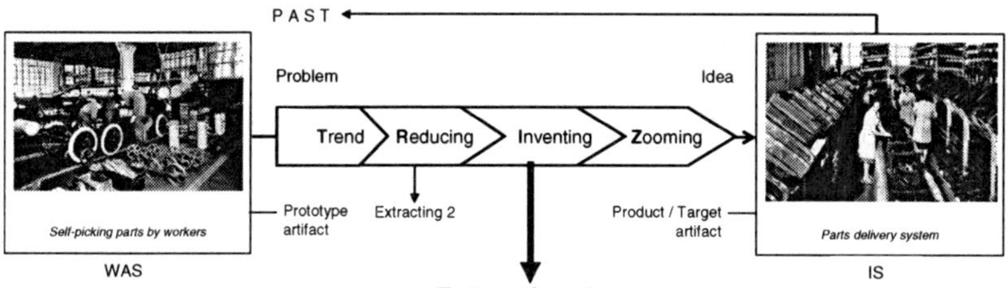

Extracting-1

LC*	No.	Navigator	Substantiation for the Extracting
+	02	**Preliminary action**	Prepare the parts in advanced so that the workers do not need to pick them.
++	07	**Dynamization**	Make the parts "moveable" to the location of the workers.
+	11	**Inverse action**	Instead of having the workers to pick the parts by themselves, the parts are delivered to the workers.
++	18	**Mediator**	Use another object to transfer all the required parts to the location of the car assembled.
+	29	**Self-servicing**	The parts are "coming by itself" to the location of the workers.

Extracting-2

Standard Contradiction: Having parts storage separated from the work station requires simple construction of work station area, but the workers had to pick all the parts required all the time.

Radical Contradiction: The parts storage should be exist in the work station so that the workers could easily pick the required parts, but the storage should also not be exist so it would not make the work station area to be crowded.

FIM: X-resource, together with available or modified resources, and without making the object more complex or introducing any negative properties, guarantees attainment of the following **IFR**:
[*low assembly time*].

fig. 12.22. Extracting: Delivery of Parts

TREND

In a Craft Production System, the automobile was built in a simple static work station and workers had to pick all the parts (wheels, springs, motor, transmission, generator, etc.) from store and put it in the work station. This activity was time consuming and became major contributor to the high assembly cycle time.

PROBLEM : How could we reduce the assembly time ?

Workers picked and put parts to their work station

REDUCING

FIM: X-resource, together with available or modified resources, and without making the object more complex or introducing any negative properties, guarantees attainment of the following **IFR**: [*low assembly time*].

Standard (Technical) Contradiction

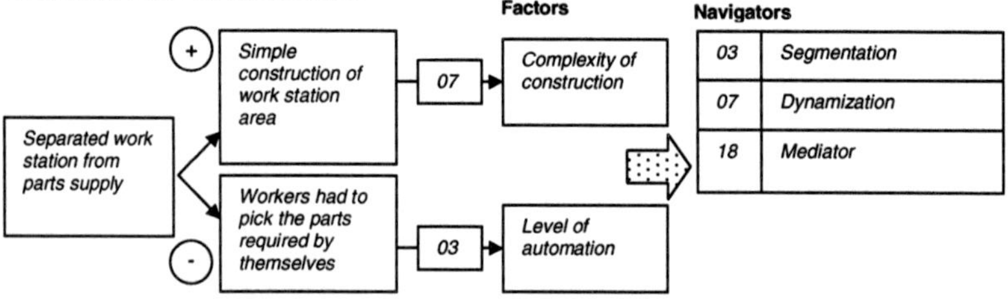

Radical (Physical) Contradiction

The parts storage	⇨	***should be exist in work station**, so that the workers could easily pick the required parts*	&	***Should not be exist**, so it would not make the work station area to be crowded*

INVENTING

Group of navigators to model the idea for the 'delivery of parts' concept :
07. Dynamization :
Make the parts 'moveable' to the location of the workers.
18. Mediator :
Use another object to transfer all the required parts to the location of the car assembled.

Parts delivery system

ZOOMING

Have the contradictions been removed? – Yes. – ~~No~~.

Super-effects: Yes. It eliminated the transportation time required for workers to pick the parts that would reduce the assembly process time.

Negative effects: Needed a good coordination between people in the store and people in the work stations in terms of number, types and time of parts need to be delivered. Also it required material handlers and system to deliver the parts.

BRIEF DESCRIPTION

The "Delivery of Parts": This was an important concept that introduced by Henry Ford that inspired the modern supply chain management. With this concept, Ford could reduce the assembly cycle time from hours to minutes. Material handlers were employed to deliver parts from store to the working station, so the workers could stay and focus to his work. Modern automobile companies currently use conveyors to move the parts to the location of the vehicles, in correct type, correct quantity and correct time.

fig. 12.23. Reinventing: Delivery of Parts

12.1.5 Division of Assembly Line

The craft production of automobile was depended on the skill of the workers. The skilled workers or craftsmen were required to perform major activities in the manufacturing and assembly processes from raw materials until finished vehicles.

The Prototype – Single-craftsman Operation

The early automobile design had already comprised of thousands of parts and hundreds of activity steps. At that time, only skillful craftsman were able to performed most of the complex tasks. However, it took a long time to finish a vehicle since he needed to change his work type and adapt with changes of tools from time to time.

In the beginning of Model T production in 1908, the vehicle body was still supplied by a craft manufacturer like Fisher Body Company.

The parts of the vehicle such as motor, transmission, generator, springs and wheels would also be assembled on a chassis by craftsman. This activity took a whole day to complete. The assembly line cycle time at that time was 514 minutes, or 8.56 hours (s. Womack). This kind of performance did not satisfied Ford's expectation for producing cars in massive output rate to achieve a state of economies.

fig. 12.24. Car frame assembly process performed by single craftsman

The Result of Contradiction Analysis

The "plus-state" of the "single-craftsman operation" can be determined as in A_S-Matrix as **number 02 (universality, adaptability)**, because it only required one worker to perform all the activities from the beginning to the end of the assembly process. The "minus-state" of the "single-craftsman operation" can be determined as in A_S-Matrix as **number 09 (ease of manufacture)**, because the worker required time to change his working pattern and to adapt with changes of tools.

Inventing Idea

Combining the plus and minus state factors above, the results will show several points in A_S-navigators. The target is to reduce the long assembly time due to the worker need some time every time he changes his type of work and tools. The new alternative will make the worker need not to change his work and tools but all the activities to assemble the cars are still carried out. After combining the A_S-Matrix 02 and 09, the idea navigators resulted :

- **Navigator 03 (segmentation)** – according to this navigator, it is required to disassemble the assembly process into several steps.

- **Navigator 11 (inverse action)** – according to this navigator, instead of assemble the car alone, the process can be performed by several workers.
- **Navigator 12 (local property)** – according to this navigator, it is necessary to change the structure of the assembly process so that it has different steps with different functions.
- **Navigator 31 (unite)** – according to this navigator, the segmentation of the assembly process can be supported by uniting several process steps that are quite similar or working on the same parts.

The Product-artifact (Solution) – Division of Assembly Line

> Place the tools and the men in sequence of the operation so that each component part shall travel the least possible distance while in the process of finishing.[159]
>
> *Henry Ford*

The history of managing complexity in the organization begins with Adam Smith[160], who published The Wealth of Nations in 1776, where he introduced concept of "division of labor" – the specialization of labor as a way of workers to achieve greater productivity by breaking large jobs into small tasks.

Another development in increasing productivity and efficiency was developed by Frederick Winslow Taylor. His principle in scientific management was known as *Taylorism* and suggest the importance of task discretion and providing training, detailed instruction and supervision to increase the performance of the workers.[161]

Ford followed above principles by applying standardization of work content and job fragmentation. Thus, not only parts were interchangeable, but labor had become replaceable too because they became merely an appendage of the production apparatus.

The impact of Taylorism at Ford replaced the previously tacit knowledge owned by craftsmen by a more complex but explicit system of clearly defined specific tasks as standards. By segmenting tasks into smaller units, job contents were reduced and could be more controlled.

After implementing the practice of delivery of parts into the workstation and reaching perfect part interchangeability, Ford decided to divide the assembly line into several subassembly divisions so that each worker would only perform a single task and move from vehicle to vehicle around the assembly hall.

---

[159] McNairn, W. and McNairn, M. (1978) *Quotations from the Unusual Henry Ford*. – Quotamus Press, Redono Beach, CA

[160] Adam Smith (1723 – 1790) – a Scottish economist and social philosopher. Cited as the father of modern economics and capitalism. Author of *The Theory of Moral Sentiments* and *An Inquiry into the Nature and Causes of the Wealth of Nations*; http://www.encyclopedia.com/topic/Adam_Smith.aspx#4

[161] Montgomery, D. (1989) *The Fall of the House of Labour: The Workplace, the State, and American Labour Activism, 1865-1925*. – Cambridge University Press, Cambridge, UK

There were divisions of workers who performed activities of fitting and assembling, parts supplying, repairing and checking. There were also foremen whose tasks were to check the activities of the workers and informed the line condition to the managers.

By August 1913, after the implementation of division of assembly line, the production cycle time for the average Ford assembler had been incredibly reduced from 514 minutes to 2.3 minutes (99.55% improvement) (s. Womack).

By 1915[162], when the assembly lines at Highland Park had already run in its full capacity, there were more than 7,000 assembly workers there. The workers spoke in more than fifty languages and many of them could barely speak English. This was a proof that division of assembly line implementation in Ford's company was very extreme that the workers need not to communicate each other in order to finished the products.

With dividing the assembly line, the workers could be concentrated on a specific task and tools that improved their working efficiency. The production cycle time would decrease and the productivity increased.

On the other hand, craftsmen did not like this concept since it reduced their role and responsibility in the production process. The company then could hire ordinary workers, that was cheaper, to do most of the assembly activities since the tasks became simpler. It also required only few minutes of training to prepare the workers in performing their tasks.

fig. 12.25. Workers on a flywheel assembly line at the Ford Motor Company's Highland Park plant, Detroit, in 1913. Each worker completed a different part of the component

fig. 12.26. Workers on a Ford assembly line working on the top and bottom ends of the engines (1946)

162 Raff, D (1987) *Wage Determination Theory and the Five-Dollar Day at Ford.* – Ph.D. Dissertation, Massachusetts Institute of Technology

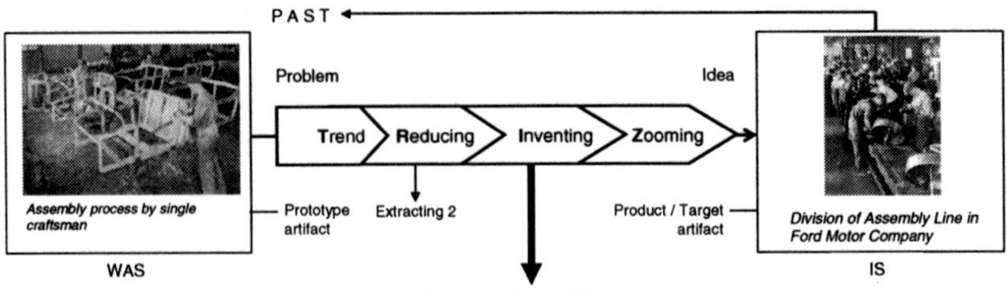

Extracting-1

LC*	No.	Navigator	Substantiation for the Extracting
+	02	**Preliminary action**	The assembly process needs to be prepared in advanced so that many workers can assemble the product in parallel and in series.
++	03	**Segmentation**	Disassemble the assembly process into several steps.
+	08	**Periodic action**	Transition from a continuous assembly process (by one worker) to several periodic ones (by many workers).
++	11	**Inverse action**	Instead of assemble the car alone, the process can be performed by several workers.
++	12	**Local property**	Change the structure of the assembly process so that it has different steps with different functions.
+	13	**Inexpensive short-life object as a replacement for expensive long-life one**	Low wage-easily replaceable workers as a replacement for high wage-difficult to find craftsmen.
++	31	**Unite**	The segmentation of the assembly process need to be supported by uniting several process steps that are quite similar or working on the same parts.

Extracting-2

Standard Contradiction: Assembly process by using single worker who built the vehicle from start to end was simple since it only requires one worker to perform all the activities, but it required long assembly time since the worker needed to adapt continuously with different type of tasks and tools.

Radical Contradiction: The assembly process steps should be few so that the worker could focus on less activities, but they should also be many since it was the nature of automobile assembly process to have many processing steps.

FIM: X-resource, together with available or modified resources, and without making the object more complex or introducing any negative properties, guarantees attainment of the following **IFR**:
[*low assembly time*].

fig. 12.27. Extracting: Division of Assembly Line

TREND

One typical characteristic of Craft Production System in automobile industry was that the whole vehicle was usually assembled by one fitter. The worker needs to have different types of skilsl to perform many kinds of activities. He needed to change his working pattern and to adapt with changes of tooling. This was the reason why craftsmanship required long period of time and skilled specialist.
PROBLEM : How could we reduce the assembly time ?

Assembly process by single craftsman

REDUCING

FIM: X-resource, together with available or modified resources, and without making the object more complex or introducing any negative properties, guarantees attainment of the following **IFR**: [*low assembly time*].

Standard (Technical) Contradiction

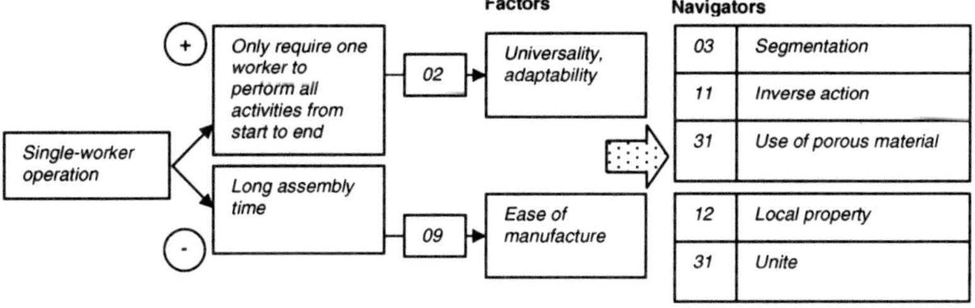

Radical (Physical) Contradiction

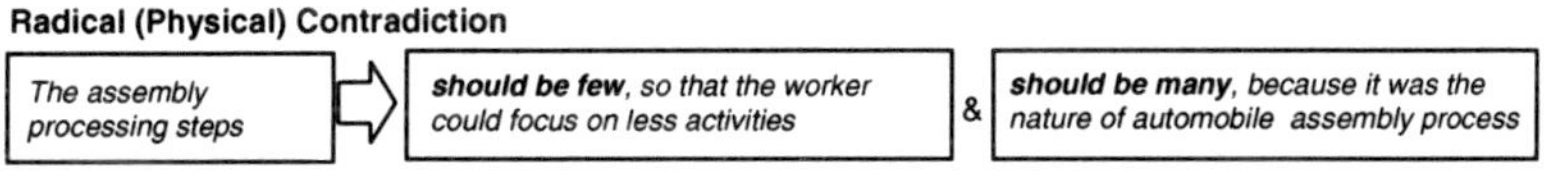

INVENTING

Group of navigators to model the idea for the 'division of assembly line' concept :
03. Segmentation :
Disassemble the assembly process into several steps.
11. Inverse action :
Instead of assemble the car alone, the process can be performed by several workers.
12. Local property :
Change the structure of the assembly process so that it has different steps with different functions.
31. Unite :
The segmentation of the assembly process need to be supported by uniting several process steps that are quite similar/working on the same parts.

Division of Assembly Line in Ford Motor Company

ZOOMING

Have the contradictions been removed? – Yes. – ~~No~~.

Super-effects: Yes. It reduced the time required to assemble an automobile due to each worker could focus on one specific task. Skilled craftsmen could also be replaced by ordinary workers.

Negative effects: Needed a good management in distributing the tasks.

BRIEF DESCRIPTION

The "Division of Assembly Line": The concept was rooted from Adam Smith's concept of 'division of labour'. Henry Ford revolutionized this concept to support Mass Production System by dividing the assembly line into several stations. This would increase the productivity, because complete familiarity with a single task meant the worker could perform it faster. This basic concept was still used by every manufacturing and assembly companies.

fig. 12.28. Reinventing: Division of Assembly Line

12.1.6 Moving Assembly Line

In a Craft Production System, the automobile company provided segmented areas for craftsmen to build the vehicle from scratch until finish. Single craftsman perform his job in one area from making frame, installing engines, tires, and fit all the parts and the body of the car.

The Prototype – Static Work station

The static workstation concept was a common sense in early automobile industry since the vehicle was mainly built by one worker only. The vehicle was built from scratch in one place, and it was the duty of the craftsman to put all the parts.

Until 1913, Ford's assembly line still followed the traditional Craft Production, where the vehicle was built in one stationary area. Although he had successfully implemented practices of supplying the parts into the workstations and fragmenting the assembly line, the workers still need to move from one vehicle to another in order to perform his tasks. It was a common sense of practice at that time because people could move easier and faster than the material.

However, Ford noticed weakness from the existing concept: walking activity, even for a small distance, took time. The workers also need to bring his tooling with him and jam-ups as faster workers overtook the slower workers in front of them.

This problem need to be resolved since the time wasted actually reduced the productivity of the workers.

fig. 12.29. Ford's Body assembly room at Highland Park Plant, Detroit, in 1910-1911. The vehicle bodies were put on a static support while the components were brought and assembled into them

fig. 12.30. Another typical of static workstation in a craft company of automobiles

The Result of Contradiction Analysis

The "plus-state" of the "stationary workstation" can be determined as in A_S-Matrix as **number 07 (complexity of construction)**, because in a static workstation it required only a simple construction to support the car that was being assembled. The "minus-state" of the "stationary workstation" can be determined as in A_S-Matrix as **number 03 (ease of manufacture)**, because the worker required time to move from one station to another.

Inventing Idea

Combining the plus and minus state factors above, the results will show several points in A_S-navigators. The target is to reduce the long assembly time due to the motion time required by the workers to move between stations. The new alternative will make the assembled cars are still in fixed position to enable the workers to do necessary activities. After combining the A_S-Matrix 07 and 03, the idea navigators resulted: **Navigator 07 (dynamization)** – according to this navigator, it is required to make the object to be able to move, instead of having the workers to move between stations.

Product-artifact (Solution) – Moving Assembly Line

> Along about April 1, 1913, we first tried the experiment of an assembly line...I believe that this was the first moving line ever installed. The idea came in a general way from the overhead trolley that the Chicago packers used in dressing beef.[163]
>
> *Henry Ford*

The moving assembly line was adopted from standard production process techniques which were already established in both the flour milling and meatpackers industries.[164]

As early as 1790[165], Oliver Evans[166] applied the principle of process flow in the flour milling industry. He recognized that the input of the material (grain) to the mill was one of the most important ways to control the speed of production. This early practice was then perfected by meatpacking industry started in 1850 by installing moving overhead trolley in slaughter houses to eliminate slow human handling of carcasses.

It was Ransome Eli Olds, owner of the Olds Motor Works in Detroit, who first introduced the basic principle of the moving material into the automotive industry in 1901.[167] Always on the hunt for more efficient and lower cost, in 1913 Ford started experimentation of the moving assembly principle in his plants.

Ford's aim of using a moveable assembly line was to reduce the amount of travel workers had to undertake to get from one workstation to the next: "If the workers were going to work like machines, Ford engineers concluded that the entire factory had to work like machine, that the success of assembly line production depended on efficient supply of materials and parts to workstations" (s. Biggs).

The development towards the moving assembly line at Ford could divided into three phases. In 1906, Ford had experimented with work slides at the Bellevue plant.

---

163 McNairn, W. and McNairn, M. (1978) *Quotations from the Unusual Henry Ford.* – Quotamus Press, Redono Beach, CA

164 Clarke, C. (2005) *Automotive Production System and Standardisation: From Ford to the Case of Mercedes Benz.* – Physica-Verlag, Heidelberg

165 Biggs, L. (1996) *The Rational Factory, Architecture Technology and Work in America's Age of Mass Production.* – The John Hopkins University Press, Baltimore, London

166 Oliver Evans (1755 – 1819) – an American inventor

167 Gartman, D. (1986) *Auto Slavery: The Labour Process in the American Automobile Industry, 1897 – 1950.* – Rutgers University Press, New Brunswick and London

In spring 1913, belt conveyor was installed in the flywheel magneto operation at Ford's Highland Park plant. In August 1913, after completing the implementation of division of assembly line, Ford started to implement this concept into the final assembly line (s. Clarke).

Ford's moving assembly line consisted of two strips of metal plates mounted on a belt rolled along the factory. Thus, with this solution, instead of having the workers moved the vehicle from station to station, the conveyor moved the vehicle past the stationary workers.

This innovation could reduce cycle time of car assembly process from 2.3 minutes to 1.19 minutes. The overall assembly time could be reduced from 12 hours and 8 minutes to 2 hours and 35 minutes by October 1913, and to 1.5 hours only by April 1914. The difference was get due to reduction of walking activity of the workers.

The production output also soared, from 12,000 units in the first year of production to 734,811 units in 1916 – contributed nearly half of the U.S. automobile production.

The moving assembly line allowed for a constant flow of production through the plant in which each station was manned by a worker placing standardized parts in a standardized manner onto the moving standardized chassis. Three years after its implementation, the moving assembly line had become standard practice in all U.S. automobile manufacturers.

fig. 12.31. Workers assembled automobiles in Ford's moving assembly line

fig. 12.32. Moving assembly line at the final assembly station (Henry Ford's automobile factory in Highland Park, Detroit, 1913)

fig. 12.33. Workers working on Model T body in a moving assembly line

fig. 12.34. Moving assembly line increased Ford Model T output

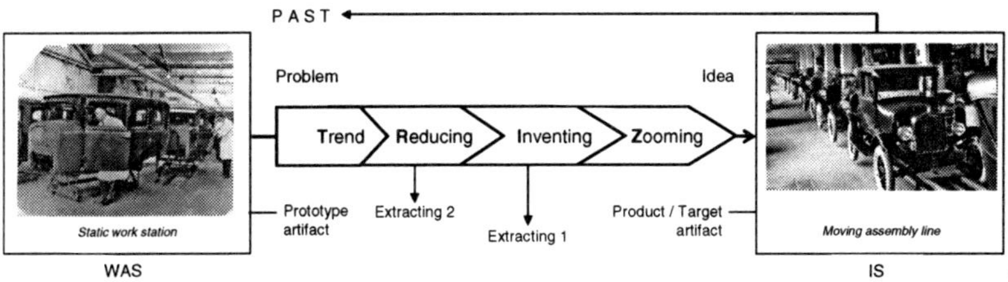

Extracting-1

LC*	No.	Navigator	Substantiation for the Extracting
+	01	**Change in the aggregate state of an object**	Change the "degree of flexibility" of the assembly line.
+	02	**Preliminary action**	The assembly line need to be prepared to reduce workers movement.
++	07	**Dynamization**	Make the object to move so that the workers need only to stand-by in their position.
+	08	**Periodic action**	Multiple usage
+	11	**Inverse action**	Instead of having the workers to move between stations, the object should be moved.
+	12	**Local property**	Change the structure of the work station so that they can deliver the vehicle being built from station to station.
+	18	**Mediator**	Use some equipment or tools to deliver the vehicle being built between stations.
+	19	**Transition into another dimension**	The work station is shaped so that it can move.

Extracting-2

Standard Contradiction: Static work station required only a simple construction, but it consumed a long time for workers to move from station to station.

Radical Contradiction: The position of the assembled cars should be stationary so the worker could do the job on it, but it should also be moveable so that the worker needed not to move between stations.

FIM: X-resource, together with available or modified resources, and without making the object more complex or introducing any negative properties, guarantees attainment of the following **IFR**:
[*low assembly time*].

fig. 12.35. Extracting: Moving Assembly Line

TREND

In a Craft Production System, the assembly process of automobile was performed in a static work station. When this concept was combined with 'division of assembly line' (different workers performed different tasks), it had a weakness. The assembly time was not in line with the expectation since the workers needed time to move from one station to another.

Static work station in a Craft Production System

PROBLEM: How could we reduce the assembly time due to the motion time of workers to move between stations?

REDUCING

FIM: X-resource, together with available or modified resources, and without making the object more complex or introducing any negative properties, guarantees attainment of the following **IFR**:

[*low assembly time*].

Standard (Technical) Contradiction

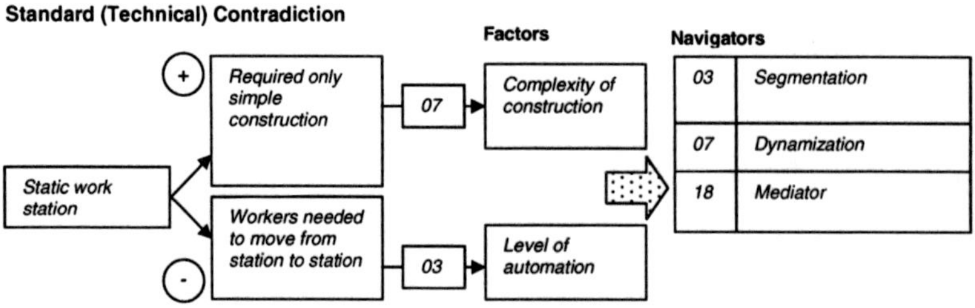

Radical (Physical) Contradiction

The position of the assembled cars	⇒	***should be stationary**, so the worker could do the iob in it*	&	***should be moveable**, so that the worker needed not to move between stations*

INVENTING

Group of navigators to model the idea for the 'moving assembly line' concept:
07. Dynamization :
Make the object to move, instead of having the workers to move between stations.

Moving assembly line with conveyor used in Mass Production System

ZOOMING

Have the contradictions been removed? – Yes. – ~~No~~.

Super-effects: Yes. It decreased the motion time required for workers to move between stations.

Negative effects: Required to re-design the factory and cost to install the conveyor.

BRIEF DESCRIPTION

The "Moving Assembly Line": This concept was introduced by Henry Ford in the spring of 1913, at his Highland Park plant in Detroit. Ford's moving assembly line consisted of two strips of metal plates mounted on a belt rolled along the factory. The device was quite similar to the long rubber belt conveyor that now serve as walkways in some airports. Thus, with this conveyor, instead of having the workers moved the car from station to station, the conveyor brought the car past the stationary workers. This innovation cut the cycle time of assembly process from 2.3 minutes to 1.19 minutes.

fig. 12.36. Reinventing: Moving Assembly Line

12.2 From Mass Production to Lean Production System

In this section, the case studies will focus on the innovations occurred in the evolution of Mass Production System to Lean Production System.

12.2.1 Quick Changeover

One of the main reasons that Mass Production was able to developed so extensively in the U.S. during in 1900s was that the American market at that time was more homogenous than the market in other parts of the world, for example in Europe. This caused the concept of product standardization could endure. With the huge amount of market demand and low claim of product varieties, it was easier for Ford to design a production system that served huge quantities of cars in a limited number of models. These are the background why companies that adopted Mass Production System used dedicated machines for their production lines.

The Prototype – Dedicated Machines

In the Mass Production era, the automobile companies used massive and expensive press lines that designed to produce more than a million units of parts in a year. The dies weighed several tons each, and they need to be aligned in the press with absolute precision. A slight misalignment would produce wrinkled parts and a serious misalignment could make the sheet metal melted in the die, obligating expensive and time-consuming repairs (s. Womack).

To avoid these problems, specialists were needed to performed the die changes, which could take a whole full day. To overcome this, the companies often stamp specific parts in a dedicated presses for months or years, without changing dies.

In Ford Motor Co., the strategy was even more extreme. Ford dedicated machines to all of the parts, even for the simple ones. For an example, by 1914, there were 15,000 machines—among them specially designed milling machines—at Ford's assembly line. With this strategy, no die changes would be needed and the production will not be interrupted by changeover.

This strategy was supported by the typical of U.S. market at that time that did not demand many models. That was the reason why Ford's dedicated machine strategy could be successful for decades with its only Model T that has been sold to more than 15 millions units until 1927 – which produced only in black color, since 1914, because black paint is cheaper, more durable, and dried faster.[168]

However, this strategy demanded too much capital investment since the machines and dies could not be used to produce other variants. When Ford released the new Model A in 1927 and stopped the production of Model T, the previous Model T's machines and dies were obsolete and need to be scrapped.

When Toyota tried to adopt the Mass Production to their plant in Japan after the World War II, they realized that they could not implement the principle of dedicated machines.

This was due to Japanese car market was smaller, and the customer demanded various type of cars – that forced the company to do frequent die changes.

---

168 McCalley, B. W. (1994) *Model T Ford: The Car That Changed the World*. – Krause Publications, Iola, WI

The problem was, with existing machines and die construction at that time, it would take a long time for specialists to perform die changes and it would not be feasible to produce a variant in small quantities. Therefore, U.S. production equipments were not appropriate for Japan because they developed for the Mass Production that handle fewer type of products, larger lots, and enormous sales. Moreover, with dedicated machines and dies, it would require too much capital – something that Toyota could not afford. Hence, a breakthrough needed to be done to solve this problem.

fig. 12.37. Dedicated stamping press machine for car fender in Ford's River Rouge Plant, 1927

fig. 12.38. A stamping press machine weighed 225 tons that produced 2,700 fenders per day, one fender at each stroke. It was such an accurate machine that the fenders required no further finishing touches before enameling

The Result of Contradiction Analysis

The "plus-state" of the "dedicated machines" can be determined as in A_S-Matrix as **number 10 (ease of use)**, because the machines were ready to use and did not need specialist that required long time to do die–changes in every set-up. The "minus-state" of the "dedicated machines" can be determined as in A_S-Matrix as **number 02 (universality, adaptability)**, because each machine could not be used for producing other type of parts. This would increase the investment cost.

Inventing Idea

Combining the plus and minus state factors above, the results will show several points in As-navigators. The target is to avoid high investment cost required to provide dedicated machines to produce many parts. The new alternative will give possibility to have fewer machines that can produce many type of parts in the same speed and quality as previous concept. After combining the A-Matrix 10 and 02, the idea navigators resulted :

- **Navigator 03 (segmentation)** – according to this navigator, it can be considered to raise the degree of disassembly of the machines.

- **Navigator 07 (dynamization)** – according to this navigator, it is advised to disassemble the machine with its dies so they can be moveable among each other.

The Result-artifact (Solution) – Machines with Quick Changeover

With regards to the problem faced by Toyota, Taiichi Ohno had an idea to develop die construction with simple die-change techniques so that the changeover activity could be done frequently in a short time. He introduced roller principle to move die in and out of machine and simple adjustment to avoid long time and skilled workers to perform the die change.

Using a few used American presses, Ohno perfected his technique for quick changeover/setup. By the late 1950s, he had reduced the time required to change dies from one day to three minutes and eliminated the need for die-change specialists (s. Womack). The setup on a bolt-forming machine at Toyota Motors could be reduced from eight hours to 58 seconds.[169] These examples were typical kinds of improvements achieved using Single Minute Exchange of Die (SMED), or Quick Changeover principle.

There are two types of setup operations:

- Internal Setup (IED – Internal Exchange of Die): setup activities that can be performed only when the machine is stopped, such as removing old dies and mounting new dies.
- External Setup (OED – Outer Exchange of Die): setup activities that can be completed without stopping the machine, such as transporting dies to or from storage.

The key principle of SMED method to decrease the setup time will be:

1. distinguish internal and external setup activities;

2. convert internal setup to external setup;

3. improve activities in both categories.

fig. 12.39. Toyota's stamping press machines with quick die change concept. Right side is Danly machine and left side is Komatsu machine

By shortening setup time with adopting SMED in the production line, some further benefits can be expected:

- increase machine operating time – increase productivity of the machine;
- can process material in small lots that can reduce finished goods inventories and inventory between processes;
- flexibility of production in responding to fluctuating demand.

---

169 Shingo, S. (1989) *A study of the Toyota Production System from an Industrial Engineering Viewpoint*. – Productivity Press, Cambridge, MA

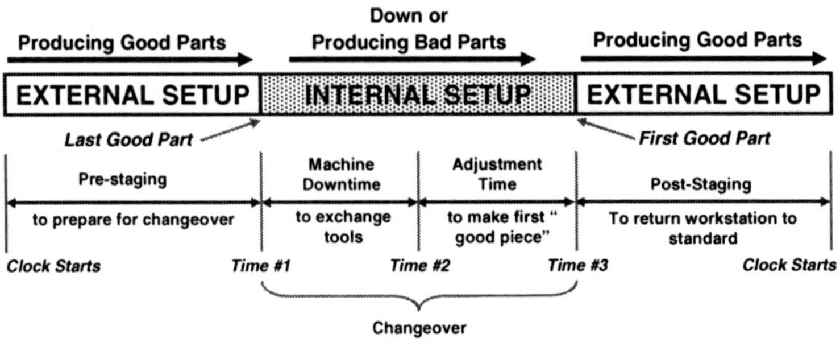

fig. 12.40. Principles of setup in a Single Minute Exchange of Die (SMED) concept

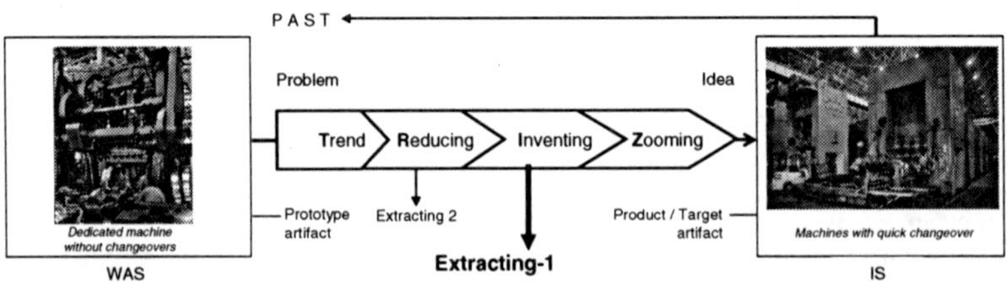

LC*	No.	Navigator	Substantiation for the Extracting
+	02	**Preliminary action**	The machines and dies shall be prepared in advanced so that the machine can handle several dies with quick changeover.
++	03	**Segmentation**	Raise the degree of disassembly of the machines / dies.
+	05	**Separation**	Separate the unnecessary activities from the changeover process.
++	07	**Dynamization**	Disassemble the machine with its dies so that they can be moveable each other.
+	08	**Periodic action**	Transition of the machine from a continuous function to a periodic one.
+	11	**Inverse action**	Instead of having one machine to produce one type of parts, the machine can be modified to be able to produce many types of parts with simple changeover.
+	12	**Local property**	Change the structure of the machine and dies so that the machine can handle many dies with simple changeover.
+	13	**Inexpensive short-life object as a replacement for expensive long-life one**	Use inexpensive machine and dies as replacement of expensive ones.
+	15	**Discard and renew of parts**	Use the same machine to serve many dies. The out-of-date dies can be discarded and replaced with new dies.
+	20	**Universality**	Use universal machine to serve many dies.

Extracting-2

Standard Contradiction: The use of dedicated massive and complex machines had an advantage of readiness in use since they did not need any set-up and specialist for die-change, but they could not be used for producing other parts.

Radical Contradiction: The press die machine should be common to accommodate many types of parts, but it should also be unique so that it could produce parts with high precision.

fig. 12.41. Extracting: Machines with Quick Changeover

TREND

Stamping press in Ford plant, River Rough, 1927

In a Mass Production System, the construction of the stamping press dies were massive, complex and expensive. In order to make another type of part, specialists require a long time (one full day) for set-up (die-change) to ensure absolute precision. To overcome this issue, companies usually used dedicated press sets for every type of part. This required high investment cost that could only be covered by high production rate. This solution became not feasible when the market volume was low or demanded high variation of car types.

PROBLEM : How could we avoid the high-cost-dedicated press dies ?

REDUCING

FIM: X-resource, together with available or modified resources, and without making the object more complex or introducing any negative properties, guarantees attainment of the following **IFR**:

[*universal & simple stamping press machine*].

Standard (Technical) Contradiction

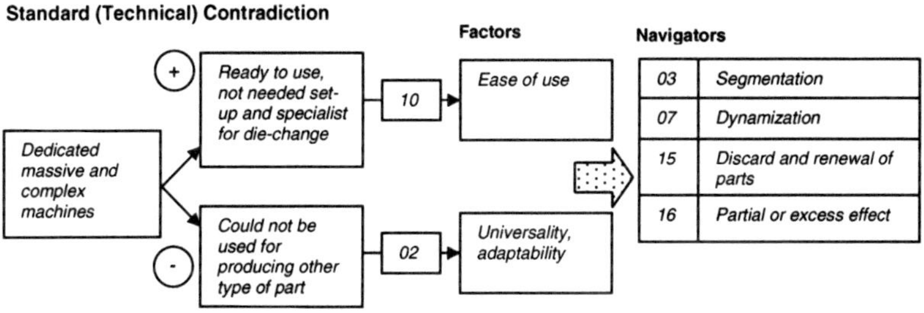

Radical (Physical) Contradiction

The press die	⇨	***should be common**, to accommodate many types of parts*	&	***should be unique**, so it could produce parts with high precision*

INVENTING

Stamping press in Toyota plant, 1970s

Group of navigators to model the idea for the 'Quick Changeover' concept :
03. Segmentation:
Raise the degree of disassembly of the machines/dies.
07. Dynamization:
Disassemble the machine with its dies so that they can be moveable among each other.

ZOOMING

Have the contradictions been removed? – Yes. – ~~No~~.

Super-effects: Yes. It could avoid the use of many dedicated press dies and reduced the set-up time.

Negative effects: Initial cost for providing the press dies with this concept.

BRIEF DESCRIPTION

Quick Changeover / Single Minute Exchange of Die (SMED): This concept was developed by Toyota to overcome problems of low market volume and low capital in providing dedicated press dies. The machines were made so that they could produce several types of parts with simple die-change set-up activity in a short period of time. In 1973, Toyota had reduced the time required to change dies to three minutes and eliminate the need for die-change specialists.

fig. 12.42. Reinventing: Machines with Quick Changeover

12.2.2 Continuous Flow

Mass Production System focuses on providing affordable products for the "masses". In order to achieve this, it must be supported by another principle: economies of scale – lower cost and price that were achieved with greater output. As prices were lowered, more people could afford to buy the products, resulting in greater sales and therefore greater production, even lower costs and prices, and so on.

The Prototype – Batch-and Queue Process

In Mass Production, it is necessary to produce high amount of products and also keep high amount of inventory. Machineries were made larger in pursuit of scale economies. The massive expense of the machinery encourages batch production and buildup of huge work-in-process and finished goods inventories (even if there is no customer to buy it). These inventories appeared as assets on company balance sheet, despite the enormous amounts of cash they absorbed.

According to Mass Production System, batch production is necessary to reduce the changeover frequency as minimum as possible, since the more frequent the changeover, the lower the available production time and the higher the labor cost.

As a matter of fact, batch production brings several disadvantages[170]:

a. It creates batch delays.

Whenever parts are processed in a batch, the entire batch, except for the one piece being processed, is delayed "in storage" – in either an unprocessed or a processed state until all pieces in the batch have been processed. This is called as "waiting-in-process". Little attention has been paid to such delays because they are typically concealed within processing times. The only reason for increasing batch size is the assumption that this will compensate for delays caused by long changeover. The truth is that batch delays prolong the production cycle time considerably.

b. It creates waste of transportation.

Batch production brings consequence of transportation of pile of parts from one station to another. Usually, improving transportation means replace manual handling with tools such as trolleys, forklifts, etc., which actually only improves the work of transport. Transport only increases costs and never adds value. Typically, labor cost in production activity consist of 45 percent of processing, 5 percent each of inspection and delays, and 45 percent of transport. Even when the manual transport is mechanized, transportation costs are just shifted from manual to mechanical – an investment without return. With this in mind, the elimination of transport through layout improvement is required.

c. It creates quality problem.

Batch production also creates quality problem because similar defect are usually replicated throughout the batch before it can be caught – usually at the end of line.

---

170 Shingo, S. (1989) *A study of the Toyota Production System from an Industrial Engineering Viewpoint.* – Productivity Press, Cambridge, MA

In 1970s, Bumper Works, located near Danville, Illinois, U.S., was a typical batch-and-queue factory. Bumper Works made chrome and painted steel bumpers in a variety of styles for pickup trucks. It made large batches of each type of bumper – typically a month's worth – before shifting production to next model and delivered the products to new-car dealers and repair body shops.

Because large batches were considered normal at that time, it was not important that it took sixteen hours to changeover Bumper Works' stamping presses. Batches of raw materials also came in large amount and they were considered unavoidable. Thus, Bumper Works had a warehouse in its plant to store tons of sheets of steel as supplies.

And because the chroming company – performing the key step in the middle of the production process – also worked in a batch mode, Bumper Works piled up semi-finished bumpers in its intermediate goods warehouse until there was an enormous batch and then shipped them to the chroming company all at once.

When the chroming company shipped the materials back in a batch, they were run through a final assembly operation (to install inner reinforcing bars, attachment brackets, and cosmetic coverings), stored once more in a finished goods warehouse, and sent in a batch to the customer according to a predetermined schedule.

In 1985, Bumper Works was signed on as a supplier for a small volume of Toyota business in North America, and in 1987 won a sole-source contract for the bumpers on the new version of Toyota's small pickup. The owner of Bumper Works, Shahid Khan – a successful immigrant from Pakistan who started his career as production worker at Bumper Works – wanted to place his company as Toyota's sole bumper supplier for North America.

fig. 12.43. Example of mass production with large batch of inventory. These were jeeps produced for WWII allies

fig. 12.44. Batch production and transport of SsangYong New Actyon in Sollers-Far East plant, Korea. Sollers is a Russian automobile company headquartered in Moscow

But there was one problem: Bumper Works' production system was still a classic Mass Production with batch-and-queue. This concept would cause Toyota need to wait for fulfillment of each product type.

Bumper Works would also deliver a huge finished goods products to Toyota, that will force Toyota to increase its inventory area and cost. Internally, the batch-and-queue process also gave burden to Bumper Works since it needed to provide huge area for storing inventories.

The inventories would also consume huge amount of inventory cost and capital.

Toyota then took Khan and his senior managers for a trip to Japan in 1989 and walked them through showcase lean suppliers. There, they learned for the first time the lean concept of Toyota to overcome the disadvantages of the batch-and-queue process.

The Result of Contradiction Analysis

The "plus-state" of the "batch-and-queue process" can be determined as in A_S-Matrix as **number 36 (power)**, because less power was required due to less changeover and transporting frequency between stations. The "minus-state" of the "batch-and-queue process" can be determined as in A_S-Matrix as **number 25 (loss of time)**, because the parts needed to wait until the batch was completed before it could go to next station for the sequential process.

Inventing Idea

Combining the plus and minus state factors above, the results will show several points in A_S-navigators. The target is to avoid long waiting time of the whole batch to be finished. The new alternative will give possibility to have shorter period to finish and deliver parts to next (internal and external) stations / customers. After combining the A_S-Matrix 10 and 02, the idea navigators resulted:

- **Navigator 02 (preliminary action)** – according to this navigator, it is needed to prepare the production line in advanced so that each part can be transferred to next process individually.
- **Navigator 40 (uninterrupted useful function)** – according to this navigator, it is necessary to eliminate idle running and interruption of the parts processing activity.

The Result-artifact (Solution) – Continuous Flow Process

After World War II, Taiichi Ohno and his technical collaborators, including Shigeo Shingo[171], concluded that the real solution for the disadvantages of batch-and-queue process was to create continuous flow in small-lot production when only few units of products were needed, not millions.

[171] Shigeo Shingo (1909 – 1990) – a Japanese industrial engineer, one of the world's leading experts on manufacturing practices and the Toyota Production System

Ford actually had introduced a piece of continuous flow in his assembly line, especially with the moving assembly line concept. But the parts supplied to the assembly line are produced entirely in large lots. At Toyota, the manufacturing and assembly processing both were performed as one-piece flow operations.

Toyota created a comprehensive system in which various parts, whether they processed within the plant or supplied from outside, were produced in small lots and flew directly into final assembly. This became one of the fundamental principles of the Toyota system and a significant difference between Ford and Toyota.

Production cycles could be reduced significantly by eliminating batch delays by using one-piece flow concept. Yet, one-piece lot transport involved increase of transportation from one process to the next, a problem solved by improving layout, such as placing succeeding processes next to one another. After layout was improved, more efficient transport means needed to be provided, such as conveyors. This allows materials to flow smoothly between processes that reduced production cycles and cut transport time dramatically.

In May 1990, Toyota sent several lean *sensei* (master in the TPS system) from its Operation Management Consulting Division to Bumper Works as personal tutors. The group was established in 1969 by Taiichi Ohno to promote lean thinking within Toyota and its supplier companies.

The first thing the lean *sensei* noted at Bumper Works was the massive inventories and batches. The production line did not "flow" at all. Since immediate right-sizing the massive stamping presses to permit continuous single-piece flow was not possible, the first solution was to reduce their changeover times and shrink batch sizes. Changeover times were then down from sixteen hours in the mid-1980s to around two hours.

The next important activity was re-layout of the production area. Transportation between stations were minimized in order to have a continuous flow of materials, from the input of raw material until the output of finished goods. The concept of one-piece flow was implemented as maximum as possible. By this concept, metal sheets as raw material flowed directly from the receiving dock to the blanking machine, which cut the steel into required size. The blank sheets then went immediately to the adjacent cell of stamping presses for giving shape. Next, they were shipped at frequent intervals in small batches to the outside chroming company and returned to the welding shop adjacent to the stamping presses. There, several parts of the bumpers were welded together. Finally, the bumper went straight to the shipping dock just in time for scheduled shipment.

So, basically the production lines were made to flow continuously without stoppage to wait for the completion of the batch.

fig. 12.45. A view of the continuous assembly line at the new Toyota plant outside St. Petersburg, Russia

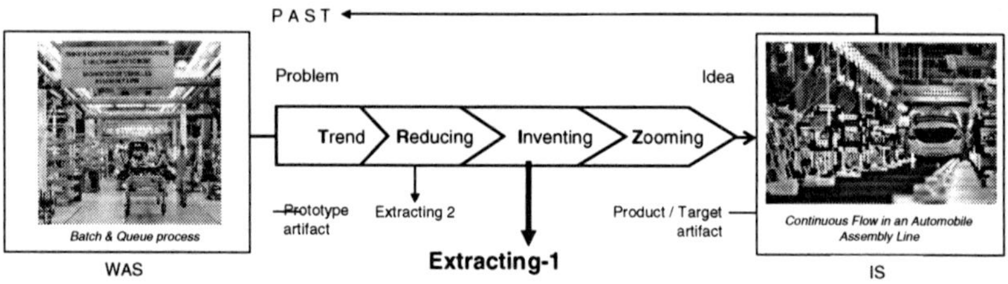

LC*	No.	Navigator	Substantiation for the Extracting
+	01	**Change in the aggregate state of an object**	Transition of the assembly process from "batching" to "continuous"
++	02	**Preliminary action**	Prepare the production line in advanced so that each part can be transferred to next process individually.
+	07	**Dynamization**	Make the "queue" material to "move"
+	11	**Inverse action**	Instead of having the assembly process in batch, the process can be performed in continuous one.
++	40	**Uninterrupted useful function**	Eliminate idle running and interruptions.

Extracting-2

Standard Contradiction: Mass production with "batch" process use less power for transportation between station, but the parts needed to wait for completion of the whole batch before they can be moved to next station.

Radical Contradiction: The product stock should be exist so that the part would not be transported individually between stations, but it should also not exist so it would not cause waiting time.

FIM: X-resource, together with available or modified resources, and without making the object more complex or introducing any negative properties, guarantees attainment of the following **IFR**:
[*production with less waiting time*].

fig. 12.46. Extracting: Continuous Flow Process

TREND

In a Mass Production System, it was common that the parts processed in a batch of quantity within a station before being delivered to the next station. The larger the batch, the longer a unit had to wait for the rest of the batch to be completed before it could go to the next process. This queue-time was considered as waste and represented time lost that was not value-added in the eyes of the customer.

Batch Transferring

PROBLEM: How could we reduce this waste of waiting time ?

REDUCING

FIM: X-resource, together with available or modified resources, and without making the object more complex or introducing any negative properties, guarantees attainment of the following **IFR**: [*production with less waiting time*].

Standard (Technical) Contradiction

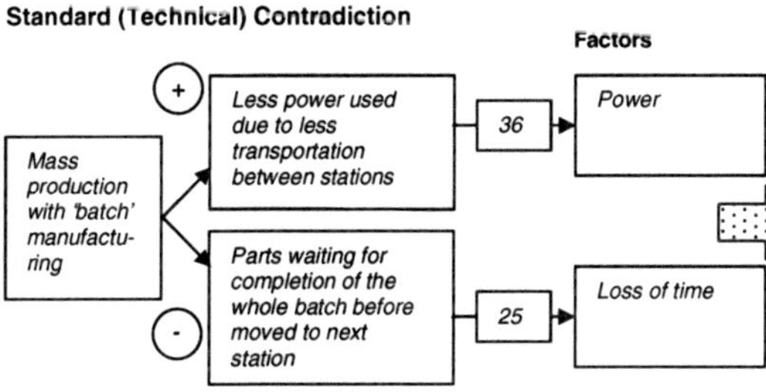

01	*Change in the aggregate state of an object*
02	*Preliminary action*
20	*Universality*
40	*Uninterrupted useful function*

Radical (Physical) Contradiction

The product stock	⇨	***should be exist**, so that the part would not be transported individually between stations*	&	***should not be exist**, so it would not cause waiting time*

INVENTING

Group of navigators to model the idea for the 'Continuous Flow' concept:
02. Preliminary action: Prepare the production line in advanced so that each part can be transferred to next process individually;
40. Uninterrupted useful function: Eliminate idle running and interruptions.

Continuous Flow in an Automobile Assembly Line

ZOOMING

Have the contradictions been removed? – Yes. – ~~No~~.

Super-effects: Yes. It reduced the waiting-in-process.

Negative effects: Factory re-layout to achieve continuous flow.

BRIEF DESCRIPTION

Continuous Flow: This is one of important strategies required in Lean Production System. The paradigm aim was to achieve one-piece flow where a single discrete unit of product flowed from process to another. In effect, the batch quantity was one. This strategy required a good integration of equipments and management of workers. Continuous flow and *kanban* were the tools required to achieve the success of Just in Time.

fig. 12.47. Reinventing: Continuous Flow Process

12.2.3 Production Leveling

Processing in large batch is one characteristics of Mass Production that is considered necessary to reduce the changeover time and by that increase the effective production time and output. The batch process itself gives consequence of batch scheduling, in which large batch of certain type of product is produced within one period of time before the line is changed to another type of products.

The Prototype – Batch Scheduling

Most of Bumper Works' customers were usually ordering massive batches – one-month lots to be delivered by the last day of the month. This caused Bumper Works to create its production schedule in large batches. For example, if there were orders for the end of the month for 2,000 of Bumper A, 2,000 of Bumper B and 4,000 of Bumper C – with 20 days of working per month – Bumper Works will produce Bumper A within the first 5 days of the month, Bumper B in the next 5 days, and Bumper C in the last ten days. They did that in order to avoid frequent changeovers that could take sixteen hours and consumed the production time.

This batch scheduling – according to Toyota – gives some disadvantages. It would not suit to Toyota's production system that processed many variations of product type in small batches in short period of time, in order to avoid inventory of many kinds of raw materials and finished goods.

The scheduling would also not be flexible in fulfilling the Toyota's after sales service division that frequently need a specific bumper to be delivered immediately during repairing customers' car.

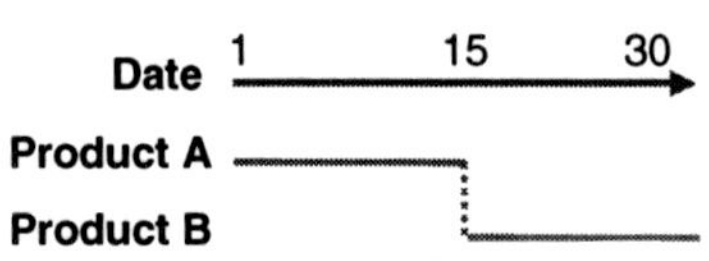

fig. 12.48. Batch production scheduling in Mass Production

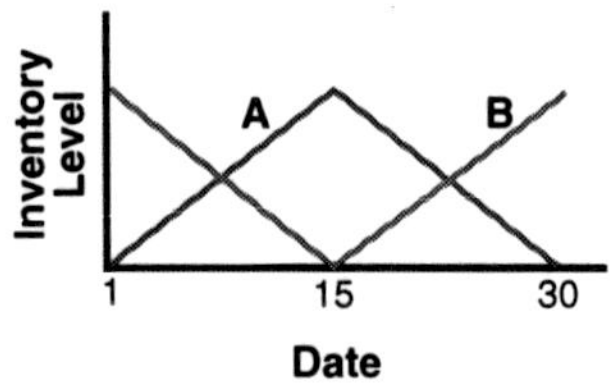

fig. 12.49. Finished goods inventory level in Mass Production

The Result of Contradiction Analysis

The "plus-state" of the "batch scheduling" mentality can be determined as in A_S-Matrix as **number 09 (ease of manufacture)**, because it did not require frequent changeover of the machines and change of the raw material and parts. The "minus-state" of the "batch scheduling" can be determined as in A_S-Matrix as **number 26 (quantity of material)**, because it had weakness of high inventory of one kind of product and shortage of the other one. This made the company could not be flexible in adapting the changes in customer demand.

Inventing Idea

Combining the plus and minus state factors above, the results will show several points in A_S-navigators. The target is to enable the company to have balance in its ability to provide products demanded by customer. The new alternative will give the company possibility to change its production line more frequent without consuming long time for changeover. After combining the A_S-Matrix 09 and 26, the idea navigators resulted :

- **Navigator 03 (segmentation)** – according to this navigator, it is necessary to "disassemble" the batching scheduling into smaller ones.
- **Navigator 06 (periodic action)** – according to this navigator, the continuous batching scheduling need to be transformed into several periodic scheduling.

The Result-artifact (Solution) – Mixed-model Production Leveling

Current market condition forces companies to produce different models continuously and limits the fluctuations in scheduled production requirements. This can be done by leveling the volume and mix the product types at the assembly line that needs support of the ability to produce small batches in more frequent changeovers and smaller WIP (work-in-process) inventories.

Production leveling is one of the prerequisites for having good Lean Production. The intent of production leveling is to maintain the availability of many types of products in order to have smooth fulfillment of customer orders and sequence them over time (Art of Lean, Inc., 2011).

The principle of mixed-model production leveling is by producing the batch volume of each product types in a shorter period of time, i.e. everyday. This will need more frequent changeovers, but with implementation of SMED, this will give no problems.

Production leveling for mixed products has advantages in eliminating work-in-process and finished goods accumulation by producing small batches in small period of time. Moreover, it can respond rapidly to the fluctuation in customer demand without having need to put huge amount of finished goods inventories and facilitates planning by letting the workers know at the beginning of the process what the average load will be.

Starting in 1992 – under Toyota's guidance – Bumper Works decided to prepare its daily schedule using Toyota's production leveling techniques. Shahid Khan's production managers would take the orders for the next month, let's say 2,000 units of Bumper A, 2,000 units of Bumper B, and 4,000 units of Bumper C. They would add them up (8,000 units) and divide by the number of working days in a month (say, twenty) to discover that Bumper Works would need each day to make 100 of Bumper A, 100 of Bumper B, and 200 of Bumper C each day. This would require three changeovers of the blanking and stamping machines everyday.

Because of its previous capability of quick changeover and continuous flow, it became possible for Bumper Works to implement this mixed-model production leveling in its assembly line. The new system gave Bumper Works the ability to balance its ability in dealing with fluctuation in customer demand in short notice without having to provide huge inventories of finished goods and raw materials. Thus, the inventory cost and area could also be reduced.

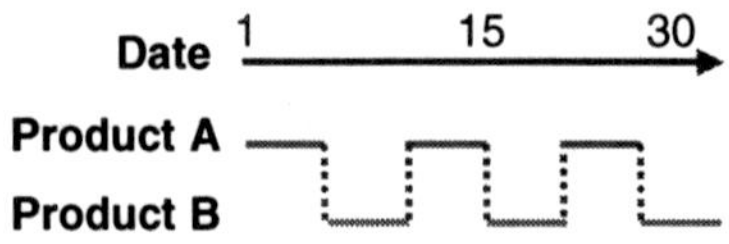

fig. 12.50. Production scheduling in TPS

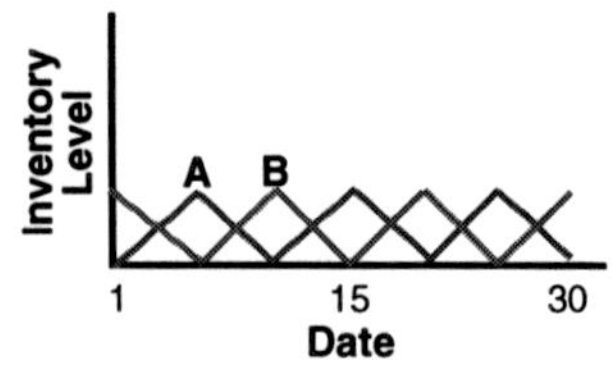

fig. 12.51. Finished goods inventory level in TPS

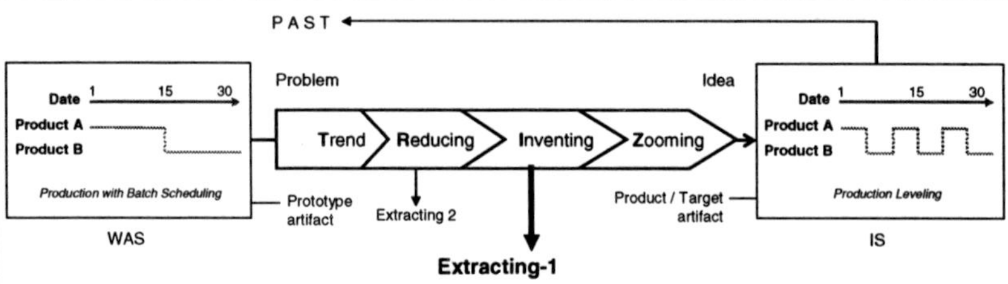

LC*	No.	Navigator	Substantiation for the Extracting
+	01	**Change in the aggregate state of an object**	Change the degree of flexibility of production schedule from long batching period to smaller one.
+	02	**Preliminary action**	Prepare schedule and machine in advanced so that production can cover all types of products in small batching period.
++	03	**Segmentation**	"Disassemble" the batch scheduling into smaller ones.
++	08	**Periodic action**	Transition from a continuous batching schedule into several periodic ones.
+	11	**Inverse action**	Instead of producing one type of product within long period, produce many types of products within short period.

Extracting-2

Standard Contradiction: Mass production with "batch" scheduling did not require frequent changeover and change of material, but it produced high inventory of one type of product and shortage of the other one.

Radical Contradiction: The batch scheduling should be exist so that the changeover time could be minimum, but it should also not be exist so that it could be flexible to customer change of demand.

FIM: X-resource, together with available or modified resources, and without making the object more complex or introducing any negative properties, guarantees attainment of the following **IFR**:
[*even workload and inventory*].

fig. 12.52. Extracting: Mixed-model Production Leveling

TREND

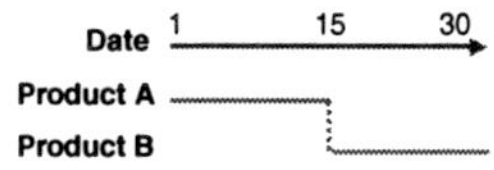

Production with Batch Scheduling

In a Mass Production System, it was common that the production was arranged to finish demand for one type of product (i.e. product A), before it moved to another product (i.e. product B). It was also common to have overtime to create stock enough to deal with future customer peak demand. Lean Production System saw this as another type of waste: *Mura* (unevenness) and *Muri* (overburden). The inventory for Product A would increase while the one of product B would decrease. Thus, the company could not have flexibility to adapt with customer change of demand. Furthermore, it would force the workers and the machines to do more than they were capable of / designed to.

PROBLEM: How could we reduce the unevenness of workload and inventory?

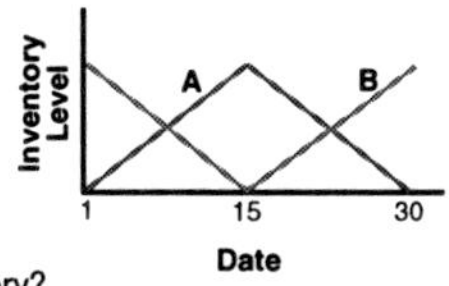

Finished Goods Inventory Level

REDUCING

FIM: X-resource, together with available or modified resources, and without making the object more complex or introducing any negative properties, guarantees attainment of the following **IFR**: [*even workload and inventory*].

Standard (Technical) Contradiction

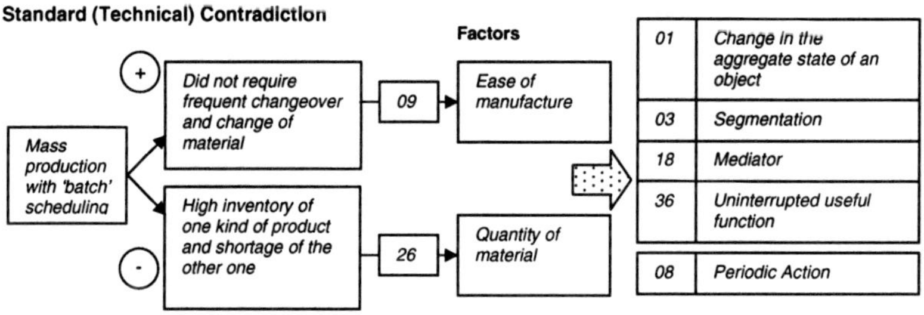

Radical (Physical) Contradiction

The batch scheduling	⇨	***should be exist**, so that the changeover time could be minimum*	&	***should not be exist**, so that it could be flexible to customer change of demand*

INVENTING

Group of navigators to model the idea for the "*Heijunka*" concept: **03. Segmentation:** Disassemble' the batch scheduling into smaller ones; **08. Periodic action:** Transition from a continuous batching scheduling into several periodic ones.

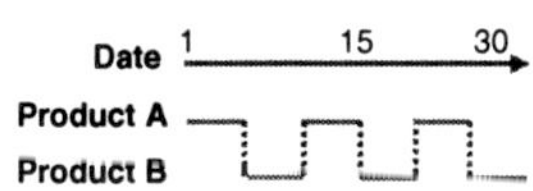

Production Leveling

ZOOMING

Have the contradictions been removed? – Yes. – ~~No~~.

Super-effects: Yes. It could increase flexibility in dealing with customer change of demand without increasing inventories and overburden in the workers/machines.

Negative effects: Required machines with quick changeover and high reliability.

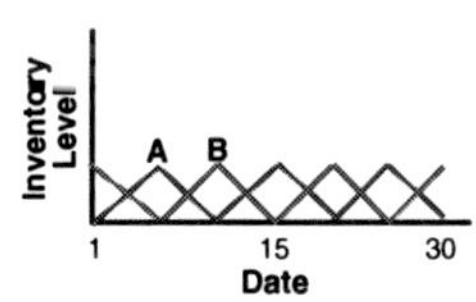

Finished Goods Inventory Level

BRIEF DESCRIPTION

Heijunka: It is a Japanese term that means 'production leveling/smoothing'. This concept was developed by Toyota to have an even workload and inventory. The concept contained demand leveling, volume leveling, product leveling or mix of them. With this strategy, the company could be flexible in dealing with fluctuation in customer demand without having to have overburden in its stations/workers in certain period of time. This strategy required good implementation of SMED (Single Minute Exchange of Die) and Continuous Flow.

fig. 12.53. Reinventing: Mixed-model Production Leveling

12.2.4 Poka-yoke

In a Mass Production System, one of the characteristics is the high output rate of production. Supported with the moving assembly line and division of labors, the line can produce output in high speed.

The Prototype – Inspection and Rework

With output rate as its main concern, it become something in common for a company that adopt Mass Production to focus on the quantity, and put aside the concern of the quality of the products.

During its peak production in 1910s, there were hardly any inspections of finished automobiles in Ford's Model T assembly line at Highland Park. There was no engine checking until the vehicle was finished, and no products were ever road-tested. The company did not deliver the product at the highest quality, mostly because there was no urgency for it at that time. The cosmetic failures such as gaps in the fender panels, stumbled engines or electrical problems did not bother the customers.

However, as the market competition became tighter, the demand of the quality of the vehicles was increased and automobile companies usually put Final Inspection Station at the end of the assembly line to check for quality problems. This was because the companies wanted to keep the production line running to achieve the output target without having to stop for errors.

The defects then multiplied and accumulated from station to station and cause burden for the workers in the Final Inspection and Rework Stations. Because the errors did not tried to be fixed, the errors keep continued and created defects in subsequent products.

Defective products needed to be inspected, reworked or even scrapped and it required allocation of man-hour and product replacement, which meant additional waste of cost.

fig. 12.54. Reworking activity at an automobile assembly line

The Result of Contradiction Analysis

The "plus-state" of the "inspection and rework" concept can be determined as in A_S-Matrix as **number 22 (speed)**, because each station would be able to run in high speed and did not need to stop for inspecting products for defects. The "minus-state" of the "inspection and rework" can be determined as in A_S-Matrix as **number 05 (precision of manufacture)**, because the defects produced in each station would be accumulated and give burden to the Final Inspection Station and consume a lot of effort and cost for rework.

Inventing Idea

Combining the plus and minus state factors above, the results will show several points in A_S-navigators. The target is to make it possible to inspect and monitor the quality of the products in each process so that the defective product will accumulated in the Final Inspection and give burden to the workers in the Rework Station. The new alternative will give possibility to the system to do monitoring without sacrificing the productivity. After combining the A_S-Matrix 22 and 05, the idea navigators resulted:

- **Navigator 02 (preliminary action)** – according to this navigator, the machine need to be prepared by implementing partial change in it so that it can detect any defects or non-conformances.
- **Navigator 29 (self-service)** – according to this navigator, the machine need to serve itself with auxiliary to detect defect / non-conformance and stop automatically.

The Result-artifact (Solution) – Poka-yoke System

Lean Production System considers quality as one of its goal. In order to achieve highest quality of products (zero defect), lean company shall realize that defects must be prevented from occurring. To do so, inspections need to be done – not merely to find defects after it happened – but it must prevent defects.

For a complete elimination of defects, 100% inspection must be adopted. Sampling inspection is not enough, because it can not guarantee product quality. However, the traditional 100% inspection in the Final Inspection Station is not suitable to be implemented since it will cause accumulation of products in that station, increase the waiting time and throughput time of a product.

Inspection by workers also not in line with TPS' principle of *jidoka* (autonomation / pre-automation / automation with human touch), where workers need to be separated from machines through the use of mechanisms to detect production abnormalities.

There are said to be twenty-three stages from purely manual work to full automation (s. Shingo). To be fully automated, a machine must be able to detect and correct its own operating problems. Since it is very difficult and not feasible to build an equipment that can correct its error by itself, then it is preferable to limit the machine capability to pre-automation stage, where it can detect its error automatically by using a simple mistake proofing device (*poka-yoke*).

Ninety percent of the result of full automation can be achieved at relatively low cost if machines are designed to merely detect problems and leave the correction of problems to the workers.

Successive, self, and source inspection can all be achieved through the use of *poka-yoke*. *Poka yoke* can achieve 100% inspection through mechanical or physical control.

There are two ways in which *poka-yoke* can be used to correct mistakes:

- Warning type: when the *poka-yoke* is activated, a buzzer sounds or a lamp flashes to alert the worker.
- Control type: when the *poka-yoke* is activated, the machine shuts down.

The warning *poka-yoke* still have weakness since it allows error to continue if workers do not respond to the warning. The control *poka-yoke* is the strongest corrective device because it shuts down the process until the error has been corrected.

Arawaka Shatai supplied car doors for Toyota. One of the production processes was to put a board covered with leather as a back lining plate for the door, which was attached by 20 retainers. Workers sometimes forgot to attach one or two retainers that caused defects on the products. They were advised to be more careful and the rate of defects dropped for a while but then returned to previous level.

Since this appeared to be a recurring problem, twenty proximity switches were installed in the equipment as warning *poka-yoke*. If a retainer was left off, the equipment would stop and a buzzer sounds to alert the workers about the problem.

With the use of the *poka-yoke*, the problem could be eliminated and the defect dropped to zero. In this case, a contact type *poka-yoke* performs 100% successive inspection.

fig. 12.55. Proximity switch: a sample of poka-yoke device

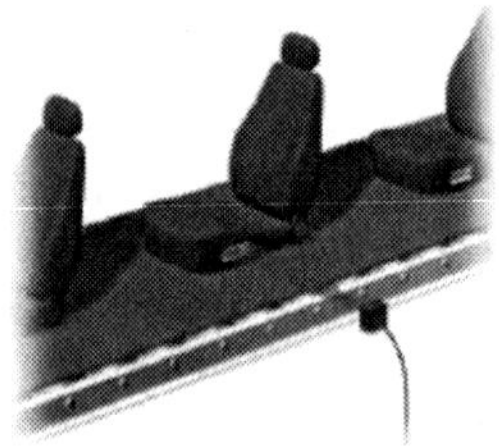

fig. 12.56. Barcode Reader, a poka-yoke system used in automobile industry to check completeness of the components need to be mounted to the passenger seats (belt buckle, side air bag, passenger check module, seat weight sensor and seat position sensor)

fig. 12.57. Toyota 4Runner: award recipient for 2011 Most Dependable Vehicles for Midsize Crossover/SUV), based on J.D. Power surveys

fig. 12.58. A result of jidoka (build-in quality) concept by Toyota: Lexus GS, award recipient for 2010 Initial Quality Study for Midsize Premium Car, based on J.D. Power surveys to provide feedback on quality during the first 90 days of new-vehicle ownership

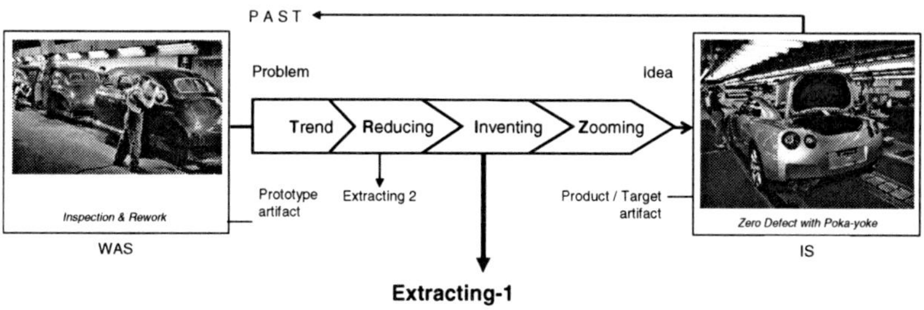

LC*	No.	Navigator	Substantiation for the Extracting
++	02	**Preliminary action**	Prepare the machine by implementing partial change in it so that it can detect any defect or non-conformance.
+	04	**Replacement of mechanical matter**	Use of devices with electrical / magnetic / electromagnetic / optical principle as sensors.
+	12	**Local property**	Change the structure of the machine s that there is one part of the machine that can detect defect automatically.
+	18	**Mediator**	Use another object to transmit an action of detecting defective parts or non-conformance process.
+	21	**Transform damage into use**	Use the "damaging factor" of stopping machine (stop production) for a useful effect (eliminate the cause of the problem).
+	28	**Previously installed cushion**	Install an additional object in the machine that can detect failure of process / defect on products.
++	29	**Self-servicing**	The machine service itself with auxiliary to detect defects/non-conformance and stop automatically.
+	36	**Feedback**	Install a device that can maintain and adjust a process to work in the given limits.

Extracting-2

Standard Contradiction: Inspection and Rework practice in the Final Station made it possible for the assembly line to run fast and achieved high productivity, but this concept would produce defects that could not be caught during process.

Radical Contradiction: The assembly line should run continuously to achieve high output, but should also be stop periodically so that the process and product can be monitored.

FIM: X-resource, together with available or modified resources, and without making the object more complex or introducing any negative properties, guarantees attainment of the following **IFR**: [*automation of inspection*].

fig. 12.59. Extracting: Poka-yoke System

TREND

Rework activity in car assembly line

In a Mass Production System, it was a common sense that the goal was to achieve high output of products. The process run in high speed and the workers did not have time and concern to check the quality of the products. Defects could only be caught in the Final Inspection Station, which caused many products needed to be reworked or scrapped. This was a kind of waste of money and material that needed to be eliminated by Lean Production System.

PROBLEM: How could we catch the defect before Final Inspection Station without stopping the machines to perform periodic inspection in every station ?

REDUCING

FIM: X-resource, together with available or modified resources, and without making the object more complex or introducing any negative properties, guarantees attainment of the following **IFR:** [*automation of inspection*].

Standard (Technical) Contradiction

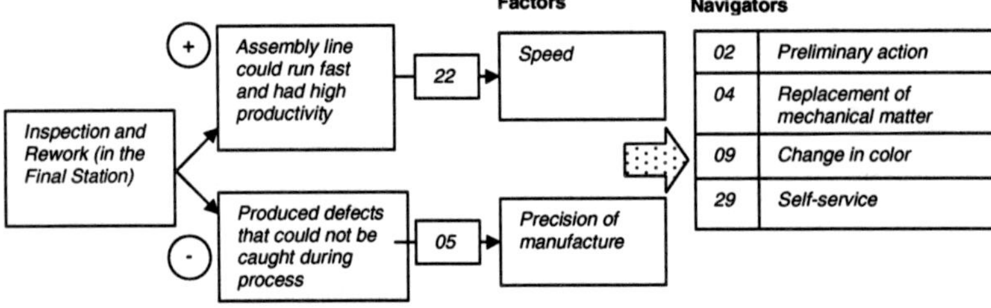

Radical (Physical) Contradiction

The assembly line	⇨	***should run continuously**, to achieve high output*	&	***should be stop periodically**, so that the process and product can be monitored*

INVENTING

Zero defect with Lean System

Group of navigators to model the idea for the '*Poka-yoke*' concept:
02. Preliminary action:
Prepare the machine by implementing partial change in it so that it can detect any defect or non-conformance.
29. Self-service:
The machine service itself with auxiliary to detect defects/non-conformance and stop automatically.

ZOOMING

Have the contradictions been removed? – Yes. – ~~No~~.

Super-effects: Yes. It increased the precision of parts manufactured

Negative effects: Cost for providing dedicated machine tools to manufacture the parts

Proximity Switch, an example of poka-yoke device

BRIEF DESCRIPTION

Poka-yoke: *Poka-yoke* is a Japanese term that means "fail-safing" or "mistake-proofing". *Poka-yoke* is a part of *jidoka* (autonomation – automation with human touch) concept introduced by Toyota. Basically it covers low cost, highly reliable devices used in machines that will stop processes in order to prevent the production of defective parts. One example of the *poka-yoke* device is proximity switch that used as sensor to detect presence of material / parts during production process.

fig. 12.60. Reinventing: Poka-yoke System

12.2.5 Andon System

The Prototype – Passing-on Defect

> We expect the men to do what they are told. The organization is so highly specialized and one part is so dependent upon another that we could not for a moment consider allowing men to have their own way. Without the most rigid discipline we would have the utmost confusion.[172]
>
> *Henry Ford*

As has been mentioned earlier in section 12.2.4, in a Mass Production system there was a tendency to pass on errors to keep the line running that caused errors to multiple endlessly. Every worker assumed that errors would be caught at the end of assembly line. Even if the worker wanted to stop the machine due to errors, he had no access or tools to inform it to the responsible line managers, who had the authority to stop the line.

This kind of discipline was assured as the managers at Ford posted the production output of each man on production board. The figures[173] were "posted hourly, and the records of those who equal or better quota set are written down in the colored crayon". This standard measure was taken to stir up "competition among workers, who performed the same operation".

So, rather than leaving his station and losing productivity, the operators preferred to pass on the defective product to next station and continue to process subsequent products. Hence, the so called "passing-on defect mentality", was commonly practiced that avoided line stoppages at any price and emphasized repair in the end of assembly line.

fig. 12.61. An example of Final inspection area in an Austin Automobile assembly line (1950s). Cars received an inspection and final polishing before went to test drive.

The Result of Contradiction Analysis

The "plus-state" of the "passing-on defect" mentality can be determined as in A_S-Matrix as **number 25 (loss of time)**, because the production process could take place continuously that would increase the productivity. The "minus-state" of the "passing-on defect" can be determined as in A_S-Matrix as **number 12 (loss of information)**, because the operators did not have chance and access to inform the defect occurred in the line. It also caused the defects multiply from station to station and created enormous amount of rework effort and cost to fix them.

---

172 McNairn, W. and McNairn, M. (1978) *Quotations from the Unusual Henry Ford.* – Quotamus Press, Redono Beach, CA

173 Porter, H. F. (1917) *Four Big Lessons from Ford's Factory.* – System, 31 (June 1917)

Inventing Idea

Combining the plus and minus state factors above, the results will show several points in A_S-navigators. The target is to enable the workers to stop the process that produce defect as immediately as possible so that the defective product will not be pass on to the next station and accumulated in the Final Inspection and Rework Station. The new alternative will give possibility to stop the machine and inform the responsible persons only when the defect is occurred so that it will not sacrifice the productivity and manpower effort to do machine monitoring all the time. After combining the A_S-Matrix 25 and 12, the idea navigators resulted:

- **Navigator 09 (change in color)** – according to this navigator, it is advised that the machine that creates problem can "change its color".
- **Navigator 18 (mediator)** – according to this navigator, it is needed to use an object to transfer information about problem in certain machine/station.
- **Navigator 35 (unite)** – according to this navigator, instead of leaving the problem to the operator in a problematic machine, the whole responsible team need to come and work on the problem.

The Result Artifact (Solution) – Andon System

At TPS, *jidoka* also means build-in quality. It means achieving highest quality during processing the materials[174]. One concept of *jidoka* – separating man's and machine's work has been mentioned earlier by the use of *poka-yoke*. The other concept of *jidoka* is to stop the work if found abnormality.

According to TPS principle, passing on defective products to the end of assembly line is also one type of waste, since there will be human effort and cost involved to rework or scrap the product. Moreover, since there is no corrective action performed to the problematic process or machine, similar defect will be produced from time to time.

To avoid these wastes, it is necessary to stop the operation or machine if there is trouble with machine operation. When problem occurs, visual control or *andon* (indicator lights) show the workers, supervisors, technicians and other related personnel where the trouble is. The troubles are then being able to be immediately communicated to everyone.

One of the important mentality to support this tool is the urgency to stop the production line and fix the problem. Top management need to commit itself to halt the machines or production lines when there is trouble. This is the key that makes *jidoka* concept with *poka-yoke* and *andon* tools become famous as autonomation – automation with a human touch.

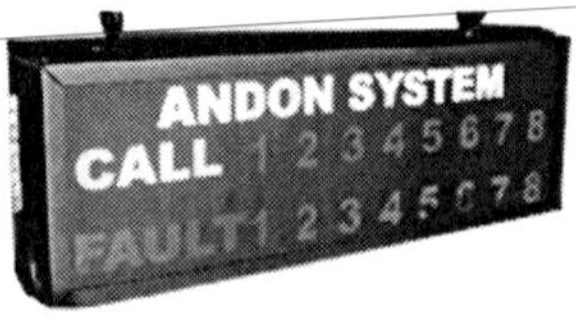

fig. 12.62. Andon system to communicate problems in machines

[174] Liker, J. K. (2004) *The Toyota Way*. – McGraw-Hill, New York

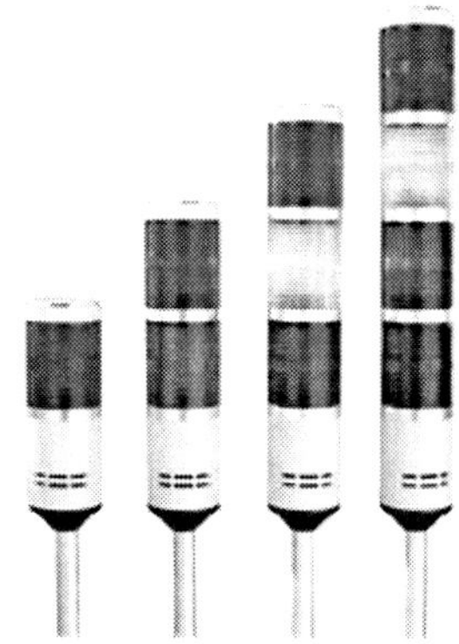

fig. 12.63. A type of andon lamp. Red colour (top) means machine error, yellow (second from top) means machine needs attention, green (third) means normal operation, blue (fourth) means short of material.

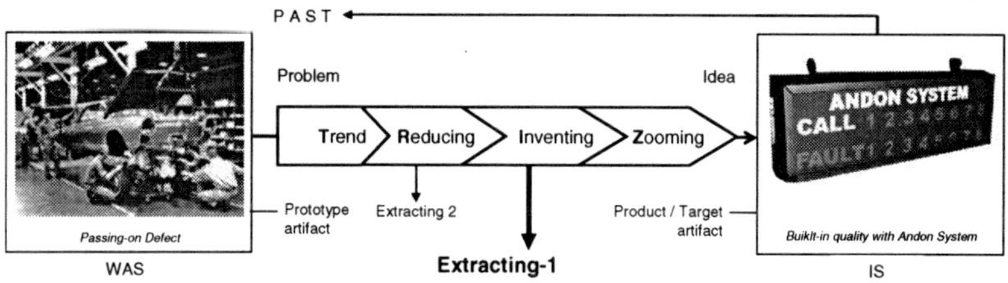

LC*	No.	Navigator	Substantiation for the Extracting
+	01	**Change... aggregate state of an object**	Change the degree of flexibility to stop the production line whenever there is problem.
+	02	**Preliminary action**	A device is provided in advance so that the worker can inform problem in the production line.
++	09	**Change in color**	Change the color of the machine / station that give problem.
+	11	**Inverse action**	Instead of implementing "passing-on defect" mentality, the company needs to encourage the workers to stop and inform for any kind of problems.
+	12	**Local property**	Change the structure of the working station to add a device that act as information board for management.
++	18	**Mediator**	Use an object to transfer information about problem in certain machine / station.
+	21	**Transform damage into use**	Use the "damaging factor" of stopping machine (stop production) for a useful effect (eliminate the cause of the problem).
++	35	**Unite**	Instead of leaving the problem to worker in a problematic machine, the whole team need to come and work on the problem.

Extracting-2

Standard Contradiction: Passing-on defect mentality made the machine can work with low stoppage time, but it gave no access and tools to inform the defect that cause high rework and scrap rate at the Final Station.

Radical Contradiction: The assembly line should run continuously to achieve high output, but should also stop periodically so that the error could be fixed.

FIM: X-resource, together with available or modified resources, and without making the object more complex or introducing any negative properties, guarantees attainment of the following **IFR**:

[*low rework rate*].

fig. 12.64. Extracting: Andon System

TREND

Since the goal in a Mass Production System was a high production output, it made the workers in each station strived to keep the line running, even if they found defects. Production line could only be stopped by the decision from senior line manager, but the workers did not have access to inform him about line problem. This caused the defects multiply from station to station. Once a defective part had become embedded in a complex vehicle, an enormous amount of rework might be needed to fix it, especially in the Final Inspection Station.

Rework activitv in car assemblv line

PROBLEM: How could we reduce the rework of products ?

REDUCING

FIM: X-resource, together with available or modified resources, and without making the object more complex or introducing any negative properties, guarantees attainment of the following **IFR**: [*low rework rate*].

Standard (Technical) Contradiction

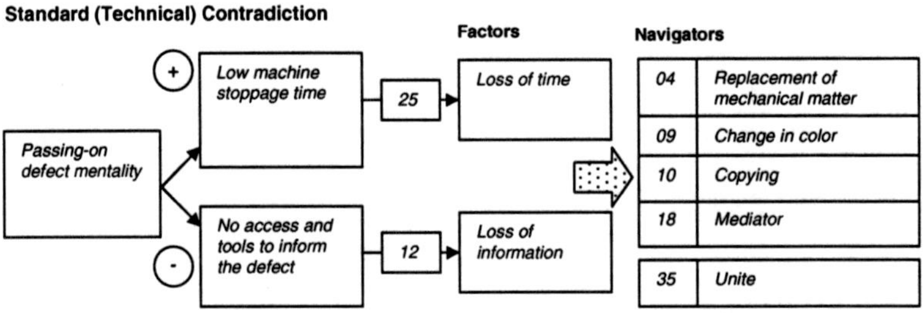

Radical (Physical) Contradiction

The assembly line	⇨	***should run continuously****, to achieve high output*	&	***should be stopped periodically****, so that the error could be fixed*

INVENTING

Group of navigators to model the idea for the 'Andon' concept:
09. Change in color:
Change the color of the machine/station that give problem.
18. Mediator:
Use an object to transfer information about problem in certain machine/station.
35. Unite:
Instead of leave the problem to worker in a problematic machine, the whole team need to come and work on the problem.

Andon System

ZOOMING

Have the contradictions been removed? – Yes. – ~~No~~.

Super-effects: Yes. It reduced the number of reworked and scrapped products. It also limited the magnitude of the problem.

Negative effects: Many line stoppages can be happened during initial implementation.

BRIEF DESCRIPTION

Andon: *Andon* is a Japanese term that means "paper lantern". It is a part of *jidoka* concept introduced by Toyota. *Andon* is a system to notify management, maintenance, and other workers of a quality or process problem. The tool used can be a signboard incorporating signal lights/alarm sound that can be activated manually by workers or automatically by the equipment itself. *Andon* gives the worker the ability to stop production when a defect is found and immediately call for assistance. Therefore, it can stop further defective products being produced and reduce time and cost to rework the products.

fig. 12.65. Reinventing: Andon System

12.2.6 Just-in-Time (JIT)

In Mass Production that adopted economic of scale principle, it is necessary to maximize output in order to reduce the production cost. This principle emphasis the maximum usage of the machines and "push" the material to the end of assembly line and encourages buildup of finished goods inventories. These inventories were considered as assets for the company, rather than accumulated waste of capital.

The Prototype – "Push" Production Control

The production process[175] with "push" concept can be analogous by the water flow in the river with many rocks (fig. 12.66). The river is the material movement, the depth of water represent the inventory and Work-in-Process (WIP), and the rocks represent the problems encountered in assembly line. The "push" concept use high inventory (high level of water) as safety buffer to deal with quality problems (defective products), long production time (due to batch processes and machine problems), long set-up time (due to complex machine), or long delivery time from suppliers.

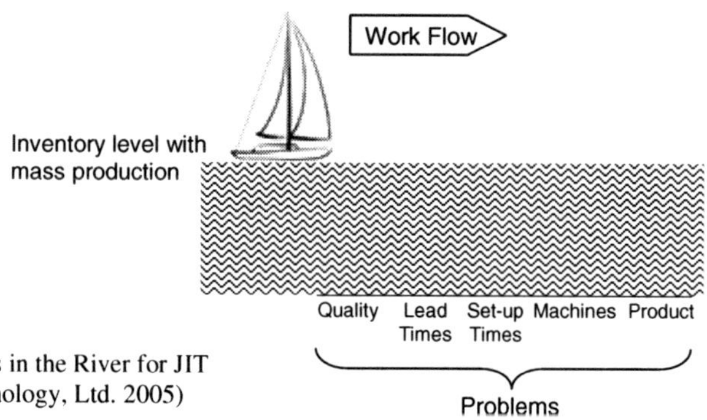

fig. 12.66. The Rocks in the River for JIT (after Tangram Technology, Ltd. 2005)

When Eiji Toyoda[176] and his managers from Toyota went to Ford's plant in 1950, to their surprised, they found that the company had many flaws. They saw the discrete process steps were based on large volumes, with interruptions between these steps causing large amounts of material to sit and wait in WIP-inventory. They saw the massive and expensive equipments.

The company tried to maintain efficiency in reducing the cost per piece, with workers keeping busy by keeping the equipment busy, resulting in a lot of overproduction and an uneven flow, with defects hidden in large batches that could go undiscovered for weeks.

---

175 Silver, E. A., Pyke, D. F. and Peterson, R. (1998) *Inventory Management and Production Planning and Scheduling* (pp. 592 – 623, 631 – 653). – John Wiley & Sons, Inc., New York

176 Eiji Toyoda (b. 1913) – a Japanese engineer, cousin of Kiichiro Toyoda (founder of Toyota Motor Corp.). He became the President (1967 – 1981) and Chairman (1981 – 1994) of Toyota

Entire workplaces were disorganized and out of control. With big trucks moving materials everywhere, the factories looked more like warehouse than an automobile plant.

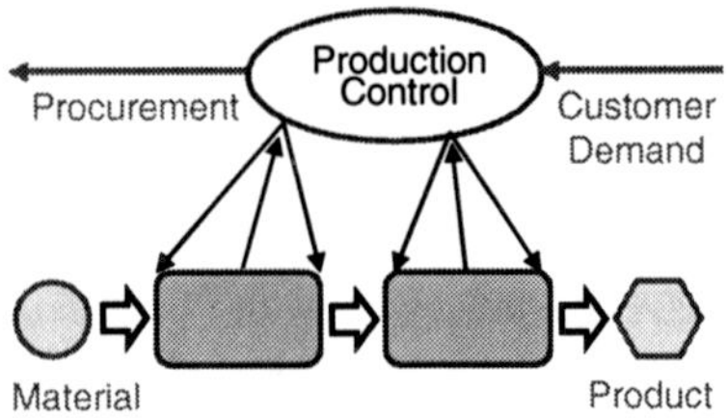

fig. 12.67. Push Production Control Principle usually adopted by Mass Production System

fig. 12.68. Unsold Dodge SUVs sat at the Atlantic Marine Terminal at the port of Baltimore, Maryland. More than 57,000 cars waiting for buyers during Q1-2009. Maryland paid USD 5.26 million for almost 10 hectares of additional car storage space near the port

The Result of Contradiction Analysis

The "plus-state" of the "push production control" can be determined as in A_S-Matrix as **number 09 (ease of manufacture)**, because each station needed to think only of their own production pace. The "minus-state" of the "push production control" can be determined as in A_S-Matrix as **number 02 (universality, adaptability)**, because this concept was not flexible in dealing with the fluctuation of the customer demand. Accumulation of inventories between stations or in the final storage could be expected.

Inventing Idea

Combining the plus and minus state factors above, the results will show several points in A_S-navigators. The target is to avoid accumulation of inventories between stations and in the final storage that consume cost and area. The new alternative will give possibility to have low stock of products without sacrificing the fulfillment of customer demands. After combining the A_S-Matrix 09 and 02, the idea navigators resulted :

- **Navigator 11 (inverse action)** – according to this navigator, instead of using "push" concept, the production can be controlled by 'pull' concept;

- **Navigator 02 (preliminary action)** – according to this navigator, the production line can be prepared in advanced so that they can adapt the change in customer demand without loss of time;
- **Navigator 18 (mediator)** – according to this navigator, a mediator is required to transfer information from customer / downstream stations to upstream stations and supplier in order to have a quick response without accumulating stocks.

The Result-artifact (Solution) – Just-in-Time

Just-in-Time (JIT) is a production planning system that first introduced by Toyota Motor Company during 1960s. It was created as Toyota's answer for the inventory management and control system that were commonly used by major automotive industries at that time, especially in the United States (s. Womack).

The JIT was said to be inspired by supermarket concept. The supermarket customers may go to the shelves and buy what they want, when they need it. The shelves then are refilled as products are sold. This concept gives an easy system for suppliers to see how many products have been taken and avoid overstocks. Thus, the most important feature of a supermarket system is that stocking is triggered and maintained by actual demand. Toyota has used this concept to create a flexible production system that is characterized as the "pull" system of order-based production.

There are two kinds of known definitions of the Just-in-Time concept:

1. JIT in production (production when it is called for): produce parts just in time; all production activities are performed when the customer requires it.

2. JIT in delivery (stockless inventory): All external supplies are delivered exactly at the moment they are required, at the correct quantity with the correct type. It can be said that the inventory is taking place on the road.

So, the objective of JIT is to produce and deliver right quantity of products with the highest quality at the right time. During this process, it is necessary to gain it with minimizing inventory and lead time, and suppress failure and defects (s. Silver, [175]). In other words, JIT system tries to pursue zero inventories, zero transaction, and zero disturbances.[177]

In connection with "river and rocks" analogy mentioned in fig. 12.66 (s. Silver), as JIT requires low inventory (low level of water, fig. 12.69), the problems (rocks) will be revealed and draw attention to be solved, otherwise the production can be stopped. This concept is in contrast with "push" system implemented by Mass Production which implement high inventory (high level of water) to cover the problems.

JIT system gives several important advantages (see Silver):

- Reduction of inventory space and cost

[177] Vollman, T. E., Berry. W. L., Whybark, D. C. and Jacobs F. R. (2005) *Manufacturing Planning and Control for Supply Chain Management*. – McGrawHill, New York

- Less material movements in/out of storage
- Reduced transactions
- Simplified manufacturing and planning control systems.
- Reduction of production throughput time
- Greater responsiveness to market demands
- Improvement of quality and reduction of quality cost.

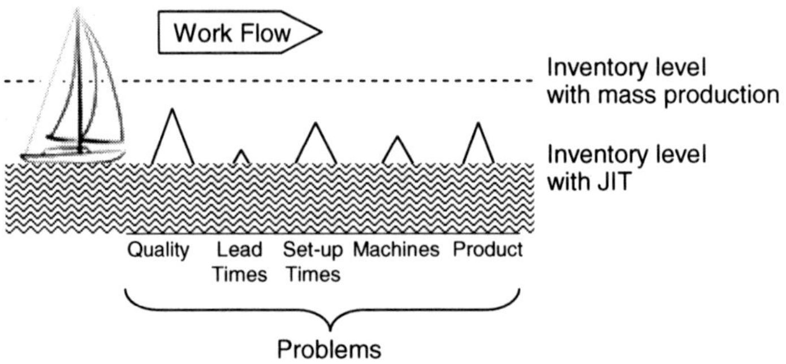

fig. 12.69. The Rocks in the River for JIT (after Tangram Technology, Ltd. 2005).

As JIT is a "pull" system, it uses a unique mechanism of information flow which is called *kanban* (Japanese term for "card"; s. Shingo). When TPS started receiving international attention in the seventies, many people were wrongly understood JIT as the "*kanban* method".

Kanban is actually tools used in JIT to help implement the "pull" principles of TPS. A *kanban* can be a variety of things, most commonly it is a card, but sometimes it is a cart, while other times it is just a marked space. In all cases, its purpose is to facilitate flow, bring about pull, and limit inventory. It is one of the key tools in the battle to reduce overproduction.[178] One important *kanban* rule requires that all materials and products be accompanied by a *kanban* card. Thus, *kanban* connected the material and information flow between working station. The downstream station gives signal for required parts to its upstream center using *kanban* cards.

The *kanban* used in the TPS serves three main functions (s. Shingo):

- identification tag: indicates type of the product;
- job instruction tag: indicates process type, quantity and delivery time required for the product;
- transfer tag: indicates from where and to where the product should be transported.

178 Wilson, L (2010) *How to Implement Lean Manufacturing*. – McGraw-Hill, New York

Basically *kanban* system can be applied in plants involved in repetitive production. *Kanban* system are not applicable in one-of-a-kind production based on infrequent and unpredictable orders.

With development in technology, the manual information system using *kanban* cards now is started to be improved by integrating the computer network system called *e-kanban*. This system gives simpler, more accurate, and more responsive flow of information between customer and suppliers, and within the organizations themselves.[179]

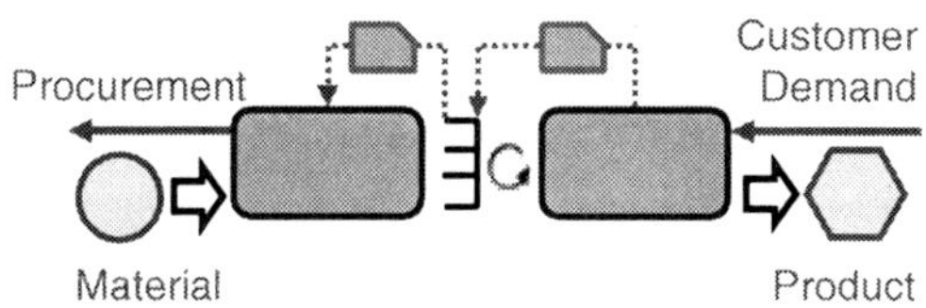

fig. 12.70. Pull Production Control Principle adapted by Lean Production System

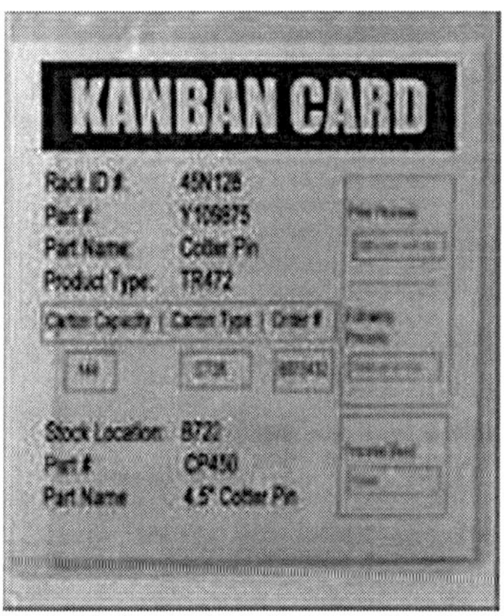

fig. 12.71. A conventional kanban card

fig. 12.72. Kanban card supplemented with barcode for simpler checking

fig. 12.73. Wireless Kanban Replenishment System, a kanban system that eliminates kanban card

179 Drickhamer, D. (2005) *The kanban e-volution.* – Material Handling Management (March), 24-26

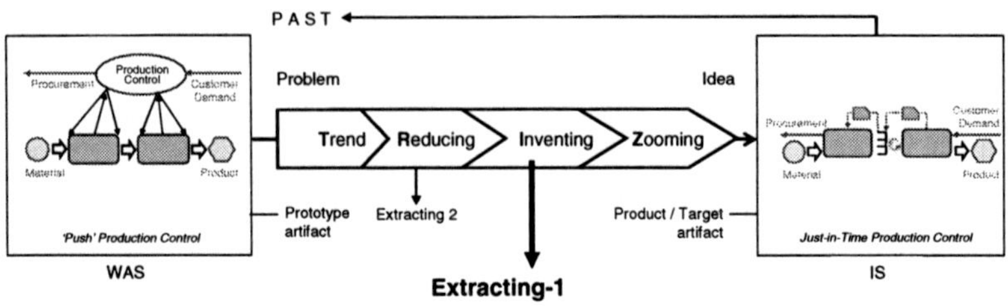

LC*	No.	Navigator	Substantiation for the Extracting
++	02	**Preliminary action**	Prepare the production line in advanced so that they can adapt the change of customer demand without loss of time.
+	05	**Separation**	Separate the "buffer stocks" unneeded by customer by producing only the required ones.
++	11	**Inverse action**	Instead of using "push" concept, the production can be controlled by 'pull' concept.
++	18	**Mediator**	Need a mediator to transfer information from customer / downstream station.
+	36	**Feedback**	Use "feedback" (order) from customer to define the quantity of products need to be produced.

Extracting-2

Standard Contradiction: Mass production with "push" concept made each station only need to think of their own pace of production, but it was not flexible in dealing with fluctuation of customer demand / downstream station condition.

Radical Contradiction: The product stock should be exist to serve customer order, but should not be exist so it would not cause problems (holding cost, etc.)

FIM: X-resource, together with available or modified resources, and without making the object more complex or introducing any negative properties, guarantees attainment of the following **IFR**:

[*production control with less inventories*].

fig. 12.74. Extracting: Just-in-Time

TREND

In a Mass Production System, every station in assembly line had to "push" each product to next station as quickly as possible. This "make-to-stock" concept was not flexible in adapting the customer demand fluctuation and condition of the downstream stations. This caused many semi-finished/finished products waited between stations / in the final storage as inventories. These inventories were considered as problems since they were restrained capital, consumed area and holding cost and they could be worn-out / damaged over time.

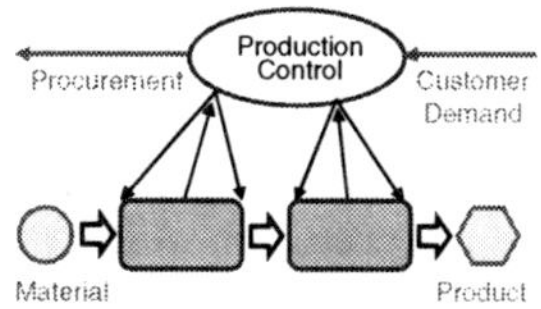

Scheme of 'Push' Production Control Principle

PROBLEM: How could we cope with problems created by "push" control?

REDUCING

FIM: X-resource, together with available or modified resources, and without making the object more complex or introducing any negative properties, guarantees attainment of the following **IFR**:

[*production control with less inventories*].

Standard (Technical) Contradiction

Mass production with 'push' concept

(+) *Each station just needed to think of their pace of production* → 09 → **Factors:** *Ease of manufacture*

(-) *Not flexible in dealing with fluctuation of customer demand/downstream station condition* → 02 → *Universality, adaptability*

Navigators	
05	*Separation*
07	*Dynamization*
11	*Inverse action*
02	*Preliminary action*
18	*Mediator*

Radical (Physical) Contradiction

The product stock ⇒ ***should be exist****, to serve customer order* & ***should not be exist****, so it would not cause problems (holding cost, etc.)*

INVENTING

Group of navigators to model the idea for the "Just-in-Time" concept: **11. Inverse action:** Instead of using "push" concept, the production can be controlled by 'pull' concept; **02. Preliminary action:** Prepare the production line in advanced so that they can adapt the change of customer demand without loss of time; **18. Mediator:** Need a mediator to transfer information from customer/downstream station.

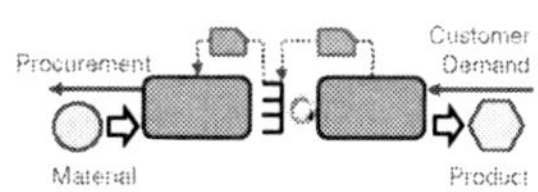

Scheme of 'Pull' Production Control with Just-in-Time Principle

ZOOMING

Have the contradictions been removed? – Yes. – ~~No~~.

Super-effects: Yes. It reduced the inventories.

Negative effects: Needed reliable machines / suppliers to ensure smooth fulfillment of customer demands.

Kanban Card, coordination tool used in Just-in-Time principle

BRIEF DESCRIPTION

Just-in-Time (JIT): JIT is a "pull" concept developed by Toyota as an answer of "push" concept that creates problem of accumulation of inventories. The system was realized by using *kanban* (visual card) that acted as mediator to inform the customer demand from downstream to upstream station. In order to support this concept, Toyota also needed to prepare its line so that the material could flow smoothly from suppliers until the finished goods area.

fig. 12.75. Reinventing: Just-in-Time

13 Selected Examples

13.1 The Super-Task of Learning and Self-Training in EASyTRIZ

And in conclusion of this course, we will deal with a number of examples whose descriptions are prepared with the use of TRIZ instruments discussed in the book.

These examples need to be preceded by a small essay on extremely complex aspects of creation which have not been satisfactorily explained by TRIZ or other sciences, such as psychology, philosophy, theory of art (any art!), theory of creation (in its currently known variants), and pedagogy.

We will now focus on perception of *beauty of the idea* and *miracle of invention.*

I cannot remain indifferent when I encounter thousands of large and small artifacts, both well known and obscure, whether when visiting world museums or while browsing web sites dedicated to the history of inventions or modern novelties. I admire any, even the least significant, successful solutions in everything: in household items, in construction, in transportation vehicles, in military systems – generally speaking, in any branch of engineering, and naturally in art – any art in the broadest meaning of the word.

At the same time, I keep asking myself and all my acquaintances the same question which may seem – and truly is – rather strange. I ask this: you see a thing, so tell me what is it that you like in that thing, why do you perceive it as something *beautiful*?

The answer is usually a variation of the following: well, what do you mean, these are *beautiful* windows, plants, shapes, colors, etc. *ad nauseam.* Those who give the answer do not understand the essence of the question! So I press on: *what is it* that makes this plant *beautiful*? *What is it* that makes these windows *beautiful*? *What is it* that makes this invention *beautiful*?

And when I proffer this simple clarification, I often do not get any answer at all. It turns out that it is difficult to answer *that* question! It is difficult to express in words the *feeling of beauty*, the *feeling of wonder*! Maybe it takes preparation, maybe we even need to be trained to *understand it* and *express it*!

So *what* and *how* can be said about the *beauty of examples* provided in the previous sections of this book, and the *beauty of examples* that I will present in this conclusion?

Because all these examples are praiseworthy, and the right thing to do would be to express – without fanfare and without reference to specific points in space and time – our respect to all those who made millions of large and small *discoveries of things that were new* along the evolutionary path of our civilization, both those whose names are known and those whose names have been lost in the course of time! Virtually all those inventions were admirable – however simple they may have been! All those inventions are reflections of the inner light produced by talent. In the overwhelming majority of cases, they are testimonies of solid knowledge of the subject matter. Sometimes they are but lucky shots in the dark, sometimes they are the well-deserved fruits of courage and willpower shown by pioneers pushing ahead despite terrible privations. They are bursting to tell us thousands upon thousands of breathtaking life stories. And all these stories are as interesting as they are undisclosed, undiscovered, unpreserved!

Huge treasures produced by our civilization remain things unto themselves and cannot be inherited or reused to beget new creation.

We still have to learn both to invent and see the beauty of our ideas.

So let us try to explain things that bear explanation, and mark – to the extent possible – those that magically attract us by their cryptic beauty.

And let us start with the picture (fig. 13.1) that we already saw in the book called *ABC TRIZ.*

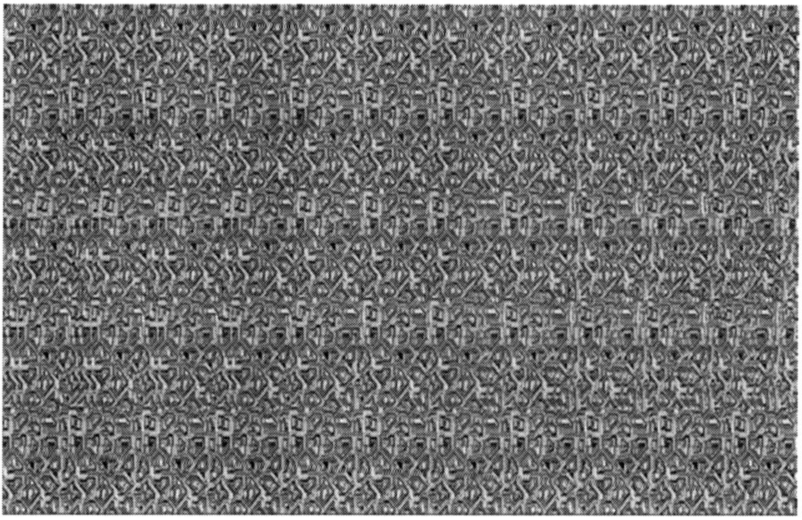

fig. 13.1. TRIZ is *Magic Vision*!

Example 13.0. ***Magic Eye***. Here is how you should look at this picture: bring it close to your face until the image becomes blurred, then start to gradually move it away looking *beyond, or through, the sheet* rather than *on its surface*. And at some point you will see a *miraculous transformation* of the image into a 3D shape!

The secret is to look *deep inside, into the nature of things, rather than on their surface*! Let us repeat the words of the great Goethe (see footnote [144]): *to place oneself at the level of the objects is to learn; to take objects in their depth is to invent.*

TRIZ works the same way!!!

When you perceive the three-dimensional depth of the image hidden in the picture, you are sure to experience the *miracle of discovering*, the *miracle of understanding*!

It may happen that your first attempts to penetrate the pseudo-3D world of the Magic Eye picture will be unsuccessful. However, the pleasure derived from watching such pictures will increase as you hone your rapid self-adjustment skills.

The same happens as you study TRIZ. First you will not perceive any breathtaking miracles behind the methods you will be learning, while your assignments will appear either too simple and uninteresting or, conversely, completely incomprehensible – just like that picture if viewed hastily and superficially.

But after you have independently completed at least several dozen assignments, after you have learnt to bravely tackle any challenging problem, *you will become initiated into the mystery and craft of TRIZ*!

And in this example we are interested not in the optical illusion, but in the analogies and metaphors that it represents. And it is important to understand the conclusion: it is not enough to know the method, you need to feel *the strength and allure of the method and the beauty of the result*!

It is important to feel that you have the power to rise above happenstance and deal with dilemmas, that you have the ability to overcome "insurmountable" difficulties, to persist and prevail. And TRIZ is all these things, too. Read an incredibly interesting story told in a book called *How to Become a Genius*[180]. It is about people who create things that are truly new, who tirelessly reiterate and reinforce eternal values. It is about Konstantin Tsiolkovsky and Friedrich Zander, about Roberto Bartini and Rostislav Alekseyev, about Andrey Tupolev and, of course, about tens of thousands of lesser known people who by right could and should have been called the "true creators of human civilization". But it is also about Richard Branson and Dean Kamen, about Burt Rutan and James Dyson – after all, it is not overnight that they became millionaires and billionaires! And this story is still not finished, because it is also about Anatoly Yunitsky whose ideas are discussed in this and other books that I wrote, and about all those who are still waiting to unveil their discoveries.

Last but not least, it is about beauty. If the feeling of beauty is nothing but emotional appreciation of harmony, efficiency, unexpectedness, then appreciation and perception are based on… information! And people who do not understand harmony and expediency cannot appreciate unexpectedness and beauty of ideas!

According to V.P. Simonov (see footnote [52]), disclosure, discovery of beauty is a function of the super-consciousness, *a property uniquely possessed by the creative brain*. The "stupid brain" is devoid of that property. Therefore, to assure further evolution of TRIZ, we will have to learn to understand beauty. Possibly, that understanding will become our "super-task", the main Ideal Final Result which will inform our creative inquiry, our ideal construction.

---

[180] G. S. Altshuller, I. M. Vertkin (1994) *How to Become a Genius. Life Strategy of a Creative Personality*. – Minsk, Belarus

Experience shows that *understanding beauty* is an acquired talent, while creating beauty requires a native talent of an artist and inventor, better still – *artistic inventor* or *inventive artist*, because inventing is both a science and an art. For our purposes, it would suffice to incorporate into the TRIZ framework, to the extent possible, the models and methods designed for mass-scale training in *understanding beauty*. I believe that it is quite possible, but we need facts and arguments, we need a diligently tilled empirical field yielding plenty of new workable concepts and patterns, instruments and outcomes.

For the time being, let us list properties pointing to the existence of a connection between creation and beauty, between things perceived at the conscious (*information content*) and unconscious (*feeling of beauty and miracle of inventing ideas*) level.

First of all, human mentality exists on many levels and "in many spaces" and, in the simplest terms, can be divided into sub-consciousness, consciousness, and super-consciousness. "Each person has an individual structure... [consisting of] three levels of mentality. All those levels are parts of such person's *inventory* assuring satisfaction of needs and their elevation in the process of historical development of mankind." [181] In this context, "*creativity requires super-consciousness.* Super-consciousness engages consciousness which, essentially, is always busy beneficiating, explicating, expanding, developing and verifying *whatever is prompted by intuition...* Consciousness assimilates connections between intuitive guesses and objective reality, and evolves knowledge and skills – experiential fruits comprising the arsenal of sub-consciousness. Super-consciousness directs inquiries, intensifies operation of consciousness (reason, logic), and mobilizes *automated skills of sub-consciousness.*

"Super-consciousness draws material for its recombinatory pursuits both in conscious experiences and in the reserves accumulated by sub-consciousness. Nevertheless, super-consciousness contains something which is truly "super", i.e. larger than consciousness *per se*. This "super" element is represented by *completely new information which cannot be directly derived from previous experiences.*

"To list everything that comprises, or may comprise, man's mental inventory, everything that transforms him into a human being is impossible, but there are *universal inventory items*, such as imagination, fantasy and logic. To have imagination is, essentially, to have *intuition* or *super-consciousness.* Since his early years, the child is enriching his consciousness... by imitating adults. He acquires super-consciousness by playing games... Game-play also develops the will – one more equally important item of human inventory.

"Game-play enhances super-consciousness because the child is required to demonstrate resourcefulness and ability to promptly make correct decisions in problem situations and boldly change existing rules. Because of its very nature, game-play dictates the need to know and follow the existing norms, and be ready to challenge

[181] Here and below, I have used ideas developed by P.V. Simonov, P.M. Ershov (1984) *Temperament, Character, Personality.* – Moscow, Science Publ. (in Russian; *Italics* are mine – M.O.)

and breach those norms. This is the technique (if we may use this term) used by super-consciousness. During game-play it is performing *"technical exercises"*, getting the child ready to use intuition when he must, for the first time in his life, create something new or make a non-trivial decision. This instrument is acquired with pleasure – therefore the risk that it may displace other instruments. That is why *productivity, effectiveness and strength of super-consciousness, one of the most important items of the human mental inventory, are determined not by super-consciousness itself, but rather by how firmly it relies on consciousness, on knowledge acquired consciously and planfully.*"

To teach TRIZ fundamentals, we need special models and methods. I am positive that the methods I call "extracting" and "reinventing" display properties typical both for game-play and inquiry and, therefore, they are aimed directly at forming super-consciousness, at nurturing and disciplining creative intuition.

There exists unity of ideas, in terms of prevailing attitudes and practical methods, between Modern TRIZ on the one hand and *theater creativity* according to K.S. Stanislavsky and *engineering creativity* according to E.I. Reghirer on the other hand.

"Stanislavsky sought to find *conscious technical ways to creative super-consciousness* of the actor… The "art of feeling" (use of templates, masks – M.O.) was replaced by the *physical action method* which laid the foundation for the science of theatrical art and created prerequisites for science-based theatrical education. According to Stanislavsky, while creating a scenic image the actor must unwind … the chain in the opposite direction: *from action to its origins, because only action can be directly controlled by consciousness*" (see footnote [181]).

We witness the same process in MTRIZ reinventing (section *6 Reinventing*) – during the reinventing exercise artifact analysis proceeds strictly from the RESULT to the PROTOTYPE in accordance with the following algorithm:

Meta-Algorithm of Re-inventing Z-I-R-T

Stage 1: Review the properties and construction of the result-artifact under analysis, and compare them to the properties and construction of the selected prototype-artifact – this corresponds to the *Zooming* stage in "direct" MAI T-R-I-Z;

Stage 2: Extract transformation models objectively participating and realized in the result-artifact – this corresponds to the ***Inventing*** stage in "direct" MAI T-R-I-Z;

Stage 3: Extract contradictions – this corresponds to the ***Reducing*** stage in "direct" MAI T-R-I-Z;

Stage 4: Describe the original problem situation for the prototype-artifact – this corresponds to the ***Trend*** stage in "direct" MAI T-R-I-Z.

In fact, all reinventing stages are performed in reverse order compared to MAI T-R-I-Z; accordingly, we can define and call the Meta-Algorithm of Reinventing either "MARI T-R-I-Z" or "MAI Z-I-R-T". The opposition of MAI T-R-I-Z and MARI T-R-I-Z can be schematically presented as follows (fig. 13.2).

The same goes for the *"conscious technical methods"* proposed by E.I. Reghirer[182] for training researchers and inventors: "Training of talents still remains individual. Musicians, sculptors, directors usually adopt creative experiences shared by universally recognized accomplished creators. The use of this training format is unavoidable because of the small number of trainees... As a result, experience is transferred from generation to generation in the same way as *folklore*: this is the most ancient form of education... In modern times it is only natural to try and find a more productive method involving proactive transfer of learning."

For almost 50 years, TRIZ has been disseminated *the "folklore" way*. Efficiency of this way for mass-scale education is highly questionable. First, not every "creator" is a good educator. Second, "universally recognized creators" are few and far between. Third, there are as many TRIZ versions as there are teachers (there are no standard models). Fourth, TRIZ school was never finalized; accordingly, there are no "new" trainers, especially those capable of implementing a "mass-scale" training program.

As a result, to organize such training, Modern TRIZ introduces a different technology which can be described as *efficient, standardized, and remotely based.* E.I. Reghirer points out that it is possible to train researchers by... *reinventing*[183] inventions made by other researchers! Idea: organization of training by "repetition of inquiry" *with known final result*!

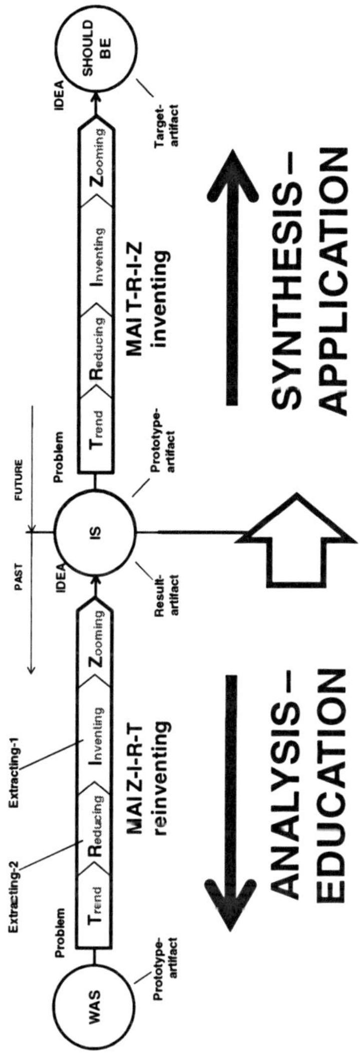

fig. 13.2. "Direct" MAI T-R-I-Z: Invention = Synthesis-Application; "Reverse" MAI Z-I-R-T: Reinventing = Analysis-Education

[182] E.I. Reghirer (1969) *Development of Researcher Abilities.* – Moscow, Science Publishers, 230 pp. (in Russian; *Italics* are mostly mine – M.O.)

[183] Here I am extending the term *reinventing* to the method developed by E.I. Reghirer, as in the English language this word can mean both "reinvent" and "rediscover"

Then "activity… of the *repeat researcher* is in no way different from activity of the *primary researcher.* To use that circumstance, it is only necessary to ensure that the trainee does not know in advance the solution found by others – moreover, the trainee should not know whether such solution actually exists, or that it is possible under the given conditions.

"This principle – which I call *discovery of the known* – can be used as the basis of researcher training, and such training can commence long before the young specialist can be directly involved in real-life research work.

"Of course, only those who are particularly gifted and tenacious can master … the *autodidact method* – very few people can learn to swim when they are thrown into the open sea – many would perish as a result of such "training".

"To … apply the *discovery of the known* method, all you have to do is … divide the process of inquiry into its typical segments (compare this to MAI T-R-I-Z! – M.O.), and thereby create an opportunity to exercise each segment … without completing the entire cycle – which must be done when ordinary, unproductively cumbersome methods are used to teach methods of inquiry."

This stage-by-stage approach is implemented when extracting and reinventing on the basis of the Meta-Algorithm of Invention T-R-I-Z are used for assisted and independent learning of Modern TRIZ idea generation techniques. Discussion of certain objective differences from the approach proposed by E.I. Reghirer lies beyond the scope of this book.

Before ending this section, I would like to share another idea which deals with learning TRIZ fundamentals.

G.S. Altshuller pointed out that TRIZ *changes human thinking.* But that is tantamount to changing the entire personality, i.e. the man himself. Changing his values, his motivations. Clearly, here we deal with a phenomenon[181] "that K.S. Stanislavsky called a *super-task.* In the same way a plant grows out of the seed, wrote Stanislavsky, a novel grows out of individual thoughts and feelings experienced by the writer… Let us … call this main, chief, comprehensive objective which attracts to itself all tasks without exception … the *super-task of the novel. True art must teach us how to consciously incite in ourselves our unconscious creative nature and make it engage in super-conscious organic creation.*"

The notion of *beauty* is undoubtedly one of the fundamental ideals of superconsciousness. TRIZ can and must teach its proponents to see the beauty of the world around them and to improve that world – and, by the same token, to keep fighting and avoid the trap of narcissism.

I would like to conclude with a wonderful prediction[184] by one of the most *talented creators* of TRIZ who refused to be restricted by TRIZ boundaries (in the early and mid 1970-es!) and instead expanded the prescient space to universal scientific systems – and who most unfortunately had to terminate his TRIZ inquiry too early (for a good reason).

---

[184] R.P. Povileyko *Engineering Creative Art.* – Moscow: Znanie (Knowledge) Publishers, *Novelties in Life, Science and Techology. Technology* Series, 1977, 4. – 64 pp. (in Russian)

"Unwritten ethics of the strictly academic approach to publication of research findings have found their way to the modern technical literature: publications consist of a brief description of the research, its results, workable schemes and constructions developed by the researcher, his main conclusions and recommendations.

Meanwhile the thinking behind these conclusions and recommendations, the researcher's guesses, doubts, trials, errors, alternative variants, original discoveries – all these things are conspicuously absent... In scientific publications the analytical section is compressed into a summary detailing approved research methods, *while in engineering publications this section is not present altogether.*" And things have not changed since then – see Section 11. Energy of Progress.

"And now let us place on the same shelf in our memories a Nartov copying lathe and a modern numerical control lathe. Yesterday machine engineering had one set of possibilities and requirements, today these things have completely changed. Yesterday the Nartov lathe was perfection, today the modern NC lathe is by far the better of the two.

"Both artifacts – in their inimitable ways – assure that their form and content are in harmony. But this harmony is further emphasized and reinforced by the sparkling talent of their creators. Their constructions are impeccable – subject to the need to streamline production costs taking into consideration the psychological, physiological and creative requirements posed by the fact that they are to be attended by man who, together with the machine, participates in the process of production – and all this leads to the emergence of a machine which can justly be called *beautiful* at the given level of historical development.

"It is not by accident then that *after thorough, detailed, thoughtful familiarization* with such works of technical art as the Nartov lather or the NC lather we are unwittingly overwhelmed by a *feeling of endless admiration inspired by the mastery, talent and commitment of those who created these machines*, and then we call such machines *wonderful, magnificent, beautiful.*

"That is why I am positive that in the future museum of human culture, in the *Hall of Beautiful Things*, there will be a place for Leeuwenhoek lenses and an electronic microscope, a stool adorned with *Khokhloma* wood painting and an automobile, a Stradivarius violin and a laser plant, a lathe built by the talented Russian inventor Andrey Nartov and a modern precision numeric control lathe[185]."

13.2 Inventions. People. Society

To the extent possible, some of the examples discussed below are accompanied by brief biographical accounts of the authors of the relevant inventions. I resolved to provide such accounts because I could not present reinventings of the artifacts created by those famous people and say nothing about themselves. I hope those cameos will shed additional light on the *wonderful creators'* discoveries and their talent.

---

[185] R.P. Povileyko (1964) *Technics and Esthetics.* – Novosibirsk, Novosibirsk Electrical Engineering Institute, 62 pp. (in Russian)

13.2.1 "Checkered" Ideas

In the sections to follow, I am going to present examples of reinventing covering a broad variety of artifacts and situations, some extremely simple and well-known, others more obscure and complex. In any event, I tried to simplify the descriptions to illustrate the basic idea of each invention, leaving out non-essential technical details, because the purpose of these examples is to serve as benchmarks that you can use to design your own examples.

May your reading be interesting – and I wish you success with your own reinventing examples!

I would like to start with a simple "craft-level" example – always keeping in mind that in this case the craftsman was not without talent! And the solution is beautiful!

Example 13.1. Chimney Stack "Stopper"

This wonderfully efficient invention may be not as simple as we would want it to be, but so far nobody has come up with a better solution!

PROBLEM. A large-diameter stack (say, 1 or 2 meters) cannot be left without some cover or, to put it simply, "stopper".

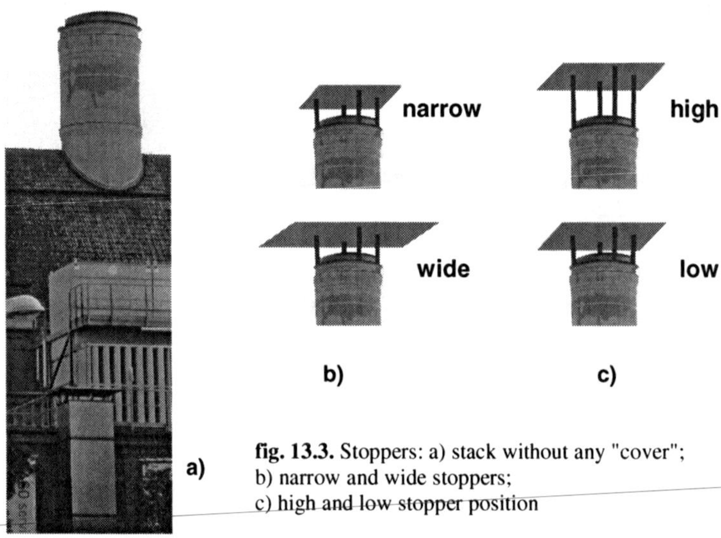

fig. 13.3. Stoppers: a) stack without any "cover"; b) narrow and wide stoppers; c) high and low stopper position

Standard contradiction for the initial problem situation according to fig. 13.3a: the stack must have a stopper to protect it from precipitation (rain, hail, snow), but a *large stopper* can be thrown off by strong wind and, therefore, will have *low reliability*.

Radical contradiction for the construction option shown in fig. 13.3b: the stopper must be *narrow* to ensure free movement of the smoke exiting the stack, and it must be *broad* to reliably protect the stack from precipitation.

Radical contradiction for the construction option shown in fig. 13.3c: the stopper must be (installed) *high* to ensure free movement of the smoke exiting the stack, and it must be (installed) *low* to reliably protect the stack from precipitation.

The **OZ** consists of the mouth of the stack and the stopper.

General contradiction: how can we make a fail-safe stopper preventing precipitation from getting inside the stack?

SOLUTION. A wonderful solution (fig. 13.4) was devised by an unknown but very talented master. This solution combines several crafts and could have had several "fathers": a blacksmith and a mechanic, a maker of armor (helmets) and a maker of moving mechanisms (carts, windmills, weather vanes, well winches, forged grilles and industrial lathes).

fig. 13.4. "Helmets" on factory chimney stacks!

Complete reinventing according to Meta-Algorithm Z-I-R-T

ZOOMING

The helmet consists of the "headpiece" (1), the weather vane (2), the spindle (3) on which the entire construction rotates, and the support (4) fitted inside the mouth of the stack. Regardless of the direction of the wind, the helmet turns so that the backside of the "headpiece" "faces" the wind and reliably protects the mouth of the stack from any precipitation!

The stronger the wind, the more securely the helmet places itself in the best possible position!

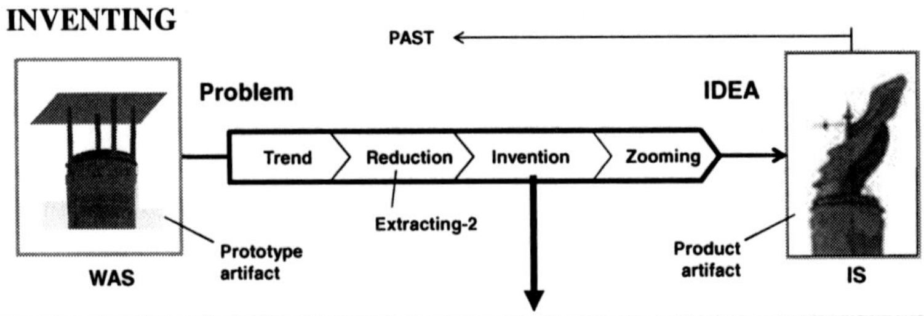

EXTRACTING-1		
07	**Dynamization** - make an object moveable that is otherwise fixed	The helmet is rotated by the wind, while the "old" stopper was immobile.
03	**Segmentation** - disassemble an object into individual parts	The previous "monolithic" stopper is divided into several modified components: the "headpiece" to protect the mouth of the stack from precipitation, the "weather vane" to enable rotation of the "headpiece" by the wind, etc.
04	**Replacement of mechanical matter** - replacement of static fields with dynamic ones, from unstructured fields to fields with a specific structure	A new mobile construction has been implemented instead of the old immobile construction by introducing new energy transfer routes and new part movement trajectories.
10	**Copying** - replace an object or a system of objects with copies	The weather vane idea has been copied (by analogy) to harness the wind as the driving force of the entire construction.
12	**Local property** - every object should exist under conditions that correspond best to its functions	This navigator (together with navigator 03) "works" as follows: all new parts work under conditions that best correspond to their functions, i.e. the "headpiece" protects, the weather vane rotates, the spindle holds the construction together and sets the axis of rotation, the support firmly fixes the construction in the mouth of the stack.
21	**Transform damage into use** - use damaging factors, especially damaging influences from the environment to achieve a useful effect	The wind BY ITSELF puts the "headpiece" into the position where it most efficiently protects the stack from precipitation!
28	**Previously installed cushion** - increase the relatively low security of an object with safety measures in advance	The device must in advance be prepared to protect the mouth of the stack from any changes in the direction and strength of the wind, and from any precipitation – snow, rain, hail, possibly sand or other large particles.
34	**Matryoshka** - an object runs through a hollow space in another object	The spindle is inserted into the "headpiece" and the support, and the support is inserted into, and fixed in, the stack.
35	**Unite** - unite similar objects or objects for neighboring operations	All new parts that were created by segmenting the simple original "monolithic" construction (navigator 03) are now united into a new monolithic mobile construction.

fig. 13.5. Extracting-1 for the "helmet"
("Direct" sequence of MAI Z-I-R-T stages is retained)

Clearly, such sets of models (fig. 13.5) are usually yielded by the MITO method. In simplified modeling, it is also possible to supplement the models from the selected cluster with requisite dominant models by applying the BICO method to just one standard contradiction.

Such modeling will be totally objective, and it will have nothing to do with "adjusting" real life to suit a preconceived explanation.

REDUCING

Let us perform Extracting-2, and develop several formal variants ("formulas") for the informal contradictions presented above.

Standard contradictions:

SC1: 13 External damaging factors **VS** 02 Universality, adaptability = 01, 21, 28, 31

Here the problem factor is reflective of the insufficient "adaptability" of the existing stopper to changing exposure to external damaging factors.

SC2: 02 Universality, adaptability **VS** 21 Shape = 03, 07, 27, 32

SC3: 02 Universality, adaptability **VS** 07 Complexity of construction = 04, 07, 14, 27

SC4: 02 Universality, adaptability **VS** 10 Ease of use = 03, 07, 15, 16

Here problem factors are varied with respect to the original idea that if we modify the existing stopper so as to increase its "adaptability" (trend), it will create problems with shape (if the old shape is retained, the problem has no solution!), complexity of construction (increased complexity is likely to be the "fee" for getting rid of the problem), and ease of operation (for example, some component may have to be controlled by an operator – and we definitely do not need that).

Maxi-FIM: the stopper BY ITSELF always provides the mouth of the stack with fail-safe protection from precipitation, even in strong winds!

Using the MITO method, we can easily "rank" recommended transformation models selected from the A-matrix for all formal standard contradictions described above: 07^3, 03^2, 27^2, 01, 04, 14, 15, 16, 21, 28, 31, 32.

TREND

Known stack covers do not provide reliable protection from precipitation, especially in strong winds. The simplest cover ("stoppers") is made from a piece of metal sheet fixed on supports above the upper edge of the stack.

How can we make a fail-safe stopper preventing precipitation from getting inside the stack?

And to finalize the results obtained, let us present the inventing process in the Meta-Algorithm of Invention (MAI) format (fig. 13.6).

ATTENTION !

It should be noted that several important navigators present in the Extracting-1 table (fig. 13.5) and absent from the resulting MITO list in fig. 13.6 were included as supplements based on the findings of objective examination of models at the extracting stage! Those navigators were not "found" by adjusting and selecting "required" factors to arrive at "required" A-matrix cells with "required" navigators!

TREND

A large-diameter stack (say, 1 or 2 meters) cannot be left without some cover or, to put it simply, "stopper". Known stack covers do not provide reliable protection from precipitation, especially in strong winds. The simplest cover ("stopper") is made from a piece of metal sheet fixed on supports above the upper edge of the stack. How can we make a fail-safe "stopper" preventing precipitation from getting inside the stack?

REDUCING

Maxi-FIM: The operating zone by itself ensures attainment of the **IFR**
[*fail-safe protection of the stack from precipitation, even in strong winds!*]

Standard Contradiction 1

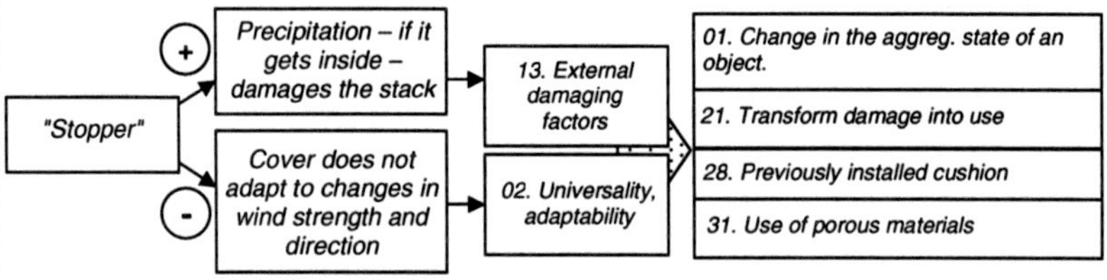

SC2: 02 Universality, adaptability VS 21 Shape = 03, 07, 27, 32

SC3: 02 Universality, adaptability VS 07 Complexity of construction = 04, 07, 14, 27

SC4: 02 Universality, adaptability VS 10 Ease of use = 03, 07, 15, 16

MITO Ranging: 07^3, 03^2, 27^2, 01, 04, 14, 15, 16, 21, 28, 31, 32

INVENTING

Dominating models: 07, 03, 04, 21 and 28.

Additional dominating navigators: 10, 12, 34, 35!

Key idea: *Join "copies" of two systems: helmet and wind vane! Make the "stopper" in the form of a "hard hat" safely covering the edge of the stack, and top the "hard hat" with a wind vane. Under the influence of the wind, the vane will be turning the "hard hat" so that it "faces" precipitation with its closed backside!*

ZOOMING

Have the contradictions been removed? – Yes. – ~~No~~.

Super-effects: *self-optimization of protection!*

Negative effects: *complexity of construction*

BRIEF DESCRIPTION

To protect the edge of the stack from precipitation, the "stopper" is made in the form of a "helmet + weather vane" which is turned by the wind so that the closed side of the helmet "looks" in the direction of the wind.

Objectively present transformation models: 07, 03, 04, 10, 12, 21, 28, 34, 35.

fig. 13.6. Reinventing of the "helmet-stopper" for a chimney stack

Example 13.2. Rubber-Blade Windshield Wiper

On a winter day in the end of 1902 Mary Anderson, farmer, viticulturalist and real estate agent, who had come to New York on business, had to use a trolley car. She noticed that the motorman had to stop the car from time to time so he could get out and clean the windshield of falling sleet. On her return to Alabama, where she lived and worked together with her mother and sister, Mary Anderson designed a windshield wiper which could be operated manually from inside the vehicle! In 1903 she patented that solution for 17 years. When her patent expired… all automotive companies started to manufacture such devices, while the author of the invention received nothing.

TREND

In the late 19-th – early 20th century, if the person driving a motorcar (tramway) wanted to clean the windshield, he had to stop the vehicle, get out and do the cleaning by hand. What can we do to eliminate this problem?

REDUCING

FIM: X-resource, together with available or modified resources, and without making the object more complex or introducing any negative properties, guarantees attainment of the following **IFR**:

[*the windshield can be cleaned from within the vehicle*].

Standard (Technical) Contradiction

Wiping precipitation off the windshield

(+) *Vehicle steering* → *35 Brightness of lighting*

(-) *Inconvenience and need to stop the vehicle from time to time!* → *10 Ease of use*

Factors / **Navigators**

04 Replacement of mechanical matter

08 Periodic action

10 Copying

18 Mediator

Radical (Physical) Contradiction

Windshield cleaning ⇨ *must be to make it possible to see through the windshield* & *must not be so that the operator does not have to stop and exit the vehicle*

INVENTING

The following models are used: 04, 08 and 10, as well as 18.

Solution: *Pursuant to US Patent No 743801 dated November 10, 1903, issued by Mary Anderson, it was suggested that a holder with a rubber wiper be installed on the windshield, and that the said holder be actuated from within the vehicle via a lever drive.*

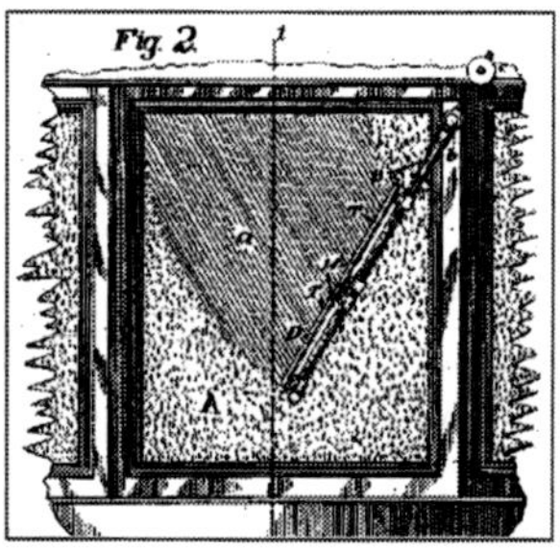

ZOOMING

Have the contradictions been removed? – Yes. – ~~No~~.

Super-effects: *It is possible to automate the drive.*

Negative effects: -

BRIEF DESCRIPTION

To increase the ease of keeping the windshield clean of snow and other precipitation, a holder with a rubber wiper is installed on the windshield. From time to time, the holder is actuated manually from within the vehicle through a system of levers.

Navigators used: 04, 08, 10 and 18.

fig. 13.7. Reinventing Mary Anderson's windshield wiper (source: www.dpma.de)

Example 13.3. Safe Safety Belts

In 1955 the American Ford became the first automotive company in the world to equip its cars with safety belts. 12 years later the Swedish Volvo followed suit. Then safety belts became a staple feature used by all car-makers in the world. By the end of 2009, three companies had announced plans to fit their cars with "inflatable" safety belts: the Japanese Honda (1997), the German Mercedes-Benz (2009), and the American Ford (2009). Still, Ford intends to become the first company to introduce such belts on a mass scale.

Extracting

Prototype	No.	Navigator	Implementation	Target Artifact
	**01**	01 Change in the aggregate state of the object	The belt has an elastic faceplate.	
	02	14 Use of pneumatic and hydraulic constructions	The elastic faceplate is inflatable.	
	03	18 Mediator	The inflatable faceplate acts as a mitigating mediator between the rigid belt and the human body.	

Reinventing

TREND *Traditional safety belts must have a high response speed and be rigid to reliably hold the human body from moving forward by inertia. However, it is safety belts that often inflict serious injuries when they are actuated upon impact, especially by strangling the passenger. What can we do?*

REDUCING **FIM:** X-resource, together with available or modified resources, and without making the object more complex or introducing any negative properties, guarantees attainment of the following **IFR**: [*elimination of risk of injury upon activation*].

Radical Contradiction (RC):

Safety belt ► rigid (to firmly hold the human body upon impact) **VS** soft (to reduce risk of injury)

Formula and solution (in general terms) of the standard contradiction (SC):

22 Speed VS 14 Internal damaging factors = 01, 05, 18, 33

INVENTING Dominant models that can be used to resolve the SC: *01 Change in the aggregate state of the object, 18 Mediator* and, additionally, *14 Use of pneumatic or hydraulic constructions.* The following changes have been made in line with these navigators: an inflatable mediator pad has been attached to the belt (model 18); compressed air is used to inflate the pad (model 14); the pad is made of elastic material, and its size increases as it is filled with compressed air (model 01).

Dominant resources used to resolve the RC: functional resource (non-injurious contact), spatial resource (new 3D shape), and material resource (use of an elastic pad and compressed air injected into the pad).

Model *05 Separation* is also efficiently used to resolve the RC, as in certain critical places the belt is given properties which are both necessary, but lie in direct opposition vis-à-vis each other. Thus the belt is *soft* where it comes into direct contact with the human body **AND** it is *rigid* elsewhere (efficient use of structural and spatial resources).

In line with model *33 Quick jump*, after the belt is actuated, the pad is filled before it is pulled taut! Resources used: temporal, material, energetical.

ZOOMING The contradictions have been removed. Super-effect: under ordinary conditions, the pad is more ergonomic and pleasant to use than the rigid belt!

"Air Belt", Honda, 1997

BRIEF DESCRIPTION

To reduce the traumatic effect of the safety belt, it is suggested that it be fitted with an inflatable pad.
Dominant navigators: 01, 14, 16 and 18.

fig. 13.8. Reinventing inflatable safety belts (source: www.membrana.ru)

Example 13.4. In 2001 *The Wall Street Journal* called the *Weego* child carrier one of the greatest inventions of the century!

Extracting

Prototype	No.	Navigator	Implementation	Target Artifact
	01	04 Replacement of mechanical matter	The familiar backpack structure has been modified.	
	02	07 Dynamization	Dynamized elements have been introduced.	
	03	12 Local property	Holes for the child's head, arms and legs; special straps and supports.	

Reinventing

TREND *In the early 1960-es, Ann and Mike Moore, a young married couple from America, worked in Africa as Peace Corps volunteers. They saw work women inTogo carrying their children on their backs, wrapping them up into large shawls or pieces of cloth. As a result, their arms remained free and could be used for work. Wouldn't it be nice to transplant that idea into America or Europe? But people raised in the western cultural tradition do not carry their babies that way. They do carry heavy things, though, for example, using backpacks. What can we do?*

REDUCING FIM: X-resource, together with available or modified resources, and without making the object more complex or introducing any negative properties, guarantees attainment of the following **IFR**: [*convenient hands-free baby-carrying device*].

Apparently, the functional idea has been borrowed from Africa, and is based on the use of a shawl. But the subsequent technical realization clearly originates from an ordinary tourist backpack! Accordingly, we select the backpack as our prototype.

Radical Contradiction (RC):

Backpack ► closed (to carry things) **VS** opened (in certain places, to carry the baby)

Formula and solution (in general terms) of the standard contradiction (SC):

10 Ease of use **VS** 21 Shape = 04, 07, 14, 15

INVENTING Dominant resources used to resolve the RC: functional resource (extension of backpack functions) and spatial resource (shape: manufacture of openings and cover flaps).

Dominant models used to resolve the SC: *4 Replacement of mechanical matter* and *07 Dynamization.* Additionally: *10 Copying, 11 Inverse action, 12 Local property* and *19 Transition into another dimension.* Changes: adjusting the shape to match the required functions (models 04 and 14); copying of functions (model 10); introduction of flaps that can be opened and closed (models 07 and 12); the device can be carried both on the back and in the front (models 11 and 19).

ZOOMING Have the contradictions have been removed? – *Yes!*

Super-effects: *1) the device can be carried both on the back (fig. 1-3) and in the front (fig. 4), 2) men will also carry the baby (they wouldn't if they had to use a shawl); 3) the device can be used to carry twins (fig. 5).*

In 1969 Ann Moore patented a child carrier. Navigators present: 04, 07, 10, 11, 12 and 19.

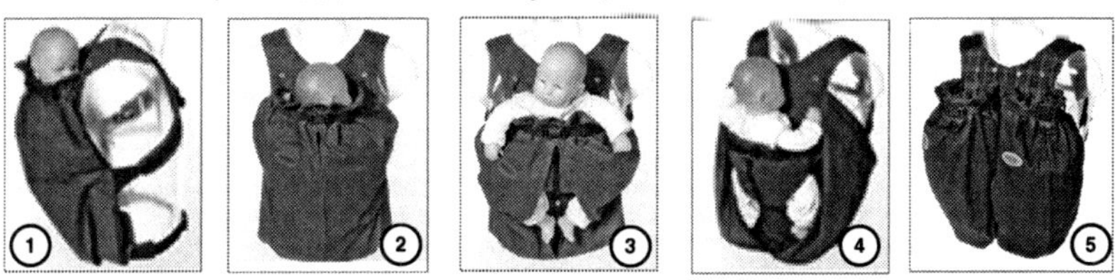

When in 1964 Ann had her firstborn, she remembered the way African women carried their children. She and her mother came up with the design, and the neighbors helped them to sew the new special backpack. Little by little, it became increasingly convenient and gained popularity. In the 1970-es, the Moore couple decided to start a company called *Weego* (from "we go"). Less than 10 years later, their child carriers became famous first in America and then throughout the world, and the company's annual turnover was measured in millions of dollars.

fig. 13.9. Reinventing Weego

Example 13.5. Duckbill? – Yes, "Duckbill". And We Are Proud of That!

The "Duckbill" idea was born in 1999, and by 2002 the company *New Impulse* opened in Zelenograd, a small city in Moscow environs, the first two retail shops offering *preliminary reservation* of goods. The operation of the network is based on several inventive ideas. The key ideas include the following: 1) organization of small discount shops where the buyers can view and order goods (with an option to purchase some goods "on the spot"); 2) creation of a *transportation network* where the goods are carried from the *central storage facility* to shops or directly to the buyer's homes; 3) creation of an *information network* through which orders literally flow into Universal Batching Terminals (UBTs or, in Russian, **UTK**s – which, together with the Russian word "**nos**it'" ("carry"), is evocative of "**utk**o**nos**" ("duckbill")), where specially trained employees pack up "client carts" for delivery; 4) use of the "reservation-delivery-acceptance-payment" scheme; NB: *payment comes only at the end of the chain*, and is effected by the client only if he is satisfied with the quality of goods and promptness of delivery! Centralized storage has excluded the need for expensive storage facilities in Moscow – creating additional opportunities in terms of lowering prices and competing with large supermarkets.

Extracting

Prototype	No.	Navigator	Implementation	Target Artifact
	01	05 Separation	Client terminals are first moved to the shop, and will soon be moved to client homes; information gathering terminals are moved to the storage facility.	
	02	11 Inverse action	Instead of the client coming to the goods, the goods are coming to the client.	
	03	35 Unite	Central storage facility	

Reinventing

TREND *The drive to increase the ease of shopping for convenience goods is a reflection of two global trends: 1) saving user time, and 2) use of information technologies based on mobile communication systems and the Internet. But what specifically can be done taking into consideration the domination of supermarket and hypermarkets – which seize the best positions, particularly in large cities?*

REDUCING **Radical contradiction (RC):** The goods ► far (in the shop) **VS** close (at home)

Formulas and solutions (in general terms) of some standard contradictions (SCs):

23 Functional time of the moveable object VS 02 Universality, adaptability = 01, 03, 11

01 Productivity VS 03 Level of automation = 01, 10, 35, 37

INVENTING ALL models chip in: *01 Change in the aggregate state of the object – redistribution of the traditional concentration of goods, information and clients; 03 Segmentation – networks comprising places where clients can order, test, and pay for the goods; 05 Separation (the model is introduced additionally based on extracting outcomes) – the required functions are separated into requisite distributed (ordering) concentrated (large "global" and small local storage facilities in retail outlets) clusters; and, naturally: 10 Copying, 11 Inverse action and 35 Unite.*

ALL resources are used to resolve the contradictions: systemic, informational, functional, temporal, spatial, material, and energetical. Still, information is the key resource, as it enables generation and collection of orders, centralized storage of stocks, optimization of transportation routes, etc.

ZOOMING The original contradictions have been removed. Super-effect: the system has good prospects (trend) of further growth, and may subsequently evolve into the Functional Ideal Model *"orders from homes, goods to homes"*.

BRIEF DESCRIPTION This is an example of efficient integration of informational resource into real life. It demonstrates the possibility of using completely different methods to create optimal structures to deliver the goods to end users.

fig. 13.10. Reinventing the idea "The goods are coming to the client, not the client to the goods"

Example 13.6. *"Tuber of the Problem", or "From Village with Love"*

"Tula Cornfield" (Tulskaya Niva, Venevo District, Tula Region) is a modern "collective farm" – "kolkhoz" is the term of preference used by its organizer and general manager, a well known industrial entrepreneur and financier Andrey Eduardovich Kazyuchits. The farm has huge plans – it intends to make "good money" on the sale of domestically manufactured agricultural produce. Yes, it imports seeds – the best of those available in the market. But the rest is all "home-made". And it IS possible and necessary to revive and strengthen the Russian countryside. To make it possible, operating efficiency must be raised to the level of industrial production, with superior product quality as the main "output", and positive emotions associated with diligent work and fitting reward as the main "inputs".

Extracting

Prototype	No.	Navigator	Implementation	Target Artifact *From Village with Love*
	01	16 Partial or excess effect	Superior quality based on the use of "thoroughbreds" and conscientious work.	
	02	29 Self-servicing	Proprietary and friendly delivery networks.	
	03	40 Uninterrupted useful function	Exclusion of blank runs (racketeers), ongoing movement towards the end user.	

Reinventing

TREND *Agricultural production – a lot of work, praying for weather, lack of city comforts, and... difficulties with selling the produce! What can be done – taking into account the almost oligarchic domination of network retailers who are trying to obliterate domestic manufacturers with exorbitant distribution costs?*

Prototype: wasteful storage of potato "the old way" – in potato pits where it often rots away. In Tula Cornfield *production is organized in a very different fashion: potatoes are washed (a), sorted (b), stored in well-ventilated facilities (c), and packaged (d) into convenient containers designed for different types of consumers.*

REDUCING Radical contradiction (RC):

The goods ► far (in the countryside) **VS** close (at home)

Formula and solution (in general terms) of the standard contradiction (SC):

10 Ease of use (for the consumer) **VS** 23 Functional time of the moveable object (quick delivery) = 12, 14, 29, 32 and, additionally: 16 and 40.

INVENTING Possible solution: organization of a proprietary of network for the delivery of goods to the END user (models 12, 29, 40); long-term storage of agricultural produce (models 14, 16, 40).

Mobilization of the informational and functional resources: network marketing and reinforcement with appropriate value-added services.

ZOOMING The original contradictions can be removed. Super-effect: the system has good prospects (trend) of further growth (and integration with other agricultural producers), and may subsequently evolve into the Functional Ideal Model *"orders from homes, goods to homes"*.

BRIEF DESCRIPTION Mobilization of the informational and functional resources (network marketing, value-added services, etc.) opens up new opportunities in terms of creation of viable structures designed to effect direct delivery of agricultural produce to the end users.

Let us wish such companies success. They will win. Not only because they have will and faith and are ready to do whatever it takes, but also because of their fortitude and humor. They write about themselves: "We are ordinary people – just like everybody else. Except that we are funny, brave, angry, persistent, devil-may-care, risky, etc." And they also write for you: "From Village with Love".

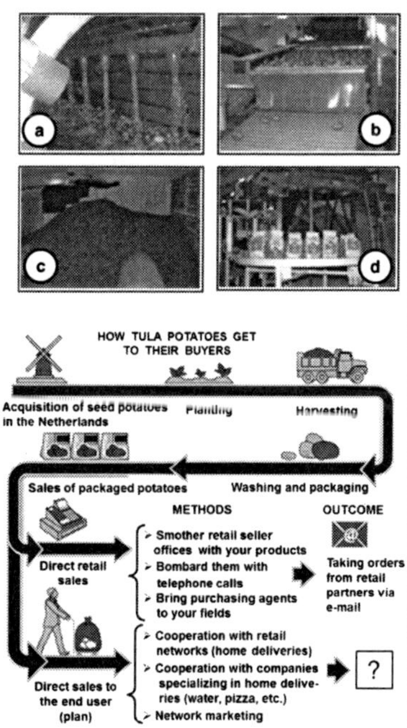

fig. 13.11. Reinventing the *Tula Cornfield* idea (sources: http://tulskayaniva.ru, http://product.ru, http://www.kommersant.ru/sf)

Example 13.7. Digital Ice Sculptures

And now let us have a look at the sculpting technology that could have been used by Michelangelo himself (see Problem 4.13). And it is definitely used by architects and artists in preparation for Ice Sculpture Exhibitions – such as the one held in Harbin, China (see Problem 1.23).

TREND There are various devices used to model the shape of new, relatively small items with the help of the so-called Rapid Prototyping.

Such devices use a special fluid which polymerizes and thickens when exposed to extreme temperatures or light. The sought item is shaped gradually, layer by layer, within the fluid.

Most Rapid Prototyping devices are rather expensive. Can we make a cheaper device, possibly at the expense of precision and durability of output (we might want to use "crude" models just to visualize the required shapes in 3D space)?

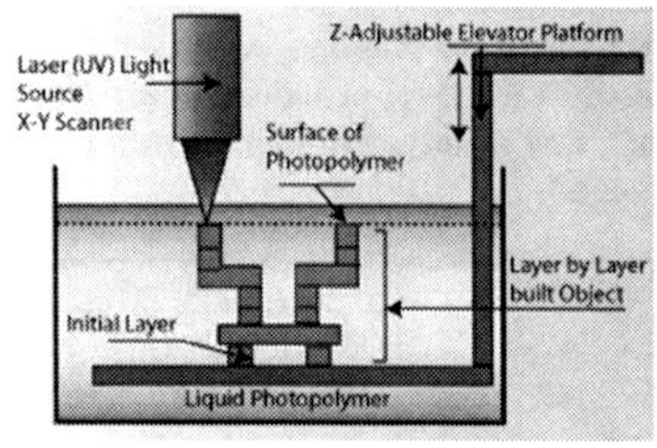

REDUCING

FIM: The X-resource in the form of particles of substance or energy is located inside the operating zone and ensures, together with the other available resources, the obtaining of the following **IFR**:

[layer-by-layer rapid growth of the model].

Formula and solution of the standard contradiction:

05 Precision of manufacture VS 10 Ease of use = 01, 03, 09, 36

Radical Contradiction:

Prototype material	⇨	must be inexpensive and easily form layers	but	must not be inexpensive and easy to use, as all known materials and devices are expensive

INVENTING

The solution is based on the material resource: the original material is replaced with a less expensive and more readily available material, such as water! Idea for the solution: the prototype is freeze-grown by application of rapidly freezing water drops.

All navigators from the cluster are used in the solution: 01 Change in the aggregate state of the object, 03 Segmentation, 09 Change in color and 36 Feedback.

Dispersal of liquid nitrogen inside the chamber brings the temperature down to, and maintains it at, minus 20-30 degrees. Water drops are applied to the model through a warmed-up nozzle. They freeze up as soon as they come into contact with the surfaces created before. Fast, simple, cheap!

ZOOMING

Have the contradictions been removed? – Yes!

Super-effects: we have devised an inexpensive Rapid Prototyping method using the most readily available material – water!

Negative effects: the "sculpture" has to be stored at sub-zero temperatures.

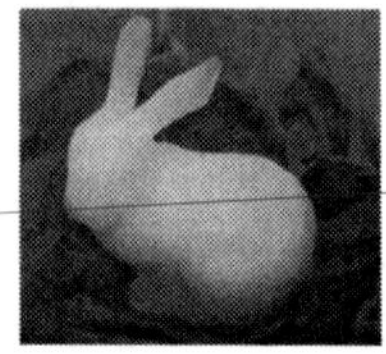

BRIEF DESCRIPTION

Scholars working the Centre for Intelligent Machines, McGill University, USA, have developed a Rapid Freeze Prototyping technology based on the use of ice – New Architecture of Phase Change. Navigators present: 01, 03, 09 and 36; also, modification of the material resource.

fig. 13.12. Reinventing the rapid ice prototyping technology
(source: www.memrana.ru)

Example 13.8. Self-Organization: By the Law of Nature

> Further technical development is unthinkable without wide systematic use of "patents" of Nature[186].
>
> *G. Altshuller*

TREND *There are known methods that can be used to obtain metal sheets with convex-concave cells (upper and lower) surfaces). The most widely used of those methods are pressure molding and broaching with pressure molding. However, they sometimes expose metal to unequal internal stresses which may lead to emergence of micro-fissures. As a result, the metal sheets are less durable and less reliable when used in the construction of roofs, walls and other elements. What can we do?*

REDUCING FIM: The X-resource in the form of particles of substance or energy is located inside the operating zone and ensures, together with the other available resources, the obtaining of the following **IFR**:

[*forming metal sheets with regular convex-concave cells without damaging sheet surfaces*].

Standard (Technical) Contradiction

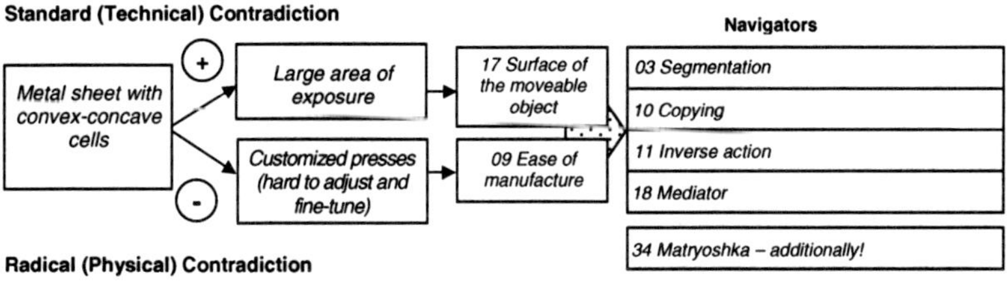

Radical (Physical) Contradiction

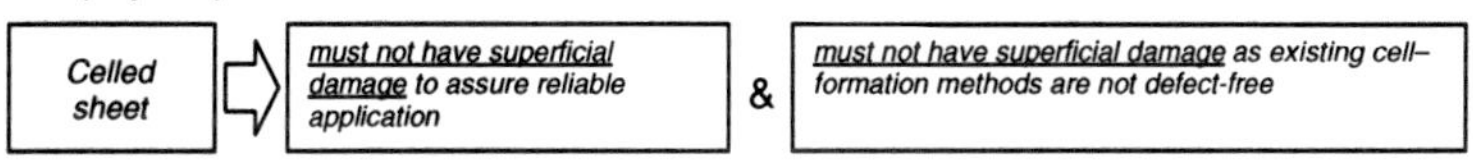

INVENTING **Using model 03:** exposure to "particles of substance" to evenly distribute pressure throughout the surface of the sheet. **Using model 10:** here inventors examined similar constructions created by Nature (fig. a) and reformulated the problem as *"self-organization"* of the cells making up the surface (b) – for example, is it possible to "copy" a tortoiseshell? **Using model 11:** instead of applying pressure "from above", form the sheet by using vacuum suction (c) – the sheet must *"shape itself"* under the influence of evenly distributed atmospheric pressure. **Using models 18 and 34:** a "mediator" has been introduced – a special instrument controlling cell growth! The "instrument", installed under the sheet (inside the machine), is creating, along the length of the sheet, (c) a support structure with a preset step!

ZOOMING Super-effects: 1) Smooth surface.
2) Equal thickness of the sheet in section view!

BRIEF DESCRIPTION

To create cellular structures (similar to those "invented" by Nature) on metal and other sheets, inventors used navigators 03, 10, 11, 18 and 34, and certain physical effects (hydro-vacuum pressing and particularly "self-organization" of cells under certain conditions).

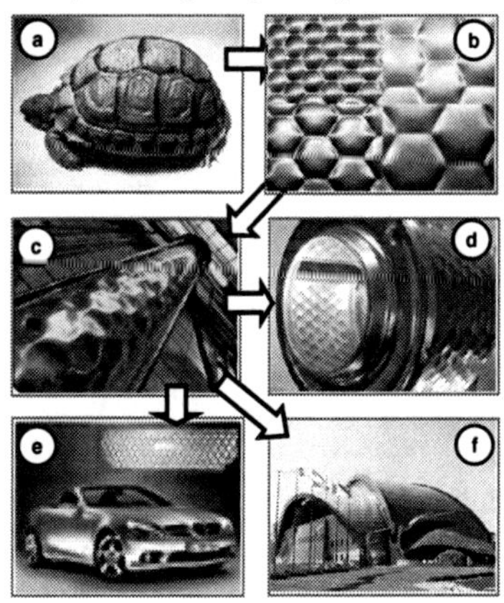

The **Dr. Mirtsch GmbH** company, Berlin, Germany (managers: a family creative collective – Professor, Dr. Frank Mirtsch, his wife Dr. S. Mirtsch, and his son Dr. M. Mirtsch) successfully uses Nature's "patents" to design new ideas and technologies enabling formation of metal and other sheets with regular cellular structures. The company is well-known throughout the world: **d)** washing machine drum from Miele; **e)** SLK motor car boots from Daimler; **f)** roof of a sports center in Odessa, Ukraine. Besides, **BioTRIZ** technologies have for many years been successfully explored by a company bearing the same name working at the English University of Bath (Professor Julian Vincent and his colleagues, Dr. Nikolay Bogatyryov and Dr. Olga Bogatyryova).

fig. 13.13. "By the Law of Nature": modeling the inventions by Professor Mirtsch

---

[186] G.S. Altshuller (1961) *How to Learn to Invent.* – Tambov, Tambov Publishers

Example 13.9. For Inventor's Day in Europe

Modern protected secret communication systems broadly use the so-called "frequency hopping" method. It is hard to believe that it was invented in 1941 (and patented in 1942) by a film actress[187] and her friend, a composer[188]. Be that as it may, communication companies waited for 20 years for the patent to expire before the method began to be used on an industry-wide basis.

In addition to that, the patriotically-minded actress once raised US$ 7 million at an auction selling War Bonds, debt securities issued by the US Government to finance its European anti-Nazi campaign. Well, the good thing is the name of the inventor was not forgotten: in honor of Hedy Lamarr, it is suggested to celebrate the Inventor's Day in Europe on November 9, which was her birthday.

TREND To avoid "leaks" when transmitting secret messages, information is encrypted. But this takes time, which may reduce efficiency of transmission. Besides, the message can be decrypted, or transmission can be disrupted (distorted or "jammed" by creating strong interference) if the "interceptor" knows, as a minimum, the frequency at which the message is being transmitted. What can we do?

REDUCING FIM: X-resource, together with available or modified resources, and without making the object more complex or introducing any negative properties, guarantees attainment of the following **IFR**: [*protection of information*].

Formula and solution (in general terms) of the standard contradiction (SC):

12 Loss of information VS 01 Productivity = 07, 11, 36

12 Loss of information VS 13 External damaging factors = 02, 03, 21

Radical Contradiction (RC):

The frequency is ► "known" (for the transmitter and the receiver) **VS** "not known" (for the "interceptor")

INVENTING Dominant models that can be used to resolve the SC using the MITO method: *02 Preliminary action, 03 Segmentation* and *07 Dynamization.* It was suggested that information be transmitted in fragments (model 03), and that transmission frequencies be changed (model 07) in accordance with the program created for the transmitter and the receiver in advance (model 02). All remaining models can also be gainfully employed: 11, 21 and 36.

Dominant resources used to resolve the RC: informational resource (frequency change management), structural resource (combination of a large number of frequencies; the original invention used 88 frequencies), temporal resource (dynamization).

ZOOMING The contradictions have been removed. Super-effect: military communication systems, as well as widespread civil applications, such as Bluetooth, Wi-Fi, WiMAX and others, use the principle of separation and change of frequencies to increase reliability of communication.

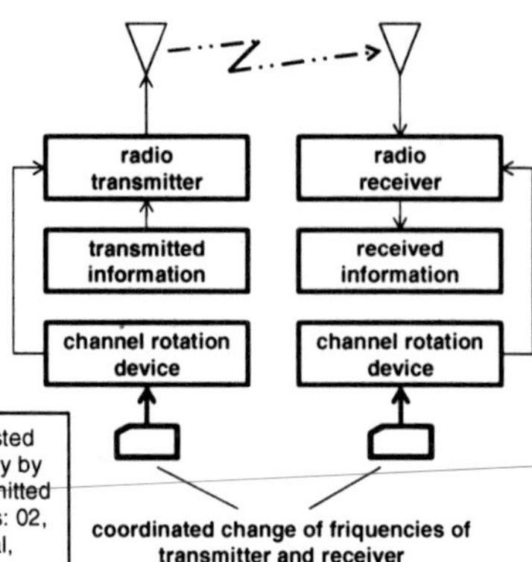

BRIEF DESCRIPTION

To protect information transmission, it has been suggested to effect synchronized change of transmission frequency by the transmitter and the receiver. The message is transmitted in fragments on different frequencies. Dominant models: 02, 03 and 07; dominant resources: informational, structural, temporal.

fig. 13.14. Reinventing US Patent No. 2 292 387

[187] Hedwig Eva Maria Kiesler (November 9, 1914, Vienna, Austria – January 19, 2000, Altamonte Springs, Florida, US) – Austrian-born American film actress popular in the 1930-es – 1940-es; scenic name – Hedy Lamarr; photo: "Inventor's Day", www.wikipedia.org

[188] George Antheil (1900, Trenton – 1959, New York, US) – American avant-garde composer and pianist; born in a family of German immigrants; photo: www.wikipedia.org

Example 13.10. Innovative Globalization from *fischer*[189]

Artur Fischer

Klaus Fischer

For the last 30 years, the family enterprise has been headed by Professor Klaus Fischer, inventor, entrepreneur *par excellence*, son of Artur Fischer

Artur Fischer would probably have agreed to transfer the Inventor's Day to the date mentioned in the previous example. The thing is, he was born on December 31, and it would be terribly inconvenient to celebrate the Inventor's Day on New Year's Eve! Meanwhile, Professor Artur Fischer is holding the "world record" in the number of patented inventions and industrial samples – in 2008 it stood at 1,121.

There is no doubt that this is a good enough reason to convert his birthday into the Inventor's Day.

And now let us try to find the "secret" of the dowel – the artifact which made its author a celebrity known throughout the world!

TREND *In the past, if we wanted to drive a screw into a brick or concrete wall, we made small wooden plugs and hammered them into pre-drilled holes. Unfortunately, sometimes the plugs crumbled and fell out of the wall. Generally speaking, it was not always possible to take the screw out of the wall and then put it in back again. Apparently, despite its apparent simplicity, plugs may break down – lose their shape, so to speak. What can we do?*

REDUCING FIM: X-resource, together with available or modified resources, and without making the object more complex or introducing any negative properties, guarantees attainment of the following **IFR**: [*tightly closing the hole into which the screw is driven*].

Formula and solution (in general terms) of the standard contradiction (SC):

07 Complexity of construction **VS** 21 Shape = 04, 07, 11, 14

Radical Contradiction (RC):

The plug is ► "hard" (to tightly close the hole) **VS** "soft" (to make it possible to drive the screw in)

INVENTION Dominant models that can be used to resolve the SC: *04 Change in the aggregate state of the object, 07 Dynamization* and, additionally, *01 Change in the aggregate state of the object.* Invention of the dowel totally agrees with these navigators: the body of the dowel is made of elastic plastic (model 01) so that it expands, and its diameter increases, as the screw goes through it (models 04 and 07).

Dominant resources used to resolve the RC: structural resource (cuts, "winglets", protrusions, etc.), spatial resource (new 3D shape), and material resource (use of elastic – but at the same time hard and durable – material) – when expanded, the dowel is tightly "wedged" in the hole.

ZOOMING

The contradictions have been removed. Super-effect: standardization and mass-scale production of mounting fittings, simplification of work!

BRIEF DESCRIPTION

Creation of a sturdy, reliable and simple device to drive screws into brick or concrete walls: standardized elastic dowel of various shapes and sizes (models 01, 04 and 07).

The dowel created by Professor Artur Fischer, founder of the fischer group, has become a truly global product.

Equally widespread and well-known are his erector sets fischertechnik and fischerTiP.

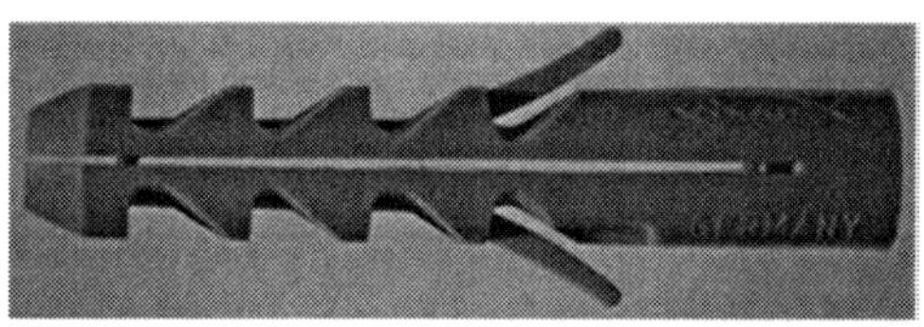

fig. 13.15. Reinventing Professor Artur Fischer's dowel

[189] Artur Fischer (born on December 31, 1919, in Tumlingen, Germany) – famous inventor and entrepreneur, creator of globally renowned products – a series of "dowels" and "fischertechnik" construction toys for children, adolescents and students, electric flash and other artifacts; photo: www.presswire.dk and www.fischer.de

Example 13.11. "Don't tell me it's impossible. Tell me you can't do it."

Yes, he[190] said that in one of his interviews (Esquire, December 2008). If that tirade were directed at himself, he could have worded it differently: "Don't tell me it's impossible. I can do it!"

TREND Any two-wheel scooter or bicycle remains stable only when in motion. In other words, these devices are "dynamically stable". However, when they stop, they fall down on their side.

There are known systems that can stabilize spatial position using liquid equilibrium detectors and gyroscopic stabilizers. However, there is "nothing to stabilize" in a bicycle transversely – the only thing we can do is install a sophisticated "scales-like" system so as to make it possible, by constantly balancing, to keep the mass center of the system strictly within the support area. But that seems too complex and completely unjustified.

Can we build a two-wheel vehicle which would be able to rapidly move along, carrying its "pilot" or "passenger" (whatever turns you on!), and keep upright without falling when forced to stop?

REDUCING

FIM: The X-resource in the form of particles of substance or energy is located inside the operating zone and ensures, together with the other available resources, the obtaining of the following **IFR**:

[*mobile two-wheel transportation vehicle with reliable self-stabilization during stops*].

Standard (Technical) Contradiction

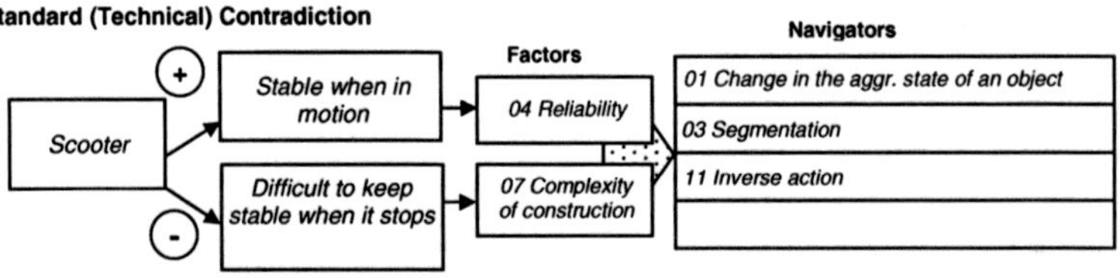

Radical (Physical) Contradiction

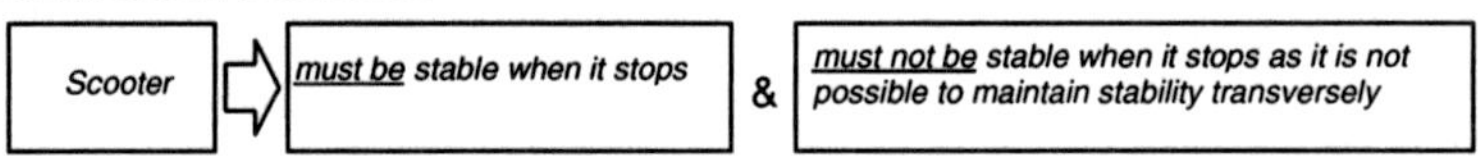

INVENTING

The first key model is Model 11: if we have no resource that we can use to adjust the position of the scooter transversely, we can do it longitudinally, while transverse stability can be assured if the wheels are placed not one after the other (as is usually the case), but coaxially so that they are attached to the edges of the scooter! Now we have spatial and functional resources that we can use for steering!

The second key idea is the introduction of an informational resource: we must use electronic devices to control the position of the mass center and keep it within the support plane!

ZOOMING The contradictions have been partially removed: we have achieved stability, but at the expense of complexity!

Super-effects: 1) This is a new mode of transportation. 2) Discovery of a new locomotion method for robots!

Negative effects: complexity of construction!

BRIEF DESCRIPTION

Due to introduction of a new informational resource and application of Navigator 11 Inverse action, we have received a new mode of transportation with two coaxially placed wheels, capable of maintaining stability by dynamic adjustment of the position occupied by the mass center of the system in the wheels' support plane. In the photo: Dean Kamen, riding his invention – Segway.

fig. 13.16. Reinventing Dean Kamen's Segway scooter
(source: http://www.wikipedia.com)

[190] Dean L. Kamen (b. 1951) – American inventor, creator of Segway, an individual transportation device fitted with a gyroscopic stabilization system

Example 13.12. P.U.M.A Is Getting Ready to Jump!

TREND

The smallest motor cars! The coveted dream of every carmaker on the planet! But simple reduction of dimensions while keeping the relative proportions of the "big" motor cars turned the "small" motor cars into something unreal, something from outer space. There was a need to create a new class of small motor cars for one, two or more passengers! Like in the fairy tale about the "the peasant's wise daughter" (Problem 1.17): it has a roof, but it's not a motor car, it has two wheels, but it's not a motor cycle! What could it be?

Seriously speaking, the main constraining factor in the minimization problem is the number of wheels. It would be kinda silly to try and reduce the number of wheels to one. Two (rather than three or four) wheels – well, that's a much better idea. But such machine already exists – it is either a motor cycle or a motor scooter. Besides, both configurations lose their specific advantages as soon as pack them inside a body with a "roof". What can be done?!

REDUCING

FIM: The X-resource in the form of particles of substance or energy is located inside the operating zone and ensures, together with the other available resources, the obtaining of the following **IFR**:

[stable two-wheel vehicle of the motor car type].

Radical Contradiction 1:

Motor car	⇨	*<u>must be</u> small in line with the urbanization development trend*	&	*<u>must not be</u> small as it looks like a grotesquely reduced "real" motor car*

RC 2: Motor car ► with two wheels (to reduce the size)
VS with more than two wheels (to assure stability)

INVENTING

The solution is based on a systemic transition: creation of the motor car of a new type – on the basis of a two-wheel self-stabilizing Segway platform!

In 2009 this idea was jointly suggested by the General Motors company and the innovator Dean Kamen within the framework of the project called P.U.M.A – Personal Urban Mobility & Accessibility. In 2010 at an exhibition in Shanghai, Segway was presented as a joint solution designed by General Motors and the Chinese company SAIC under the brand Electric Networked-Vehicle – EN-V.

ZOOMING

Have the contradictions been removed? – Yes, the acute (radical) esthetic and technical contradictions have been removed!

Super-effects: proposal of a new class of motor car boasting conceptually new design, construction and road behavior.

Negative effects: -

The new systemic solution creates a new class of motor cars, establishes new perception stereotypes and new esthetical and technical criteria that will be used to evaluate future models in this class.

fig. 13.17. Reinventing P.U.M.A and EN-V, new-class automobiles based on Dean Kamen's Segway

Example 13.13. Hand Dryer with Air "Brushes"

TREND

Photo by the author

10 years ago, on the eve of the 2nd millennium, a new sight appeared in the heart of London – the famous London Eye. Each year, millions of tourists admire breathtaking views of the Thames, the Big Ben tower, and the building of the British Parliament from a height of 135 meters. Incidentally, many of them (and they are millions, too!) visit the London Eye's bathrooms. The old hand driers sometimes scald your hands; besides, they are too slow-working, and visitors are forced to get in line to use them. Paper napkins are too expensive, and trash bins quickly get clogged up with them. What can be done?

REDUCING

FIM: The X-resource in the form of particles of substance or energy is located inside the operating zone and ensures, together with the other available resources, the obtaining of the following **IFR**:

[*quick and safe hand-drying*].

Standard (Technical) Contradiction

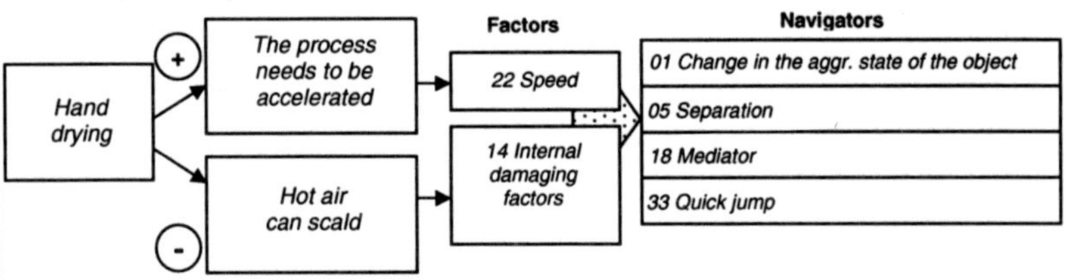

INVENTING

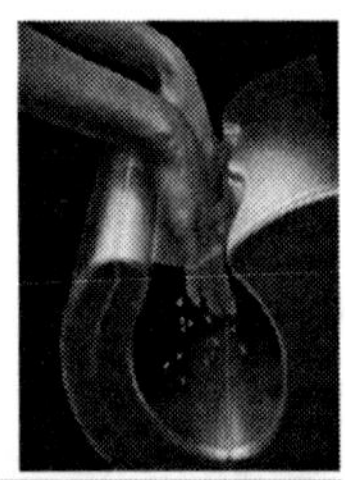

All models in the cluster are used. According to models 01 and 33, it is suggested to boost the speed of the air jets issuing from the hand drier to 640 km/h! The air jets act like two razor-sharp brushes which "shave off" residual water from the skin according to model 18; besides, they operate in line with model 05 by introducing a useful property (removal of moisture) and removing harmful properties (scalding, slow operation).

ZOOMING

Have the contradictions been removed? – Yes. – ~~No~~.

Super-effects: the device can be used to rapidly dry many various objects; the drops are gathered in a vessel below – anti-bacteriological effect; minimal energy consumption!

Negative effects: -

BRIEF DESCRIPTION

The hand drier uses high-speed air blades. One of the first uses – bathrooms at the London Eye.
Navigators used: 01, 05, 18 and 33.
Dyson Airblades™ were invented by James Dyson (in the photo).

fig. 13.18. Air Blades from James Dyson[191]
(source: http://www.dysonairblade.co.uk/)

[191] Sir James Dyson (b. 1947) – famous English inventor and billionaire entrepreneur, creator of the "cyclone" vacuum cleaner (late 1970-es) and other artifacts

Example 13.14. Solution for Problem 3.4. Fan without Propeller Blades!

TREND

For dozens of years we have known desktop (and other) fans with large, usually rubber-coated blades – noisy, always ready to give you a rap on the knuckles (quite literally) if you don't keep your hands away from it. And one more thing: the feeling produced for the air stream generated by such fan is not always pleasant – instead of an even refreshing breeze you get gusts of cold wind. What can be done?

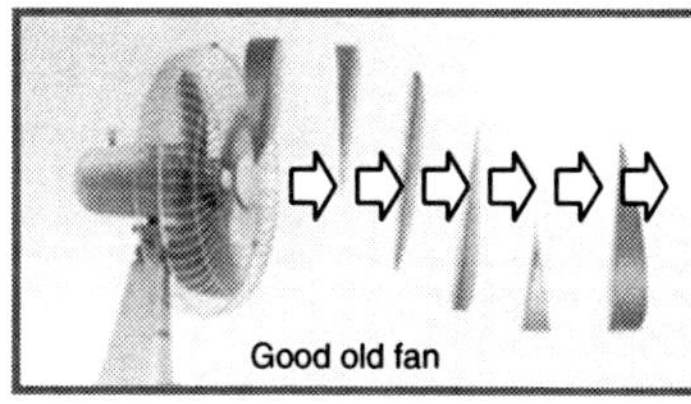
Good old fan

REDUCING

FIM: The X-resource in the form of particles of substance or energy is located inside the operating zone and ensures, together with the other available resources, the obtaining of the following **IFR**:

[*creating a safe and evenly strong air stream*].

Standard (Technical) Contradiction

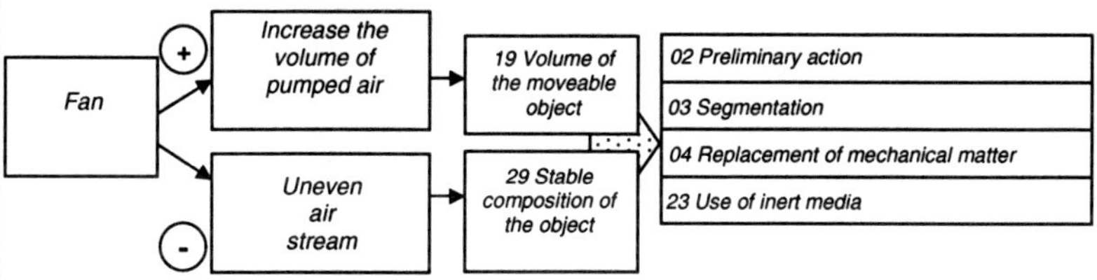

INVENTING

The key model is Model 04: However, to restructure the "field" (air stream), we need to turn to physical/technical effects. The inventor used the Coandă effect: the air or liquid jet "clings" to the nearby surface if it tangentially moves close to such surface (fig. 1)! The "work stream" (a) is pumped out through narrow fissures around the mouth of the entry plane (fig. 1) and creates (fig. 2) an additional stream (b) "sucked in" at the entry point, and another additional stream (c) "attracted at the exit point. The work stream is fed by the pump installed in the base of the fan through cavities in the round frame.

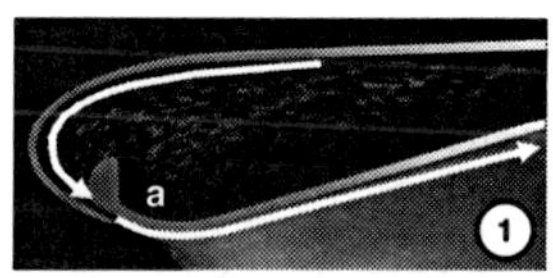

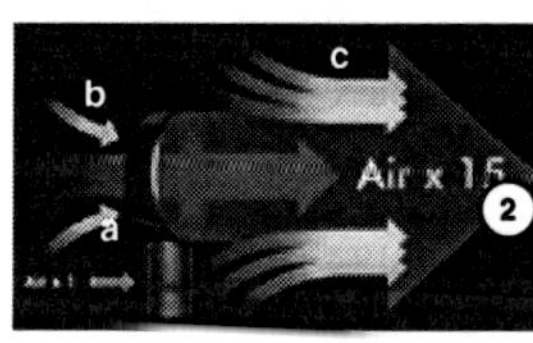

ZOOMING

Have the contradictions been removed? – Yes. – ~~No~~.

Super-effects: by the time the air taken in by the fan gets to the exit point, it is amplified 10-20 times! The stream is completely even and "smooth".

Negative effects: -

BRIEF DESCRIPTION

Based on the Coandă effect, the shape of the work stream is made cylindrical, while its volume is amplified more than 10-20 times! Key navigators: 02, 03 and 04.
Dyson AirMultiplier™ was invented by James Dyson (fig. 3).

fig. 13.19. Reinventing James Dyson's "bladeless" fan
(source: http://www.dyson.co.uk/fans)

13.2.2 Benchmark Examples: Kremlin Stars

Example 13.15. How to Make a Kremlin Star Brighter?

TREND The Kremlin stars must be seen well not only during the night, but also during the day. This means that need to be very bright. However, the light generated by the light bulb is evenly dispersed throughout the interior of the star, meaning that the rays of the star, especially their tips, are "light-starved". It is not expedient to use more powerful light bulbs, as the star may have to be cooled down, and the construction of its lighting system may become prohibitively complex. In addition to that, this will boost power consumption.

PROBLEM: how do we make a Kremlin Star brighter?

REDUCING FIM: X-resource, together with available or modified resources, and without making the object more complex or introducing any negative properties, guarantees attainment of the following **IFR**:

[the star is equally bright in the center and at the tips of its rays].

Standard contradiction (SC):

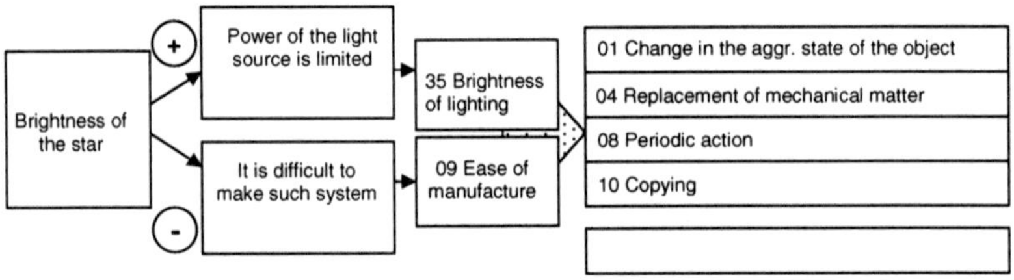

Radical Contradiction (RC):

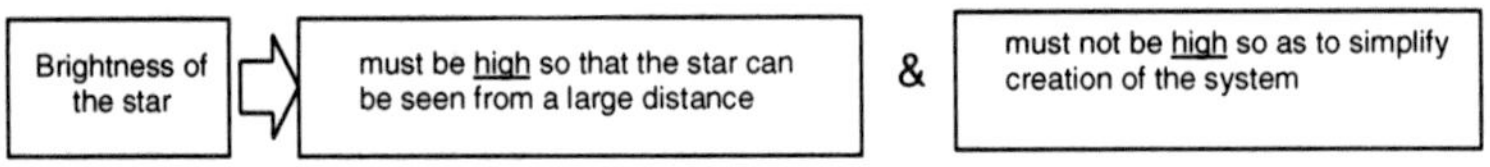

INVENTING All models work here!

In line with Model **01,b)** change concentration, Model **04,c)** replace unstructured fields with fields with a specific structure, and Model **10,b)** replace an object or system of objects with their optical copies (images), we need to project into the rays focused light streams!

Key idea: a 15-element light concentrator (with 3 focusing plates for each of the 5 star rays) is installed in the center of the star.

In line with Model **08**, the brightness of the star during the night is reduced compared to its brightness during the day.

ZOOMING Have the contradictions have been removed? – Yes. – ~~No~~.

Super-effect: reduced power consumption.

Negative effects: more complex construction.

BRIEF DESCRIPTION

A 15-element concentrator is installed around the Kremlin Star's light bulb to make it brighter. During the night, the brightness of the star is reduced by decreasing the power supplied to the light bulb.

Navigators present: 01, 04, 08 and 10.

fig. 13.20. Reinventing of the light concentrator

Example 13.16. How to Replace a Deficient Light Bulb?

TREND Like any other technical system, Kremlin Star elements have a finite useful life. In addition to that, material defects may result in unexpected failure of certain elements, for example, incandescent filaments. To increase durability, each light bulb has two filaments. The backup filament "kicks in" in the event of the main filament's failure. After that the bulb needs to be replaced. But how can this be done? It is impossible to get there through the peak of the tower – because it is too narrow, and because its interior is fitted with bearings with diameters far less than average dimensions of a human body. Besides, the star – as we well know – keeps rotating around its axis. Should we climb up 70 meters on the outside to "open" the star so that we can access its bulb?

REDUCING **Macro-FIM:** X-resource, together with available or modified resources, and without making the object more complex or introducing any negative properties, guarantees attainment of the following **IFR**:

[quick replacement of the star's light bulb].

Maxi-FIM: the lamp BY ITSELF "comes" to the repairman to be fixed up or replaced!

Standard contradiction (SC):

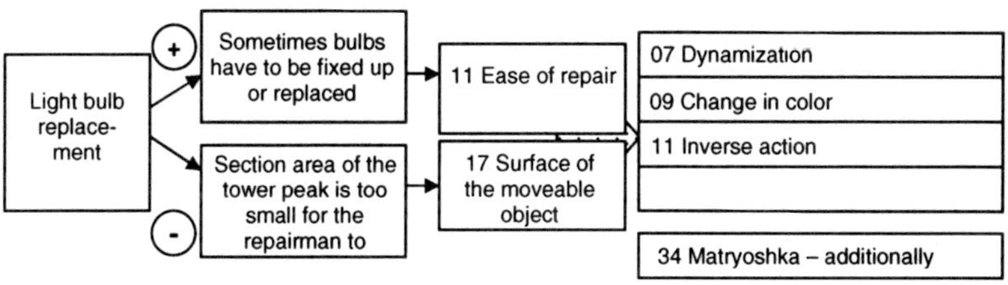

Radical Contradiction (RC):

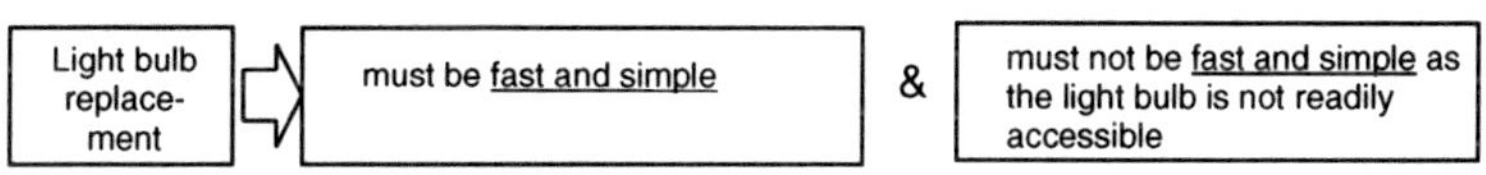

INVENTING Dominating models – 07, 11 and 34.

We are going to use Model **07,b)** disassemble the object into parts that move with respect to each other, Model **11,b)** make a moveable part of the object (or of the environment – the repairman!) fixed or a fixed part (the bulb!) moveable, and Model **11,c)** turn the object "upside down" or "inside out". We can also add Model **34,b)** one object runs through a hollow space in another object. The dominant transformations affect the structural and spatial resources.

Key idea: there is a rod that can move up and down inside the start and the peak of the tower (and that moves through the bearings)! A non-rotating light bulb is installed at the top of the rod! When the bulb needs to be repaired, it is lowered into a small room under the peak of the tower. In other words, we have a clear-cut case of 11 Inverse action: it is not the repairman that comes up to the bulb, but the bulb that goes down to the repairman! Bulb replacement takes 30-35 minutes.

ZOOMING

Have the contradictions have been removed? – Yes. – ~~No~~.

Super-effect: the bulb does not rotate, which lifts a serious problem that could have arisen in terms of supplying it with electric power.

BRIEF DESCRIPTION

The light bulb of the star is installed on a rod inside the peak of the tower which goes through a series of bearings. The rod is lowered together with the bulb into a small room under the peak of the tower. Dominant navigators: 07, 11 and 34.

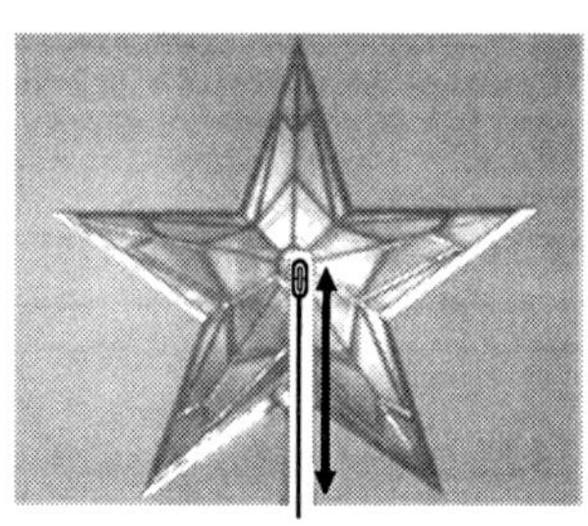

fig. 13.21. Reinventing of the star-mounted light bulb replacement system

Example 13.17. "Ruby" means *"Bright Red"*!

TREND Despite the fact that we have created a sophisticated light concentrator feeding focused light into star rays, ray tips remain insufficiently bright. As a result, in blazing sunlight they seem to be colored black! That is the effect of the light being reflected from the selenium "ruby". The star becomes truly red only when strongly lit from the inside. However, it is undesirable to use more powerful light bulbs. How can we increase the brightness of the star without resorting to more powerful light bulbs?

REDUCING

Micro-FIM: The X-resource in the form of particles of substance or energy is located inside the operating zone and ensures, together with the other available resources, the obtaining of the following **IFR**:

[the star has a bright red color regardless of external lighting].

Standard contradiction (SC):

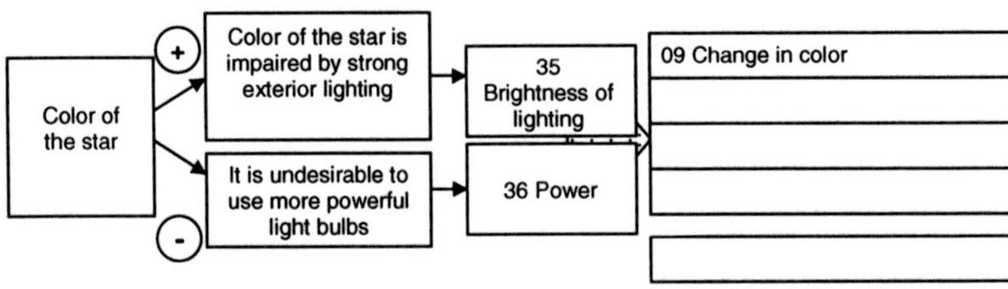

Radical Contradiction (RC):

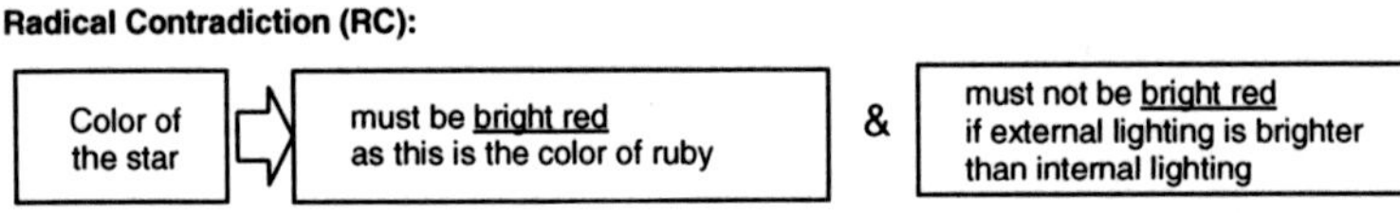

INVENTING

Key idea: In line with the only available Model **09,b)** change the level of the transparency of the object, the selenium rube plates are made with variable transparency (transparency decreases from the center of the star towards ray tips!).

Clear domination of the material resource to resolve a radical contradiction.

ZOOMING Have the contradictions have been removed? – Yes. – ~~No~~.

Super-effect: changes in glass transparency remove the need to use more powerful light bulbs.

BRIEF DESCRIPTION

To increase the brightness of internal lighting of the star's plates (compared to external lighting), selenium ruby plates have varying transparency decreasing from the center of the star to its ray tips.
Dominant navigator: 09 Change in color.

fig. 13.22. Reinventing of the method and the technical solution used to boost the brightness of the light generated inside the star without increasing the power of the internal light source

Example 13.18. Invention is *Science* + *Art*!

TREND

Powerful light provided by the bulb and the concentrator creates a contrasting play of light and shadow (or, to be more exact, insignificant – but very undesirable – local loss of brightness in certain places along the ray) on the main ruby plates. Due to uneven brightness, the color of the star becomes insufficiently homogenous. At this stage, it is already impossible to change the power of the light bulb or the construction of the concentrator. Changes in the level of transparency of ruby plates cannot remove the unwelcome color differences, either.

How can we make the lighting of the ruby plates from the inside more even?

REDUCING

FIM: X-resource, together with available or modified resources, and without making the object more complex or introducing any negative properties, guarantees attainment of the following **IFR**:

[ruby plates are homogenously colored throughout the entire length of each ray of the star].

Standard contradiction (SC):

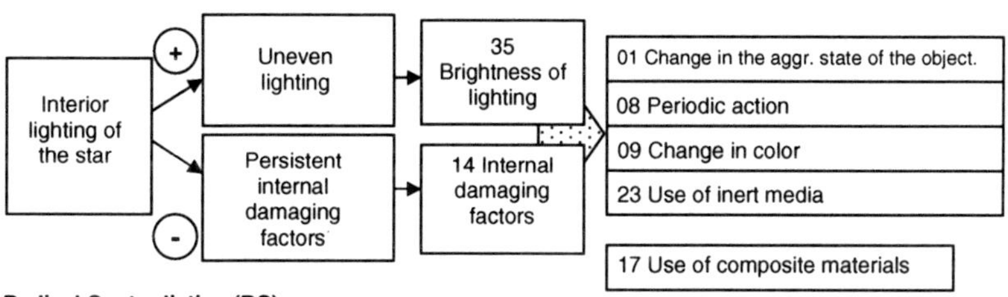

Radical Contradiction (RC):

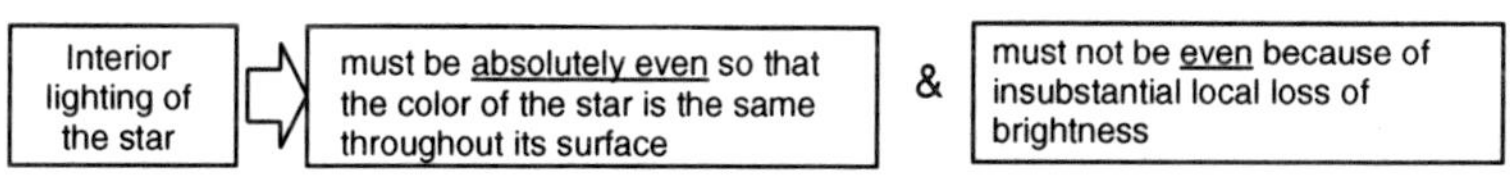

INVENTING

Key models: **01, b)** change concentration; **09,a)** change the color of the object or its environment; **b)** change the level of transparency; **17:** move from homogenous materials to composites.

Key idea: place between the light source and the ruby plates (a) thing layers of special crystal glass (b) and matte glass (c) diffusing light and removing contrasting color transitions.

ZOOMING The contradictions have been removed.

Negative effects: increased complexity of construction.

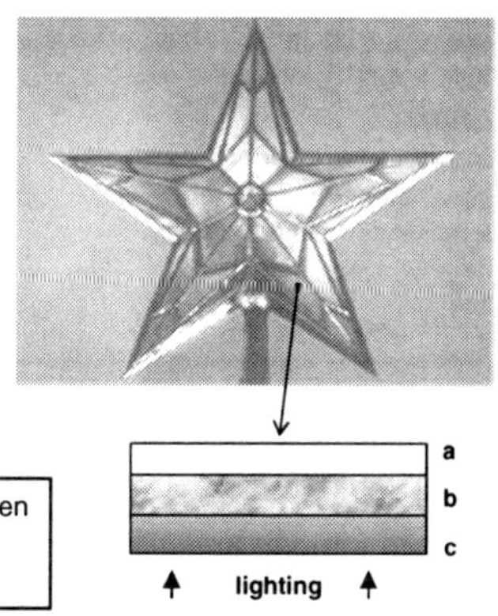

BRIEF DESCRIPTION

To make the lighting of ruby plates more even, they have been supplemented with thin layers of crystal and matte glass.

Navigators present: 01, 09 and 17.

fig. 13.23. Reinventing of the compositional structure of glass veneer on the star

13.2.3 Mysteries of the Three Elements: Water, Air, Space

Example 13.19. We Are All Aquanauts!

TREND

Known muscle-powered submersibles have a serious drawback – they require application of considerable efforts to ensure prolonged navigation at relatively high speeds. The reasons is that all propulsion units used by such submersibles to date, including screw propellers, have relatively low efficiency ratios, and low efficiency ratios translate into high loss of power.

Additional difficulties arise when an attempt is made to design hermetically closed submersible vehicles. Such vehicles are more comfortable, but their power consumption is immeasurably higher.

On the whole the power required (2.5-5.0 kW) is by about an order of magnitude higher than the normal power (0.2-0.4 kW) that can be sustained by a human being over a period of 2-4 hours.

Human power cannot be increased. What can we do?

REDUCING

FIM: The X-resource in the form of particles of substance or energy is located inside the operating zone and ensures, together with the other available resources, the obtaining of the following **IFR**:

[reduction of environmental resistance to motion and production of additional power].

Formulas and solutions (in general terms) of standard contradictions (SCs):

01 Productivity **VS** 36 Power = 01, 02, 40

01 Productivity **VS** 39 Loss of energy = 01, 02, 04, 14

Radical Contradiction:

Submersible propulsion unit	⇨	<u>must have</u> increased efficiency	but	<u>must not have</u> increased efficiency subject to existing propulsion devices, for example, screw propellers

INVENTING

Dominant models: 01 Change in the aggregate state of the object and 04 Replacement of mechanical matter.

The solution is based on the energetical resource and application of a physical/technical effect: it is suggested that the new submersible use a water propulsion unit. Outside water is sucked in by a muscle-powered pump and ejected in two places (fig. 3): (1) fore, so that the water "clings" to the sides of the submersible in accordance with the Coandă effect thereby rarefying the medium and reducing resistance to motion, and (2) aft, passing through water ducts lining the body of the submersible. As a result, the Coandă effect increases power efficiency of both propulsion units.

ZOOMING

Have the contradictions been removed? – Yes!

Super-effects: increased range of applications; high safety and efficiency

Negative effects: -

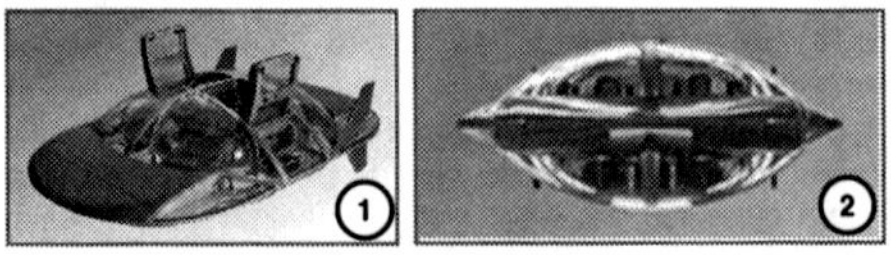

Muscle-powered submersible *Blues Space* – project developed by the *Marine Innovative Technologies* company, Saint-Petersburg, General Director: V.S. Taradonov: 1) with opened entry/exit hatches; 2) front view

BRIEF DESCRIPTION

A new high-speed muscle-powered submersible has been proposed. The craft's propulsion unit is based on the use of the Coandă effect.
Dominant navigators: 01 and 04. The energetical resource is intensified.

Submersible of the future: length – 3.5 m, height – 1.2 m, width – 2 m, weight (in the air) – about 2 tons.
Diving depth – up to 30 m, underwater speed – up to 5 knots (up to 10 km/n), number of passengers

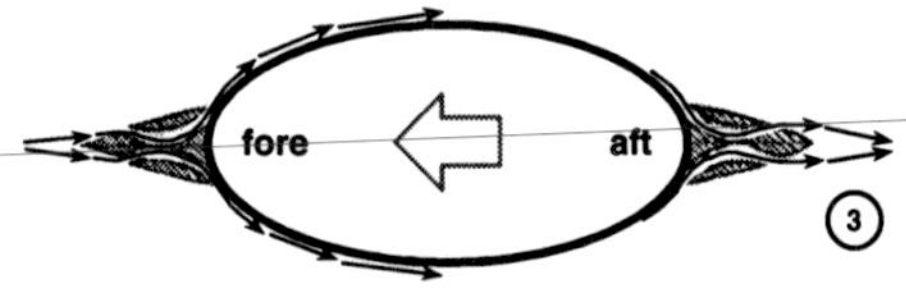

Pressure profiles: thrust production by propelling force (aft) and additional thrust production by rarefaction (fore and aft) due to the pressure difference generated by the Coandă effect

fig. 13.24. Reinventing of the *Blue Space* submersible

Example 13.20. Squall from under the Water!

TREND The need to boost the speed at which a torpedo moves under water to, say, more than 100 km/h results in a multiple increase of power requirement. This happens due to the rapid non-linear growth of environmental (water) resistance to motion. The problem affects primarily torpedoes equipped with propeller screws. What can we do?

REDUCING

FIM: The X-resource in the form of particles of substance or energy is located inside the operating zone and ensures, together with the other available resources, the obtaining of the following **IFR**:

[multiple increase of torpedo speed and moderate increase of power expenditure].

Formula and solution (in general terms) of the standard contradiction:

22 Speed **VS** 13 External damaging factors = 01, 03, 04, 36

Radical Contradiction:

Resistance of the water medium to motion of the submersible must be low (to increase speed), and it must be high (in line with natural properties of the medium).

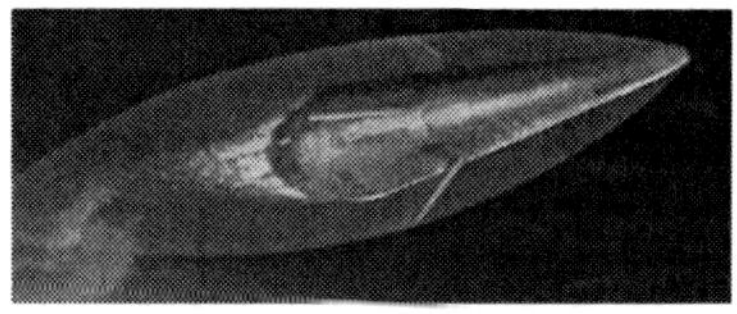

fig. 1. *Squall* torpedo in cavitational "cloud"

INVENTING Dominant models: **01** Change in the aggregate state of the object, **03** Segmentation and **04** Replacement of mechanical matter.

The solution is based on application of a special physical/technical effect and activation of the spatial and energetical resources.

1. In line with Navigator **01**, it is possible to saturate water with gaseous inclusions.

2. In line with Navigator **03** and paragraph 1, such saturation can lead to "segmentation" of the monolithic medium (water) into loosely connected "particles".

3. In line with Navigator **01** and **04**, it is necessary to seek to replace the dense liquid medium with a gaseous-liquid and then (if possible) gaseous medium! Purpose: to reduce water resistance.

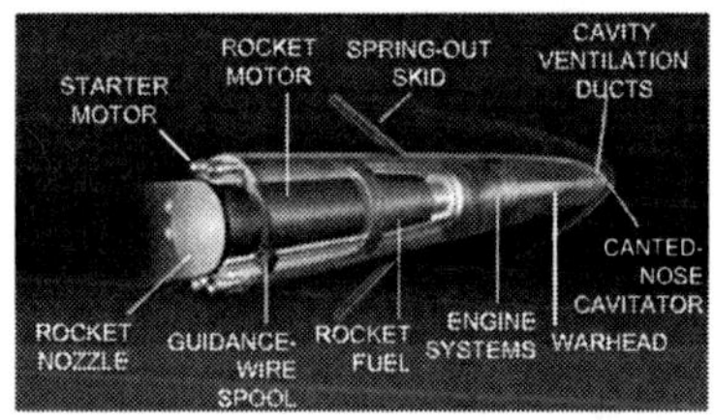

fig. 2. Construction of *Squall* torpedo, Russia

Solution: 1) creation of a cavitational cloud before the front part of the torpedo with the help of a powerful ultrasound (or any other) generator; 2) use of a reactive rocket engine.

fig. 3. Relatively close modern analog (2005): *Barracuda* torpedo, *Diehl BGT Defence*, Germany

ZOOMING

Have the original contradictions have been removed? – Yes!

Super-effects: ability to achieve under water speeds of 600-800 km/h and even supersonic speeds.

Negative effects: increased complexity of motion control. Pursuant to Navigators **05, 07, 10** and **18**, to compensate for this drawback, the torpedo is fitted with folding control fins which can redirect the torpedo by changing their tilt in the upper part of the cavitational cloud.

fig. 4. *Squall* rocket super-cavitation generator

BRIEF DESCRIPTION

In the 1960-es, Russian engineers invented the still-unsurpassed rocket torpedo *Squall* with a fore cavitational cloud generator creating a gaseous-liquid dynamic medium in front of the torpedo's bow.
NB: *Squall* can carry nuclear warheads.
Dominant specialized navigators: **01, 03** and **04**. A special physical/technical effect (creation of super-cavitation) has been used to intensify the spatial and energetical resources.

fig. 13.25. Reinventing of the *Squall* rocket torpedo

Example 13.21. "Flying Lens" from Russia (1): Ideal Integration

TREND One of the most serious "habitual" drawbacks of aircraft is that they... need air fields. The larger and faster the aircraft, the larger the air field it requires. The air field is also one of the factors constraining the aircraft's cargo lifting capacity. We need machines that can make do with a simple even field several hundred meters long. Next, if we wanted to build aircraft capable of carrying, say, 800-1,000 passengers or 100-200 tons of cargo, we could encounter irresolvable design problems. One of those problems is the "thin" wing and its attachment to the hull. Finally, wheeled landing gears simply buckle down under the heavy load. What can we do?

REDUCING

FIM: X-resource, together with available or modified resources, and without making the object more complex or introducing any negative properties, guarantees attainment of the following **IFR**:

[rapid transportation of heavy cargoes for large distances by "field" aircraft].

Formula and solution of the standard contradiction:

32 Weight of the moveable object **VS** 15 Length of the moveable object = 07, 14, 15, 32

Radical Contradiction:

Aircraft	⇨	must have high load lifting capacity and high speed, effect long-distance cargo transportation, and use small air fields several hundred meters long	but	must have low load lifting capacity and low speed not to need large specially designed and equipped air fields

INVENTING

Dominant models: 07, 14 and 32.

In line with model**14** Use of pneumatic or hydraulic constructions, we can resort to an air-cushion VTOL system. This will enable us to discard the wheeled landing gear.

In line with model **32** Counterweight: 1) to assure a short takeoff run and long super-economic flight, it is possible to use airfoil flight mode; 2) to increase load lifting capacity, it is possible to use the flying "thick" wing concept – the body of the aircraft is its wing! In other words, subject to paragraph 1, one solution integrates different aviation concepts!

In line with model **07** Dynamization, it is necessary to optimize the work of various systems at different stages of the flight.

ZOOMING Have the contradictions have been removed? –Yes! **Super-effects:** 1) top-line safety! 2) possibility to use water surface as the "air field"! 3) possibility to use natural gas as fuel – extremely high environmental friendliness!

BRIEF DESCRIPTION

In the early 1980-es in Russia, Professor L.N. Shchukin proposed an aircraft of the "thick wing" type with an air-cushion takeoff-landing system. It could fly in airfoil and aircraft modes, had high load lifting capacity and flight range, and did not require special air fields!
Key idea: integration of alternative aviation concepts. Navigators present: 07, 14 and 32.

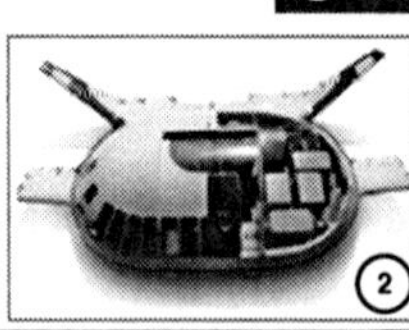

1) Lev Nikolayevich Shchukin (1931-2001), expert of the "space" design bureau created by the space engineering pioneer Sergey Pavlovich Korolyov (1907-1966), later Professor of the Moscow Energy Institute, founder of the company called **EKIP** (from the Russian abbreviation meaning *Ecology and Progress*); 2) EKIP cargo version – up to 120 tons; 3) EKIP passenger version – up to 1,200 seats. Flight altitude – up to 11,500 m. Flight speed – 600 km/h. Flight range – up to 6,000 km. Takeoff run – up to 600 m. Length – 62 m, wing span – 102 m, height – 20 m.

fig. 13.26. Reinventing of the EKIP aircraft concept

Example 13.22. "Flying Lens" from Russia (2): To Conquer the Unconquerable

TREND "Thick wing" has a large air support area and, consequently, ensures higher load lifting capacity. However, it also inherits the key drawback of all wings – emergence of vortices along the upper surface (a), at the tips, and at the trailing edge. For a long time, this problem was thought to be irresolvable. Clearly, it had to be resolved by designing the EKIP craft.

With EKIP, the problem is exacerbated by the "thickness" of the wing-body. It is easy to imagine the "storms" (vortices, turbulence) raging behind a wing which is 20 meters "thick", 20 meters wide, and has a span of about 100 meters!

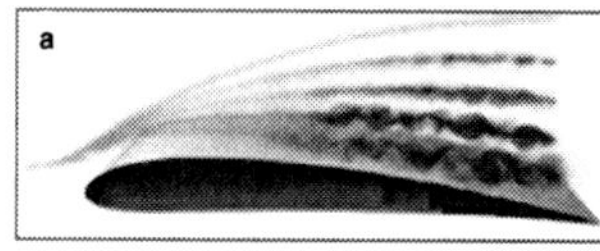

A known solution (1930-es, Germany) is to suck the bordering air layer into the interior of the wing through small holes (b). It is inefficient for large surfaces, while the growth of the number of holes has adverse effect on the wing's durability. What can we do?

REDUCING

FIM: The operating zone BY ITSELF ensures the obtaining of the following **IFR**:

[*ideal vortex-free air shape of the wing*].

Formula and solution of the standard contradiction:

02 Universality, adaptability **VS** 04 Reliability = 01, 11, 18, 32

Radical Contradiction:

Thick wing	⇨	must produce vortices because of the poor air shape of the wing	but	must not produce vortices to assure high-speed securely controlled flight

INVENTING

Dominant models: 01 Change in the aggregate state of the object and 11 Inverse action.

The solution is based on the "inverse action" principle: instead of reducing the number of holes, we must merge them into a continuous aperture running at right angles to the direction of air movement! The aperture can be made in the form of a cylindrical "pipe" from which air is sucked into the body. Such solution can be used because the craft's body has a large usable volume. Accordingly, it may be possible to make several such apertures (1 – wind-tunnel testing of the craft). This will make it possible to manage the parameters of the air layer around the surface of the craft, and obtain a virtually "perfect" air shape (2).

The presence of model 01 can be interpreted as change of pressure and movement of vortices inside the apertures. Such change leads to the emergence, on the surface of the craft, of vortex-free (laminar) flow without separation, which improves the craft's steering sensitivity and overall performance (due to reduction of fuel consumption).

ZOOMING

Have the contradictions been removed? – *Yes!*

Super-effects: 1) superior sensitivity of the craft, *inter alia*, during takeoff and landing at high angles of attack;
2) superior performance of the craft due to reduction of resistance to motion.

BRIEF DESCRIPTION

In the early 1980-es in Russia, Professor L.N. Shchukin proposed, for the **EKIP** aircraft, a system to control air layer parameters above the aft portion of the craft. The system consisted of multiple tubular apertures located transversely throughout the span of the wing-body and connected with the surface of the craft by solid suction fissures.
Navigators present: 01 and 11 (instead of limiting the number of holes, as was the case with previous **EKIP** versions, now the fissures and apertures crossing the body are made continuous).

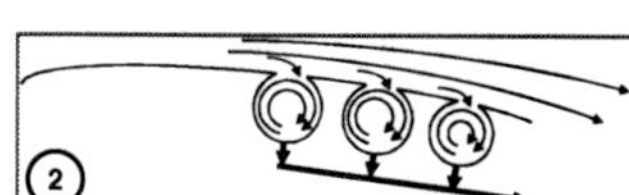

fig. 13.27. Reinventing of the EKIP body-wing

Example 13.23. Russian "Flaying Saucer" – Giant of the Aerial Ocean

TREND

Helicopters would be ideal as the means of transportation of large-size cargoes, as they can hover over the cargo pickup and dropdown areas.

Unfortunately, helicopters have limited load lifting capacity. This drawback restrains their use to put out fires, especially in high-rise buildings.

Their second drawback is the limited distance by which they can carry cargoes. This drawback prevents the use of helicopters to fly to remote areas (taigas, deserts, polar areas, seas and oceans, etc.).

What can we do?

REDUCING

FIM: The X-resource in the form of particles of substance or energy is located inside the operating zone and ensures, together with the other available resources, the obtaining of the following **IFR**:

[long-distance transportation of super-heavy cargoes for by VTOL aircraft].

Formula and solution of the standard contradiction:

32 Weight of the moveable object **VS** 30 Force = 02, 06, 27, 32

Radical Contradiction:

Aircraft	⇨	must have high load lifting capacity and move cargo by long distances	but	must have low load lifting capacity and flight range due to large own weight and cargo weight, such drawback not being sufficiently compensated by the lifting force generated by rotors in known crafts

INVENTING

Dominant models: 27 Full use of thermal expansion and 32 Counterweight.

The solution is based on the use of material and energetical resources: In line with model 32, we need to create a craft based similar to a dirigible (or aerostat) and filled with helium (this gas is not combustible, which makes it superior to hydrogen). The deficient lifting capacity must be obtained by heating the air inside the craft's body according to model 27 (this idea – in application to an all-metal aerostat (in those times, the term "dirigible" did not exist) – was voiced by K.E. Tsiolkovsky in 1885-1886)!

ZOOMING

Have the contradictions been removed? – Yes!

Super-effects: takeoff and landing do not require the use of ballast, but are performed by heating/cooling the air inside the body of the craft!

Negative effects: -

BRIEF DESCRIPTION

A dirigible- or aerostat-like aircraft has been proposed where the lifting force is generated by combined use of a lighter-than-air gas (helium – the first component) and heating of the air (the other component).

Navigators present: 27 and 32. The material and energetical resources are also intensified.

Gas-heat lens dirigible ALA Thermoplane (joint design by the Moscow Aviation Institute, Thermoplane Design Bureau, and Ulyanovsk Aviation Plant, late 1980-es)

"Craft of the future" – LOMOSKY company, Moscow-Ulyanovsk

Aircraft of the future: diameter – about 300 m, flight altitude – 5 km, flight speed – 250 km/h, load-lifting capacity – 2,000 tons.
Source: www.locomosky.ru

fig. 13.28. Reinventing of the gas/heat-powered Thermoplane dirigible

Example 13.24. Idealization According to Bartini: Annular Wing Aircraft

Aviation experts know that at high speeds wingtips start to generate considerable lift-induced drag. It is caused by the vortices emerging at wingtips as air emerges from under the wing under very high pressure. To reduce the drag, designers sometimes use "winglets" – vertical wingtip extensions.

Original RC: the wing *must have a tip* (like any physical object with terminal length) which produces air vortices, **AND** the wing *must not have a tip* to prevent emergence of such air vortices.

In 1935, as R. Bartini was designing his DAR aircraft (a long-range Arctic reconnaissance plane), he installed two coaxial propellers inside a cylindrical nacelle with profiled surface elements (fig. 13.29). Flight tests revealed the so-called "Bartini effect" – increased thrust[192]!

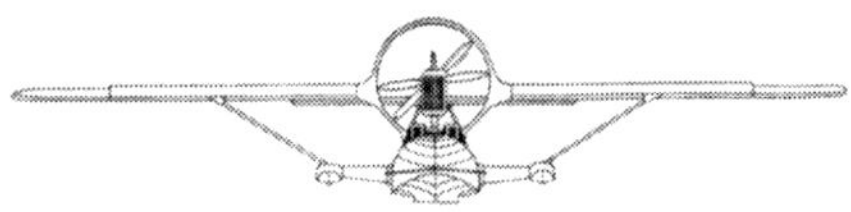

fig. 13.29. Bartini's DAR (symbolically, the Russian word "dar" means "gift"); www.navy.su

The ring in fig. 13.29 can be regarded as a part of the wing. Now, here is an idea: what if such ring encompassed the entire plane! This would produce a solution according to Bartini – join wingtips to form an "endless" wing", and you have an annular wing aircraft, or a "ringplane"! And such aircraft were built (fig. 13.30,a,b), but the idea did not catch on.

However, recently this severed thread of evolution was picked up again. In 1988 two Belarus fellow thinkers – aviation mechanic A.A. Narushevich and pilot A.L. Gushchin – built and tested a "ringplane" (fig. 13.30) with an oval wing! In 2006 they were joined by military pilot, former Chief Navigator of Belarus Air Forces A.M. Anokhin and Professor L.I. Grechikhin. L.I. Grechikhin believes (www.popularmechanics.ru) that "aircraft with classical wings will soon exhaust themselves. It is really very simple: the larger the aircraft, the heavier and sturdier its wings, the more difficult it is to maintain their rigidity. On the other hand, an oval wing which produces the same lifting force is two times lighter." Keeping in mind that technology evolves along a spiral-shaped path, we wish these enthusiasts luck with testing new materials and technologies, but especially with finding… investors!

fig. 13.30. "Ringplane" from Belarus and its prototypes

a) M.V. Sukhanov project (1936-1943), Moscow Aviation Institute; b) Heinkel Lerche ("Lark") VTOL, 1944; c) Belarus "ringplane" nicknamed "Potatoplane" ("Bulbolyot" – in Russian pronunciation) by its creators, 1988; "Potatoplane" during take-off d) and in flight e)

[192] I.E. Chutko (1978) *Red Aircraft*. – Moscow, Politizdat (Political Publishers; in Russian)

Example 13.25. Visible Is Incredible: Is This a Working *Pepelatz*!?

fig. 13.31. *Pepelatz* samples from the planet *Plyuk-213*

Most of those who saw the wonderfully sarcastic space sci-fi adventure movie *Kin-Dza-Dza!* (Director: Georgiy Danelia, b. 1930) had a good laugh at the expense of strange aircraft trudging over the surface of the planet *Plyuk-213*, apparently employing some "anti-gravitation" principle. Well, we do know at least one of their secrets: they used... matches! Probably to "fire their engines". Those aircraft (fig. 12.31) were called *pepelatz*.

G. Danelia explained in one of his interviews that that name came from the Georgian word *pepel* which means *butterfly*.
We will discuss "anti-gravitation" in our last Example 13.38 *The Lord of the Rings and the Strings*. At this juncture I suggest that we dwell on something which is simpler – but equally impressive (see video clip at www.aesir.com). Incidentally, in ancient German and Scandinavian mythology, the word *Aesir* refers to Gods, supreme celestial beings, or even entities. Know what I mean, huh?

Anyway, their *pepelatz* had to be able to fly the appropriate way. And it appears that the English firm AESIR has built such aircraft (fig. 13.32).

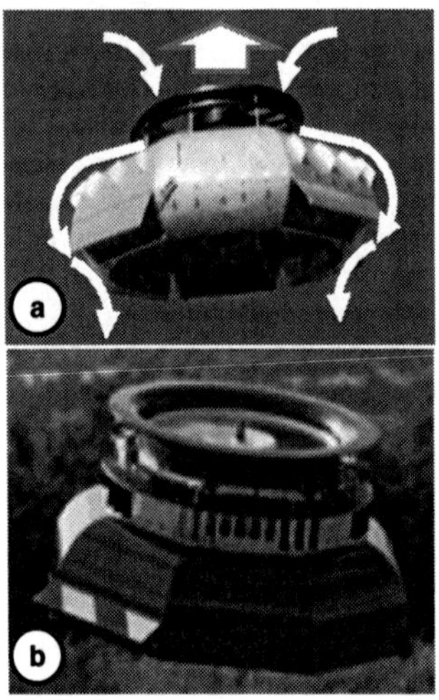

fig. 13.32. *Pepelatz* developed by AESIR
a) test model;
b) Embla, diameter 60 cm

OK, here goes – **Jocular RC:** the aircraft *must have* a propulsion unit (to be able to fly) **AND** it *must not have* a propulsion unit (so that we all have absolutely no idea how on earth (or in the air ☺) it moves!).

And indeed – we see it, but we do not believe it: *aparent = incredible*[193]! Yes, it has a rotating propeller at the top. If you do not know the answer, it may initially appear that this aircraft is an air-cushion vehicle or a "normal" helicopter whose rotor is built into a ring! But then you start to have your doubts: there is just the one rotor, but the body of the "helicopter" is not getting any "reverse" rotation! And if it were an air-cushion vehicle, it would not be able to rise so high above the ground!

Very serious answer: the aircraft goes up because in accordance with the Coandă effect there is a difference between pressure levels *above* and *under* its body! The rotor creates an air jet which flows down along the body of the aircraft and "clings" to its curved surface. The aircraft steering and stabilization functions are performed by deflecting flaps (visible in the photos).
Well, that Cambridge IS a mysterious place! Because it is the birthplace of both the host of the *Apparent – Incredible* TV show and the ASIER version of the *Pepelatz*!

---

193 *Apparent – Incredible*, a famous Russian science TV show which went on the air on February 24, 1973, and for 37 years running (a record!) has been hosted by Professor Sergey Petrovich Kapitsa (b. 1928, Cambridge, England) – outstanding scholar and science promoter, son of P.L. Kapitsa (1894-1984), 1978 Noble Prize winner

Example 13.26. Self-Organization – Learn from Nature!

TREND There are known methods of parachute-assisted descent, in the Earth's atmosphere, of various cargoes, ranging from large (military equipment) to small (scientific instruments). The parachute has a number of drawbacks. First, the "unmanaged" deployment of the main parachute – it opens up only under the influence of the oncoming air (due to the drag produced by the pilot chutes) – which is not sufficiently reliable. Second, the effect of sidewise winds in the atmosphere during the descent. As a result, the parachute may "collapse", the lines may twist or tangle up, or other dangerous developments may ensue. If the descent takes place in the atmosphere of other planets, for example, on the Mars where the atmosphere is more rarefied than on the Earth, the drawbacks of the parachute-assisted descent described above become even worse. What can we do?

REDUCING FIM: The X-resource in the form of particles of substance or energy is located inside the operating zone and ensures, together with the other available resources, the obtaining of the following **IFR**:

[safe form and structure of the deceleration system ("brake") regardless of the weight of the cargo].

Radical Contradiction (RC):

The "brake" stabilizes ► by itself (like the parachute, but this is not reliable) **VS** forcefully (like the wing, but this is complex)

Formulas and solutions (in general terms) of standard contradictions (SCs):

To achieve deceleration at high speeds, the "brake" must have a large surface area –

SC1: 22 Speed **VS** 17 Surface of the moveable object = 14, 15, 25

At high speeds, the shape of the working surface of the "brake" can become disrupted –

SC2: 22 Speed **VS** 21 Shape = 01, 06, 07, 15

When the speed of descent decreases, the "brake" is exposed to damaging atmospheric factors –

SC3: 22 Speed **VS** 13 Internal damaging factors = 01, 03, 04, 36

To decelerate a heavy cargo, the "brake" must have a large surface area –

SC4: 32 Weight of the moveable object **VS** 17 Surface of the moveable object = 14, 15, 19, 30

Increasing the working surface of the "brake" reduces its reliability –

SC4: 17 Surface of the moveable object **VS** 04 Reliability = 14, 39

MITO Ranging: 14^3, 15^3, 01^2, 03, 04, 06, 07, 19, 25, 30, 36, 39

INVENTING Taking into consideration the entirety of requirements as described above, to resolve the RC in the first approximation we may need to consider the analogs created by Nature, such as falling leaves or maple-tree samaras, or existing in sports, such as shuttlecocks in badminton. These structures have high sustainability to external exposures, and are capable of recovering their orientation in flight. Under the influence of the air, they always turn with their heavy part down, assuring that the center of gravity of the cargo (fig. a) is almost always located at the lowest point of the "construction". This solution also matches model ***19*** *Transition into another dimension.* What we have is a flying roly-poly!

In line with recommendations provided by models ***04*** *Replacement of mechanical matter,* ***07*** *Dynamization* and ***14*** *Use of pneumatic or hydraulic constructions,* it is possible to make the body of the "brake-shuttlecock" inflatable. So, the first step in reinventing is: *the use of a self-organizing construction, life a leaf or a shuttlecock.* During a protracted descent from the orbit, it would be advisable to protect the "brake" from overheating by using a similar first "stage" (fig. b), which is consistent with models ***03*** *Segmentation* and ***15*** *Discard and renewal of parts.* The landing module (fig. c) is inserted into the main stage in accordance with model ***34*** *Matryoshka.*

ZOOMING Super-effect: possible use for saving people caught in high-rise building fires.

BRIEF DESCRIPTION

A "shuttlecock" pneumatic deceleration device (PDD) has been suggested to effect descent of cargoes from the orbit to the surface. The device has two inflatable bodies: one for deceleration in an overheated environment, and one for soft low-speed landing. The reinventing relies on models 14, 15, 01, 03, 04, 07 and 19, and on model 34 Matryoshka.

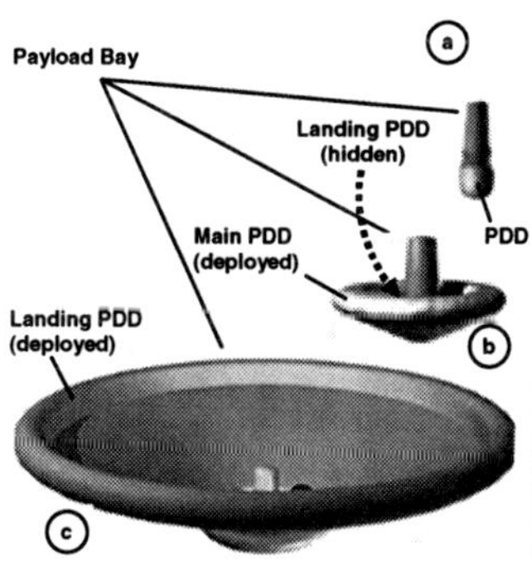

fig. 13.33. Reinventing of the inflatable deceleration "shuttlecock" for landing on the Mars (The original concept was designed at the Babakin[194] Research Center under the Lavochkin[195] Research and Production Association; a similar project was later developed by the NASA Langley Research Center)

---

194 Georgiy Nikolayevich Babakin (1914-1971) – outstanding Soviet designer of spacecraft, including the world's first artificial satellites ("sputniks") and descent modules for Moon, Venus and Mars landings

195 Semyon Alekseyevich Lavochkin (1900-1960) – outstanding Soviet aircraft designer, creator of famous fighters used during the Second World War, the world's first supersonic cruise missile *Burya* (Tempest), anti-aircraft rockets, sputniks and inter-planetary space stations

Example 13.27. Dynamization-1: Transformation Model and "Law of Development"

The history of aviation is an inexhaustible source of creative models – models that can be used to make inventions and innovations. This was noted by G.S. Altshuller: the most irreconcilable contradictions – speed VS weight, maneuverability VS simplicity of construction, etc. – are played out in the field of aviation science. And it is aviation designers that have tested many ideas that have since become classical, and that have blazed the trail to resolving many "standard" conflicts by using the same methods.

One of the most efficient methods of doing that is "dynamization" of construction. It is a manifestation of an important system evolution law according to which system manageability can be improved only when the parameter being managed can be modified within a sufficiently wide range of values. One of the best examples illustrating that law is evolution of the aircraft wing construction. For apparent reasons, military applications of aircraft have often been the main driving force behind such evolution.

Alexander Lippisch (1894-1976)

Adolf Busemann (1901-1986)

We have reviewed similar examples in the previous sections. Now we will limit ourselves to several key events of the wing dynamization history which affected two of the numerous realizations of the dynamization method. These two examples will deal with two outstanding inventions by aircraft designer Professor Alexander Lippisch[196].

In the mid 1930-es, when maximum aircraft speeds did not exceed 300-400 km/h, the German aerodynamicist Professor Adolf Busemann[197] published a mathematical substantiation of the supersonic flight theory asserting that swept wings would be ideal for such flight. However, his computations also showed that in low-speed flight such wing was inferior to the usual "rectangular" wing.

There arose an "irreconcilable" *radical* contradiction:

Wing sweep angle must be ⇨ *large* for supersonic flight & *small* for low-speed flight

In 1942 A. Lippisch patented a solution of that contradiction – which at the time did not seem too promising, but was interesting in terms of the technical possibilities that it opened – a variable-sweep wing. Unfortunately, with the war under way there were many other tasks to attend to, and the idea was not implemented until 1944. Among other things, since 1939 A. Lippisch had been the "brain father" of Messerschmitt-163, the world's first rocket-powered fighter aircraft which had exceeded the historical 1,000 km/h speed limit as early as in 1941. He was also involved in designing VTOL aircraft.

---

196 Alexander Martin Lippisch (1894, Munich – 1976, Cedar Rapids, USA) – German aerodynamics engineer who researched such pioneering ideas as "flying wing", "delta wing", "ground effect", and "variable-sweep wing", worked on vertical takeoff-landing aircraft and designed the first jet aircraft; in 1945 under Operation Paperclip he was taken to the United States where he designed a number of experimental and serial aircraft.

197 Adolf Busemann (1901, Lübeck, Germany – 1986, Boulder, USA) – German supersonic aerodynamics pioneer, creator of the "swept wing" idea and the "area rule", the classical design technique used to optimize performance metrics while designing supersonic aircraft; taken to the USA under Operation Paperclip; photo: www.scientistsandfriends.com/aerodynamics.html

Wilhelm Messerschmitt (1898-1978)

On a more intriguing note, in 1942, in an aerodynamic laboratory in Adlershof, Berlin's south-eastern suburb (the Modern TRIZ Academy is now situated at a distance of some 300 meters from that historical building), his team started wind-tunnel testing of a prototype variable-sweep wing aircraft Messerschmitt P-1101[198].

In early 1945 an almost completed Messerschmitt P-1101 was captured by a special US team ordered to find the secrets (and experts!) of Lüftwaffe and other branches of the German armed forces. Messerschmitt P-1101 inspired Bell engineers to design Bell X-5, the world's first aircraft that could change its wing sweep in flight (fig. 13.34). Its maiden flight took place in 1951.

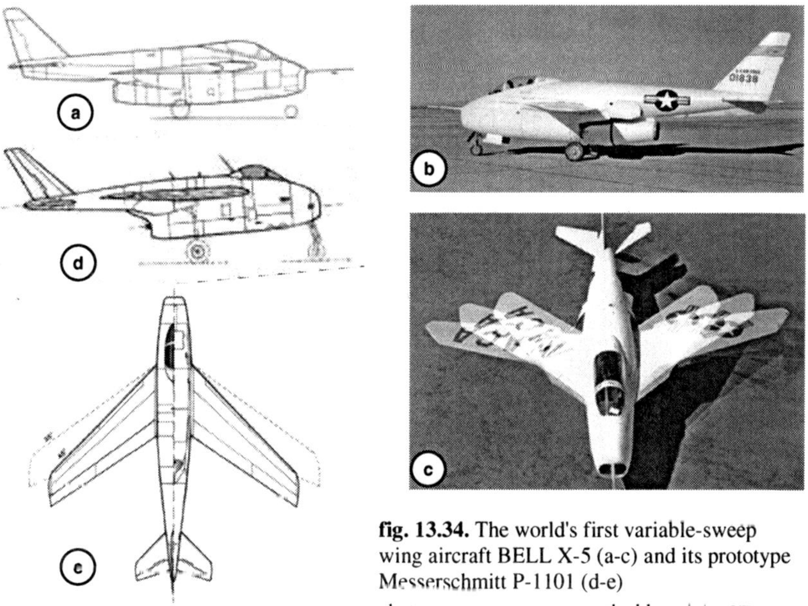

fig. 13.34. The world's first variable-sweep wing aircraft BELL X-5 (a-c) and its prototype Messerschmitt P-1101 (d-e)

photo: www.nasa.gov, www.the-blueprints.com

The fighter MiG-23 became one of the world's first serially manufactured variable-sweep wing aircraft. It was originally designed by famous aircraft designers A.I.Mikoyan[199] and

[198] Wilhelm Emil Messerschmitt (1898, Frankfurt am Main – 1978, Munich) – outstanding German aircraft designer, creator of Messerschmitt Bf 109, the most produced fighter in history, and a number of other excellent combat and experimental aircraft. Wilhelm Messerschmitt refused to go to the USA, and was sentenced to two years in prison because his enterprises had used PoW labor during World War II. In 1960-1968, he built the experimental German jet fighter EWR VJ 101, the first VTOL aircraft to break the sound barrier.

[199] Artyom Ivanovich Mikoyan (1905-1970) – one of the pioneers of jet aviation in the USSR, designer of Mig-1 and Mig-3 aircraft which took part in World War II, and of MiG-15 – MiG-25 aircraft, head of Design Bureau 155 (later named after him; current name: RSK Russian Aircraft-Building Corporation "MiG"

M.I.Gurevich[200] (the famous MiG abbreviation is based on the first letters of their last names). Design and testing of the aircraft were completed in 1967 by G.A. Sedov[201], prominent designer and test engineer.

MiG-23 boasted the following parameters: maximum supersonic speed – 2,500 km/h (more than 2M); maximum speed near ground – 1,350 km/h (higher than sound speed!); flight range – 2,000 km; flight ceiling – 18,000 m; landing speed – 240 km/h; takeoff run – 500 m; landing run – 750 m.

Artyom Ivanovich Mikoyan (1905-1970)

Mikhail Iosifovich Gurevich (1893-1976)

Grigoriy Aleksandrovich Sedov (b.1917)

Aircraft dimensions: length – 16.7 m; height – 5.77 m; wing span in "swept" and "spread" positions – 7.8 m and 14.0 m, respectively. Fig. 13.35 shows two wing positions together with aerial snapshots of one of the world's fastest birds – swift: And in conclusion – standard reinventing (fig. 13.36) for the variable-sweep wing.

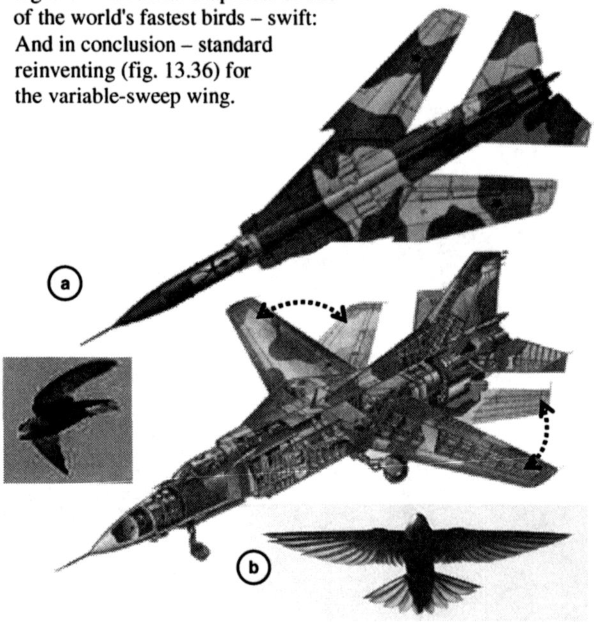

fig. 13.35. Fighter MiG-23
www.rusplane.ru, http://avs-lt.livejournal.com/90881.html

Having completed this example, the most attentive readers may have retraced their steps to Example 2.30. *B-1 and Tu-22M3 Bombers* and Example 2.31. *Su-24 Attack Aircraft*. Still, the variable-sweep wing is becoming part of history, as it is being replaced by other wing types (see the following example) based on increasingly sophisticated applications of the dynamization method and becoming closer and closer to the analogs created by Nature.

[200] Mikhail Iosifovich Gurevich (1892-1976) – outstanding Soviet aircraft engineer, co-head of Design Bureau 155 in 1940-1964

[201] Grigoriy Aleksandrovich Sedov (b. 1917) – outstanding Soviet test pilot and aircraft designer, Chief Designer of Mikoyan Design Bureau in 1970-1998

TREND

The faster fighter planes fly, the higher their pre-landing speed. Accordingly, with longer landing runs (AND takeoff runs) they require longer runways. What can be done to resolve this problem?

REDUCING

FIM: The X-resource in the form of particles of substance or energy is located inside the operating zone and ensures, together with the other available resources, the obtaining of the following **IFR**:

[shorter runway].

Standard (Technical) Contradiction

Fighter aircraft

- (+) Higher speed → Factors: 22 Speed
- (-) Landing on shorter runway → Factors: 16 Length of the fixed object

Navigators

- 07 Dynamization
- 19 Transition into another dimension
- 25 Use of flexible covers and thin films

Radical (Physical) Contradiction

Fighter aircraft ⇨ must have high speed when in flight & must have low speed when landing

INVENTING The key model is 07 **Dynamization:** a) the characteristics of the object (or its environment) must changed so as to be optimal at every step of the work procedure; b) disassemble the object into parts that are moveable with respect to each other; c) if the object is fixed, make it moveable – this model calls for the making of variable sweep wings.

The IFR according to Bartini is clearly present in the solution of the radical contradiction:

sweep angle ► "large" **AND** "small"

ZOOMING The contradictions have been removed: the new aircraft will have shorter takeoff and landing runs, and will be able to use shorter runways. Drawback: increased complexity and weight of the aircraft.

BRIEF DESCRIPTION

Due to the application of the variable sweep wing (transformation model implemented: 07 Dynamization), the fighter can take off and land using shorter runways.

Extreme wing positions at various sweep angles: a) 16°; b) 72°. Intermediate position – 45°.

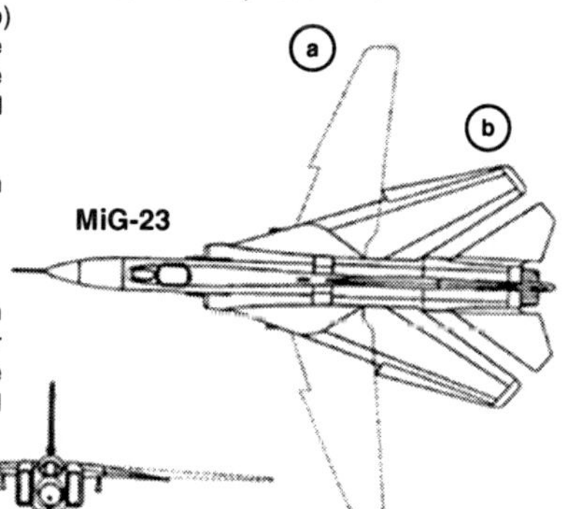

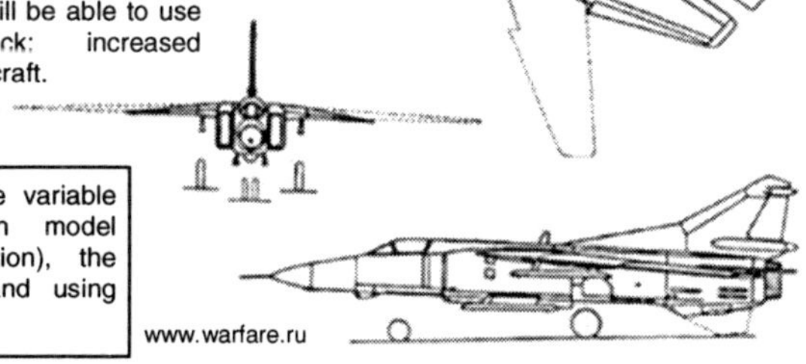

www.warfare.ru

fig. 13.36. Reinventing of the MiG-23 fighter aircraft variable-sweep wing

Example 13.28. Dynamization-2: Continuous Change – "Law of Nature"

Let us continue to resolve the contradiction discussed in the previous example.

If we, acting in accordance with TRIZ recommendations, escalate the requirement imposed on the fighter aircraft at the Trend stage in the reinventing shown in fig. 13.37, the maximum requirement can be formulated as follows: the runway must be not just short, its length should be equal to "zero"! Super-IFR!

This means that the aircraft must rise into the air from a standstill position and then proceed to fly horizontally once a certain altitude has been reached. Conversely, the landing must involve transition from horizontal flight to vertical descent ending in a smooth touchdown. This is the "functional portrait" of a VTOL or V/STOL (Vertical and/or Short Take-Off and Landing) aircraft. Incidentally, some helicopters are also classified as V/STOL aircraft.

For many decades, many designers have dreamed about creating an efficient V/STOL. Our interest, however, lies in identifying certain creative principles realized in a number of well-known examples. It turns out that numerous different realizations are based on the same principle – ***dynamization*** of construction. By the same token, constructions keep being changed in an attempt to enhance efficiency.

Be that as it may, unless we are prepared to modify the physical method of thrust production, every solution by hook or by crook must use the dynamization principle.

fig. 13.37. V/STOL aircraft using the dynamization principle:

a) rotating jet engines – German EWR VJ 101 (1960-es, with the participation of W. Messerschmitt); **b)** rotating propeller engines – American Bell XV-15 (1970-es); **c)** rotating flaps – German Dornier Aerodyne E-1 conceived by A. Lippisch (late 1960-es); **d)** rotating nozzles – English Hawker P-1127 (1960), predecessor of the famous Harrier (photo by the author, Science Museum, London, 2003)

So there is no doubt that the future "flying saucer" or "graviplane" using, as its prototype, the *pepelatz* described in Example 13.25 will be based on a completely new, currently unknown operating principle.

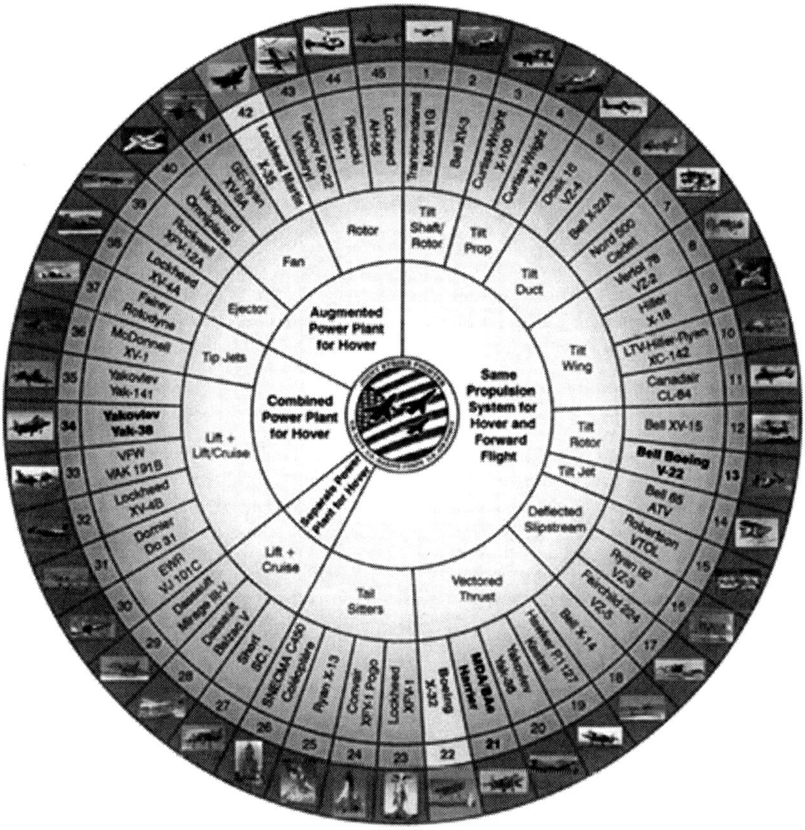

fig. 13.38. The "wheel of evolution" of serial V/STOL aircraft from early 1960-es to mid 1990-es

Source: www.vstol.org

This gorgeous diagram illustrating the flights of creative engineering fancy shows that most projects are initiated by, and developed for, the military. And this means that we are wasting huge financial and material resources, squandering away our common heritage accumulated over the course of history of the human civilization (the price of one of the "items" is disclosed in the following example).

And this is only a small portion of military projects, both completed and ongoing. It is a colossal "black hole" devouring resources that would many times over suffice to resolve burning environmental and other similar issues facing the planet.

Example 13.29. Dynamization-3: Variable Permanency – "Law of Nature"

There is another important TRIZ law – all systems "seek idealization" (see Chapter 9. *System Evolution Models*). But while constructions keep improving, the original fundamental principle, the key functional idea remains the same. It is a system evolution invariant.

This TRIZ idea is fully supported by experimental research. In this section we will discuss evolution of two invariant ideas – "flying wing" and "delta wing" (fig. 13.39).

Unfortunately, it does not cover all interesting ideas (some of them are scattered throughout the other sections of the book, though). Presentation of the two "lines" is not intended to preserve the strict evolutionary sequence, and certain models may feature attributes typical for both "lines".

So this illustration is but a rough sketch of the general pattern, but that pattern is extremely important, as it reflects the basic principle (see section *9 System Evolution Models*) in accordance with which evolution of an idea lasts in time, can produce branches which then "converge" and/or "diverge", and may assume cyclical and helical shape[202].

And each "point" on this evolutionary trajectory is a step along the path of development of the original systemic idea, i.e. an invention.

Reinventing of such examples and evolution lines is likely to produce new models and methods that can be used to create innovative and inventive solutions. However, to the best of my knowledge, such work was initiated (but never completed) only within the framework of several projects started by large companies.

The absence of a concept and language for standard presentation of extracting and reinventing results remains one of the most significant obstacles.

The extracting and reinventing models and examples discussed in this book lay the foundation for developing a paradigm that would enable accumulation by major producers and industrial research centers of creative knowledge objectively present in any innovation, any invention, any new product.

Creation of databanks storing information about efficient extracting and reinventing examples would change the entire organization of basic TRIZ/creative theory training. This is the key operating area of the Modern TRIZ Academy which has the concepts and the know-how required to develop such databanks.

And our Academy invites TRIZ experts and other interested researchers and designers to engage in cooperative projects aimed at developing new practice-oriented technologies and products.

TRIZ – the systemic creation theory – continues to evolve...

---

[202] See also *Helical Technical Evolution Models in Projective Projects*, *Morphic Fighter Wing of the Future: Forgotten Beginnings*, and other works by the well-known TRIZ expert Yu.E. Danilovsky (www.metodolog.ru/category/2/23).

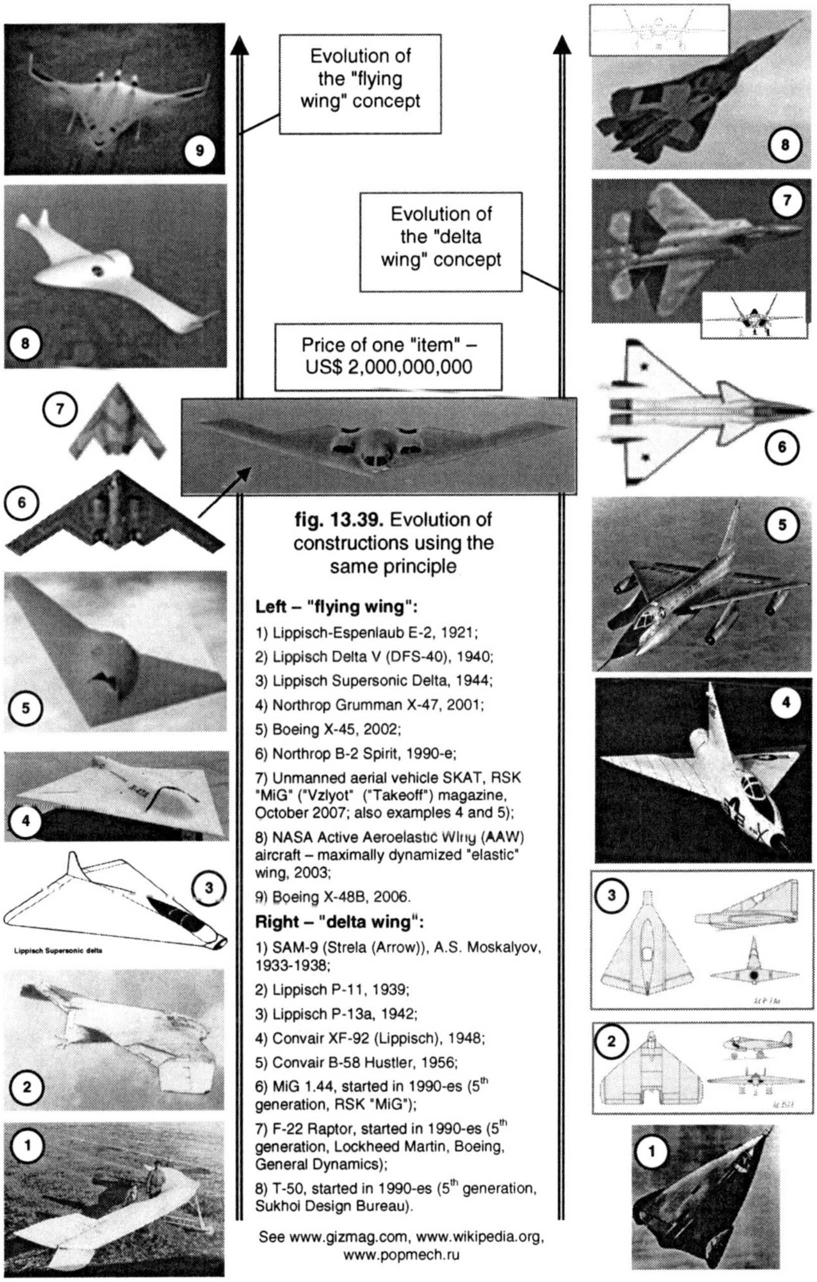

fig. 13.39. Evolution of constructions using the same principle

Left – "flying wing":

1) Lippisch-Espenlaub E-2, 1921;
2) Lippisch Delta V (DFS-40), 1940;
3) Lippisch Supersonic Delta, 1944;
4) Northrop Grumman X-47, 2001;
5) Boeing X-45, 2002;
6) Northrop B-2 Spirit, 1990-e;
7) Unmanned aerial vehicle SKAT, RSK "MiG" ("Vzlyot" ("Takeoff") magazine, October 2007; also examples 4 and 5);
8) NASA Active Aeroelastic Wing (AAW) aircraft – maximally dynamized "elastic" wing, 2003;
9) Boeing X-48B, 2006.

Right – "delta wing":

1) SAM-9 (Strela (Arrow)), A.S. Moskalyov, 1933-1938;
2) Lippisch P-11, 1939;
3) Lippisch P-13a, 1942;
4) Convair XF-92 (Lippisch), 1948;
5) Convair B-58 Hustler, 1956;
6) MiG 1.44, started in 1990-es (5th generation, RSK "MiG");
7) F-22 Raptor, started in 1990-es (5th generation, Lockheed Martin, Boeing, General Dynamics);
8) T-50, started in 1990-es (5th generation, Sukhoi Design Bureau).

See www.gizmag.com, www.wikipedia.org, www.popmech.ru

Example 13.30. The Fight for Speed Is Not Over!

TREND Attainment by aircraft of supersonic speeds disrupts their pitch stability. When an aircraft breaks the sound barrier, the nature of the airflow as it passes along the wing changes, and the center of pressure (point of application of the net lifting force) shifts back, producing a negative pitching moment. The solution based on redistribution of fuel in the aircraft's fuel tank system was invented in the late 1950-es by the outstanding Soviet air designer V.M. Myasishchev (1902-1978) while he was working on the supersonic strategic bomber M-50. That solution was also used in the fighter-bomber Su-24 developed in the early 1970-es by the young designer M.P. Simonov (b. 1929). However, the system has complex construction and is difficult to control. What can we do?

M-50

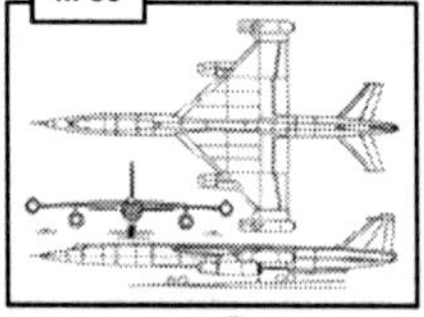

REDUCING

FIM: X-resource, together with available or modified resources, and without making the object more complex or introducing any negative properties, guarantees attainment of the following **IFR**: [simplification of the aircraft's pitch control system].

Formula and solution (in general terms) of the standard contradiction:

01 Productivity **VS** 13 External damaging factors = 01, 11, 18, 21

INVENTING

M-56

Dominant models: 11 Inverse action and 21 Transform damage into use. Interesting interpretations are also possible for models 01 Change in the aggregate state of the object and 18 Mediator.

It is possible to use model **11,a)** instead of the action prescribed by the conditions of the assignment, complete a reverse action, and model **21,b)** eliminate a negative factor by combining it with other negative factors: create higher aerodynamic resistance in the fore part of the aircraft by using forward horizontal surfaces! This solution was proposed for the strategic bomber M-56 (1960, speed up to M=2.5-3.25).

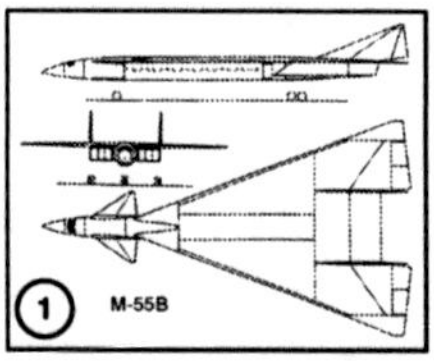

At transonic speeds those surfaces work as "weathervanes" and do not affect the steering of the aircraft. At supersonic speeds, however, they are fixed at a certain angle (this arrangement is called "canard", which is the French word for "duck"), while pitch and altitude are controlled by the elevons installed at the trailing edge of the wing.

Historically, forward horizontal surfaces were first used in 1902-1912 by the American aviation pioneers brothers Wilbur Wright (1867-1912) and Orville Wright (1871-1948).

ZOOMING Super-effect: possible multiple uses.

BRIEF DESCRIPTION

Special horizontal surfaces have been introduced to prevent emergence of the negative pitch moment when "large-ticket" aircraft break the sound barrier. Navigators present: 11 and 21.

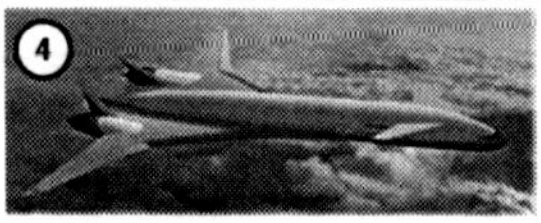

1) Supersonic passenger aircraft **M-55B** (1959): length – 57 m; wing span – 30 m; number of passengers – up to 120; speed – 2M; flight range – 4,000 km; number of engines – 6.

2-3) Mikhail Petrovich Simonov, General Designer of the Sukhoi Designer Bureau since 1983, is presenting the supersonic passenger aircraft **S-51** (2009): number of passengers – up to 50-70; speed – 2M; flight range – 9,000 km; number of engines – 4.

4) Many famous companies are involved in the current technological race to create economic high-speed passenger aircraft. Boeing project **Sonic Cruiser** (2001): number of passengers – 100-300; speed – 0.95M; flight range – 16,000 km; number of engines – 4.

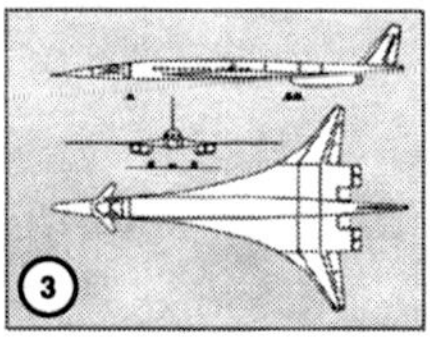

fig. 13.40. Reinventing of canard surfaces

Example 13.31. Even Faster!

TREND Creation of one-stage hypersonic aircraft (OGLA – Russian abbreviation) involved negotiating powerful physical barriers which gave rise to acute contradictions. In particular, it was necessary to resolve the engine problem, as the use of, say, turbojet engines is virtually impossible at supersonic speeds, while the use of jet engines is inefficient at transonic speeds. What can we do?

REDUCING **FIM:** The X-resource in the form of particles of substance or energy is located inside the operating zone and ensures, together with the other available resources, the obtaining of the following **IFR**:

[variable thrust production in different speed ranges].

Formulas and solutions (in general terms) of standard contradictions (SCs):

SC1: 22 Speed **VS** 32 Weight of the moveable object = 04, 05, 11, 30

SC2: 22 Speed **VS** 34 Temperature = 04, 05, 25, 26

SC3: 22 Speed **VS** 36 Power = 01, 05, 08, 30

SC4: 22 Speed **VS** 37 Energy use of the moveable object = 01, 07, 30, 32

MITO Ranging: 05^3, 30^3, 01^2, 04^2, 07, 08, 11, 25, 26, 32

INVENTING **Dominant models:** *05 Separation, 30 Use of strong oxidants, 01 Change in the aggregate state of the object, and 04 Replacement of mechanical matter.* Naturally, the solution is based on the physical/technical effects supporting the appropriate navigators.

	Navigator	Recommendation	Implementation
01	**05 Separation**	single out properties that are critically necessary	at each stage of the flight, use engines whose parameters are best suited for such stage: turbojet engine + propulsive jet engine + liquid propellant jet engine
02	**30 Use of strong oxidants**	b) replace enriched air with oxygen	use of oxygen-hydrogen fuel
03	**01 Change in the aggregate state of the object**	change the properties of substances used	1) use of liquid hydrogen; 2) use of variable air pressure to enable propulsive jet engine operation in a broad range of supersonic speeds
04	**04 Replacement of mechanical matter**	b) transition from fixed fields to fields that change in time	fuel composition and working medium change depending on the flight stage
05	**07 Dynamization**	a) characteristics of the object must be optimal at each step of the operating procedure	see paragraph 01

In the early 1970-es, the Tupolev Design Bureau suggested to create the OGLA on the basis of a combined propulsion unit which consisted of: 1) economical turbojet engines working in the M=0-2.5 speed range; 2) broad-range propulsive jet engines accelerating the aircraft to M=20-25 (!); 3) liquid propellant jet engines increasing acceleration to orbit velocity and enabling orbital maneuvering. In the mid 1980-es, the project was dubbed Tu-2000 in two modifications (fig. 1): space bomber and supersonic passenger aircraft.

ZOOMING Have the contradictions been removed? – *Yes!*

Super-effects: *possible multiple uses.*

BRIEF DESCRIPTION

To build an OGLA, it was proposed to use a combination of several engines, for example: turbojet engine + propulsive jet engine + liquid propellant jet engine.

Dominant navigators: 01, 04, 05, 07 and 30; supporting physical/technical effects are also in play.

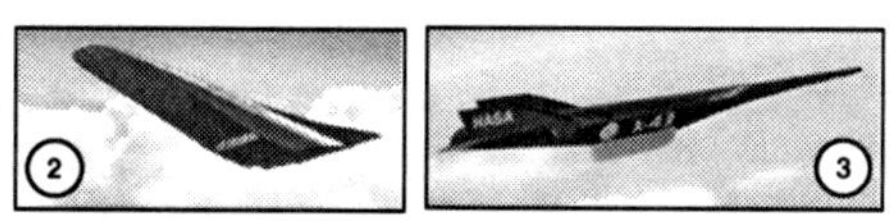

Other projects from the mid 1980-es: 2) **X-30 NASP** (with the participation of McDonnell Douglas, Rockwell International, Rocketdyne, Pratt & Whitney, etc.); 3) **X-43 Hyper-X** (Boeing, MicroCraft Inc, Orbital Sciences Corporation, GASL etc.). Modern project: 4) **A2 Lapcat** (2008) from Reaction Engines, England (NB: engine configuration similar to that used in the hypersonic strategic bomber M-50 designed by V.M. Myasishchev, 1950-es).

fig. 13.41. Simplified reinventing of one-stage hypersonic aircraft

Example 13.32. Spirals of History

The following three examples are using three levels of creation and extraction of DISC model components.

1. Axiological-Motivational Modeling

In the early 1960-es, the USA and the USSR waded into an all-out competitive race seeking to create new military rocket-and-space technologies. By then the USA had almost completed the project envisaging development of a manned spaceplane X-20 that would be placed into an exoatmospheric trajectory by rocket (fig. 12.43), and then would be able to correct that trajectory using its own engines. Eventually it would land as an ordinary airplane.

G.E. Lozino-Lozinsky

(www.buran.ru)

Such aircraft are highly maneuverable, and can be a serious threat for both space-based and surface objects. The fact that the military began to train 7 space pilots, including Neil Armstrong[203], testified to the importance attached to the project.

The Soviet Government instructed Mikoyan Design Bureau to develop a system that would counterbalance the US spaceplane. G.E. Lozino-Lozinsky[204] was appointed the Chief Designer of the project. And he and his colleagues did develop such system which was later named SPIRAL (fig. 13.42).

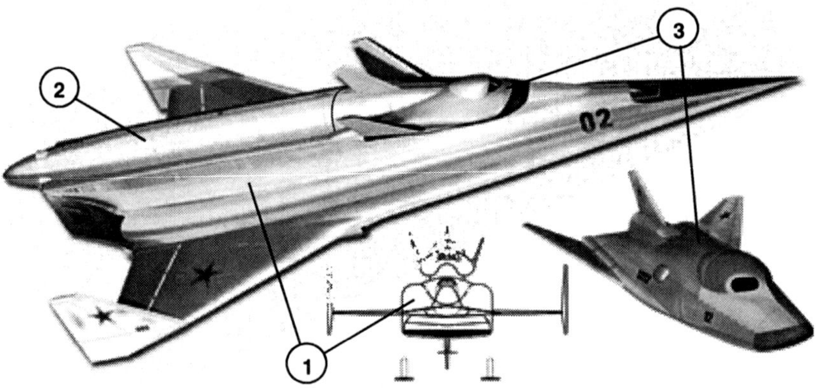

fig. 13.42. Aero-orbital system SPIRAL
1 – hypersonic accelerator plane (HAP), 2 – two-stage booster rocket (TSBR), 3 – orbital plane (OP) which creators nicknamed *Lapot'* (a Russian word meaning *bast shoe*) because of its shape

[203] Neil Alden Armstrong (b. 1930) – American astronaut, first man to set foot on the Moon during the Apollo-11 mission

[204] Gleb Evgenyevich Lozino-Lozinsky (1909-2001) – one of the leading Soviet aerospace designers, member of the Mikoyan Design Bureau since 1941, Chief Designer of **MiG-31**, General Designer of Research and Production Association "Molniya" ("Lightning"), lead designer of the **BURAN** and **MAKS** projects

2. Systemic-Applied Modeling

Rocket launches consume huge amounts of energy. Besides, they are "attached" to certain launching pads. The latter fact makes it difficult to put the spaceplane into orbit at an arbitrarily selected time. More flexibility in this respect is offered by airplane-based launching platforms.

F.A. Zander (see footnote [68]) suggested the use of accelerator planes in the Earth atmosphere as early as in 1924. Subsequently similar ideas were developed by a number of other designers.

The following projects, in particular, present considerable historical and technical interest: pre-WWII *Silbervogel* (Silver Bird, fig. 13.44) and post-WWII German projects (fig. 13.45 and fig. 13.46) *Sänger I* (Junkers, early 1960-es) and *Sänger II* (Messerschmit-Bölkow-Blohm, late 1960-es and prototypes) developed by the German designer Professor Dr. E. Sänger and subsequently used as prototypes for *X-20 Dynamic Soaring*, *X-34*, *X-37* and *Space Shuttle*.

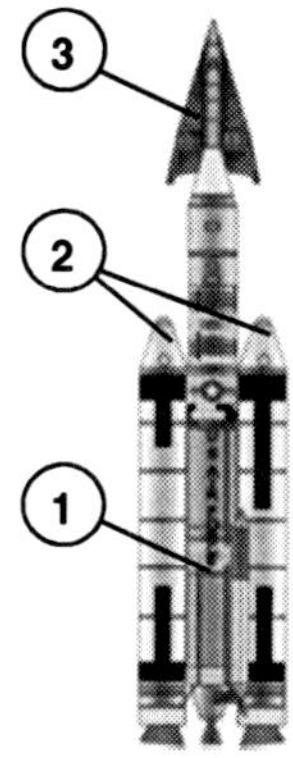

fig. 13.43. Rocket Orbital System X-20 Dynamic Soaring (Boeing, 1959-1963)

1 – booster rocket Titan III C, 2 – additional fuel tanks, 3 – OP

fig. 13.44. Project Silbervogel (1935-1941 and after 1945) by Professor Dr. E. Sänger

1) 3-kilometer monorail trestle with a 30° tilt and a sled carrying 2) the booster rocket and 3) the orbital bomber plane (OP)

www.daviddarling.info/encyclopedia

Eugen Sänger (1905-1964), first President (1951) of the International Astronautical Federation (IAF)

3. Teleological-Creative Modeling

Whether by chance or not, but the title assigned to the SPIRAL project and system is naturally evocative of the helical trajectory of evolution of technical systems. Indeed, pursuant to F.A. Zander, a two- or three-stage aerospace system could have been realized only at the current level of development of its main components.

The three-component system shown in fig. 13.50 was suggested as the key conceptual solution. The spaceplane could also be launched into orbit by the Soyuz (Union) booster rocket.

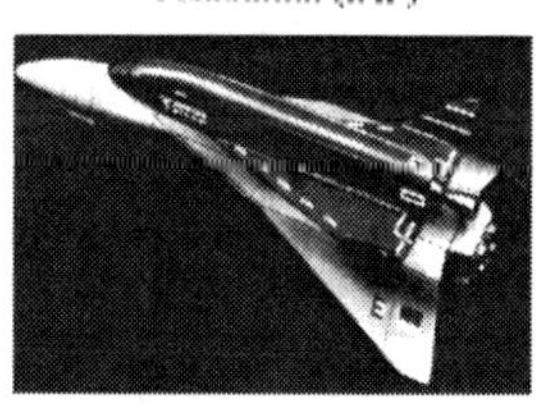

www.astronautix.com

fig. 13.45. Project Sänger I

fig. 13.46. Project Sänger II

Comparison to the rocket prototype enables us to engage in resource-based DISC modeling (fig. 13.47) and extract the following meta-models from the SPIRAL project concept: **D** – *Differentiation*, **I** – *Integration*, **S** – *Separation*, and **C** – *Combination.*

More detailed modeling is possible upon expansion of DISC meta-models with the help of the DISC Catalog (sections *9.4 DISC Meta-Model*). Then it becomes possible to extract dominant models and resources present in the project.

Meta-Model	Dominant Navigators	Dominant Resources	Description
D	**01** Change in the aggreg. state of the object, **04** Replacem. of mech. matter, **07** Dynam., etc.	**Functional**	Selection of the HAP depended on determination and optimization of many functional parameters.
D	**01** Change in the aggregate state of the object, **07** Dynamization, **08** Periodic action, **17** Use of comp. materials, etc.	**Material**	It was necessary to design new materials to protect the OP surface from exposure to excessive heat.
D	**01** Change in the aggregate state of the object, **26** Phase transitions, **30** Use of strong oxidants	**Material**	Selection of fuel was critical: 1) HAP – first kerosene, later liquid hydrogen; 2) TSBR – oxygen-hydrogen fuel; 3) OP – gradual implementation of liquid fluorine as the oxidizer.
D	**02** Preliminary action, **04** Replacement of mechanical matter, **05** Separation, **07** Dynamization, **08** Periodic action, **19** Transition into another dimension, etc.	**Systemic, Functional**	The reference orbit range was 130...150 km by altitude and 45°...135° by inclination north and so uth for launches originated in the territory of the USSR; the flight program had to be completed over the course of 2-3 orbit passes (with the 3rd orbit pass used to land the craft). The rocket propulsion unit installed on board the OP was designed so as to ensure change of orbit inclination by 17° for the *reconnaissance plane* and *interceptor plane*, and by 7°...8° for the rocket-carrying *strike plane* (with reduced fuel load). The *interceptor plane* was also designed to perform the following complex maneuver: change the orbit inclination by 12° and concurrently increase its altitude to 1,000 km.
I	**05** Separation, **10** Copying, **12** Local property, **15** Discard and renewal of parts, **18** Mediator, **19** Transition into another dimension, **32** Counter-weight, **34** Matryoshka, **35** Unite	**Structural, Spatial, Temporal**	To create a highly efficient complex subject to the absence of sufficiently powerful aviation booster engines, it was decided to *integrate* into a single structure the following three components: the OP, the acceleration plane, and the additional booster rocket system.
I	**05** Separation, **07** Dynamization, **11** Inverse action, **12** Local property, **20** Universality, etc.	**Systemic, Functional**	Integration of the orbital maneuvering regime with the aircraft landing regime.
S	Transformation into a super-system	**Systemic**	Creation of the OP, and of the entire system as a whole, resulted in the emergence of a new type of aerospace systems.
S	Transformation into a super-system	**Systemic**	Possibility to use the accelerator plane as a hypersonic passenger or cargo plane.
C	**05** Separation, **07** Dynamization, **12** Local property, **20** Universality, etc.	**Systemic, Functional**	The OP was designed to perform multiple functions: reconnaissance plane, jamming plane, fighter-interceptor plane (space-based and aerial targets), strike plane (surface and sea-based targets).
C +**I**	**02** Preliminary action, **04** Replacement of mechanical matter, **05** Separation, **08** Periodic action, **35** Unite, etc.	**Structural, Energetical**	Inasmuch as the energy of aviation engines was not sufficient, it was expedient to use rocket engines as additional boosters.

fig. 13.47. DISC modeling for the SPIRAL project (fragments)

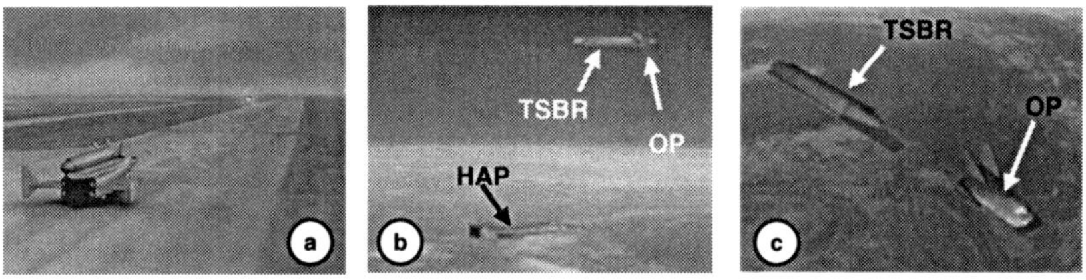

fig. 13.48. Fragments of the animation clip about the **SPIRAL** system
a) system before launch; b) separation of the OP and TSBR from the HAP;
c) separation of the OP from the TSBR (www.buran.ru)

Advantages of the SPIRAL system: 1) 3.0-3.5 times reduction of orbit launch costs per one kilogram of payload compared to rocket systems using the same fuel components; 2) spacecraft launches in a wide range of directions; 3) efficient OP maneuvering not only in space, but also during descent and landing; 4) rapid launch of the OP into any point in the orbit; 5) minimization of the number of air fields required to support the system; 6) aircraft-type landing at night and in difficult weather conditions at the air field assigned to, or selected by, the crew during any of the three orbit passes; 7) unprecedented survivability and guaranteed rescue of the crew at any point during the launch or in flight, whether in atmosphere or in space.

The SPIRAL project laid the foundation for the development, in the 1970-es and 1980-es, of the launch system made up of the *Energy* rocket and the *Buran* (Snowstorm) space shuttle (fig. 13.49). On November 15, 1988, for the first time in history, that extremely complex system performed an unmanned launch of *Buran* which then orbited the Earth and successfully landed exactly at the designated spot!

The ideas realized in the SPIRAL project were further advanced in the unprecedented MAKS Multipurpose Aerospace System (fig. 13.50). MAKS orbit launch costs per one kilogram of payload were even lower than those achieved by SPIRAL, beating rocket launch costs by 10-20 times!

That colossal potential represents invaluable heritage that is useful for, and one day will be claimed by, mankind.

fig. 13.49. Launch of the *Energy-Buran* complex, November 15, 1988, Baikonur Cosmodrome

fig. 13.50. Multipurpose aerospace system **MAKS** (Research and Production Association "Molniya" ("Lightning'), 1994 variant; carrier plane An-225 "Mriya" is shown with retracted landing gear legs)

Example 13.33. *Screw it, let's do it!* [205]

Be innovative.
Nothing is impossible.
Think creatively.
The system is not sacred.
Find another way.

Richard Branson

news.bbc.co.uk

Sir Richard Charles Nicholas Branson (b. 1950), famous British entrepreneur, billionaire, founder of the air carrier *Virgin Atlantic Airways* (1984) and several other companies, including *Virgin Galactic* (2004), holder of world and other records in flying air balloon and crossing the Atlantic Ocean (on a hot-air balloon in 1986, and on a speed vessel called *Virgin Atlantic Challenger II* in 1987).

Elbert Leander "Burt" Rutan (b. 1943), outstanding American innovator and designer of high-efficiency light aircraft, including *Model 76 Voyager* (1986 world record: 9-day non-stop flight around the world); *Virgin Atlantic Global Flyer* (2005 world record: 3-day non-stop flight around the world, etc.); *White Knight One* and *White Knight Two* carrier planes; *SpaceShip One* and *SpaceShip Two* spaceplanes, etc.

James Stephen Fossett (1944-2007), famous American aeronaut and yachtsman; billionaire; he completed the first solo around-the-world air balloon flight – it took him only two weeks, from June 19 to July 3, 2002; he also was the sole pilot of *Virgin Atlantic Global Flyer* in 2005.

Richard Glenn "Dick" Rutan (b. 1938), Burt Rutan's brother and companion, famous aeronaut; in particular, in 1986 he co-piloted the *Voyager* together with Jeana Yeager.

13.33,a. *SpaceShip*

1. Axiological-Motivational Modeling

For me, one of the best examples of commitment, faith, dedication, tenacity, enterprise, professionalism, and creativity is the separate and joint life story of three individuals: Richard Branson, Burt Rutan, and Steve Fossett. There are other people standing by their side, people such as Dick Rutan, Burt's brother – whom I cannot afford the coverage they deserve both because of space restrictions and because their lives – marked with talent, creativity and courage – call for a separate account.

These people turned a new page in the history of human civilization by starting a new type of business – mass-scale space tourism[206]! According to the plans hatched by the *Virgin Galactic* company, in 2012 it will start selling short suborbital "jumps" complete with experiencing the feeling of weightlessness and observing the Earth from a height of about 120 kilometers.

By mid April 2010, the number of "tickets" reserved for those trips exceeded 300!

These people are creating new types of activity which means they are shaping the new values that will gain currency in the 3rd millennium! The results that they have achieved are so precious that they may be regarded as the cornerstone of a new "super-system". They have blazed a trail for many creative people who want to make things better – in their professions, in their countries, on their planet – the ONLY planet available for improvement!

Well, at least for the time being…

---

[205] From book by Richard Branson (2007) *Screw It, Let's Do It: Lessons in Life and Business (Expanded)*

[206] There are already 8 space tourists – people who went into space on Russian Soyuz rockets and spent 1-2 weeks on the International Space Station.

2. Systemic-Applied Modeling

Objectives: Space tourism requires a system which has unprecedented effectiveness, reliability and safety. And, of course, it needs to have irresistible appeal similar to that of an "amusement park ride", as this is consistent with human curiosity and constant urge to experience something new and unusual. And a 120-km "jump" offers just that.

Price: The price must be high enough to make the project profitable and ensure an adequate return on investment, and it must be low enough to make it affordable for potential clients. The current price of a Soyuz/ISS "space tour" is of the "sky-is-the-limit" variety peaking out at US$ 20-30 million, or US$ 2-3 million per day! Very few people can afford that. Clearly, for a large-scale debut – even with sufficient initial demand – this is unacceptable.

Reliability and safety: Similarly to the first paragraph, success depends on the technical characteristics of the system. It was here that the *Scaled Composites* company, headed by B. Rutan since 1982, focused its main efforts. In the early 2000-es it enlisted the support of some of influential people, including Paul Gardner Allen (b. 1953), a billionaire and one of the founders of *Microsoft*, who agreed to become the company's sponsor in 2001.

Scaled Composites struck gold in 2004, when it won a US$ 10 million prize for two successful private (non-governmental) space launches conducted within one week! This became possible due to the amazing qualities of the *SpaceShipOne (SS1)* spaceplane which was "thrown" into space by an equally amazing mothership *White Knight One (WK1).*

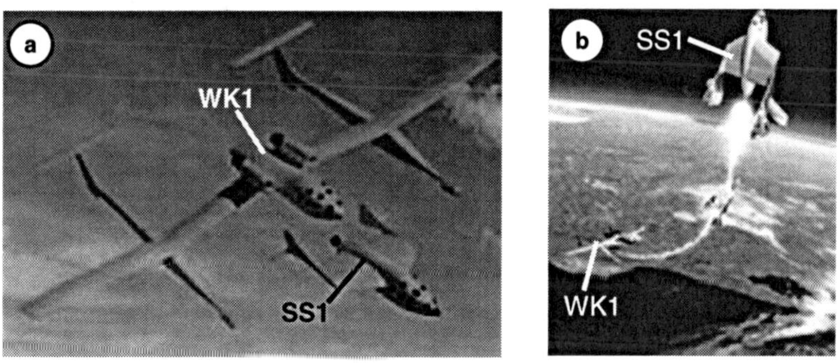

fig. 13.51. SS1 and WK1 in the same film frame:
www.scaled.com: a – separation of the spaceplane from the carrier; b – SS1 rocket-type launch

Development: In 2005 Richard Branson and Burt Rutan founded the *Virgin Galactic* company to promote space tourism! Today the "invitation price" of the space ticket is US$ 200,000. This is 10 times less than the cost of a ride to the International Space Station (ISS). However, if we factor in the duration of the space stay, the price also assumes "cosmic" proportions: even when recalculated for a two-hour flight (with only 6-10 minutes of weightlessness at the altitude of 110 km), the cost per hour skyrockets to a staggering US$ 100,000, which is tantamount to US$ 2.4 million per day! However, this does not diminish the enthusiasm of those who are prepared to lay out that kind of money to go to space – if only once in their lives!

The following sections deal with innovative technical solutions realized in the new system.

3. Teleological-Creative Modeling

3.1. Evolution of Carriers: from *White Knight One* and *Virgin Atlantic Global Flyer* to *White Knight Two*

Resource	Description	Dominant Meta-Model
01. Systemic	1. Creation of a new aerospace system (ASS) type: aircraft-aircraft. Trends: 1) increased "jump" altitude; 2) creation of a spaceplane for quasi-orbital around-the-earth flights, possibly using aerodynamic "ricochet" from the Earth's atmosphere! 3) transportation of other OPs, such as X-37B. 2. Creation of a new ASS application and a new business segment: space tourism. Trend: participation in construction of space hotels. 3. Carrier: can be used as a testing laboratory. 4. Spaceplane: can be used for pilot and astronaut training.	*Separation*, among other things, to ensure adaptation to new applications
02. Informational	1. From the very beginning and over the course of many years, designers have been using extensive mathematical modeling of components and of the entire construction as a whole. 2. Absence of hyper-speeds enables the use of comparatively less heat-resistant materials. 3. Simplification of the carrier management concept (minimization of elements).	*Differentiation*
03. Functional	1. Transformation of hull functions vis-à-vis the fuel tank. 2. The wings perform only one function – that of aerodynamic lifting surface (there are no control elements on the wings! – NB: we would like to note the apparent presence of certain transformation models, including *inverse action* and systemic *separation*).	*Differentiation, Integration*
04. Structural	1. Structural evolution with varying numbers of hulls and engines is shown in fig. 12.53 – optimization of structure to match existing functionality. 2. WK1 and WK2 use structural "unification" and "segmentation" (attachment and detachment of the spaceplane). 3. Wings and hulls have "sections" to divide fuel into portions.	*Combination, Differentiation, Integration*
05. Temporal	1. Fuel stored in the "sections" is consumed "centripetally" – from wing/hull tips towards the "center". 2. Absence of hyper-speeds enables the use of comparatively less heat-resistant materials.	
06. Spatial **WK1**: length – 17 m; wing span – 25 m. **WK2**: length – 24 m; wing span – 43 m.	1. Spatial "separation" of the hulls and the central position of the "cargo" and "power" hull redistributes the load borne by the wings so that the wingtips have the lightest "burden". 2. The fact that the spaceplane is suspended under the wing places the center of gravity of the system in the optimal position. 3. The wings are W-shaped ("inverted gull wings") (see Example 5.13. Stal-7: Steel Gull) – the wing's resistance to breaking is increased. 4. With tail surfaces installed on two "separated" hulls, it is easier to install the engines without unduly increasing the length or height of the aircraft. 4. In WK1 the spaceplane's tail surfaces are positioned so that they cannot be affected by the hot gas jet produced by the engines. The same is true for WK2 tail surfaces. 6. There are many portholes of different diameters instead of one "cockpit", which improves pilot view without impairing the aircraft's durability.	*Differentiation, Integration*
07. Material	1. The weight of the aircraft is drastically reduced due to the use of carbon-based composites. WK1 empty weight is about 1 ton (!), while fully fueled it weighs 4 tons; for WK2 these parameters are 3.5 tons and 13 tons, respectively. Carbon fiber-reinforced plastic is ten times more durable than steel, and much lighter than standard aluminum-based alloys! 2. The carriers use kerosene. The fuel of choice for the spaceplane is non-explosive and non-poisonous ("caoutchouc" + nitrous oxide).	*Differentiation, Integration*
08. Energetical	1. It may well be that both WKs are the world's slowest jet airplanes. Accordingly, despite the large wings, their engines are powerful and light! 2. Power supply: lithium batteries. 3. Across-the-board weight reduction.	*Differentiation, Integration*

fig. 13.52. Resource-based solution modeling (fragments)

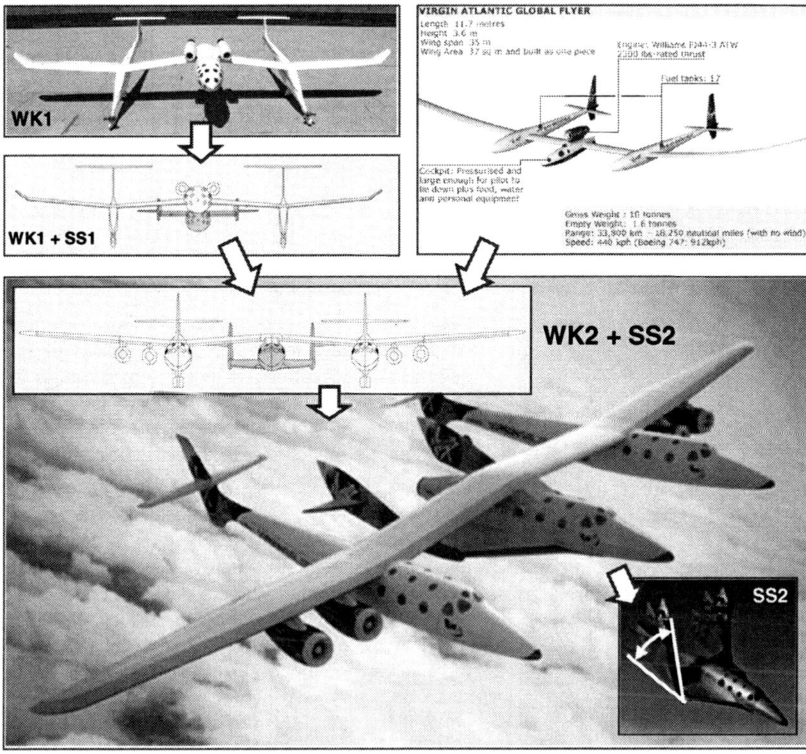

fig. 13.53. Evolution of carriers WK1 and WK2 for spaceplanes SS1 and SS2

Let us note that the carrier WK2 received its own "name": **VMS "Eve".** This is another manifestation of creativity – this time in playing with names and meanings: VMS stands for *Virgin Mothership*, but it is also a dedication to Richard Branson's mother. Her name is Evette Branson or, for short, Eve, which also means the day preceding, or leading up to, an event of some kind. Ultimately, the name proves to be a symbiotic merger of poetry and realism.

And in conclusion of this section, here is one more symbolic trait describing the flight of the carrier before launching the spaceplane: the carrier climbs to an altitude of 15 km directly above the air field *spiraling up*! Then the spaceplane detaches itself from the carrier and goes up almost vertically (at the angle of 84° with accompanying launch acceleration lasting for about 1.5 minutes). So what we have is a typical *"evolutionary spiral"* ending in a *"peak"* – a rocket-like launch! Wanna take a ride?

3.2. Evolution of Spaceplanes: from *SpaceShipOne* to *SpaceShipTwo*

Let us consider one more creative solution dealing with the descent of the spaceplane after it has completed the "jump". Taking into account the fact that by then all fuel has been used up…

the spaceplane simply falls down **from a height of 120-140 km!**

It employs aerodynamic maneuvering and glides down to the air field from which it started its journey!

This is the reason for the spiral assent – *the spaceplane should not move two-dimensionally away from the launch site.* It is a very creative solution of the "radicallest" contradiction – to be both "far" from, and "close" to, the launch site.

At the highest "point" of its flight trajectory, the spaceplane has a speed of more than 3M (three times the speed of sound), while by the end of the fall, some 15 minutes later, it slows down to less than 1M. The external surface of the spaceplane heats up to almost 600°C! This is where heat-resistant composite materials come in handy!

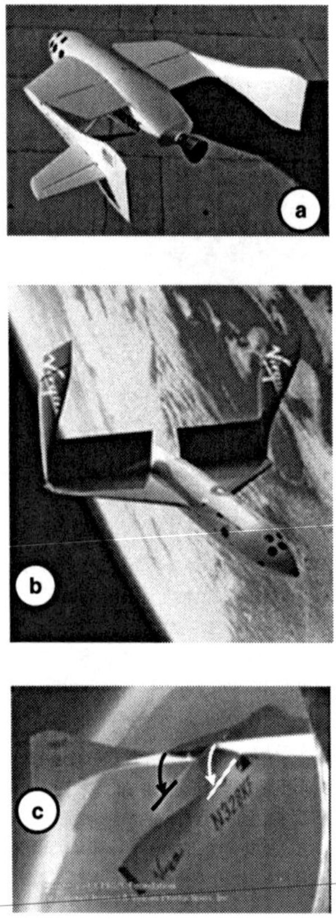

fig. 13.54. Rotating wings of the SS1 spaceplane (see also fig. 13.53)

Deceleration in increasingly dense atmospheric layers is achieved through correct orientation of the body of the spaceplane – it falls, so to speak, in its "normal" position, almost as it stands on the ground. But how can this be? The spaceplane looks like an arrow, and its empennage must inevitably turn it and plunge it into a nosedive!

The solution (fig. 13.54) objectively uses one of TRIZ models, namely, *Dynamization*: at the beginning of the descent, the wings of the SS1 spaceplane turn "up", and its "tail surfaces" keep it in the "right" position! At an altitude of some 20 km, the wings revert to the position used in horizontal flight, but the machine continues its descent along a rather steep trajectory, starting its horizontal glide only when it has gone down to 3-5 km. In fig. 13.54,a you can clearly see the line separating the immobile and mobile parts of the wing. In fig. 13.54,b SS1 is shown with its wings turned up. In fig. 13.54,c the wings start to turn into the gliding position. Similar "dynamization", as implemented in the SS2 spaceplane, is shown in fig. 13.53.

"Dynamization" of the wing pursues two objectives: *deceleration* and *protection of passengers from excessive g-force* by placing their bodies in the seats "back down".

SS1 has a length of 8.5 m and can seat 3 people: 2 passengers and 1 pilot; SS2 has a length of 18 m and can seat 8-9 people, including 6-7 passengers and 2 pilots.

The SS2 spaceplane was also given its own name: **VSS "Enterprise"**, or *Virgin Space Ship Enterprise*.

Here the word "enterprise" fully describes everything inherently present in the new business initiated by the partners: *daring undertaking, initiative, audacity, resourcefulness* and, more directly, *business, firm, company.*

3.3. Where Would We Be Without "Dynamization" and "Matryoshka"?

And how should people be placed in their seats throughout the duration of the flight? During the steep rise, the nose of the spaceplane looks almost directly up, and to cope with acceleration the astronauts must be "sitting". Then they will be looking straight ahead, while the g-force will be pressing down their backs. During the controlled "fall", the spaceplane will be oriented "horizontally". Consequently, the astronauts must be "lying" instead of "sitting". *This is a radical contradiction!* One of the simplest solutions from the TRIZ viewpoint is dynamization! And such solution is implemented in the spaceplane! And, naturally, designers used the famous Matryoshka: look at how densely and cozily the people are "packed" into their seats, while the seats themselves are embedded in their holding niches (fig. 13.55).

fig. 13.55. Dynamized astronaut seats:
a – launch; b – deceleration;
c – before landing

13.33,b. *Model 76 Voyager*

Inventive solutions are present in all machines developed by Burt Rutan while he worked for Scaled Composites. Let us perform a TRIZ composition analysis of the legendary *Voyager* airplane on which Dick Rutan and Jeana Yeager for the first time in history flew around the Earth without a single intermediate landing or refueling! Over the course of 9 days (plus 3 minutes and 44 seconds) from December 14 to December 23, 1986, they covered more than 40 thousand kilometers at an average speed of about 190 km/h.

It had taken 5 years (since 1981) to design the airplane (fig. 13.56). Its most impressive feature is its structure and engine configuration. The airplane was originally intended for an around-the world flight without landing or refueling. By that time Burt Rutan had already tried his hand at designing light aircraft made of composite materials with one or two pusher-propeller engines, including *Beechcraft Model 2000 Starship* (fig. 13.57).

fig. 13.56. *Voyager*
(length 8.9 m, wing span – 33.8 m, height – 3.1 m, empty weight – 1 ton, loaded weight – 4.4 t)

fig. 13.57. *Beechcraft "Starship"*
(length 14 м, wing span – 16.6 m, configuration: "canard" with fore-empennage)

The key idea was to *lengthen the wing* to a point where its length to width ratio would be at least 30! Composite materials made it possible for such wing to be light and durable.

For a long flight, it appeared sensible to have two engines:

1) higher overall efficiency of the engines: both would work at the initial stage (when the plane is fully loaded with fuel), and only one (the cruise engine) would continue to work later, when the airplane became lighter by the weight of the fuel it had consumed;

2) higher reliability and survivability of the aircraft, for example, if one of the engines failed;

3) higher survivability and safety of the aircraft, for example, if it were to bypass a thunderstorm front or fly over mountains.

If we use the familiar one-hull aircraft as the prototype, the following radical contradiction becomes immediately apparent:

1) it is impossible to place the two engines side by side on the wing, as soon (when only the cruise engine remains in operation) thrust will be produced asymmetrically;

2) it is impossible to place the two engines coaxially along the length of an extended hull, as because of the back propeller it will be necessary to make the landing gear higher (and, accordingly, heavier)!

Possible solution: make the hull shorter; however, a long flight requires a substantial amount of fuel, which makes the shortening of the hull impossible! So, here goes – **RC1**: the hull must be *short* (to enable coaxial installation of the engines), and it must be *long* (to take on more fuel).

The solution devised by Burt Rutan is anything but standard – he decided to use the triple-hull scheme! The contradiction was resolved in space and in structure: two additional hulls provide ample fuel storage space, and the central hull with the cockpit could be shortened! Tail surfaces could be divvied up between the additional hulls. Supportive specialized navigators: *03 Segmentation* (one hull is divided into three), *05 Separation, 10 Copying, 12 Local property, 16 Partial or excess effect* (three hulls instead of one), *21 Transform damage into use* (the plane can take on more fuel), *34 Matryoshka* (the central hull is placed between ("inside") the two additional hulls), *35 Unite, 37 Equipotentiality* (there is no need for a higher landing gear).

And there are three positive super-effects: 1) the landing gear can be hidden inside the additional hulls, 2) there are additional straddled wing supports, and 3) the tail surfaces remain outside of the operating zone of the front propeller.

Unfortunately, the new construction also had a major drawback: contradirectional movement of the additional hulls might generate unacceptable moment of torsion affecting the thin wings! **RC2**: the additional hulls *must be* (to store more fuel and enable the use of a short central hull), and they *must not be* (to avoid unacceptable exposure of wings to torsion action).

Here is the solution invented by Burt Rutan: supplement the three-hull construction with an additional front wing joining the three hulls! The decisive role was played – again! – by the spatial and structural resources. Supportive specialized navigators: *03 Segmentation* (the wing is divided into two wings – the main wing and the additional front wing), *05 Separation, 07 Dynamization* (the front wing is dynamized for pitch control), *10 Copying, 11 Inverse action* (horizontal surfaces are placed in the front in line with the "canard" configuration), *12 Local property, 19 Transition into another dimension, 21 Transform damage into use, 32 Counter-weight* (additional lifting surface), *35 Unite, 37 Equipotentiality, 39 Preliminary counteraction.*

Super-effects: 1) there emerges a "canard" configuration, with pitch control being transferred to the front surfaces, and 2) the overall rigidity of the system is increased, which helps to avoid torsion action and reduce the swaying (sagging) of wingtips.

As a result, the *Voyager*'s fuel consumption turned out to be as low as 0.9 liters per 10 kilometers! Just about as much as in an average-sized motor vehicle driving through the countryside! Meanwhile, the *Voyager*'s takeoff weight is 5 times more than the weight of such motor vehicle, and the *Voyager* flew at an average speed of 190 km/h instead of just rolling along the highway.

Ten years later, with new composite materials coming into play, it became possible to ensure sufficient durability of the wing without resorting to the "canard" configuration. The record-breaking 2005 aircraft *Virgin Atlantic Global Flyer* (fig. 13.53) and the carriers WK1 (fig. 13.51 and fig. 13.53) and WK2 (fig. 13.53) now have traditional engine configurations – only now the are using jet engines.

13.33,c. *New Turn of the Spiral for the White Knight Carriers*

Coincidentally, this page is being written exactly on the day of the long-announced X-37B micro-shuttle launch with the help of the booster rocket *Atlas-5* (April 24, 2010).

fig. 13.58. WK1 test flight with micro-shuttle X-37B
a) X-37B; b) X-37B in test flight under WK1 (2005-2006),
c) detachment from WK1

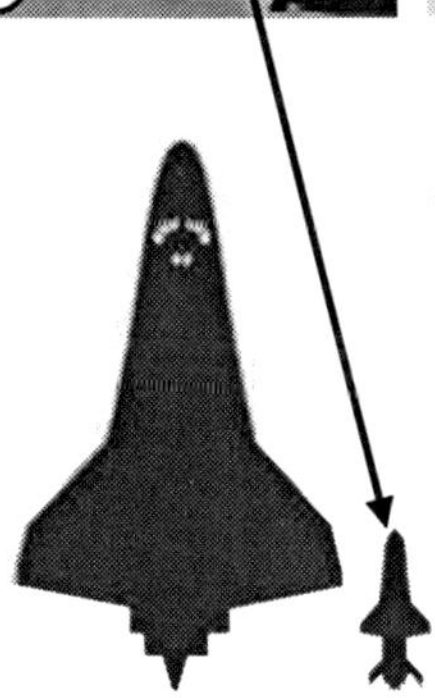

fig. 13.59. Micro-shuttle X-37B **vs** Space Shuttle Orbiter

X-37B parameters: length – 9 m, wing span – 4.6 m, weight – 5 tons;

Space Shuttle Orbiter parameters: length – 37 m, wing span – 24 m, weight – 80 tons

This Boeing project is a new turn of the technological spiral compared to the *X-20 Dynamic Soaring* project (see example 13.32. *Spirals of History*), mostly because of the construction of the micro-shuttle resembling that of the Space Shuttle (fig. 13.59). Other exceptional features include ability to engage in free flight, ability to perform in-orbit tasks for an extended period of time (up to 270 days), and ability to land automatically. This reminds of the unprecedented flight and landing of the Soviet *Buran* (see Example 13.32) on November 15, 1998.

Let us note at this point that the WK1 carrier (fig. 13.58) was used to test the X-37B craft. Our prediction that WK carriers can be successfully used to launch orbital spacecraft is already coming true, only instead of SS1 or SS2 the role of the spaceplane is played by the micro-shuttle (using the *Combination* and *Separation* models).

Example 13.34. NEVA – Superliner of the Future from Saint-Petersburg, Russia (DISC Modeling Sketch)

An amazing inventive concept was suggested in the late 1980-es by Vladimir L'vovich Freistadt (Hypersonic Systems Scientific Research Company, Saint-Petersburg) for a one-stage hypersonic aircraft (Russian: OGLA) *NEVA*[207]. Estimated flight speed at an altitude of 100 km is approximately 7M, i.e. about 7.5 km/s.

The author already performed TRIZ modeling for that machine in his first book[208] which was written in 1999 (and translated into German and English in 2001-2002)

Creators of the OGLA encountered a multitude of extremely complex problems. These two were the worst: 1) overheating of the craft's body cover; and 2) creation of a powerful universal engine that could be used for flying both in dense and rarefied atmospheric layers.

Analysis shows that the following resources dominate the resolution of these problems: *functional resource* (operating principle), *material resource* (fuel, refrigerants), and *energetical resource* (power supply). Dominance is assumed by the following two meta-models: *Integration* – the need to implement alternative principles in the same propulsion device, and *Differentiation* – radical alignment of processes and process parameters in various systems of the craft.

The authors of the OGLA *NEVA* concept "joined the unjoinable": excessive harmful heat with preparation of fuel for the "burn" and ... refrigeration (!) of the craft's body cover, huge power demand with enormous power supply present in the oncoming ionized air stream captured by the ram jet engine.

Magnetic-Plasma-Chemical Engine

At small altitudes, the ram jet engine works quite routinely, as the pressure of the oncoming air stream is sufficient to ensure continuous consumption of the fuel (kerosene). The higher the altitude, though, the more ionized the air stream becomes. At this point, the first idea kicks in (models *04 Replacement of mechanical matter, 11 Inverse action* and *29 Self-servicing*): the ion stream induces current in the first magnetohydrodynamic (MHD) system (see fig. 13.60) working in the MHD generator mode, and that energy is sufficient to feed all OGLA systems.

Then comes the turn of the second fundamental idea (models *04 Replacement of mechanical matter, 21 Transform damage into use* and *30 Use of strong oxidants*): the air stream at the exit point of the MHD generator is slowed down and, accordingly, has higher pressure – and this is exactly what we need at the entry point of the ram jet engine combustion chamber! But the authors do not stop there: incandescent ionized jet stream (plasma) existing the ram jet engine combustion chamber is further accelerated by the second MHD system working in the MHD accelerator mode!

But where is the "chemistry" to make it a "plasma-chemical" engine?

---

[207] V.L. Freistadt, A.L. Kuranov, E.G. Sheikin, *Use of MHD Systems in Hypersonic Aircraft. – Technical Physics Journal*, 1998, Vol. 68, No. 11, p. 43-47 (in Russian)

[208] See the first publication of this example in: Orloff, M. (2002; 3 editions) *Grundlagen der klassischen TRIZ: ein praktisches Lehrbuch des erfinderischen Denkens für Ingenieure.* – Berlin, Heidelberg, Springer (VDI-Buch), 352 pp.; and further: Orloff, M. (2003; 2 editions) *Inventive Thinking through TRIZ: A Practical Guide.* – New York, Springer, 352 pp. See also: Orloff, M.A. (2005; 4 editions) *Classical TRIZ Foundations.* – SOLON Press, Moscow, 416 pp. (in Russian)

Fuel

In the combustion chamber, kerosene is heated up, evaporated and oxidized (burned). To avoid wasting energy on heating the fuel, kerosene is pumped through the double wall of the OGLA cover. As a result, the cover is cooled... and the fuel is preheated! Dominant models: *02 Preliminary action, 07 Dynamization, 11 Inverse action, 17 Use of composite materials, 18 Mediator, 21 Transfer damage into use* and *26 Phase transitions.*

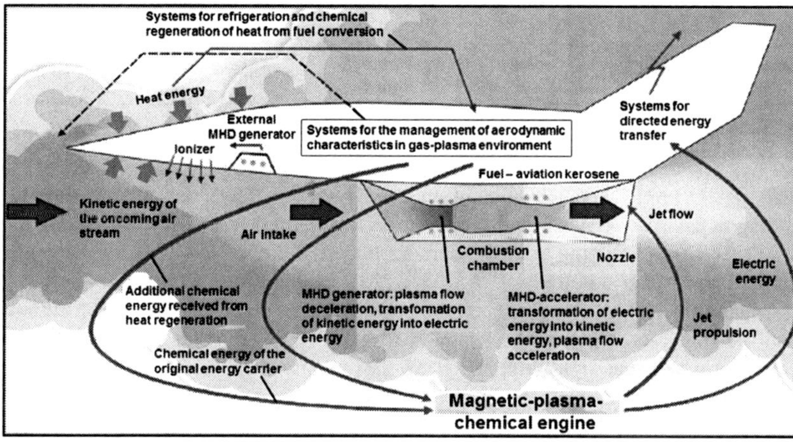

fig. 13.60. Conceptual structure[209] of the one-stage hypersonic aircraft *NEVA*

In addition to that, fuel contains water, and thermo-chemical destruction of that water (models *03 Segmentation, 16 Partial or excess effect, 30 Use of strong oxidants*, etc.) in the presence of a special catalyst yields atomic hydrogen which accelerates fuel consumption 5 times compared to liquid hydrogen engines!

Let us also note the presence of the *Differentiation* and *Integration* meta-models in the merger of two sets of "anti-systems" – MHD generator with MHD accelerator, and MHD ionizer (with the participation of a powerful laser assembly) with MHD generator.

Finally, let us note the possibility of a similar (i.e. creating a "system"-"anti-system" pair) application of *Differentiation* and *Integration* to produce a gas-plasma environment around the craft in order to reduce resistance while moving at hypersonic speeds – see Example 12.20 *Squall from under the Water!* Due to the use of the MHD generator, OGLA *NEVA* has sufficient (and colossal!) power supply capacity for that purpose.

Example 13.35. Evolution of Aerospace Systems (DISC Modeling Sketch)

Evolution from subsonic airplanes to aerospace aircraft can be presented by a chart[210] (fig. 13.62) and a comparative table resembling a morphological matrix (fig. 13.61).

---

209 This illustration is borrowed from this wonderful work: V.M. Fomin, Yu.P. Gun'ko, I.I. Mazhul', *Evolution of Jet Airplanes.* – Popular Science Magazine *Nauka* (Science), 2005, 3 (6), p. 40-53

210 Key ideas and fragments of a similar scheme have been borrowed for illustrative purposes from: V.M. Fomin, Yu.P. Gun'ko, I.I. Mazhul' *Evolution of Jet Airplanes.* – Popular Science Magazine *Nauka* (Science), 2005, 3 (6)

Stage / System	Rocket-Rocket (RR)	Rocket-Aircraft (RA)	Aircraft-Rocket (AR)	Aircraft-Aircraft (AA)
launch	**R**	**R**	**A**	**A**
acceleration	**R**	**R**	**R**	**A**
space flight	**R**	**R**	**R**	**R**
landing	**R**	**A**	**A**	**A**
Example	**Soyuz**	**Buran, Shuttle**	**SPIRAL**	**Neva**

fig. 13.61. Aerospace system types

Each system type has its advantages and disadvantages. Thus, due to their superior landing capability, RA, AR and AA systems have an edge on RR systems. By the same token, AA systems tend to be more efficient due to lower energy use and minimal atmospheric pollution.

Apparently, AA systems present the most attractive alternative for the future! That translates into "non-rocket" launch from the surface of the Earth! Does that mean that we are ready to relegate to the past what we have been so proud of for an entire century? No, rocket launches still make sense – if made from outside the Earth's atmosphere, from the orbit of the Earth's satellite, the Moon, etc. Accordingly, the experience that we have accumulated is still valuable.

What next? Various events occurring in the course of evolution of aviation and space systems can be interpreted in terms of certain DISC models. We are interested in the future of, say, AA systems. Near-earth space exploration will be favoring one-stage systems in line with the *Integration* meta-model. Integration is unthinkable without *Differentiation*, without perfect alignment and improved manageability and efficiency of all resources. The key trend for AA systems has always been – and will remain – development of the energetical resource: the engine (based on integration of alternative systems) and the aircraft power supply system as such.

Cargoes that AA systems will be able to lift off the surface of the Earth are relatively small, say, up to 1,000 tons! At the time, we do not have even that – and the road there is not going to be easy! Still, this is a "must-happen" trend for the development of circumterrestrial civilization. We need new clean sources of energy. We need new engines and propulsion units, we need... "anti-gravitation"! First for small mass, then for large mass...

Let us conclude this section with a small jocular advertisement. I have thought up several "bold" comparisons: ***EASyTRIZ*** is the ***SpaceShip*** for all creative personalities! Their first "suborbital jump"! And this: ***Modern TRIZ*** is like ***NEVA*** – a system with a completely different engine system and organization!

In any event, I wish you success in mastering Modern TRIZ!

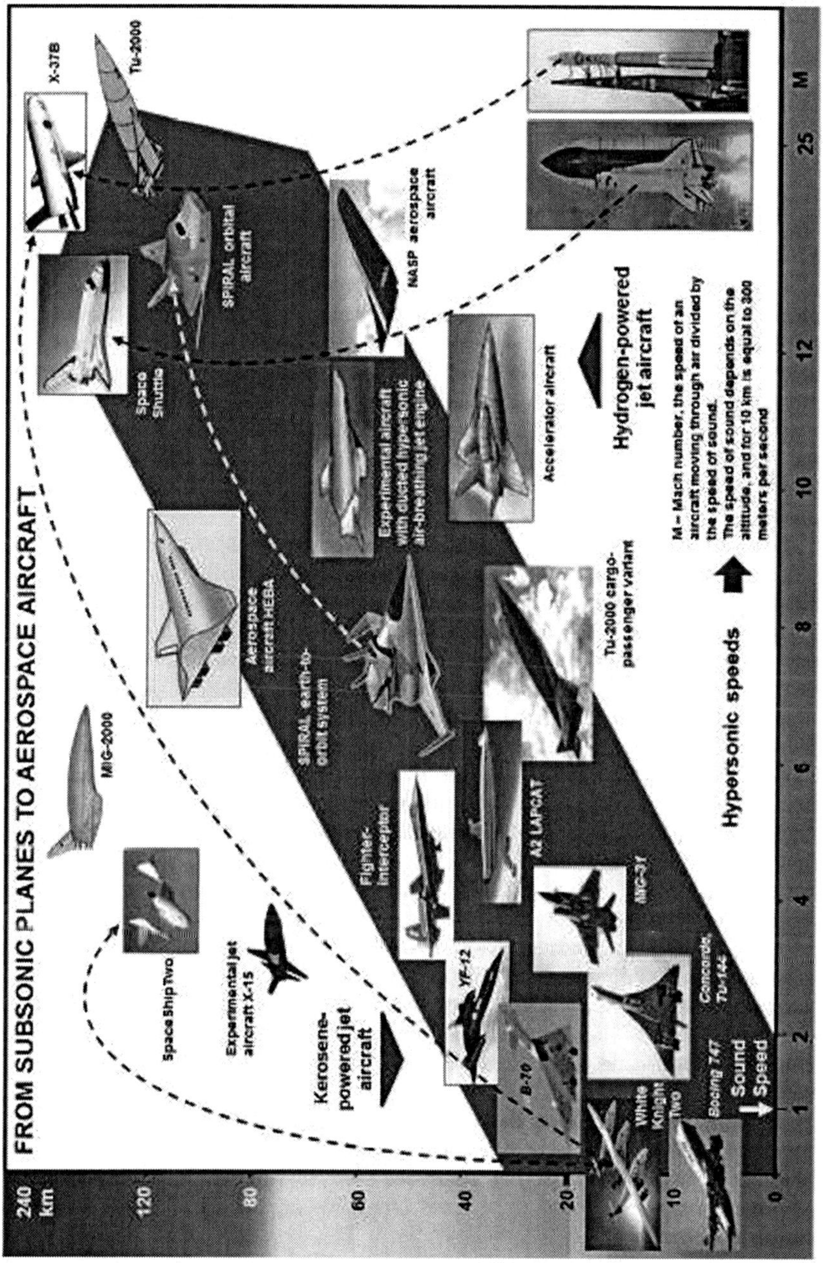

fig. 13.62. Sketch of evolution from subsonic airplanes to aerospace aircraft

13.2.4 Steps of Evolution

Example 13.36. World's First Mechanized Lathe Support

Timeframe: 1712 – 300 years ago!

Inventor: Andrey Konstantinovich Nartov

– Russian turning craft pioneer, [211]

19-year-old innovator who was ¾ of a century ahead of his time.

My first encounter with the Nartov copying lathe occurred in the autumn of 1963 as I was doing a training course called *"Metal Cutting. Machines and Instruments"*. The course was taught by Stanislav Antonovich Baltsevich, our tutor at the Minsk Polytechnic where I was majoring in *Control and Measurement Instruments.* I was reveling in kinematic configuration diagrams for various types of machines, admiring the paths by which energy was traveling from gear to gear. I was simply overwhelmed by the power stored in spring-loaded mechanisms, by the fathomless secrets lurking inside magnetic and electromagnetic elements and optical devices, and by many other things which amounted to technical discovery miracles.

Day after day we went to the workshops to try our hands at various machines: lathes, mills (horizontal, vertical, rotary-table), planes, drills… I loved music, mostly classical, played the chess, went in for boxing, skiing and track-and-field events – and got seriously high ratings and some cups, too. I also read wherever and whenever I could. I already mentioned Jack London and Ernest Hemingway. But there were also dozens of other writers and poets, hundreds of books. From time to time I went to the philharmonic. And cinema! Ah, those wonderful movies that came to us from Europe and the USA! The tears that I shed watching *Rocco e i Suoi Fratelli* (Rocco and His Brothers) with Alain Delon and Annie Girardot – because of the sympathy I felt for them, and because I was painfully aware that I could not help them, even though being a boxer, just like Rocco, I certainly knew how to exact vengeance for all his hurts and losses. Anyway, life was anything but boring…

I believe that this digression is not redundant. I also believe that it should not to be read superficially. Because it also speaks of the beauty dwelling in ideas and solutions. Or, to be more exact, of perceiving – and preparing to comprehend – that beauty. It all happened, and it was inseparable from the themes that our tutor discussed with us: life, people, creation, modesty, dignity, courage, solidarity, empathy, camaraderie… The list goes on… And the things he said have stayed with me all my life.

Anyway, once he told us about Andrey Konstantinovich Nartov. As we were gaping at the impeccable schematic drawings that were magically emerging on the blackboard, our tutor was saying: *Hear and remember forever the names of those who made our Motherland strong and invincible regardless of which bastards sought to destroy us.*

---

[211] Andrey Konstantinovich Nartov (1693-1756) – great Russian mechanic, inventor of mechanical support, automatic copier for the turning machine, and many other things

And when he told us about the life that Nartov led next to Peter the Great[212] who valued him and lavished gifts on him, and then about the indignities and heartaches that he suffered at the hands of Russian and foreign officials after the czar's death, the tutor's choice of words ("bastards") did not any longer seemed frivolous or overly rude.

Today you can learn Nartov's life story from many sources, but I heartily recommend a thin book[213] in the "old", "classical" style that our tutor told us to "have a peep at". I am reproducing here two of Nartov's inventions based on that book. OK, Invention Number One.

Support and Copying Lathe

Fig. 13.63 and fig. 13.64 show lathes that were used in the late 17th – early 18th centuries.

The turner (fig. 13.63) pushed the pedal 9 down with his foot, and rope 7 which was wound around the part 8 turned it (the part). The turner cut the part using the knife 11 which he pressed to the part with both hands. The knife rested against, and could move along, the carriage 6.

When the turner let go of the pedal, the elastic bow 10 pulled the rope up, raising the pedal and resetting it in preparation for the next work cycle. While the part turned in the reverse direction, the knife was removed from it, and no cutting was done. In other words, half the cycle was wasted on the reverse motion of the part.

Using such lathe (fig. 1364) was tiresome. It was very difficult to make identical parts. The work required strong hands, a good eye, and copious experience.

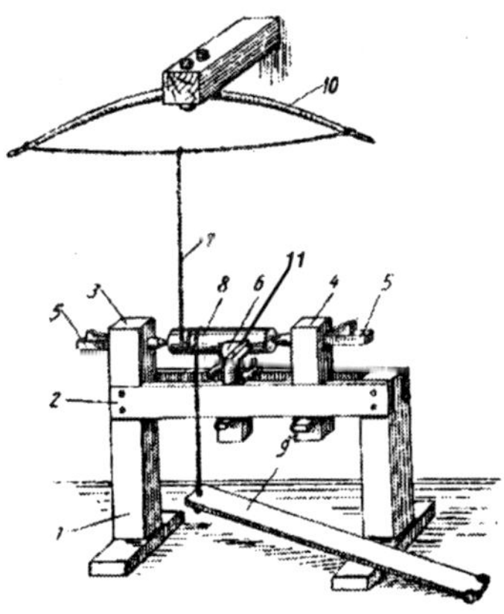

fig. 13.63. Foot-operated bow lathe

fig. 13.64. Turner working at a bow lathe

---

[212] Peter the Great (1672-1725) – Moscow Czar from the Romanoff dynasty (since 1682) and first Emperor of Russia (since 1721), reformer, founder of the Russian Empire

[213] Professor A.S. Britkin, Engineer S.S. Vidonov (1950) *A.K. Nartov, Great Mechanical Engineer of the 18th Century*. – Moscow, State Scientific and Technical Publishing House (Mechanical Engineering), 145 pp. (in Russian)

Turning metal parts was particularly difficult due to the need to exert great physical efforts.

In fig. 1364, you can also see various knives lying on the frame of the lathe and hanging on the wall.

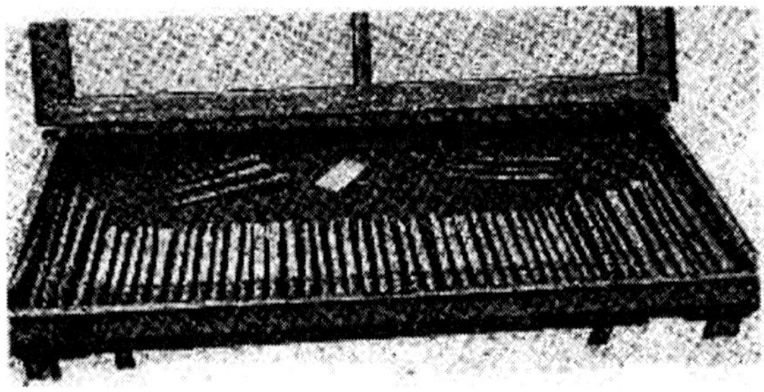

fig. 13.65. Set of turner and locksmith instruments owned by Peter the Great

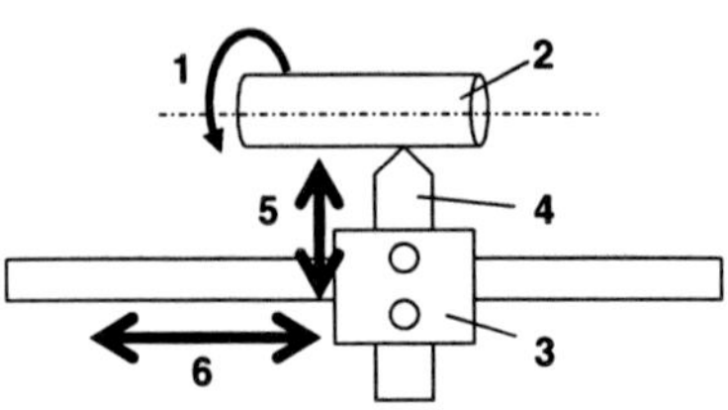

fig. 13.66. Mechanized support designed by A.K. Nartov

First (13.66), Nartov invented the mechanical gear drive which ensured continuous rotation 1 of the part 2, and the support 3 – a device which supported the knife 4 and moved it to and from the part 5 and along the part 6.

That invention was made 75 years before the first similar device was patented abroad.

To enable exact reproduction of identical parts, Nartov invented a mechanism guiding knife movement with the help of an ingenious copying device.

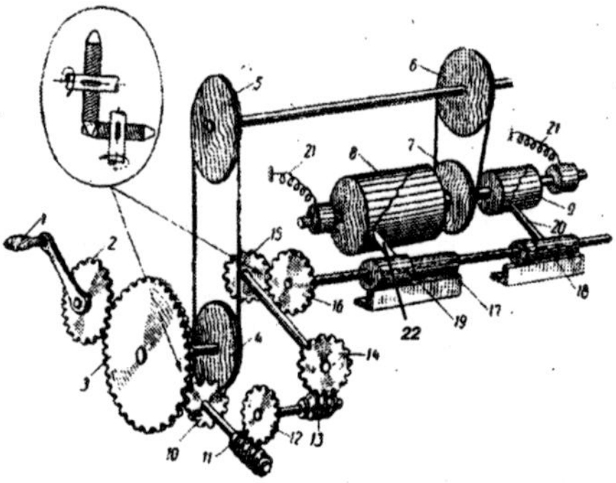

fig. 13.67. Copying lathe designed by A.K. Nartov

Fig. 13.67 shows the copying lathe developed in 1718-1729.

The finger 22 rigidly fixed to the spring-loaded support 19 moves along the copier 8. All movements of the support 19 are transmitted to the knife support 18 and, accordingly, to the knife 20 working on the part 9.

The finger/copier movements are synchronized with the knife/part movements. Generally speaking those movements are different, but the theory behind the copier and discussion of its construction lie beyond the scope of this book.

Copier profiling makes it possible to program intricate knife trajectories and make parts with very complex shapes.

Copiers may be made on the basis of different principles using different kinematic schemes. On the whole, they guide knife movement in accordance with a rigid program fixed as a unity of "special" surfaces and lines making up the complex "integrated" copier shape.

Fig. 13.68 shows a more modern copying lathe support designed with the participation of S. Yakovlev, Nartov's disciple, in the late 1720-es (the copier and the part are presented as rough sketches, without going into excessive detail; besides, I corrected an error made by the draftsman who prepared the drawings for the cited book – M.O.). The gear system clearly visible in the drawing became the "forefather" of all future lathe gearboxes.

> The operating principle of modern lathes has not changed since Nartov times – for a staggering **300 years**!

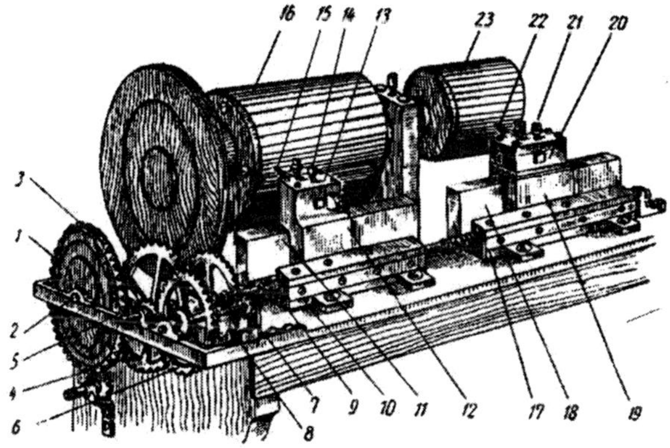

fig. 13.68. Support of the Yakovlev-Nartov copying lathe

And now let us perform the basic procedures designed to retrieve inventive models from artifacts – extracting and reinventing.

Support Reinventing According to MAI Z-I-R-T

ZOOMING

In fact, we have already done the zooming. Today it has become absolutely clear that Nartov's inventions are superior not only to the archaic bow lathes, but also to many machines which use the same operating principle (support, copier, gearbox) but were made abroad 75 or even 150 years later.

INVENTING

Here we will demonstrate TRIZ transformation models extracted from Nartov's solutions in comparison with known historical prototypes – bow lathes (fig. 13.69).

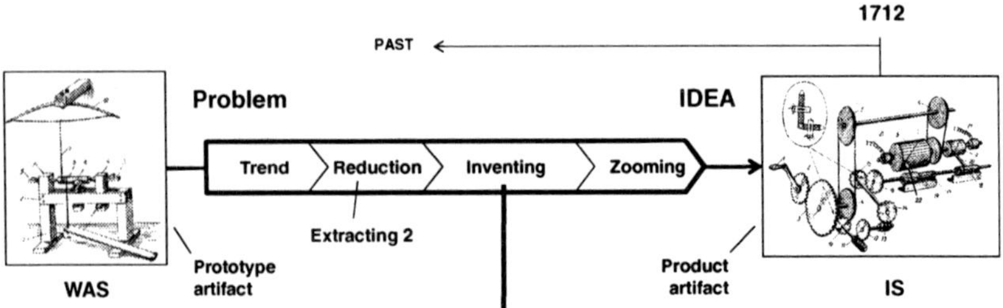

EXTRACTING – 1		
04	**Replacement of mechanical matter** - use the fields with a specific structure instead of the unstructured fields	1) Piece rotation – efficient ways (structural schemas) are designed to transfer energy using conventional gears; 2) Energy required to move the support is transferred using a) rack-and-pinion mechanism and, later, b) screw mechanism.
10	**Copying** - replace an object or a system of objects with copies; use here a change in measure (blow-up or reduce the copy)	1) Turner actions are copied by the support – man is replaced by a machine (holding, directing and moving the knife); 2) Lathing profile is copied using a special "copy template – finger" assembly with the finger support connected to the knife support.
12	**Local property** - every object should exist under conditions that correspond best to its functions	The use of certain gear ratios ensures synchronization of all movements: turning of the piece, movement of the copy template and copy template finger, movement of the knife.
18	**Mediator** - use another object to transfer or transmit an action	1) The supports act as intermediaries between the man and the knife (in non-template controlled lathes, the movement of the support and the knife is controlled by man) and, later, the piece; 2) Intermediary transmission mechanisms create efficient parameters governing the movement of all lathe parts.
19	**Transition into another dimension** - an object is shaped so that it can move or is placed not only in a linear fashion, but also in two dimensions, meaning on a surface. It is also possible to improve the transition from a surface to a three-dimensional space	1) The knife can move not only in the narrow cutting area on the surface of the supporting table, but also along the entire piece; 2) With the help of the copy template, the knife can create 3D profiles! 3) Gear wheels can be engaged not only in one plane, but also in multiple orthogonal planes using transformation of rotational motion into progressive motion (rack-and-pinion drive or screw drive of the support) and complex spatial engagements (worm gears in reduction units).
20	**Universality** - an object has several simultaneous functions so that other objects are not needed	The support holding the knife performs all major functions associated with moving the knife – lengthwise movement along the piece, advancement to required cutting depth, complex-trajectory cutting movement.
34	**Matryoshka** - an object runs through a hollow space in another object.	1) "Bolt-nut" transmission for bidirectional support movement – forcing screw goes through the nut embedded into the support; 2) Complex "gear boxes" hidden inside the frame.
35	**Unite** - unite similar objects or neighboring operations	With the help of the support, individual machining operations along the entire length of the piece which require numerous transitions and tool adjustments coalesce into a continuous machining process.
40	**Uninterrupted useful function** - complete a job without interruptions where all parts work continuously at full capacity; eliminate idle running and interruptions.	1) There is no more need for idle reverse turns to reset the foot drive. 2) The copying system has enabled the "almost automated" and "almost continuous" machining of the blank until it is transformed into the finished product.

fig. 13.69. Extracting-1 for Yakovlev-Nartov copying lathe

Other transformation models – standard and basic (dealing with physical/technical and other effects) are discussed at more advanced training stages. In this book, however, we must stick to the announced subject-matter – specialized primary transformations.

REDUCING

Extracting-2: Let us formulate the main standard and radical contradictions inherent in the prototype – the bow lathe.

Examples of Standard Contradictions:

SC1. The knife must perform *complex movements*, but that is *inconvenient* for the worker and *calls for application of strenuous physical efforts.*

SC2. The knife must follow *the same complex trajectory* to ensure manufacture of identical parts, but *precision of positioning* of the knife and *precision of the preset cutting parameters* depend on the worker and *cannot be high.*

Examples of Radical Contradictions:

RC1. Rotation of the part *must be continuous* to increase productivity, and it *must be discrete* because the foot drive needs to be reset.

RC2. The movement of the knife along the part *must be continuous*, and it *must be discrete* to enable repositioning of the carriage which supports the knife.

IFR: rotation of the part must be continuous, and the knife must continuously move along the part following a preset trajectory.

Maxi-FIM: the lathe BY ITSELF performs all movements required to obtain the IFR.

TREND

For many centuries, cutting of rotating parts was not mechanized, let alone automated. Construction of lathes had to be modified to increase productivity of labor, improve precision and enable reproduction of identical parts (serial production), minimize reliance on experience and ability of the turner, and decrease labor intensity of work.

It was necessary to invent new lathes using new principles guiding part movement, and new mechanisms to implement those principles.

At that point in history the young talented master Nartov met Peter the Great who was interested in new technologies and sought to design machines and products that would surpass everything Europe had to offer at the time.

Peter the Great was going to create the invincible army and navy that would protect Russia from its enemies – and then expand its possessions and transform it into an empire.

In conclusion of the first reinventing, let us have a look at its general outcome (fig. 13.70) in the MAI T-R-I-Z format.

TREND

Foot-powered dead center lathes have major drawbacks: low productivity due to massive time loss on idle running; low machining accuracy and, as a consequence, low reproducibility of pieces; and high power loss on manual lathe knife operation. Can we improve the lathe?

REDUCING

Maxi-FIM: The operating zone by itself ensures attainment of the **IFR**

[continuous rotation of the piece in the working direction and continuous cutting of the piece using new mechanical components of the lathe!].

Standard Contradiction 1

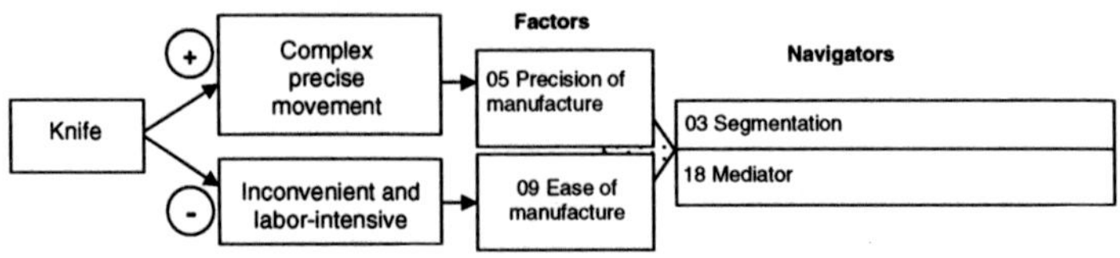

SC2: 06 Precision of measurements VS 08 Complexity of inspection and measurements = 04, 09, 10, 18

SC3 (instead of RC1): 01 Productivity VS 03 Level of automation = 01, 10, 35, 37

SC4 (instead of RC2): 25 Loss of time VS 03 Level of automation = 01, 04, 18, 25

MITO Ranging: 18^3, 01^2, 04^2, 10^2, 03, 09, 25, 35, 37

Radical Contradiction

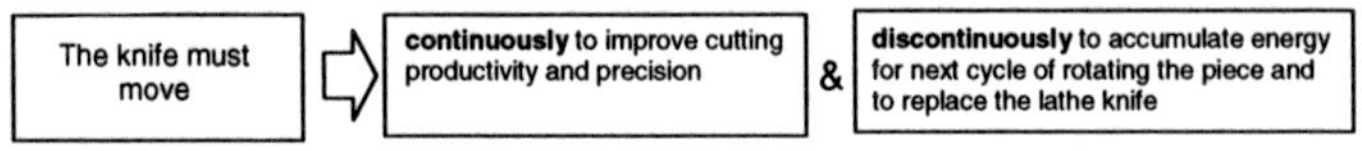

INVENTING

Dominating models – 18, 04 and 10.

Key ideas:

1) according to model 18, the knife must be fixed in a holder resting on the lathe-bed so that the turner does not have to hold it with his hands when cutting;

2) according to model 04, it is necessary to enable the holder to move lengthwise back and forth along the entire length of the piece and transversely for the required cutting depth – create holder guides; it is also necessary to modify power feed to the piece to enable its continuous rotation;

3) according to model 10, it is necessary to create a mechanism directing the movement of the holder with the knife so that the knife BY ITSELF gradually creates the required profile for the turned piece.

END RESULT: continuous rotation of the piece is powered by a common source through a series of connecting gears; the knife is fixed in a special holder which is connected, through rack-and-pinion drives, screw drives and other gearing mechanisms, to a device guiding its movement; the device is made in the form of a similar holder fitted with a finger moving along the copy template – a body of a certain form designed to obtain required knife movement.

ZOOMING

Have the contradictions been removed? – Yes. – ~~No.~~

Super-effects: revolution in lathe-building trailblazing the way to future automated machines

Negative effects: none

Automated movement diagram

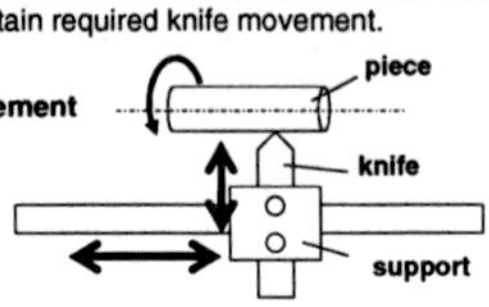

BRIEF DESCRIPTION

To increase work productivity and precision, reduce labor-intensity and improve reproducibility, the lathe is fitted with a continuous rotation mechanism and a special knife holder (support) whose movement is controlled by a copying mechanism. The copying mechanism consists of a similar support with a finger which moves along the template shaping the knife movement trajectory.

Objectively present transformation models: 04, 10, 18, as well as 12, 19, 20, 34, 35 and 40.

fig. 13.70. Reinventing of support and Nartov copying lathe

Example 13.37. Nartov's "Katyusha"[214]

This Nartov's invention is interesting not only because of its "combat qualities", but also because of the creative ideas underpinning its technical solutions (fig. 13.71). And it rightfully occupies a prominent place in the exposition of the world's largest Military Historical Museum of Artillery, Engineering Corps and Signal Corps dating back more than 300 years (!) to the first "Zeughaus"[215] at the Saint Peter and Saint Paul Fortress in Saint-Petersburg created in 1703 by order of Peter the Great to collect and store "ancient", "memorable" and "curious" artillery pieces "for remembrance and glory eternal".

TREND

By the end of the 17th century, the military already used multi-barrel firing systems featuring as many as 25, or even 105, rifle-type barrels, including systems mounted on rotating "drums" with small cannons (mortars) where "free" barrels could be reloaded with minimal time loss. However, such systems had fixed firing angles, while "drum" systems took too much time to clean and reload. How can we increase the firing rate?

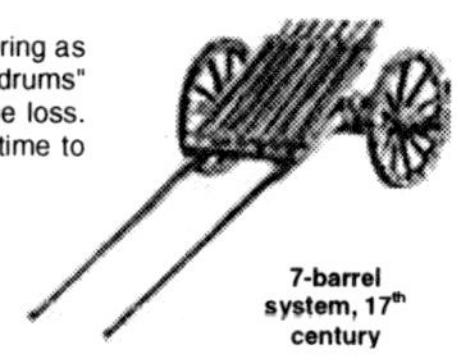

7-barrel system, 17th century

REDUCING

Maxi-FIM: The operating zone by itself ensures the obtaining of the following **IFR**:
[increasing the firing rate and accuracy of multiple-mortar systems].

Standard Contradiction 1

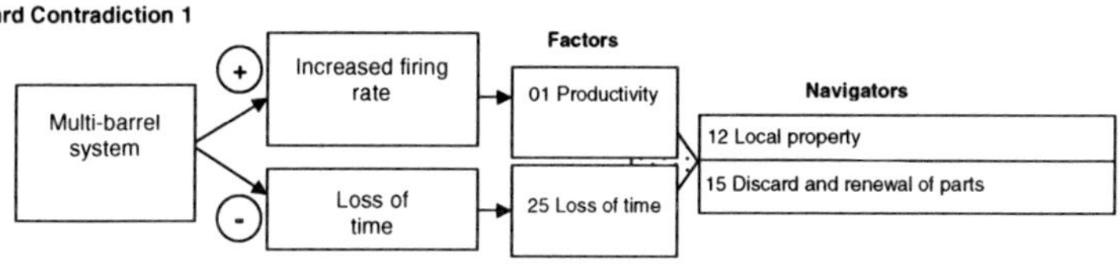

01 Productivity **VS** Shape (inefficient reloading) = 02, 15, 17, 22

MITO Ranging: 15^2, 02, 12, 17, 22

INVENTING

Dominating models – 15, 02, 12 and 22. "Composite sketch" of the TRIZ solution:

1) according to model 15, used parts of the object should be immediately replaced during operation;

2) according to model 02, objects must be prepared in advance so that they can be put to work from the best position and are available without loss of time for delivery;

3) according to model 12, every part of the object must exist under conditions that correspond best to its functions;

4) according to model 22, it is necessary to use rotating motion.

END RESULT: 44 small mortars are installed in a radial configuration on an oak-wood disc. The mortars are united into groups of 5-6 to be ignited and fired together. Cleaning and reloading have become simpler due to easier access to the mortars. The firing angle can be adjusted by tilting the disc (to do that, it is necessary to turn a screw at the butt-end of the support).

ZOOMING

Have the contradictions been removed? – Yes. No.

Super-effects: -

Negative effects: rapid deterioration of the carriage on which the system is mounted.

Rapid-fire Nartov battery (44 mortars)

BRIEF DESCRIPTION

To increase firing rate and accuracy, 44 mortars are installed on a rotating disc platform. The mortars are joined into groups of 5-6 for salvo firing.
Objectively present transformation models: 02, 12, 15 and 22.

fig. 13.71. Reinventing of the 44-barrel Nartov battery

[214] "Katyusha" ("Little Katie") is the popular nickname of the multiple rocket launcher that was developed in the late 1930-es by the Soviet RNII (Research Institute for the Study of Reactive Motion) and successfully used by the Soviet Army during WWII

[215] Zeughaus is the German word for "arsenal" or "armory"

Example 13.38. "The Lord of the Rings and the Strings"

> Creative personalities succeed not owing to, but in spite of, society. In spite of society of today. For the sake of society of tomorrow.[216]
>
> *G.S. Altshuller*

> Out of a hundred geniuses ninety-nine die unrecognized and the hundredth is faced with huge difficulties in making a break-through.[217]
>
> *Rudolph Diesel*

1. The Becoming of the "Lord of the Rings"

Mankind can be proud of the rockets it has created over the course of its history – from the Chinese arrow rockets to giant modern booster rockets launching interplanetary spaceships and carrying cargoes into the near-earth space. In this book, we have spoken at length about outstanding inventions improving rocket technologies.

However, at the dawn of the space era, amid triumphant reports of Soviet and American rocket launches, a high school student from Belarus challenged the expediency and the "eternal glorious future" of rockets as the main instrument of masterng the near-earth space! Incredible but true!

It was the year of 1955. A six-year-old boy who knew nothing of Konstantin Tsiolkovsky, let alone Wernher von Braun or Sergey Korolyov, started to build "rockets" before the world had even heard of Yuri Gagarin, the first man to fly into space. For "fuel" he used pieces of plastic combs burning as bright as gunpowder, or scraps of film left over by the projectionist operating the "cinemobile" that came each week to show movies at the village community center. But best of all the boy liked the long macaroni-like tubes of German smokeless gunpowder that came from a huge arsenal destroyed by guerillas at the railroad station of Posudovo, some 8 kilometers from the boy's native village of Kryuki lost among the endless bogs of Belarus Polesye (Woodland). It was dangerous – the site of the arsenal was pockmarked with crosses topping the graves of other curious children who had blown themselves up trying to disassemble mines or shells they had found lying around.

In 1962, soon after the historical first manned spaceflights, the boy's family moved to Dzhezkazgan, Kazakhstan, to work on virgin lands. The fact that his new home was close to the Baikonur Cosmodrome only increased the boy's interest in rockets. He had to make his rocket and rocket engine models from paper and

---

[216] From: G.S. Altshuller, I.M. Vertkin (1994) *How to Become a Genius. Life Strategy of a Creative Personality.* – Minsk, Belarus

[217] Rudolph Diesel (1858-1913) – great German engineer, inventor of the "Diesel" engine; died during the Antwerp to London voyage on *Dresden* ferry. Quoted from *The 7 Golden Rules of Creativity. Imagination at the Heart of Business* by Gottlieb Guntern, *Business Digest*, 112, October 2001; a more extensive quote is given in *Genetics of the Genius* by V.P. Efroimson (1908-1989) – a famous Soviet geneticist

glue – the only materials available to schoolchildren in those times. He also managed to manufacture smoke powder from components sold in ordinary drugstores.

When he was in the 8th grade at school, his three-stage rockets could go up several kilometers. He also had his "astronauts" – gray mice which came down to earth on paper parachutes. By the time he finished school, he had acquired superior knowledge of physics and mathematics, taken part in numerous physical and mathematical contests for schoolchildren, examined the works of Konstantin Eduardovich Tsiolkovsky, and undertaken his own analysis of the rocket method of space exploration. Being a huge sci-fi fan, he had no doubt that the future of mankind was there, up among the stars...

But he was bitterly disappointed. It turned out that, if you factored in the energy expended on manufacturing fuel components and the cost of jettisoned disposable elements of construction, the energy efficiency ratio of the booster rocket was way below one percent. For comparison, the energy efficiency ratio of a modern steam engine is 15%. Technical achievements of space science – excessive power of the booster rocket engines (about 100 million HP), high reactive jet velocity (many times faster than the fastest bullet) and high temperature of combustion products (higher than the temperature inside a Martin furnace) – not only failed to inspire the young man, but chagrined him.

He found out that the aggregate weight of payload launched into near-earth orbits (about 300 km) by the mankind over the course of many years amounted only to several hundred tons. The same amount of cargo could be transported on the Earth over the same period of time and by the same distance of 300 km... by just one horse pulling just one cart! In other words, all achievements of modern space exploration were backed up by one exorbitantly priced golden, no, diamond "space cart" and one diamond "space horse" with an efficiency ratio lower than that of a steam engine by an order of magnitude.

Thus the boy destroyed his own youthful dreams about the bright future of mankind – about space cities, orbital plants and power stations. It was simply impossible to build and maintain them using just one "space cart".

The boy became a student of the Tyumen Engineering and Construction Institute where he majored in *Road Engineering (Automobile Roads and Constructions)* and acquired in-depth knowledge of resistance of materials, structural mechanics, physics and higher mathematics. He continued his studies, and soon realized that the most environmentally friendly way to launch things into space from the surface of the planet was for the propulsion unit to use ... the internal forces of the space transportation system!

Ideal functional model! An unexpected application of the Baron Münchhausen principle! It is rumored that the inventive baron pulled himself and his horse out of the bog by his own hair made into a pigtail. This principle is very efficient in that the "Münchhausen-Horse" transportation system rose up without bearing on, or interacting with, its immediate environment, i.e. had absolutely no negative effect on that environment.

Unfortunately, inasmuch as Münchhausen did not attend modern school, he did not know that he had breached one of the conservation laws stating that it is impossible to move the mass center of a system in space using internal forces of such system.

So, we have an *irreconcilable radical contradiction*: in order to make a launch, we need a "point of support" on the Earth – the rocket is "supported" first by the launch pad and then by the gas jet issuing from the engine – and we need the mass center of the object being transported to be moved into the near-earth space; on the other hand, to exclude interaction with the environment, the "point of support" must not be situated on the Earth, and the mass center must not be moved into space!

Apparently, the problem has no solution.

But the student did find a solution – and a very simple one: the spacecraft must be made in the form of an elastic ring encompassing the Earth, and then that ring must be spun at a speed exceeding circular velocity! Then the ring will overcome gravitation, and begin to rise and stretch under the influence of centrifugal forces!

By the time the ring reaches an altitude of 300 km, its diameter will have increased only by 600 km, or approximately by 5%. The mass center of the system will coincide with the mass center of the planet – i.e. it will remain on, or rather in, the Earth! And the "point of support" is simply not there! There is only the ring spin trajectory, but it does not need to interact with the Earth to pull of the launch trick!

The miracle accomplished by this unique unprecedented idea is more impressive than all the magic worked by all the wizards in *The Lord of the Rings*[218]! It is the miracle of a self-organizing system which is based on the laws and forces of Nature – and which remains in harmony with Nature! The real "Lord of the Rings" is the 19-year-old inventor Anatoly Unitsky.

The more so because student Unitsky invented not only the method to launch just one ring, but the concept of intermittently launching and landing multiple rings which are realigned in flight so as to occupy different orbits in space.

2. The Rise of the "Lord of the Strings"

In the early 1970-es, Unitsky returned to his native Belarus, and in 1973 he graduated from the Belarus Polytechnic Institute (University) in Minsk. He continued to fine-tune the engineering and design aspects of the Planetary Transportation Vehicle (PTV) concept. In 1986 he became a member of the Cosmonautical Federation of the USSR, and in 1988 hosted the First International Research Conference *Rocket-Free Industrialization of Space: Problems, Ideas, Projects*!

Working on his project, Unitsky had to deal not only with numerous fierce critics, but also with direct resistance from the KGB both in Moscow and in Minsk. Having detected a "round peg", the "square-hole" society unleashed a war on a person who freely shared his talent and his inventions.

The main argument of Unitsky's critics was that construction around the Earth of a trestle which was to serve as the PTV "launching pad" would require too

---

218 *The Lord of the Rings* is the title of the well-known and repeatedly cinematized fantasy novel by the English writer John Ronald Reuel Tolkien (1892-1973), published in London in 1954-1955

much construction materials. He proceeded to improve the trestle design. The streamlined construction would take about as much concrete as had been used to build the Sayano-Shushenskaya Dam on the Yenisei River, and as much metal as had been used to build every hundredth motorcar on the planet!

And then Unitsky made his second invention/discovery which was equally unique and unprecedented: the trans-planetary ring needed an exceptionally smooth and straight track – and only the STRING could be used as the prototype for such structure!

Only a self-organizing system!

String constructions are universal and widely used by the Creator. For example, tubular bones in our skeleton are so durable because they are made based on the same templates as strings: some of their fibers are stretched, while others are compressed. It is not inconceivable that all material world and our entire Universe are made of miniscule quantum strings.

In 1995 Unitsky published a limited edition of his first scientific monograph *String Transportation Systems: On the Earth and in Space.* While PTV continues to evolve as the main avenue of future industrialization of space, development of string bearing structures has resulted in the invention of string bridges and trestles, string runways, string high-rise and seismically-resistant buildings, string vacuum glass, and many other string technologies.

Let us draw conclusions (www.stringtransport.com, www.stt21.ru).

PTV can lift into space, in one go, millions of tons of cargo and, if necessary, millions of passengers.

Technically speaking, PTV is a torus with a diameter of about 10 m. Inside the torus, there runs an annular vacuum channel housing an endless (annular) rotor suspended inside magnetic field and not touching the walls. When a linear actuator accelerates the annular rotor running inside the vacuum channel (i.e. around the Earth) to circular velocity, the rotor will become weightless. Additional acceleration will create excessive centrifugal force which will be trying to increase the diameter of the ring. If that force exceeds the weight of the construction, each point of the rotor will commence vertical ascent, and some two or three hours later the ring, noiselessly increasing its diameter and, accordingly, stretching out, will occupy a circular orbit at the altitude of, say, 300 km. Incidentally, the ring will lift the payload without ever touching it: the payload will be evenly distributed along the length of the ring and attached to it by magnetic fields. When the speed of rotation of the ring in orbit is decreased, the ring will gradually descend and "land" to the launching trestle on the Earth.

String technologies can make a breakthrough in road and bridge construction technologies, and in many other areas. For example, why not build a string bridge across the Bering Strait to connect the African, Eurasian and Pan-American continents into a transportation mega-system (preferably, of the string variety)?!

Let us wish the "Lord of the Rings and the Strings" – the man whose ideas have come [?] years before their time – speedy realization of his dreams and implementation of his projects!

fig. 13.72. Medium-sized high-speed passenger modules

fig. 13.73. A.E. Unitsky preparing a transportation module for the wind-tunnel test (Saint-Petersburg, 2001)

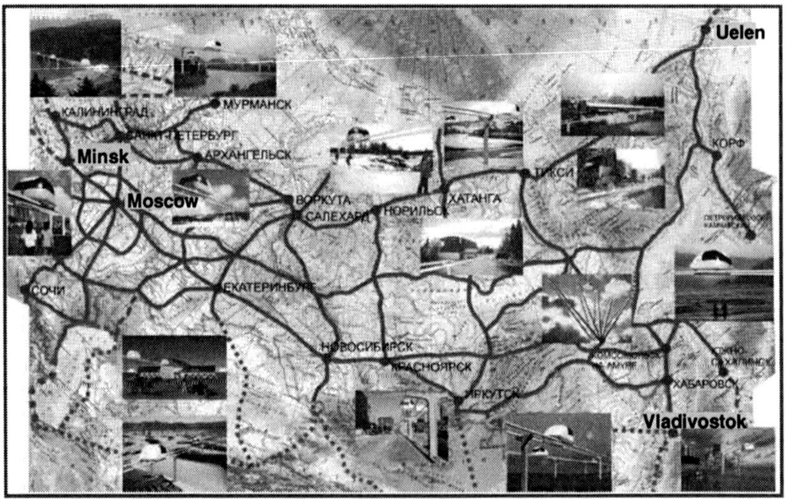

fig. 13.74. Trans-Russian String Transportation System exiting onto the Bering Strait (2003 project)

14 Reference Materials

14.1 A-matrix

14.1.1 Plus- and Minus-factors

#	Factor
01	productivity
02	universality, adaptability
03	level of automation
04	reliability
05	precision of manufacture
06	precision of measurement
07	complexity of construction
08	complexity of inspection and measurement
09	ease of manufacture
10	ease of use
11	ease of repair
12	loss of information
13	external damaging factors
14	internal damaging factors
15	length of the moveable object
16	length of the fixed object
17	surface of the moveable object
18	surface of the fixed object
19	volume of the moveable object
20	volume of the fixed object
21	shape
22	speed
23	functional time of the moveable object
24	functional time of the fixed object
25	loss of time
26	quantity of material
27	loss of material
28	strength
29	stabile structure of the object
30	force
31	tension, pressure
32	weight of the moveable object
33	weight of the fixed object
34	temperature
35	brightness of the lighting
36	power
37	energy use of the moveable object
38	energy use of the fixed object
39	loss of energy

14.1.2 A-matrix table

Problem-factor (minus-factor) → Trend-factor (plus-factor) ↓		productivity	universality, adaptability	level of automation	reliability	precision of manufacture	precision of measurement	complexity of construction	complexity of inspection and meas.	ease of manufacture	ease of use	ease of repair	loss of information	external damaging factors
	#	01	02	03	04	05	06	07	08	09	10	11	12	13
productivity	01		03.01 04.27	35.37 01.10	03.01 02.30	09.03 06.02	03.02 15.04	37.19 04.18	01.06 13.05	01.04 05.18	03.04 34.08	03.09 02.29	11.07 36	21.01 11.18
universality, adaptability	02	01.04 20.27		13.15 01	01.11 32.18	27.12	01.35 03.02	07.14 27.04	03	03.11 31	07.15 03.16	03.16 34.24	27.11	01.28 09.31
level of automation	03	35.37 01.10	13.24 03.01		28.13 09	04.10 06.36	04.10 02.15	07.18 02	15.13 29	03.10 11	03.37 15.12	03.01 11	01.38	05.38
reliability	04	03.01 14.30	11.01 32.18	28.11 13		28.09 03	09.12 28.36	11.01 03	13.17 04	04.02 24.17	13.19 17	03.28	02.04	13.01 05.17
precision of manufacture	05	02.06 09.23	01.11 18	10.04 06.36	28.09 03		10.04 18.11	10.05 06	02.36	18.03	03.09 01.36	29.02	18.34 29	10.04 02.26
precision of measurement	06	02.15 04.09	11.01 05	04.05 02.15	35.28 03.36	10.09 27		13.01 02.15	10.18 09.04	20.01 29.06	03.11 19.15	03.09 11.28	34.10 01.21	04.18 21.10
complexity of construction	07	37.19 04	14.07 04.27	07.03 18	11.01 03	10.18 09	05.10 02.15		07.02 27.04	13.10 03.11	13.39 10.18	03.11	34.08 09.03	21.08 14.17
complexity of inspection and meas.	08	01.06	03.07	15.33	13.17 04.32	27.34 24.09	10.18 09.04	07.02 27.04		35.04 28.14	05.35	37.10	01.38 13.21	21.08 14.04
ease of manufacture	09	01.03 02.04	05.11 07	32.04 03	39.04 13.38	37.20 11.07	03.01 37.06	13.10 03	20.04 28.03		05.35 11.16	01.03 28.39	09.18 06.16	18.05
ease of use	10	07.03 04	07.15 03.16	03.15 37.12	19.13 32.17	03.09 01.36	29.11 05.15	09.29 37.19	03.02	05.35 37		37.10 03.09	24.02 13.21	05.29 04.23
ease of repair	11	03.09 02	34.03 24.16	15.01 34.11	28.02 03.16	29.02	02.05 11	01.03 11.28	01.24 19.11	03.01 28.02	03.37 10.07		12.39 11.10	01.02 05.16
loss of information	12	11.36 07	18.35 39.17	01	02.04 36	27.24 09.34	09.03 27.34	20.11 18.24	01.38	09	13.21	05.02		21.02 03
external damaging factors	13	21.01 11.18	01.28 21.31	38.12 15	13.18 05.17	10.04 02.06	04.38 36.10	21.08 14.17	21.08 14.17	18.01 05	05.29 04.23	01.02 05	21.02 05	
internal damaging factors	14	21.01 06.23	07.19 24.18	05	18.05 17.23	24.19 15.10	12.38 10	08.03 31	05.33 13.03	24.17 18.34	24.18 22.20	03.18	02.33 14	01.18 34
length of the moveable object	15	22.24 04.14	22.07 03.16	19.18 10.16	02.22 14.17	02.04 14.27	04.09 24	03.08 10.18	01.03 10.18	03.14 19	07.14 01.24	03.04 02	03.18	03.07 19.18
length of the fixed object	16	25.22 34.10	03.01	02.11	07.14 04	05.09 02	09.04 12	03.10	10	07.19 13	05.29	12	18.10	03.06
surface of the moveable object	17	02.10 15.05	07.25	22.25 04.36	14.39	05.09	10.04 09.12	22.03 11	05.26 10.06	11.03 10.18	07.19 11.16	07.11 02.03	25.10	21.38 04.03
surface of the fixed object	18	02.07 19.34	07.16	36	09.01 17.24	05.14 06.26	10.04 09.12	03.06 26	05.01 25.06	17.16	16.24	16	25.16	13.05 23.01
volume of the moveable object	19	02.20 05.15	07.14	01.15 16.18	22.03 17.28	29.04 05.16	29.10 04	10.03	14.10 24	14.03 17	07.11 25.37	02	05.21	21.33 13.01
volume of the fixed object	20	01.27 02.05	31.20 11.09	02.03 11.18	05.01 16	01.02 29	19.10 09.31	03.31	05.19 10	01	34.10 18.19	03	09.22 18	15.23 08.13
shape	21	19.10 15.02	03.07 14	07.03 09	02.17 16	09.25 17	04.09 03	16.14 03.04	07.11 23	03.09 19.04	09.07 10	05.11 03	19.34 07.09	21.03 05.01
speed	22	04.02 11	07.02 10	02.06	28.01 13.04	02.04 09.29	04.09 03.18	02.04 24.15	12.15 13.16	01.11 32.03	09.04 11.37	15.05 04.13	11.10	03.04 01.36
functional time of the moveable object	23	01.19 22.08	03.01 11	20.02	28.05 11	12.13 16.17	12	02.24 14.07	08.14 23.01	13.03 24	37.13	14.02 13	02	21.07 38.04
functional time of the fixed object	24	40.02 16.30	05	03	15.13 20.17	02.10 18	02.10 18	35.02 24	29.15 20.01	01.02	03	03	02	19.03 17.38
loss of time	25	02.18 24.11	01.04	18.04 01.25	02.25 24	18.10 04.06	18.15 04.09	20.14	06.04 09.02	01.04 15.24	24.04 02.15	09.03 02	18.10 04.09	01.06 15
quantity of material	26	11.14 12.13	07.12 14	32.01	06.12 04.17	38.25	12.05 04	12.11 13.02	12.13 14.06	14.03 01.13	01.14 02.29	05.09 02.29	18.04 01	01.38 14.31
loss of material	27	04.01 02.36	07.02 05	01.02 06	02.14 23.01	01.02 18.31	16.15 31.04	01.02 04.18	01.06 02.11	07.15 38	09.04 05.18	05.01 15.13	09.03 02.12	38.21 25.17
strength	28	14.01 02.22	07.12 09	07	28.12	12.13	12.13 16	05.11 04	13.12 07.17	28.12 02.09	09.17 04.05	13.28 12	04.10 02	06.01 27.03
stabile structure of the object	29	36.01 17.12	01.25 15.05	03.32 01	01.17 18	06	11	05.01 21.10	01.21 23.36	01.08	09.01 25	05.01 02.16	18.02 09	01.18 06.25
force	30	12.04 01.27	07.19 06.40	05.01	12.01 11.33	04.14 27.26	01.02 36.18	10.01 02.06	26.27 02.08	07.27 06.03	03.04 12.29	07.03 28	09.34 02	03.01 17.06
tension, pressure	31	02.22 01.27	01	01.18	02.11 08.01	12.01	20.04 29	08.03 01	05.26 27	03.01 16	28	05	04.05 34.18	21.05 27
weight of the moveable object	32	01.12 18.27	14.35 07.32	10.01 06.08	12.28 03.13	04.01 10.06	04.13 01.10	10.25 26.15	04.14 10.09	13.04 03.26	01.12 05.18	05.13 04.28	02.18 01	21.33 06.13
weight of the fixed object	33	03.04 07.01	08.07 14	05.10 01	02.04 32.12	02.03 01.19	06.10 04	03.02 10.23	29.04 19.07	04.03 09	20.11 03.09	05.13 04.28	02.07 01	05.08 21.27
temperature	34	07.04 01	05.06 13	10.05 08.16	08.01 12.02	18	09.08 18	05.19 16	12.13 01.31	10.13	10.13	24.02 16	39.18	21.38 01.05
brightness of the lighting	35	05.29 16	07.03 08	05.10 02	01.03	12.09	28.07 09	20.09 11	09.07	08.01 04.10	04.10 08	07.19 11.16	03.20	07.08
power	36	04.01 15	08.19 15	04.05 19	08.18 10.31	09.05	09.07 05	40.08 25.15	08.01 16	10.02 15	10.01 02	01.05 02.15	02.08	08.21 31.05
energy use of the moveable object	37	37.04 01	07.19 11.16	09.05	08.33 28.13	12.35 04.32	12.03 09	05.14 13.04	01.30	04.10 25	08.01	03.07 19.04	11.18 34.21	03.01 20.13
energy use of the fixed object	38	03.20	01.11	02.05	02.26 36	04.24 06	04.01	04.24	08.01 16.29	03.24	11.18	01.19	05.02 34	02.05 21.27
loss of energy	39	04.02 14.01	07.22 11.31	05	28.02 01	18.10 09.04	09	34.36	01.12 07.36	02.01	01.09 03	05.08	08.02	33.21 01.05
	#	01	02	03	04	05	06	07	08	09	10	11	12	13

Problem-factor (minus-factor) ➡ / Trend-factor (plus-factor) ⬇		internal damaging factors	length of the moveable object	length of the fixed object	surface of the moveable object	surface of the fixed object	volume of the moveable object	volume of the fixed object	shape	speed	functional time of the moveable object	functional time of the fixed object	loss of time	quantity of material
	#	**14**	**15**	**16**	**17**	**18**	**19**	**20**	**21**	**22**	**23**	**24**	**25**	**26**
productivity	**01**	01.21 06.23	06.24 04.30	25.34 22.10	02.10 15.31	02.01 19.34	05.20 15.02	01.27 02.05	22.02 15.17	01.12 18	01.02 05.06	40.02 16.30	12.15	01.30
universality, adaptability	**02**	18.24 28	01.03 14.05	03.01 16	01.25 14.34	07.16	07.01 14	07.31 16.12	07.27 03.32	01.02 22	11.03 01	05.16	01.04	12.01 07
level of automation	**03**	05	22.11 04.19	36	19.22 11	11.10 24	01.11 16	10.11 18.31	07.09 03.11	04.02	20.39	02.16 11	18.04 01.25	01.11
reliability	**04**	01.05 17.10	07.39 22.24	07.14 04.28	19.02 22.16	09.01 17.24	12.02 22.18	05.01 18	01.03 16.28	33.01 28.04	05.01 12.29	15.13 20.17	02.25 24	33.04 17.12
precision of manufacture	**05**	24.19 15.10	02.04 14.27	05.09 02	04.38 14.09	05.14 06.26	09.04 05	29.02 01	09.25 17	02.04 09	12.13 17	18.04 15	09.10 04.06	09.25
precision of measurement	**06**	12.38 23.02	04.10 35.16	09.04 12.16	10.04 09.12	10.04 09.12	09.11 20	10.18 11.31	20.04 09	04.11 09.18	04.20 09	02.10 18	18.15 04.09	05.20 09
complexity of construction	**07**	08.03	03.08 10.10	10	22.03 11.16	20.26	15.10 20	03.16	14.11 04.07	15.02 04	02.24 04.07	02.11	20.14	11.12 13.02
complexity of inspection and meas.	**08**	05.33	16.19 10.18	10	05.11 06.19	05.23 25.16	14.03 24.16	05.06 10.31	13.11 03.23	12.24 16.01	08.14 29.23	29.15 20.01	06.04 09.39	12.13 14.06
ease of manufacture	**09**	33.18 23	03.14 11.19	07.19 13	11.03 10.37	16.17	11.14 03.17	01	03.04 11.13	01.11 32.03	13.03 24	01.16	01.04 15.24	01.36 03.18
ease of use	**10**	31.18	03.19 11.37	03.24	03.19 11.16	06.16 07.23	03.16 01.07	24.06 23.31	07.15 14.04	06.11 15	14.12 32.29	03.16 29	24.04 02.15	37.01
ease of repair	**11**	07.37 04	03.04 02.29	12.06 31	07.11 09	16.29	29.05 01.28	03	03.11 05.24	15.39	28.14 04.13	03	09.03 02.29	05.04 02.29
loss of information	**12**	02.33 21	03.10	10	25.10	25.16	34.31	05.21	24.09	10.09	02	02	18.10 04.09	18.04 01
external damaging factors	**13**	11.18 19.24	19.03 23.24	03.06	21.03 38.04	13.05 23.01	21.36 27.01	15.23 08.13	21.03 12.01	33.21 01.04	21.07 38.04	19.03 17.38	01.06 15	01.38 14.31
internal damaging factors	**14**		19.07 16.21	22.18	19.05 06.23	21.03 17	19.05 17	25.06 01.24	01.03	01.04 12.36	07.21 38.31	33.23 16.21	03.21	12.18 23.03
length of the moveable object	**15**	19.07		19.19 11.25	07.19 24	34.07 03.24	34.19 04.01	19.31 08.24	03.32 02.14	11.24 32	08	02.01 03.08	07.05 14	14.01
length of the fixed object	**16**	01.31	12.03 24		24.19 01	19.34 02.17	25.34 07	01.32 05.22	11.22 07.34	22.24 31	01.14 31	03.17 01	25.14 22	24.31 29.22
surface of the moveable object	**17**	19.05 06.23	22.07 06.24	22.19 24.11		03.24 12.18	34.22 19.24	22.11	35.15 14.24	14.25 24.15	20.12	03.12 08	10.24	14.25 20.11
surface of the fixed object	**18**	21.03 17	12.03	10.34 39.23	24.31 34		22.34 11	10.11 24	24.34 10	10.35	11.01	05.02 08.25	02.01 24.06	05.06 17.24
volume of the moveable object	**19**	19.05 17.03	03.34 01.04	34.07 24	03.34 24.19	22.34 31		01.22 18	03.07 14.24	14.24 30.15	20.01 24	25.31 03	05.20 15.02	14.25 34
volume of the fixed object	**20**	25.06 01.24	08.22	01.32 05.22	22.24 25.11	34.11 07	11.04 05.34		34.05 01	17.05 04	08.03 07.15	01.15 30	01.16 09.06	01.12
shape	**21**	01.03	14.15 35.24	11.22 02.34	35.15 24.02	19.22 09.24	22.24 07.21	34.05 01		01.07 15.06	22.10 39.29	25.11 35.21	22.02 15.19	26.21
speed	**22**	05.18 01.33	11.22 32	19.07 25	14.25 15	22.19 24	34.14 15	04.34	01.07 06.15		12.08 01.35	12.11	11.02	02.08 14.30
functional time of the moveable object	**23**	33.23 10.21	05.08 39	37.39 08	12.19 08	39.37 08	02.05 08.25	02.25 37	22.10 04.29	12.01 35		02.40 24	40.02 04.06	12.01 02.17
functional time of the fixed object	**24**	21	19.17 08	03.17 01	01.06 08.22	01.19 12.34	01.12 11.24	01.15 30	15.00 11.34	14.24 22.11	18.04		04.40 02.16	12.01 01
loss of time	**25**	01.21 06.23	07.05 14	25.18 22.35	10.24 35.16	02.01 19.24	05.35 15.02	01.16 09.06	24.02 15.19	04.10 02.24	40.02 04.06	04.40 02.16		01.30 06.16
quantity of material	**26**	12.01 17.23	14.22 01.08	07.31 01.04	07.22 14	05.06 17.24	07.40 14	01.30 31.03	01.22	01.14 15.04	12.01 02.17	12.01 31	01.30 06.16	
loss of material	**27**	02.03 15.14	22.14 02.23	02.04 18	01.05 02.31	02.06 23.31	03.14 25.26	12.23 06.31	11.01 12.35	02.11 04.30	04.13 12.06	13.16 06.30	07.06 01.02	20.12 09.19
strength	**28**	07.01 21.05	03.07 32.01	07.22 04.10	12.15 17.14	39.17 04	02.07 22.34	03.22 19.07	02.25 01.17	32.11 10.22	13.12 10	18.10 24	14.12 04.02	14.02 13
stabile structure of the object	**29**	01.17 13.23	11.07 03.04	27	05.28 11	23	04.02 08.23	15.04 01.17	21.03 06.24	38.07 04.06	11.13 02.01	23.12 01.36	01.13	07.09 01
force	**30**	11.12 26.18	19.08 39.26	04.02	08.02 07	03.06 26.27	07.39 37.27	05.26 06.27	02.01 17.15	11.04 07.37	08.05	05.02 10	02.27 26	22.14 06.26
tension, pressure	**31**	05.38 13.06	01.02 26	01.03 22.16	02.07 26.04	02.07 26.27	20.01 02	01.18	01.24 07.02	20 01.26	08.12 13	22.01 19.05	27.26 24	02.22 26
weight of the moveable object	**32**	21.01 31.23	07.32 14.15	07.19 37.14	14.19 30.15	04.03 31.24	14.05 17.04	17.05 24.34	02.22 01.17	05.32 07.30	35.15 31.01	02.04	02.01 40.04	12.10 06.31
weight of the fixed object	**33**	01.21 03.23	19.24	02.03 14.01	34.01	01.25 11.05	22.11 12	35.01 22.05	11.02 14.22	01.19 25	01.02 37	05.13 08.20	02.40 01.10	08.20 06.10
temperature	**34**	21.01 05.18	07.08 39	07.08 39	12.01 23.06	01.30	15.23 17.06	01.20 24	22.21 08.09	05.04 26.25	08.11 23	08.06 26.17	01.04 33.06	12.19 25.23
brightness of the lighting	**35**	01.08 09.23	08.09 16	22.01	08.09 10	19.24 03	05.11 02	22.11 02	09.25	02.11 08	05.08 20	20.02 04	08.03 10.19	03.08
power	**36**	05.01 06	03.02 01.27	03.01 24	08.30	19.09 11.30	01.20 30	25.20 29	14.22 05.17	07.01 05	08.01 02.30	16	01.40 02.20	24.15 08
energy use of the moveable object	**37**	05.01 20	37.04	05.08 07.04	07.08 29	01.05 19.07	01.11 06	12.15 01.34	37.05 14	32.07 01	04.01 20.06	04.11 02.08	01.30 08.06	15.36 16.06
energy use of the fixed object	**38**	08.21 06	19.24 37	19.39 16	11.37 18	19.16 08	11.08 04	01.06	34.18	03.04	08.24 04	19.04 10	02.17 34.39	12.01 31
loss of energy	**39**	33.01 05.21	34.05 20.11	20.30 34	07.10 19.25	19.34 25.06	34.06 36	34	24.18	16.01 30	33.01	19.31	02.06 09.34	34.06 29
	#	**14**	**15**	**16**	**17**	**18**	**19**	**20**	**21**	**22**	**23**	**24**	**25**	**26**

Problem-factor (minus-factor) / Trend-factor (plus-factor)		loss of material	strength	stabile structure of the object	force	tension, pressure	weight of the moveable object	weight of the fixed object	temperature	brightness of the lighting	power	energy use of the moveable object	energy use of the fixed object	loss of energy
	#	**27**	**28**	**29**	**30**	**31**	**32**	**33**	**34**	**35**	**36**	**37**	**38**	**39**
productivity	**01**	04.02 01.36	14.04 02.06	01.12 21.23	04.07 02.26	02.27 22	01.10 18.27	04.13 07.12	01.33 04.02	10.19 08.03	01.40 02	01.02 30.08	03	04.02 14.01
universality, adaptability	**02**	07.02 05.11	01.12 09.20	01.25 22	07.19 40	01.16	03.20 07.32	08.07 14.16	13.05 12.01	20.21 10.03	08.03 14	08.01 14.11	16.03 37	06.07 03
level of automation	**03**	01.02 06.35	29.11	06.03	05.01	11.01	04.10 06.01	04.10 01.02	10.05 08	32.09 08	04.05 13	05.09 11	01.03 05.11	36.04
reliability	**04**	02.01 14.23	28.04	12.03 18.05	32.04 02.12	02.18 01.08	12.32 02.17	12.02 32.04	12.01 02	28.09 11	33.28 10.31	33.28 13.08	26.36	02.28 01
precision of manufacture	**05**	01.31 02.18	12.13	25.06	04.08 15.26	12.01	04.09 11.06	10.01 13.39	08.10	12.09	09.05	09.05	02.18 35.40	11.09 05
precision of measurement	**06**	02.16 31.04	04.20 09	09.01 11	09.05	20.04 09	09.01 10.04	04.01 29.10	20.08 04.18	20.03 09	12.20 09	12.20 09	18.16 24	10.09 13
complexity of construction	**07**	01.02 04.14	05.11 04	05.21 19.08	10.16	08.03 01	10.25 15.26	05.10 01.23	05.19 11	18.19 11	40.08 25.15	13.05 14.04	04.02 11	02.01 11.05
complexity of inspection and meas.	**08**	03.06 02.18	13.12 07.04	28.21 23.25	26.04 17.08	01.26 27.09	13.10 04.11	20.11 04.03	12.13 01.16	05.18 10	08.03 16.02	01.30	08.01 16	01.12 07.08
ease of manufacture	**09**	07.15 38	03.12 02.09	28.11 03	01.37	01.08 03.27	04.14 07.16	03.13 26.11	13.10 06	04.18 13.03	13.03 37.18	04.10 13.03	03.24	08.01
ease of use	**10**	04.09 05.18	09.17 12.04	09.01 25	04.11 01	05.09 37	29.05 11.07	20.11 03.29	10.13 11	11.19 03.18	01.15 05.02	03.11 18	18.37 11	05.08 11
ease of repair	**11**	05.01 15.13	03.28 05.39	05.01	03.28 02	11	05.13 01.28	05.13 01.28	24.02	07.03 11	07.02 09.05	07.03 04.16	03.07 11.16	07.03 09.08
loss of information	**12**	34.19 12.11	18.31 22.17	25.10 18.35	11.03	18.21	02.18 01	02.01 35	21.03 18	08	02.08	03.40 08	18	08.02
external damaging factors	**13**	38.21 08.17	06.01 27.03	01.18 25.06	11.01 23.06	21.05 27	21.33 13.23	05.21 11.18	21.38 01.05	03.08 09.11	08.21 31.05	03.18 20.13	02.05 21.27	33.21 01.05
internal damaging factors	**14**	02.03 15	07.01 21.05	01.17 13.23	01.04 03.17	05.38 13.06	08.21 07.23	01.21 03.23	21.01 05.18	08.18 23.09	05.01 06	05.01 20	08.21 06	33.01 05.21
length of the moveable object	**15**	24.14 36.02	32.01 14.15	03.32 07.15	19.02 24	03.32 01	32.07 14.15	03.19 07.24	02.07 08	09	03.01	32.01 18	24.03 18.07	34.05 01.23
length of the fixed object	**16**	02.04 18.01	07.22 04.10	23.27 01	04.02	03.22 01	25.31 32.17	01.04 17.14	12.01 30.06	12.29	37.32	01.18 33.11	25.31 37.11	20.04
surface of the moveable object	**17**	02.01 05.23	12.07 17.22	28.05 11.23	08.25 01.05	02.07 26.04	05.19 14.04	12.31 05	05.07 16	07.09 08.11	08.02 09.06	08.09	19.08 35	07.19 25.10
surface of the fixed object	**18**	02.22 06.23	17	05.30	03.06 01.26	02.07 26.27	22.31 24.11	25.05 22.06	01.23 30	03.18 01.09	19.09	08.11 04	01.17	19.34 25
volume of the moveable object	**19**	26.23 15.02	39.22 07.34	04.02 03.23	07.01 26.27	20.01 26.27	05.10 14.17	31.17 10	15.23 02.06	02.11 05	01.20 11.06	01	30.38 08	34.07 11.16
volume of the fixed object	**20**	02.23 01.15	39.22 19.07	18.04 01.17	05.06 27	18.01	31.25 24.08	01.02 08.22	01.20 24	01.18 04.34	25.20	01.08 31.11	01.17	01.17
shape	**21**	01.14 12.35	25.22 02.17	38.03 06.24	01.02 27.17	15.07 02.22	32.02 14.17	07.02 10.12	21.22 08.09	11.07 09	24.20 05	05.20 15.22	22.07 31	22
speed	**22**	02.11 04.30	32.12 10.22	04.38 03.06	11.04 07.08	20.06 30.17	05.04 11.30	03.11	04.25 26.05	02.11 08	08.01 30.05	32.07 01.30	01.08	22.40 08.01
functional time of the moveable object	**23**	04.13 12.06	13.12 02	11.12 01	08.05 16	08.12 13	08.35 15.31	31.15	08.01 23	05.08 24.01	08.02 01.30	04.20 01.06	20.06	02.18 01
functional time of the fixed object	**24**	13.16 06.30	01.39	23.12 01.36	19.17 39	19.24 17	31.24 07	20.13 08.16	08.06 26.17	17.18 .34	16	11.17 18	01.17	02.17 15.18
loss of time	**25**	01.06 02.23	14.12 04.06	01.12 21.35	02.27 26.35	27.26 24	02.40 27.01	02.40 10.35	01.14 33.06	03.08 10.19	01.40 02.20	01.30 08.06	03	02.35 06.09
quantity of material	**26**	20.12 02.18	22.01 15.02	07.05 19.17	01.22 12	02.26 22.12	01.20 06.31	13.10 06.01	12.19 23	01.04 25.31	01	15.14 16.06	12.01 31	34.06 29
loss of material	**27**		01.04 31.17	05.22 25.17	22.07 06.17	12.26 27.02	01.20 36.17	01.20 21.09	33.26 23.31	03.20 11	04.13 06.30	01.06 18.35	04.13 37.31	01.13 05.31
strength	**28**	01.04 31.17		11.19 01	02.06 12.22	02.12 06.17	03.32 17.07	17.10 13.03	25.02 17	01.08	02.10 01.04	08.01 02	01	01
stabile structure of the object	**29**	05.22 25.17	19.39 07		02.01 33.16	05.01 17	33.01 05.23	10.23 03.17	01.03 09	09.12 13.07	09.01 13.31	11.08	13.24 14.06	22.05 23.20
force	**30**	32.01 17.35	01.02 22.13	01.02 33		06.33 28	32.03 27.06	06.11 03.04	01.02 33	11.08 01.18	08.01 06.27	08.19 02	03.16 26.27	22.07
tension, pressure	**31**	02.26 12.27	39.06 12.17	01.38 05.17	26.01 33		02.26 27.17	11.14 02.06	01.23 08.05	18.37 21.01	02.01 22	22.18 02.27	19.24 37.18	05.26 29
weight of the moveable object	**32**	35.01 12.31	04.13 06.17	03.01 08.23	32.02 06.27	02.26 27.17		08.01	20.14 24.30	08.03 09	37.26 06.31	01.37 15.31	01.03 04	20.05 15.08
weight of the fixed object	**33**	35.32 11.25	04.05 02.13	10.23 03.17	32.02 08.01	11.14 02.06	17.31 03		04.08 09.21	01 08.09	07.08 06.21	12.19	06.08 04.03	06.08 04.07
temperature	**34**	33.26 14.31	02.25 21.17	03.01 09	01.02 12.33	01.23 08.05	26.21 20.30	21.01 09		09.25 33.16	05.22 19.29	08.07 12.19	01.12	33.19 01.30
brightness of the lighting	**35**	11.03	01.08	09.12 13	10.08 20	25.01 37	08.03 09	05.01 09	09.01 08		09	09.03 08	09.01 03.07	08.16 03.20
power	**36**	04.13 06.30	10.02 04	01.09 07.31	10.05 26.01	21.02 01	32.26 30.31	08.10 19.13	05.22 19.29	16.20 08		16.20 08.27	08.07 16.03	02.01 30
energy use of the moveable object	**37**	01.18 06.35	35.08 39.01	08.11 19.18	16.10 33.05	36.22 29	37.06 04.31	04.37 31.11	08.18 12.22	05.07 08	20.08 27.06		07.11	37.21 07.18
energy use of the fixed object	**38**	04.13 06.31	01	13.24 14.06	26.27	19.39 01.17	04.11 32.08	08.39 20.13	01.08 33.26	08.05 01.09	35.08 11.01	05.08 11.35		01.31 03.18
loss of energy	**39**	01.13 05.27	10	22.05 23.20	26.30	24.11	07.20 08.04	08.20 06.39	08.30 34	03.11 09.07	12.30	01.08 12	01.08 24	
	#	**27**	**28**	**29**	**30**	**31**	**32**	**33**	**34**	**35**	**36**	**37**	**38**	**39**

14.2 A-catalogs (short reference versions)

14.2.1 40 primary transformation models (navigators)

#	Navigator
01	change in the aggregate state of an object
02	preliminary action
03	segmentation
04	replacement of mechanical matter
05	separation
06	use of mechanical oscillations
07	dynamization
08	periodic action
09	change in color
10	copying
11	inverse action
12	local property
13	inexpensive short-life object as a replacement for expensive long-life one
14	use of pneumatic or hydraulic constructions
15	discard and renewal of parts
16	partial or excess effect
17	use of composite materials
18	mediator
19	transition into another dimension
20	universality
21	transform damage into use
22	spherical-shape
23	use of inert media
24	asymmetry
25	use of flexible covers and thin films
26	phase transitions
27	full use of thermal expansion
28	previously installed cushion
29	self-servicing
30	use of strong oxidants
31	use of porous materials
32	counter-weight
33	quick jump
34	matryoshka (nested doll)
35	unite
36	feedback
37	equipotentiality
38	homogeneity
39	preliminary counter-action
40	uninterrupted useful function

14.2.2 As-catalog

01. change in the aggregate state of an object	a) this includes transitions into "pseudo-states" ("pseudo-liquid") and into transitional states such as the use of the elastic properties of solid objects as well as simple transitions such as from a solid to a liquid state; b) changes in concentration or consistency, in the degree of flexibility, in temperature, etc.
	(01-02) The German firm ZELTEC has developed a transparent cone to extract clean water from the ground **(navigator 01).** The cone is placed on the ground and solar rays heat the surface. Water in the ground evaporates, precipitates on the walls of the cone and flows into a collector placed around the cone at its bottom.

02. preliminary action	a) previous necessary (partial or complete) change of an object; b) prepare objects in advance so that they can be put to work from the best position and are available without loss of time.
	(02-02) Nets are stretched around sheep in advance to collect wool that falls from the animals upon application of a special chemical preparation **(navigator 02).**

03. segmentation	a) disassemble an object into individual parts; b) make it possible to disassemble an object; c) raise the degree of disassembly (reduction into parts) of an object.
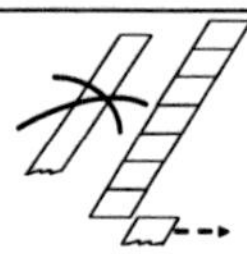	(03-02) In order to regenerate the cutting function of a knife, its blade is formed out of several segments that can be removed when they wear out. Small indentations are cut into the blade for this purpose **(navigator 03).**

04. replacement of mechanical matter	a) replace mechanical schemes with optical, acoustic, or olfactory schemes; b) use of electrical, magnetic, or electromagnetic fields for the interaction of objects; c) replacement of static fields with dynamic ones, from temporally fixed to flexible fields, from unstructured fields to fields with a specific structure; d) use of fields in connection with ferric-magnetic particles.
	(04-02) The German firm "Microtech" has made micro-miniature "submarine" for inspection and treatment of blood-vessels. The external rotating magnetic field actuates the "propeller" **(navigator 04).**

05. separation	separate the "incompatible part" ("incompatible property") from the object or - turned completely around - include the only really necessary part (necessary property) into the object.
	(05-01) A device is integrated into a wheelchair that is equipped with an electrical motor, an accumulator and a guidance system in order to ensure a motorized forward movement of the chair (**navigator 05**).

06. use of mechanical oscillations	a) cause an object to vibrate; b) raise the frequency of the vibrations up to and including ultra-high frequencies if the object is already in motion; c) use of the resonating frequency, application of quartz vibrators; d) use of ultra-sound vibrations in connection with electromagnetic fields.
	(06-01) An American firm has produced special ultrasonic whistles which are mounted on the front bumper. At speeds over 50 kilometers per hour under influence of the air stream, the whistles emit ultrasonic signals which are frightening for animals, but inaudible for people **(navigator 06).** The higher the speed, the "louder" the signal.

07. dynamization	a) the characteristics of an object or an environment are changed to optimize every work procedure; b) disassemble an object into parts that are moveable among each other; c) make an object moveable that is otherwise fixed.
	(07-01) A new dynamic solution was found to more securely anchor yachts and improve their movement especially in shallow waters: the keel can be folded and turned using a hinge in the hull with flexible components **(navigator 07).**

08. periodic action	a) transition from a continuous function to a periodic one (impulse); b) change the periods if the function already runs that way; c) use the breaks between impulses for other functions.
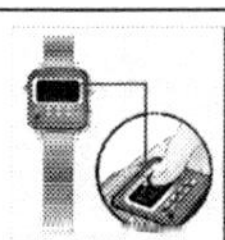	(08-02) The Japanese firm "Seiko" has produced an electronic watch which "falls asleep" (ceases to show time) if the watch remains motionless for more than three days. Internal memory continues to keep time, and the watch "wakes up" at a touch and starts to show the correct time right away **(navigator 08).**

09. change in color	a) change the color of an object or its environment; b) change the level of the transparency of an object or its environment; c) use color supplements to observe objects or processes that are difficult to see; d) add lighting if this kind of supplements is already in use.
	(09-02) The French firm DAITEM proposed a new system to protect apartments and cars can be protected from thieves. The system uses a device which fills the apartment or car with dense thick milky-white smoke upon detecting unauthorized entry **(navigator 09).** The device is activated when the burglar alarm system is tripped off. The smoke does not damage either the apartment or the car.

10. copying	a) use a simplified and inexpensive copy instead of an inaccessible, complicated, expensive, inappropriate, or fragile object; b) replace an object or a system of objects with optical copies; use here a change in measurement (blow-up or reduce the copy); c) if visible copies are used, they can be replaced with infra-red or ultra-violet copies.
	(10-01) A system has been developed to copy aquariums with many different species of fish. The system consists of a display, for example, a computer monitor installed behind an aquarium with thin glass walls that imitates the movement of air bubbles and the natural movements of plants. The fish and their entire world are projected with video recordings on the computer screen (**navigator 10**). The combined planes create an illusion of existence of the fish in a big aquarium.

11. inverse action	a) instead of an action prescribed by the conditions of an assignment, complete a reverse action (heat an object instead of cooling it); b) make a moveable part of an object or the environment fixed or a fixed part moveable; c) turn an object "upside down" or around.
	(11-02) The American firm "Kinesis" has developed a keyboard with left- and right-hand keys sunk in two separated recesses **(navigator 11).**

12. local property	a) change the structure of the object (the external environment, external influences) from the same to a different one; b) different parts of an object have different functions; c) every object should exist under conditions that correspond best to its functions.
	(12-02) It has been suggested that pipes made of readily fusible material filled with fire extinguishing liquid be installed above electrical cables to automatically "locate" and suppress electrical fires **(navigator 12).**

13. inexpensive short-life object as a replacement for expensive long-life one	replace an expensive object with a group of inexpensive objects without certain properties, for example, long life.
	(13-01) It has been suggested to create and apply small and inexpensive unmanned planes to record meteorological data above oceans **(navigator 13).**

14. use of pneumatic or hydraulic constructions	use gaseous or fluid parts instead of fixed parts in an object: parts that can be blown up or filled with hydraulic fluid, air-cushions, hydrostatic or hydro-reactive parts.
	(14-02) The Japanese firm "Mugen Denko" produces an inflatable life vest for motorcyclists **(navigator 14).**

15. discard and renewal of parts	a) parts that have fulfilled their task and are no longer part of an object should be disposed of (dissolved, evaporated, etc.); b) used parts of an object should be immediately replaced during work.
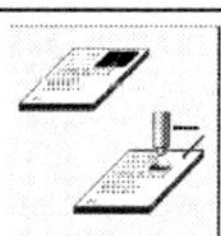	(15-02) The Japanese firm "Toshiba" has developed computer printer ink which becomes almost completely colorless when heated **(navigator 15).**

16. partial or excess effect	when it is difficult to achieve the desired effect completely, we should try to achieve a bit less or a bit more. This can make the task much easier.
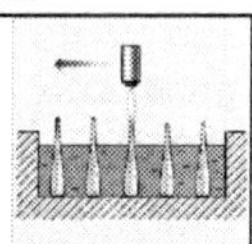	(16-02) Medicinal ampoules can be sealed off quickly by dipping them into cooling liquid that protects them from strong thermal field except for the tip which subjected to intense heating (**navigator 16**).

17. use of composite materials	move from homogeneous materials to combinations.
	(17-02) In England, a multi-layered curtain has been proposed. It contains pores of different sizes providing mechanical filtration of sound waves so that the aggregate pass band of a set of filters approximately corresponds to the spectrum of sea surf **(navigator 17).**

18. mediator	a) use another object to transfer or transmit an action; b) temporarily connect an object with another (easily separable) object.
	(18-02) Single-cell micro-organisms that start to glow in moving water were introduced into salt water pools to investigate the movements of dolphins (mediators according to **navigator 18**). The water flowing around the dolphins' bodies can then be recorded with high precision video cameras.

19. transition into another dimension	a) an object is shaped so that it can move or is placed not only in a linear fashion, but also in two dimensions, meaning on a surface. It is also possible to improve the transition from a surface to a three-dimensional space; b) do construction on several floors; tip or turn the object on its side; use the back of the space in question; c) optical rays that strike a neighboring space or the back of the present space.
	(19-02) Measuring with a normal measuring tape often means that the tape is used in different directions, either horizontally, vertically, or at various angles. This means the measurements are made in different dimensions (**navigator 19**). This makes it possible to construct closely connected three dimensional representations of a room.

20. universality	an object has several simultaneous functions so that other objects are not needed.
	(20-02) A robot with universal learning abilities has been developed at the TU Berlin to dismantle complex household appliances and industrial facilities that are no longer in use (solution according **navigator 20**).

21. transform damage into use	a) use damaging factors, especially damaging influences from the environment to achieve a useful effect; b) eliminate a negative factor by combining it with other negative factors; c) support the damaging factor until it is no longer causes damage.
	(21-01) Wood by-products (bark, shavings and leftovers) could be used in large power plants. The production of extra electrical energy for saw mills or neighbouring facilities could compensate for the damage caused by large amounts of wood waste that collects in them **(navigator 21)**.

22. spherical-shape	a) change from linear parts of the objects to curved ones, from flat surfaces to spherical ones; from parts shaped like cubes or parallelepipeds to round structures; b) use rollers, balls, and springs; c) change to turning movements by using centrifugal force.
	(22-01) Scientists from the University of Potsdam (Germany) have developed a special stand on the basis of a rotating disk **(navigator 22).** The disk is fitted with protrusions that can be used as hand grips and foot rests. The axis of rotation of the disk can also change position.

23. use of inert media	a) replace a normal medium with an inert one; b) complete a process in a vacuum.
	(23-01) Powdered dry ice has for a long time been used to disperse clouds; now German meteorologists in Germany have successfully used it to disperse fog at an airfield **(navigator 23).**

24. asymmetry	a) move from a symmetrical shape of an object to an asymmetrical one; b) increase the degree if the object is already asymmetrical.
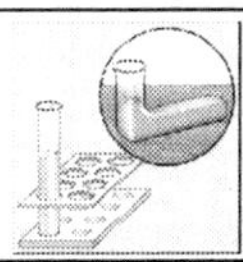	(24-01) A reagent tube needs to be developed with curved form and flat foundation so that it can stand on a table without a special frame; this is expected to simplify the use of such devices (**navigator 24**).

25. use of flexible covers and thin films	a) flexible covers and thin layers are used in place of the usual constructions; b) isolate objects from the external world with flexible covers or thin layers.
	(25-01) German entrepreneur Peter Aschauer suggested a new rescue device – an avalanche airbag made of bright orange nylon **(navigator 25).** The airbag is carried in a small backpack and inflated by compressed nitrogen from a small cylinder activated by the user when he finds himself in danger of boing buried under snow.

26. phase transitions	full use of phenomena that occur during phase transitions such as a change in volume, radiation or absorption of warmth, etc.
	(26-01) To enable rapid cooling of injured spots, the French firm "Crionic Medical" offers gun-shaped portable devices; when the trigger is pulled, the gun emits a cloud of dry ice **(navigator 26).**

27. full use of thermal expansion	a) full use of the expansion (or reduction) of materials when heating them; b) use of materials with different coefficients of heat expansion.
	(27-01) The Firm Sealed Air Corporation (USA) has developed highly elastic polyethylene bags which come in all sizes. Mechanical or thermal impact triggers off production of polymer foam which is then evenly distributed throughout the interior of the bag **(navigator 27).**

28. previously installed cushions	increase the relatively low security of an object with safety measures in advance.
	(28-02) Parachutes in the passenger section could help save passengers in plane crashes (**navigator 28**). The "superfluous" weight of cargo sections, wings and engines could be separated and discarded so that the burden on the parachutes is reduced (as is their own weight). This could reduce the operating expenses attributable to such system.

29. self-servicing	a) the object services itself with auxiliary and repair functions; b) reuse waste (energy, material).
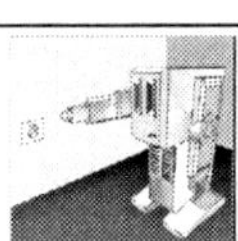	(29-01) Robots have a program that instruct them to search for a specified or any nearby electrical socket in order to charge their accumulators (**navigator 29**).

30. use of strong oxidants	a) replace normal air with an enhanced stream; b) replace an enhanced stream with oxygen; c) influence air or oxygen with ionizing rays; d) use of oxygen with ozone; e) replace ionized or ozone-oxygen with ozone.
	(30-01) The Swedish firm "Mediteam" has developed a method of treatment of caries without drilling holes, but by using a special mix which dissolves the diseased substance of the tooth in less than half a minute **(navigator 30).**

31. use of porous materials	a) make an object porous or use supplementary porous elements (inserts, coverings, etc.); b) if the object already consists of a porous material, the pores can be filled with some kind of material in advance.
	(31-01) Experts of FAO (Food and Agriculture Organization of the United Nations) have developed and patented a way to sterilize coconut water without heating. The drink is filtered with micro-porous filters which remove microbes and their spores **(navigator 31).**

32. counter-weight	a) compensate for the weight of an object with its connection to another object with lifting power; b) compensate for the weight of an object using interaction with the external environment (for example, with aerodynamic or hydrodynamic forces).
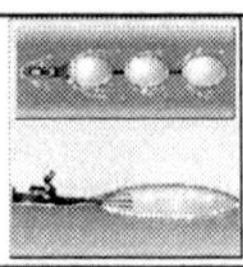	(32-02) Turkish engineers have offered a way to deliver lots of fresh water by sea transportation of long "chains" and "networks" of tanks made of polymeric film; each such tank contains 10 000 cubic meters of water. As fresh water is lighter then sea water, the tanks keep afloat **(navigator 32).**

33. quick jump	complete a process or some of its (damaging or dangerous) stages at high speed.
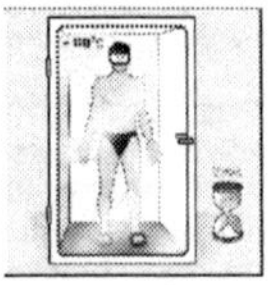	(33-02) An unusual way of treating arthritis and rheumatism is used by rheumatologists in the German city of Sendenhorst: the patient is placed in a refrigerating chamber with a temperature of minus 110-120 degrees of Celsius for 1-2 minutes **(navigator 33).**

34. matryoshka (nested doll)	a) an object is inside another object that is also inside another, etc.; b) an object runs through a hollow space in another object.
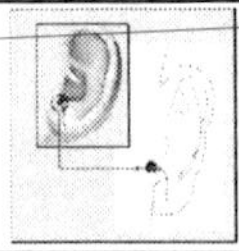	(34-02) A receiver and synthesizer are inserted into the ear in order to conveniently and discreetly hear radio programs, messages from mobile telephones and other signals (**navigator 34**).

35. unite	a) unite similar objects or objects for neighboring operations; b) temporarily unite similar objects or objects for neighboring operations.
	(35-02) 10,000 thermal microelements are installed on the underside of a watch to produce an sufficiently large electrical potential to power the watch; the microelements transform the difference between the temperature of the wrist and the temperature of the ambient environment into electricity (**navigator 35**).

36. feedback	a) create a retroactive influence; b) change a retroactive influence that already exists.
	(36-02) Siemens has developed a computer program which determines the amount of cash to be placed inside each ATM based on previous usage data (**navigator 36**).

37. equipotentiality	change work conditions so that it is not necessary to lift or lower an object.
	(37-02) To make it easier for elderly and sick people to get into and out of the bathtub, it is fitted with a door which can be opened when the tub is empty, and is tightly sealed when the tub is filled with water **(navigator 37).**

38. homogeneity	objects that interact with the object in question must be made from the same material (or from one with similar properties).
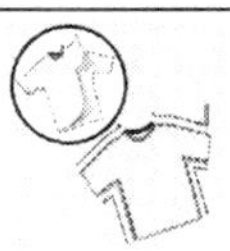	(38-01) In the Welding Institute in Cambridge (Great Britain), a shirt made from pieces of a synthetic fabric has been welded by laser **(navigator 38).**

39. preliminary counter-action	if the conditions of a task require an action, then a opposite action should be taken in advance.
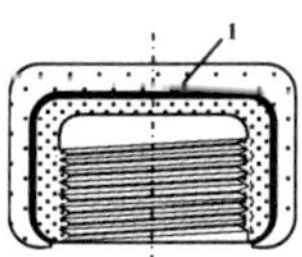	(39-02) English engineers form Salford University have developed a special cap for medicine tubes that cannot be opened any more after the expiration date (**navigator 39**).

40. uninterrupted useful function	a) complete a job without interruptions where all parts work continuously at full capacity; b) eliminate idle running and interruptions.
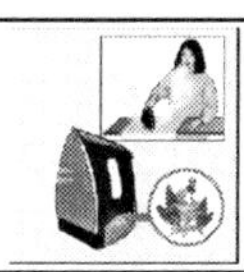	(40-02) The American firm "Procter & Gamble" has patented a way of treatment by inhalation during ironing by filling a steam generator of the iron with water containing several drops of medicine **(navigator 40).**

14.2.3 Af-catalog (simplified)

1. Separation in space	One part of system space has property A, while another part of system space has property not-A.
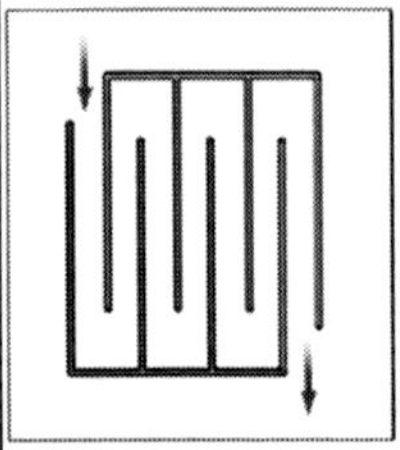	(**QUEUE ORDERING**) To physically eliminate the possibility of a chaotic crowd emerging in front of a building (museum, theater, exhibition hall), the area is reorganized by erecting dividing constructions which form a narrow corridor which people can negotiate only walking one after another (**transformation in space**).

2. Separation in time	During one time interval, the system has property A; during another time interval, it has property not-A.
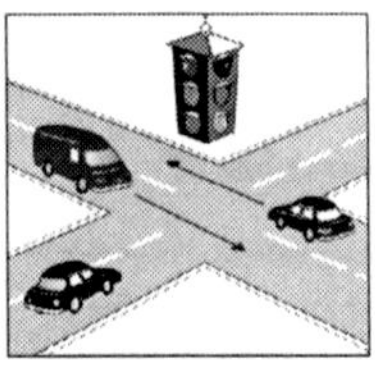	(**SAME-LEVEL ROAD CROSSING**) To physically eliminate the possibility of vehicles colliding as they traverse traffic routes at the same level, conflicting traffic flows are divided in time by using special signal systems (such as traffic lights or traffic wardens) which regulate the order of movement (**transformation in time**).

3. Separation in structure	Some elements of the system have property A, while other elements or the system as a whole have (has) property not-A.
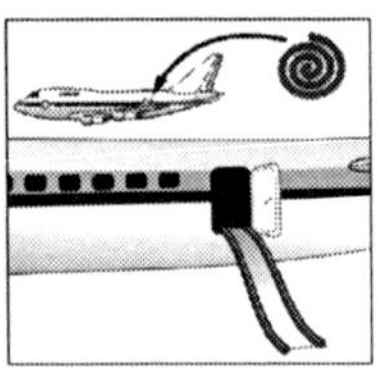	(**EMERGENCY ESCAPE CHUTE**) To create an emergency escape chute that is durable and "inflexible" when in use and rolled-up and "flexible" when not in use, it is made in the form of several joined elastic pipes which are inflated by compressed air (**transformation in structure**).

4. Separation in material	For one purpose, the material has property A; for another purpose, it has property not-A.
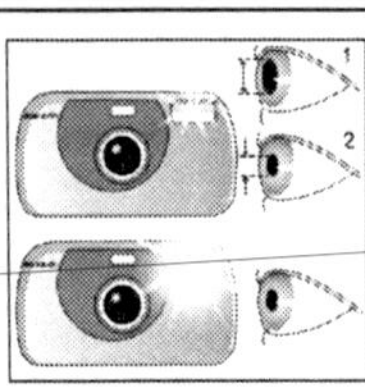	(**PRE-FLASH**) To prevent the "red-eye effect" when making flash-assisted photographs, many cameras generate one or several short smaller flashes before setting off the main flash (**transformation in material and energy**); as a result, pupils contract, and less light from the main flash reaches the retina.

14.2.4 Afs-catalog (simplified)

Principle of transfor-mation	Connection to As-navigators
1 separation of the conflicting properties in space	**05 removal:** remove the disruptive part, emphasize the part needed. **10 copy:** use of simplified and inexpensive copies. **19 transition to another state:** increase the freedom of an object, use construction in several layers, use lateral and other surfaces. **22 spherical-shape:** transition to curved surfaces and shapes, use of wheels, balls, or springs. **24 asymmetry:** transition to asymmetrical shapes, increase asymmetry. **25 use of flexible covers and thin layers:** use flexible covers and thin layers instead of normal constructions. **34 matryoshka:** store an object in another one in stages, place an object in the hollow space of another one.
2 separation of the conflicting properties in time	**02 previous effect:** run the necessary effect partially or completely; arrange the objects so that they can go to work faster. **07 dynamization:** make an object (or its parts) moveable, optimize the characteristics of the process (of an object) in every stage. **08 periodic effect:** transition from a continuous effect to a periodic one, change the periodicity, interruptions in the effect. **18b transmission:** connect an object with another (easily removable) one for a specific time. **28 previously installed cushion:** consider possible disruptions in advance. **33 quick jump:** accelerate a process strongly so that damages don't even appear. **35b connect:** temporally connect similar or neighboring operations with each other. **39 previous opposite effect:** an opposite effect must be completed in advance to run the primary effect. **40 uninterrupted useful function:** eliminate idle time and interruptions so that all parts of an object function at full capacity.
3 separation of the conflicting properties in structure	**03 separation:** disassemble an object into its parts, increase the degree of "disassembly". **11 opposite:** take the opposite action from the one that is apparently given by the conditions at hand. **12 local property:** transition from a similar to a different structure so that every part can complete its function under the best condition. **15 disposal and regeneration of parts:** a used-up part can be disposed of or regenerated during its function. **18a transmission:** use a part to transmit an action. **35a connect:** connection of similar objects or objects for neighboring operations with each other.
4 separation of the conflicting properties in material	**01 change of the aggregate state of an object:** change the concentration or consistency, full use of properties like the elasticity of materials, etc. **17 use of materials that consist of several components:** transition from similar materials to those consisting of several components. **23 use of inert media:** replace a medium with something inert. Let processes run in a vacuum. **26 use of phase transitions:** full use of phenomena that occur during phase transitions: changes in volume, in the radiation or absorption of heat. **27 use of heat expansion:** full use of the heat expansion of materials, the use of materials with different heat expansion. **29b self-servicing:** use of the waste of material and energy. **30 use of strong oxidants:** replace air with oxygen, influence air with ion beams, use of ozone. **31 use of porous materials:** shape an object in a porous way, fill porous parts with some kind of material. **38 similarity:** manufacture objects that influence each other from one and the same material.

14.2.5 Advanced Af-catalog "Change of Conditions – Change of Function"

#	Resource	Change of Conditions – Change of Function	
		Definition	**Example**
01	**Systemic**	Under one set of conditions, the **system** has property A; under another set of conditions, it has property not-A.	When there is no emergency, the life vest is hidden and occupies little space; when there is an emergency, it is on and filled with compressed air.
02	**Informational**	Under one set of conditions, the **information** has property A; under another set of conditions, it has property not-A.	A signal with the same parameters can switch the system on (if it was off) and off (if it was on).
03	**Functional**	Under one set of conditions, the **function** is accompanied by property A; under another set of conditions, it is accompanied by property not-A.	Low-altitude flight is accompanied by emergence of supplementary lifting force when the aircraft attains certain speed (screen effect), and is not accompanied by such effect at lower speeds.
04	**Structural**	Under a certain set of conditions, some **elements** of the system have property A, while other elements have property not-A.	To sort out small non-magnetic metal scrap, a magnetic-rheological medium is used to create conditions which modify "buoyancy" of individual elements – elements with lower specific weight float first followed by "heavier" elements, etc.
05	**Spatial**	Under a certain set of conditions, one part of system **space** has property A, while another part of system space has property not-A.	Water: 1) "hard" water around "soft" water (mixture of water and air – anti-injury "air cushion") – in a diving pool; 2) immobile water around mobile water – in a swimming pool with artificially generated water flow (to enable the swimmer to swim without turnarounds).
06	**Temporal**	Under one set of conditions during one **time** interval, the system has property A; under another set of conditions during another time interval, it has property not-A.	Encapsulated ice melts in the drink (turns into water) and cools it; then is frozen in the refrigerator (turns into ice) to be reused at a later time.
07	**Material**	Under one set of conditions, the **material** has property A; under another set of conditions, it has property not-A.	Material with Curie point: non-magnetic at one temperature; magnetic at another temperature.
08	**Energetical**	Under one set of conditions, the **energy** has property A; under another set of conditions, it has property not-A.	All other things being equal, electric charge (lightning) cannot pass through non-ionized air, while it can pass through ionized air.

14.2.6 Advanced Af-catalog

"Change of Function – Change of Purpose"

#	Resource	Change of Function – Change of Purpose	
		Definition	**Example**
01	**Systemic**	For one purpose, the **system** has property A; for another purpose, it has property not-A.	Reversible pump can be used to pump water both in and out.
02	**Informational**	For one purpose, the **information** has property A; for another purpose, it has property not-A.	Pseudo-random binary sequence appears as noise, while in reality it carries useful information (encoding, radio-location, etc.).
03	**Functional**	For one purpose, the **function** has property A; for another purpose, it has property not-A.	Polishing with iced-in abrasive: heating together with cooling.
04	**Structural**	For a certain purpose, some **elements** of the system have property A, while other elements or the system as a whole have (has) property not-A.	Individual links in a bicycle chain are "inflexible", while the chain as a whole "flexibly" encircles gear wheels.
05	**Spatial**	For a certain purpose, some part of the system **(in terms of its geometrical dimensions, shape or position, etc.)** has property A, while the system as a whole has property not-A.	The paint jet issuing from the nozzle of the pulverizer is hugged by a high–pressure air "cone" – as a result, paint particles do not leave the "cone" and cover the surface being painted more densely; super-effect: you can work without a respirator.
06	**Temporal**	For one purpose during one **time** interval, the system has property A; for another purpose during another time period, it has property not-A.	The captain's deckhouse lowers when the pleasure boat passes under a low bridge.
07	**Material**	For one purpose, the **material** has property A; for another purpose, it has property not-A.	An escape chute made of soft rubber is rolled up for storage, but when it is filled with compressed air, it becomes "hard" and can be used to evacuate the plane.
08	**Energetical**	For one purpose, the **energy** has property A; for another purpose, it has property not-A.	Live electric wire in the field: the voltage is sufficient to scare animals off, but not sufficient to injure them.

14.2.7 Table for the SITO-plus method

No.	Factor Position	01	02	03	04	05	06	07	08
01	productivity	02	01	04	03	06	05	10	30
02	universality, adaptability	01	03	07	14	16	11	05	20
03	level of automation	01	11	04	10	03	05	02	06
04	reliability	01	28	02	12	04	17	13	03
05	precision of manufacture	09	04	02	05	06	10	01	12
06	precision of measurements	09	04	20	10	01	02	11	12
07	complexity of construction	10	11	03	04	02	05	08	14
08	complexity of inspection and measurements	04	01	10	13	16	03	05	06
09	ease of manufacture	03	01	11	13	04	16	18	07
10	ease of use	03	11	05	04	09	15	29	37
11	ease of repair	03	02	05	01	28	11	07	29
12	loss of information	02	01	10	21	04	08	09	18
13	external damaging factors	21	01	05	03	38	04	06	08
14	internal damaging factors	01	21	05	03	23	06	17	07
15	length of the moveable object	03	14	01	07	02	19	24	32
16	length of the fixed object	01	04	22	03	10	02	07	12
17	surface of the moveable object	05	07	10	25	11	02	04	09
18	surface of the fixed object	05	06	01	02	17	25	16	23
19	volume of the moveable object	01	03	02	05	14	07	15	20
20	volume of the fixed object	01	05	02	22	06	08	14	15
21	shape	02	03	07	22	09	15	01	05
22	speed	04	01	11	02	08	15	30	05
23	functional time of the moveable object	08	01	02	12	13	04	05	06
24	functional time of the fixed object	01	02	03	16	17	13	15	20
25	loss of time	02	01	06	04	24	35	09	15
26	quantity of material	01	12	14	06	02	22	13	17
27	loss of material	02	01	04	06	31	05	13	18
28	strength	01	12	02	17	07	04	13	22
29	stabile stricture of the object	01	05	23	13	17	03	06	07
30	force	01	02	27	06	26	03	07	04
31	tension, pressure	01	02	26	27	05	17	22	08
32	weight of the moveable object	01	04	06	10	13	14	15	31
33	weight of the fixed object	01	02	04	08	03	05	06	10
34	temperature	01	08	05	12	21	04	06	09
35	brightness of the lighting	08	09	03	01	07	05	10	11
36	power	01	08	02	05	30	10	15	19
37	energy use of the moveable object	01	08	04	06	05	07	37	18
38	energy use of the fixed object	01	08	06	13	02	03	05	20
39	loss of energy	34	01	05	06	08	20	30	02

14.2.8 Table for the SITO-minus method

No.	Factor Position	01	02	03	04	05	06	07	08
01	productivity	01	02	04	03	07	14	15	22
02	universality, adaptability	07	03	01	05	14	11	04	08
03	level of automation	01	05	10	15	03	04	02	16
04	reliability	02	01	28	17	13	03	04	13
05	precision of manufacture	09	02	04	06	01	10	05	03
06	precision of measurements	04	09	10	12	02	03	15	18
07	complexity of construction	03	02	10	04	01	11	14	05
08	complexity of inspection and measurements	01	13	04	10	05	06	07	08
09	ease of manufacture	03	01	04	10	13	07	11	02
10	ease of use	03	01	04	09	05	07	10	11
11	ease of repair	03	07	02	05	28	01	11	13
12	loss of information	02	18	01	10	21	04	03	05
13	external damaging factors	21	01	05	03	04	38	06	08
14	internal damaging factors	01	05	21	23	06	03	17	33
15	length of the moveable object	03	07	14	04	19	22	01	08
16	length of the fixed object	10	22	01	02	03	04	07	34
17	surface of the moveable object	19	07	10	11	02	14	03	08
18	surface of the fixed object	01	06	16	17	02	05	19	23
19	volume of the moveable object	01	02	05	14	07	11	15	20
20	volume of the fixed object	01	05	06	03	15	22	24	31
21	shape	01	03	14	02	07	15	22	11
22	speed	01	04	11	15	02	07	24	30
23	functional time of the moveable object	01	12	08	04	13	16	02	20
24	functional time of the fixed object	16	01	02	03	17	05	08	13
25	loss of time	02	01	04	06	09	15	24	10
26	quantity of material	01	12	06	14	02	08	18	20
27	loss of material	02	01	04	06	05	13	18	23
28	strength	04	01	12	02	18	22	08	10
29	stabile stricture of the object	01	23	03	05	06	11	17	25
30	force	01	02	26	04	05	08	06	27
31	tension, pressure	01	02	27	26	05	06	08	12
32	weight of the moveable object	01	04	07	32	05	10	17	02
33	weight of the fixed object	01	10	13	02	03	04	08	11
34	temperature	01	08	05	02	12	23	06	04
35	brightness of the lighting	09	08	03	11	05	07	18	01
36	power	01	08	02	05	06	09	20	03
37	energy use of the moveable object	01	08	20	03	05	09	18	02
38	energy use of the fixed object	03	01	06	08	04	13	16	21
39	loss of energy	01	05	08	07	02	04	06	09

14.3 List of the Examples*

---

* Pr. – problem; Ex. – example; p – page; b – beginning; c – continuation; s – solution; digits after p, b, c, s – number of page

14.4 Terms and abbreviations

14.5 Sources of illustrations for section 12

12.1 – www.eyewitnesstohistory.com/ford.htm
12.2 – blacksmithandmachineshop.com
12.3 – www.coachbuilt.com
12.4 – www.huffreport.com
12.5 – 12.6, etc. for similar forms – Orloff, M. (2010): *Modern TRIZ. Handout material for Global Production Engineering*. – Global Production Engineering, Technische Universität Berlin
12.7 – 12.9 – www.coachbuilt.com
12.10 – lwww.flickr.com
12.13 – allworldcars.com
12.14 – paul.bert.sens.pagesperso-orange.fr
12.15 – www.automotivetraveler.com
12.16, 12.19, 12.20 – www.rarecarrelics.com
12.21 – www.austinmemories.com
12.24 – www.coachbuilt.com
12.25 – auto.howstuffworks.com
12.26 – theoldmotor.com
12.29 – 12.32 – www.rarecarrelics.com
12.33 – www.beaconschool.org
12.34 – www.motortrend.com
12.37 – mniebuhr.com
12.38 – www.flickr.com
12.39 – artoflean.com
12.40 – mdcegypt.com
12.43 – www.allposters.com
12.44 – rtdr.org
12.45 – www.theepochtimes.com
12.48 – 51 – Rabe, M. (2010): *Lean Management. Handout material for Global Production Management*. – Global Production Engineering, Technische Universität Berlin
12.54 – www.coachbuilt.com
12.55 – www.supplierlist.com
12.56 – www.keyence.com
12.57 – 12.58 – www.jdpower.com
12.61 – www.austinmemories.com
12.62 – www.dollarpaisa.com
12.63 – www.sortega.com
12.66 – www.tangram.co.uk
12.67 and 12.70 – Mertins, K. (2010): *Just in Time. Handout material for Global Production Management*. – Global Production Engineering, Technische Universität Berlin
12.68 – www.caradvice.com.au
12.69 – www.tangram.co.uk
12.71 – www.bradyid.com
12.72 – www.manufactus.com
12.73 – www.leansupermarket.com

14.6 Selected web-sites

Web-sites of Academy International for Modern TRIZ:

www.abc-triz.com
www.easytriz.com
www.modern-triz-academy.com

Key web-site in Russian: www.altshuller.ru

Key international web-journal: www.triz-journal.com

CPSIA information can be obtained at www.ICGtesting.com
Printed in the USA
LVOW070055030713

341301LV00001B/13/P

9 783642 252174